HANDBOOK OF GEOPHYSICAL EXPLORATION

SEISMIC EXPLORATION

VOLUME 18

PHYSICAL PROPERTIES OF ROCKS:
FUNDAMENTALS AND PRINCIPLES OF PETROPHYSICS

HANDBOOK OF GEOPHYSICAL EXPLORATION

SEISMIC EXPLORATION

Editors: Klaus Helbig and Sven Treitel

Volume
1. Basic Theory in Reflection Seismology[1]
2. Seismic Instrumentation, 2nd Edition[1]
3. Seismic Field Techniques[2]
4A. Seismic Inversion and Deconvolution: Classical Methods
4B. Seismic Inversion and Deconvolution: Modern Methods[2]
5. Seismic Migration (Theory and Practice)
6. Seismic Velocity Analysis[1]
7. Seismic Noise Attenuation
8. Structural Interpretation[2]
9. Seismic Stratigraphy
10. Production Seismology
11. 3-D Seismic Exploration[2]
12. Seismic Resolution
13. Refraction Seismics
14. Vertical Seismic Profiling: Principles
15A. Seismic Shear Waves: Theory
15B. Seismic Shear Waves: Applications
16A. Seismic Coal Exploration: Surface Methods[2]
16B. Seismic Coal Exploration: In-Seam Seismics
17. Mathematical Aspects of Seismology
18. Physical Properties of Rocks
19. Shallow High-Resolution Reflection Seismics
20. Pattern Recognition and Image Processing
21. Supercomputers in Seismic Exploration
22. Foundations of Anisotropy for Exploration Seismics
23. Seismic Tomography[2]
24. Borehole Acoustics[1]
25. High Frequency Crosswell Seismic Profiling[1]
26. Applications of Anisotropy in Vertical Seismic Profiling[1]

[1] In preparation.

[2] Planned.

SEISMIC EXPLORATION

Volume 18

PHYSICAL PROPERTIES OF ROCKS: FUNDAMENTALS AND PRINCIPLES OF PETROPHYSICS

by

J. H. SCHÖN

Professor for Petrophysics and Well Logging
JOANNEUM RESEARCH
Institute of Applied Geophysics
Leoben
Austria

PERGAMON

ELSEVIER SCIENCE Ltd, The Boulevard, Langford Lane, Kidlington
Oxford OX5 1GB, UK

First edition 1996
Second edition 1998

Library of Congress Cataloging in Publication Data

Schön, Jürgen H.
Physical properties of rocks: fundamentals and principles of petrophysics/by Jürgen H. Schön.
p. cm. - (Handbook of geophysical exploration. Section I, Seismic exploration: v. 18)
1. Rocks. I. Title. II. Series
QE431.6.P5S36 1995
552'.06-dc20 95-46751

British Library Cataloguing in Publication Data

A catalogue record for this book is available from the British Library

ISBN 0 08 0410081

Printed in The Netherlands

Dedicated to Monika and Peter for patience and encouragement

CONTENTS

CHAPTER 3: DENSITY OF ROCKS

CHAPTER 4: MAGNETIC PROPERTIES OF ROCKS

CHAPTER 5: NATURAL RADIOACTIVITY OF ROCKS

CHAPTER 6: ELASTIC PROPERTIES OF ROCKS

CHAPTER 8: THERMAL PROPERTIES OF ROCKS

CHAPTER 9: ELECTRICAL PROPERTIES OF ROCKS

CHAPTER 10: SOME RELATIONSHIPS BETWEEN VARIOUS PHYSICAL PROPERTIES OF ROCKS

Preface

The interpretation of geophysical data in exploration geophysics, well logging, engineering, mining and environmental geophysics needs the knowledge of the rock properties and their correlation. Growing interest in rock physics results from the need to extract more information from geophysical measurements concerning the accuracy, reliability and representative validity of the results.
The expectations result from:
➢ the expanding field of applied geophysical investigations,
➢ the increasing number, variety, and complexity of properties of interest,
➢ the "broadening" of the spectrum of rock types, which are of interest,
➢ the dramatic development concerning "input data" by the modern equipment.

We must transform this growing data volume into equivalent information about the properties of interest. From these trends, a problem and contradiction results between the rapidly growing inflow of data and the improved technical complexities of their processing, and the sometimes traditional basis of interpretation to obtain increased relevant information.
Rock physics is a key in the network of applied geosciences and related engineering disciplines - it must connect the various properties of rocks. It gives the basis for transformation or translation from measured into interesting properties. The optimal use of all relevant information is the crux of a modern interpretation.
In this way, rock physics

- is *integrated* into the general techniques, strategies, algorithms, and the complete process of exploration, and simultaneously
- is an *integrating* part of this process, because rock physics couples and connects the different disciplines.

The main subject of this book covers aspects of applied geophysics and applied geosciences. Its idea is a comprehensive and concise systematic presentation of the physical properties of rocks. The book will present fundamental results of experimental investigations and some main directions of a theoretical basis. It may be an addition to the existing data collections and excellent books regarding individual properties. The growing volume of results and papers demands a selection and the refer in many parts of this book. I apologize for this selection and briefly discussion of results in this book.

One particular idea is to show the some "classic fundamentals of our discipline rock physics" and the deep understanding of physics and geology in the work of the pioneers in rock physics ... but also to show the pleasure we get from investigation of the natural rocks and their fascinating beauty

"... to see a world in a grain of sand and a heaven in a wild flower,
hold infinity in the palm of your hand, and eternity in an hour ... "
(William Blake, 1757-1829)

Colleagues and friends helped me by reviewing parts of this book. I am especially grateful to Klaus Helbig and Frank Börner.

For indispensable help in refinement of this text I thank John F. Arestad for his careful reading and rereading of the manuscript. Many thanks, dear John for spending all the time and attention, for valuable comments and support.

Jürgen H. Schön

CHAPTER 1

Rock Physics as an Integrated and Integrating Component of Geosciences

"Petrophysics" is suggested as the term pertaining to the physics of particular rock types. ... This subject is a study of the physical properties of rock which are related to the pore and fluid distribution ..." G.E. Archie (1950), who was the pioneer in application and quantification of rock physical relations as a part of geosciences and petroleum engineering.

1.1. Introduction

Rock physics is the investigation and study of the physical behaviour and properties of the rocks concerning

- experimental / theoretical aspects and methods, and
- fundamental and practical questions connected with applications in geophysics, geology, reservoir and geotechnical engineering and related disciplines.

Scientists in rock physics today are attempting to solve problems regarding the earth's interior and evolution, and the "classic field" of reservoir characterization starting with the fundamental work of Archie (1942), as well as, problems of engineering geophysics and environmental studies. So it can be seen that this discipline also includes the study of anthropogeneous changes of rock properties.

Yale (1985) has characterized rock physics as "a highly interdisciplinary field involving geology, geophysics, geochemistry, physics, acoustics, well logging, core analysis, and petroleum, chemical and mechanical engineering" and, with a view on its application in industry, emphasizes that "... nearly all major petroleum companies conduct research in rock physics."

The main subject of this book covers aspects of applied geophysics and applied geosciences with their theoretical and practical background. The growing interest in rock physics results from the need to extract more information from geophysical measurements concerning the

- growing spectrum of parameters necessary for exploration, and
- higher accuracy, reliability and representative validity of the results.

The position of rock physics within the framework of the general approach to solving an exploration problem using geophysical methods is illustrated in Figure 1.1:

The interpretation process starts with the geophysical field result (seismogram, log, record, etc.); today, it is generally presented in digital form. The first step is to use this field result - after corrections and calibrations - to obtain a model. This model represents the spatial distribution of diagnostic properties of the applied method (see Table 1.1). For example, for electrical resistivity measurements the diagnostic property is the specific electrical resistivity. Such a distribution may be represented by a profile, a two-dimensional cross-section, a map or a three-dimensional presentation.

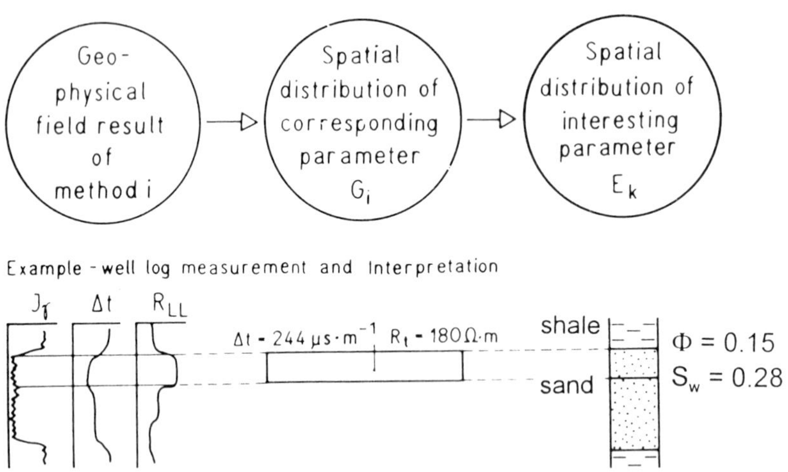

Figure 1.1. General path for solving an exploration problem using geophysical methods, illustrated by an example (simplified well log of a sand-shale-profile)
J_γ - Natural gamma intensity Δt - Acoustic log travel time R_{LL} - Laterolog resistivity

During the next stage of this process, we must transform the distribution of the corresponding properties into another distribution of properties that are significant for a particular geological or technical problem (e.g., porosity, mineral composition, mechanical strength). For this, "interrelationships" between the different properties are required. These may be derived either theoretically on the basis of models or by empirical observations.

The development of such interrelationships, which are fundamental for modern interpretation techniques and strategies, is an important subject of research in rock physics. The well-known Archie-relation (Archie, 1942) reflects this general feature:

$$R_0 = R_w \cdot F = R_w \cdot \left(\frac{a}{\Phi^m}\right) \tag{1-1}$$

where R_0 and R_w are the specific electrical resistivities of the water saturated rock and pore water - these are the corresponding properties and the result of measurements. F is the "formation resistivity factor", which is controlled by porosity (Φ); an empirical factor a is related to "tortuosity", and the empirical exponent m is called "cementation exponent". The porosity is the "target" of the interpretation.

In this procedure of "transformation" or "translation" from measured into interesting properties, all relevant additional information must be included (in the simple case of Archie's equation, this is the knowledge or assumption concerning the empirical parameters a and m). The optimal use of all information is the crux of a modern interpretation.

Table 1.1. Main geophysical methods and corresponding diagnostic physical properties

Method	Parameter(s)
Magnetometry	susceptibility, magnetic permeability
Gravimetry	density
Geoelectric, Electromagnetic	conductivity (or specific resistivity), dielectric behaviour (permittivity)
Seismic	seismic velocities and densities (impedance), attenuation
Geothermal	thermal conductivity, thermal diffusivity, specific heat capacity
Radiometric, Nuclear methods	content of radioactive elements, cross-sections for nuclear processes

In this way, rock physics

- is *integrated* into the general techniques, strategies, algorithms, and the complete process of exploration, and simultaneously
- is an *integrating* part of this process, because rock physics couples and connects the different disciplines based on interrelationships between their different diagnostic properties.

The connection is based on physical laws and considering the specific nature and features of the subject - the natural rock.

Now, we can formulate the causes of the *growing expectations* of rock physics more precisely. They result from the following:

1) The expanding field of applied geophysical investigations. This field has been extended from oil, gas and mineral exploration to other arenas of high economic, technologic, social and human importance such as engineering problems, hydrology, and environmental research.
2) This is reflected in an increasing number, variety, and complexity of parameters and properties of interest, in addition to porosity and water saturation. Of increasing interest are mechanical properties (deformation, strength, time behaviour...) and properties characterizing the hydrogeological regime. In many cases, there is a time dependency, where, future development of such properties are of interest, not just a "snapshot" of the present state.
3) This has resulted in a "broadening" of the spectrum of rock types, which are of interest. In the past the hydrocarbon reservoir rock (carbonates and sandstones, with such difficulties as the "shaly sand problem") were of dominant concern. Today, unconsolidated rocks are of interest (for engineering and environmental geophysics, hydrogeology, etc.), as well as, igneous rocks for projects such as buildings, water engineering plants, geothermal energy resources, underground sites for waste disposal, and caverns for different uses. Last, but not least, the recent development in crustal research projects (Kola deep drillhole, Kontinentale Tiefbohrung KTB, Ocean Drilling Project, etc.) have presented new results and raised new questions regarding rock physics.

Additionally, we are confronted with some "exotic" materials. There are the samples from the moon taken by the Apollo- and Luna-missions or measurements in the neighbourhood of volcanos or active tectonic parts of the earth crust (e.g. Eichelberger, 1989).

However, there are also dramatic *developments* with respect to "input data and tools" for rock physics studies:

- An explosive growth of data has been produced by modern digital equipment. This development is characterized for example by 3-dimensional seismic acquisition, spectral methods in field and borehole analysis (e.g. nuclear spectral analysis, natural gamma spectometric measurements, spectral IP, nuclear magnetic resonance imaging etc.) and other "high data rate" methods, such as full waveform acoustic log or the formation microscanner (or imager). We must transform this growing data volume into equivalent information about the properties of interest.
- Fundamental results in physics, physical chemistry, mathematical disciplines and geosciences can and must be incorporated into research and practical applications in rock physics and geophysics.

From these trends, a problem and contradiction results between
- the rapidly growing inflow of data and the improved technical complexities of their "handling" and processing ,
- and the frequently traditional basis of a complex interpretation to obtain increased relevant information.

1.2. The rocks of the earth - their composition and classification

1.2.1. Overview

"Rocks are natural mineral assemblages crystallized by rock forming processes. Rocks form the earth crust and mantle..." (Huckenholz, 1982).

The crust - approximately 1% of the earth's volume and 0.5% of its mass - is the upper part of the spherical shell structure of the earth. These main shells are: Crust (0 to about 30 km), mantle (about 30 to 2900 km), and core. The boundaries are characterized by distinct changes of chemical composition (see Table 1.2) and thermodynamic conditions. This book is primarily concerned with the properties of rocks found in the crust.

We distinguish rocks according to the origin (genesis) and composition between

- igneous rocks (magmatites, magmatic rocks),
- metamorphic rocks (metamorphites), and
- sedimentary rocks (sediments).

This is the fundamental geological nomenclature of rocks, and includes the great variety of rocks - an illustration of which is given in Figure 1.2.

Table 1.2. Estimated and calculated chemical composition (weight -%) of the earth's crust, mantle and core (after Correns, 1968; see also Huckenholz, 1982)

Element	Crust	Mantle	Core
O	47.25	43.7	
Si	30.54	22.5	
Al	7.83	1.6	
Fe	3.54	9.88	86.3
Ca	2.87	1.67	
K	2.82	0.11	
Na	2.45	0.84	
Mg	1.39	18.8	
Ti	0.47	0.08	
P	0.08	0.14	
Mn	0.07	0.33	
Ni	0.0044		7.36
Co	0.0012		0.40
S	0.031	___	5.94
Σ	99.3466	99.65	100.00

1.2.2. Igneous rocks

Igneous rocks (magmatites, magmatic rocks, eruptive rocks, eruptiva) are mineral assemblages, originating from a melt (magma) by crystallization. Three types are distinguished by their natural occurrence:

- volcanic rocks (reaching the surface),
- plutonic rocks (crystallization at greater depth), and
- dike rocks (small dikes and sills separating large rock masses).

Table 1.3. shows the average mineral composition of the crust and typical igneous rocks. This illustrates that:

- feldspars are the most abundant assemblages (about 62% in most rocks)
- quartz is the second most important mineral (about 21 %).

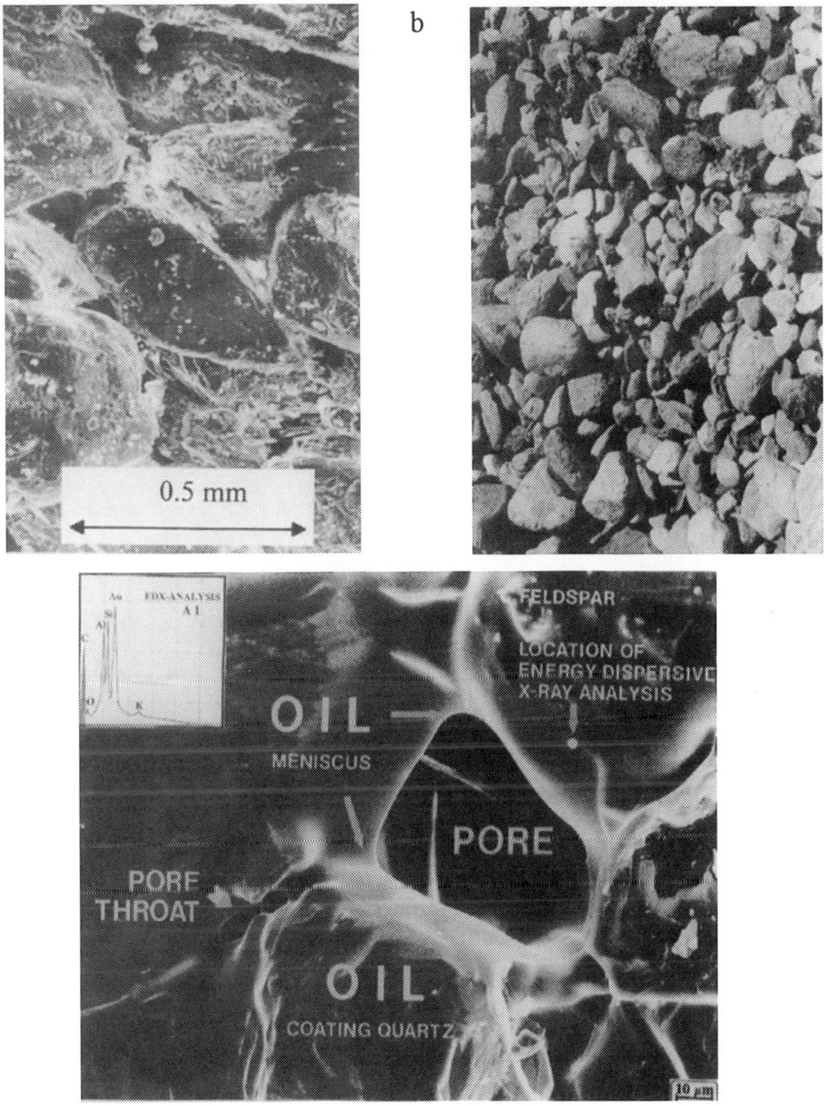

Figure 1.2. Some typical examples of natural rocks - illustrating the broad variety of species, the complexity of shape and structure ... *and the beauty of our subject of study.*
a) Sandstone, Cretaceous (Elbsandsteingebirge Germany), b) Beach sediment (Enns Austria); c)Cryo-SEM micrograph of a highly oil-saturated sandstone sample (GFE Gemeinschaftslabor für Elektronenmikroskopie der RWTH Aachen; see also Mann et al., 1994

Table 1.3. Approximate average mineral composition (volume - %) of the earth's (upper) crust and of major intrusive rocks (after Wedepohl, 1969; see also Huckenholz, 1982)

Mineral	Crust	Granite	Grano-diorite	Quartz-diorite	Diorite	Gabbro
Plagioclase	41	30	46	53	63	56
Alkalifeldspar	21	35	15	6	3	
Quartz	21	27	21	22	2	
Amphibole	6	1	13	12	12	1
Biotite	6	1	13	12	12	1
Orthopyroxene	2				3	16
Clinopyroxene	2				8	16
Olivine	0.6					5
Magnetite, Ilmenite	2	2	2	2	3	4
Apatite	0.5	0.5	0.5	0.5	0.8	0.6

1.2.3. Sedimentary rocks

"More than 50% of the earth surface is covered by a (thin) blanket of sediments which makes up about 6% of the total earth's crust" (Huckenholz, 1982).
Sedimentary rocks (sediments) are the result of weathering and sedimentation processes: Igneous, metamorphic and also previously deposited sediment rocks are broken down physically and chemically. The residual material is transported as debris and in solution and accumulates at the locus of deposition and/or precipitation.
Therefore, we can distinguish, after occurrence and origin, two main types of sedimentary rocks:

- ► the clastic or fragmental sediments (e.g. sandstones, shales) and
- ► the chemical and biochemical sediments (e.g. carbonates, evaporates).

Clastic sediments make up about 92% of the total.

Because quartz is one of the most resistant minerals in the crust it becomes enriched in sedimentary rocks (see Table 1.4.), whereas, the feldspars, amphiboles, pyroxenes, mica and olivines suffer a partial or complete disintegration. The results are residual remnants (weathering residua) and new minerals (weathering products). The last type are mainly very fine-grained phyllosilicates with grain diameter of less than 0.02 mm (clay mineral group).

Table 1.4. Mean mineral composition of sedimentary rocks after Huckenholz, 1982

Mineral	weight - %
Quartz	30
Feldspar	9
plagioclase	(1)
alcali feldspar	(8)
Mica	23
muscovite, illite	(15.5)
biotite, glauconite	(7.5)
Clay minerals	17.5
Chlorite	2
Calcite	6
Dolomite	2.5
Oxides	6.5
Apatite	< 0.5
Carbonaceous matter	0.5
Others	2.5

The most commonly used classification principle for sedimentary rocks is based on grain size (see Figure 1.3).
For nonspherical grains, the particular diameter measured depends on the technique used (for example sieve analysis, which gives a close estimate of the diameter of the minimum cross-sectional area; Jorden and Campbell, 1984). The grain size is given in mm or in "phi units", where

$$phi = -\log_2 d \qquad (1\text{-}2)$$

where d is the grain diameter in mm.

Natural sediments exhibit a grain size distribution, which can be shown by a grain size distribution curve in graphic presentation (cumulative distribution curve). The properties of a grain size distribution can be characterized by the

- median diameter *Md*: It is defined as the specific size of the cumulative curve at which 50 wt% of the particles are greater or smaller than the *Md* value.
- coefficient of sorting *So* (Trask sorting coefficient): The coefficient of sorting is defined as

$$So = \sqrt{\frac{d_{25}}{d_{75}}} \tag{1-3}$$

where d_{25} is the grain size (mm) of 25 percentile and d_{75} is the grain size (mm) of 75 percentile (cumulative wt %). A clastic sediment can be called *very-well sorted* when the coefficient *So* is <1.4, *well sorted* with 1.4 to 1.87, and *poorly sorted* with >1.87 (Huckenholz, 1982).

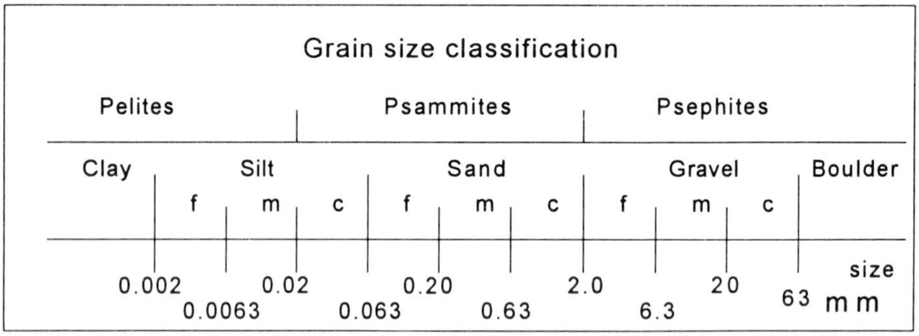

Figure 1.3. Grain size scale and classification of sediments (the grain size is defined as "diameter"); the terms psephites, psammites, and pelites are defined for more than 50 %-weight of the corresponding grain size range; *f* - fine *m* - medium *c* - coarse

It may be noted that the term "clay" describes a grain size range of particles and a specific type of minerals (phyllosilicates) with special properties (for example cation exchange).

During their transport, movement and sedimentation, rock particles or grains change and receive their shape and "face", which is described by parameters of sphericity (shape) and roundness (angularity).

Within the petrology of sedimentary rocks are applied so-called "textural properties" such as grain size distribution, median diameter, sorting, sphericity, roundness, and porosity (see chapter 2.). These are features of a detailed classification which relates to the genesis of rocks. More detailed descriptions are given in textbooks on petrography and sedimentology (e.g. Pettijohn, 1975; Tucker, 1991).

1.2.4. Metamorphic rocks

"Metamorphic rocks (metamorphites, metamorphica, schists) are the result of metamorphism. Metamorphism is the solid-state conversion of igneous and sedimentary rocks under the pressure-temperature regime" of the crust. (Huckenholz, 1982). Metamorphic transformations are caused by a change in temperature and pressure. The preexisting mineral assemblages are converted into new assemblages corresponding to thermodynamic conditions. Such temperature and pressure changes within the earth's crust are caused by different types of processes, such as:

- regional metamorphism
- contact or thermal metamorphism
- dislocation or cataclastic metamorphism.

The great varieties of metamorphic rocks are characterized, classified and named according to their mineral assemblages and textures. The main types are: Phyllites, schists, gneisses, skarns, serpentinites, felses, quartzites, granulites, eclogites, amphibolites, and marbles. The different metamorphic rock types are described in the petrographic literature. For example, a short but clear-cut description, including the chemical composition, is given by Huckenholz (1982).

The structure of most metamorphic rocks is characterized by parallel oriented elements (minerals or mineral axes, respectively) and in many cases also an orientation of microcracks and other "internal defects". This causes a typical, more or less planar cleavage of rocks like gneisses, schists, and slates. Other types of metamorphic rocks like felses and hornfelses are compact without such parallel structures.

Components or parts of the structural features of metamorphic rocks are predetermined from their premetamorphic sedimentary and magmatic source materials. However, the typical schistosity or foliation is produced by directed stress and its effects on the mineral aggregate, the orientation of the minerals and the creation of fractures, etc..

1.2.5. Physically determined classification of rocks

The geological classification described in the foregoing is based primarily on genetic features. By examining their physical aspects, there is another classification which includes physical features and properties. Initially, we consider the characterization of natural rocks, which are generally heterogeneous and structured systems consisting of different minerals with various physical properties, shapes, sizes and arrangement in space. Also included are complicated boundary effects and interactions.

These features, as well as, thermodynamic conditions generally control and determine the physical properties of rocks. A recommended classification is given in Figure 1.4..

This pattern classifies the rocks according to

- ➢ their substantial content in monomineral and polymineral rocks;
- ➢ their content of fluid components and the geometry of these components in nonporous (dense), and porous and fractured (cracked, fissured etc.) rocks;
- ➢ their character of bonding, cementation or other effects at the point of contact between the solid components (particles) in consolidated (solid "bridges" at contact) and unconsolidated (friction or physico-chemical interface forces like adhesion and capillary tension) rocks.

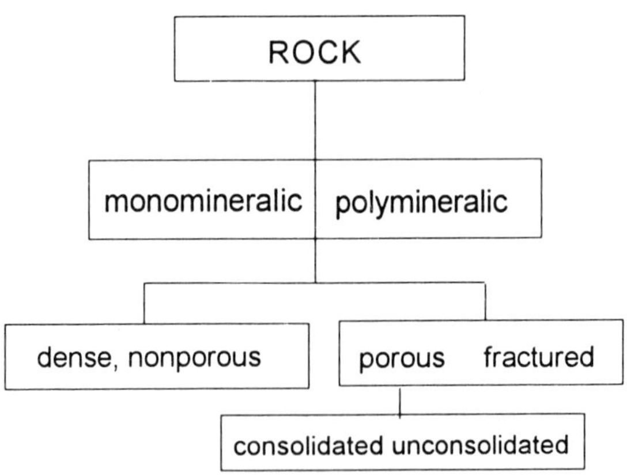

Figure 1.4. A physically determined classification of rocks

Three remarks related to problems regarding physical characterization of rocks are necessary:

1) Homogeneity and inhomogeneity, isotropy and anisotropy

Rocks are *homogeneous or inhomogeneous* in a considered volume; this pair of terms indicates whether the property is a function of local coordinates within the volume or not. In the case of homogeneity, the property is the same at every point, and it differs for inhomogeneity.

Rocks are also *isotropic or anisotropic* in a considered volume; this pair of terms indicates whether the property is a function of direction or not. In the case of isotropy, the property is equal in all directions and can be used as a scalar. In the case of anisotropy, the property depends on direction and must be used as tensor with its significant symmetry.

Gassmann (1964) has given an illustrative description of these two pairs of terms and their physical picture (Figure 1.5).

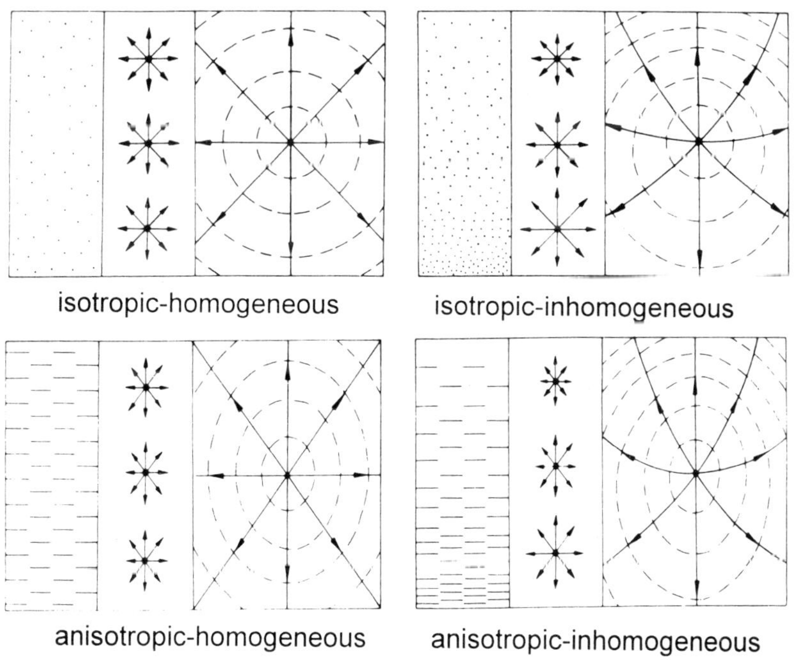

Figure 1.5. Illustration of the terms isotropy - anisotropy and homogeneity - inhomogeneity; after Gassmann (1964)

Investigations of anisotropy show the strong correlation of this phenomenon with structural-textural properties as for example in the case of seismic properties (orientation of cracks, grain and mineral anisotropy originated by depositional environment etc., see section 6.3.4.3).

The study of anisotropy may be one "key" to understanding and to quantitative characterization of the internal structure of rocks from a physical point of view. This leads to the next problem:

2) Structure and texture

The characterization and definition of the internal structure of rocks is a very difficult problem. Using a physical point of view, "structure" indicates the type of internal geometry of the rock and the geometrical arrangement of its parts in space. There is not always a clear distinction between the terms "structure" and "texture" in the geosciences.

Serra (1984) gives one definition: "When describing the texture of a rock, we look at the grains and cement, and their intrinsic properties. The structure refers to the 3-dimensional arrangement of these constituents, and their anisotropy"; but Jorden and Campbell (1984) said: "Texture deals with the size, sorting, shape, roundness, and packing arrangement of the rock solids."

3) Scaling problem

The heterogeneity of the substance "rock" includes the general problem of different scales when using various techniques in the geosciences. This fundamental problem of scaling and of the "representative volume" is illustrated with a comparison between the effective properties for a geological cross section as they are relevant (e.g. for seismics, the measuring environment of a logging tool, and a core or a thin section, respectively). In most cases, laboratory measurements in rock physics are related to the smallest scale with dimensions of only mm or cm (see Figure 1.6.).

Integrated reservoir modelling includes geological, geophysical, and petrophysical data of different scales and resolution for a consistent reservoir characterization.

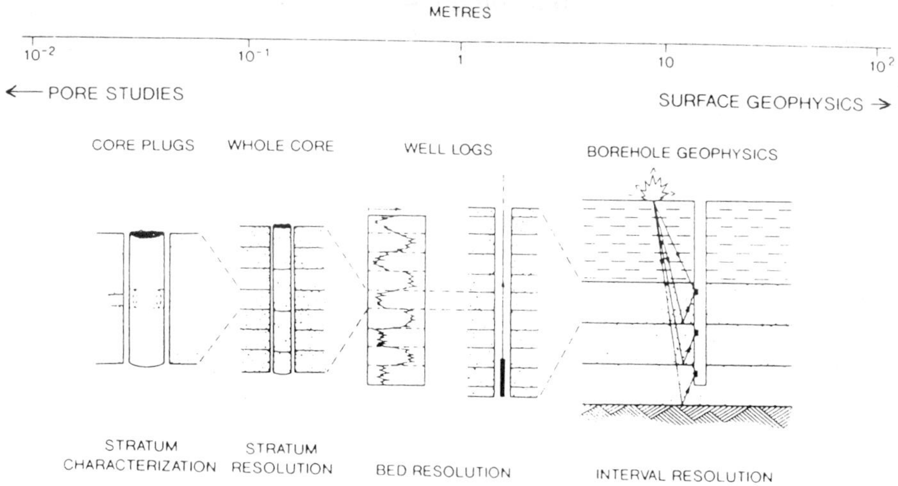

Figure 1.6. Relationship of different scales of measurement through progressively increasing resolutions (Worthington, 1991)

1.3. Results of experimental investigations - an overview

Because natural rocks are heterogeneous materials, consisting of components with different properties, it is obvious that their effective properties depend upon rock composition (volumetric fractions) and properties of the components (minerals, pore content). In most cases, the dominating difference between the properties of the solid and fluid constituents creates a distinct correlation with porosity. But also internal geometry such as grain size and its distribution, grain shape, texture and structure and bonding properties (e.g. cementation) and interface effects are influential. Additionally, these properties are usually influenced by pressure and temperature.

We should understand and use the following pattern as a thesis to take with us during our walk across the experimentally determined physical properties of rocks:

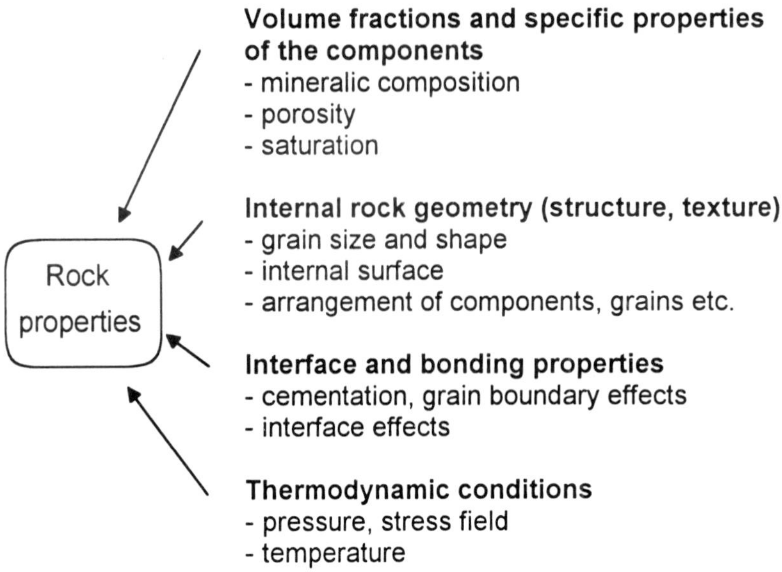

Figure 1.7. Rock properties - main dependencies

Scalar properties depending only on the first and last influencing parameters, whereas, tensorial properties are influenced by all parameters. In many cases, there is a significant correlation between internal structure and anisotropic behaviour.

Figure 1.8 shows examples illustrating some of these correlations and their influencing parameters.

1.4. Models as a basis for rock physics theories - an overview

The creation and derivation of a theory begins with a conceptual rock model. This model is the key for the mathematical formulation of relationships between the different properties and parameters of the rock model as an idealized representation of real rock.
We can summarize the function of models and of theoretically derived relations in rock physics with three points:

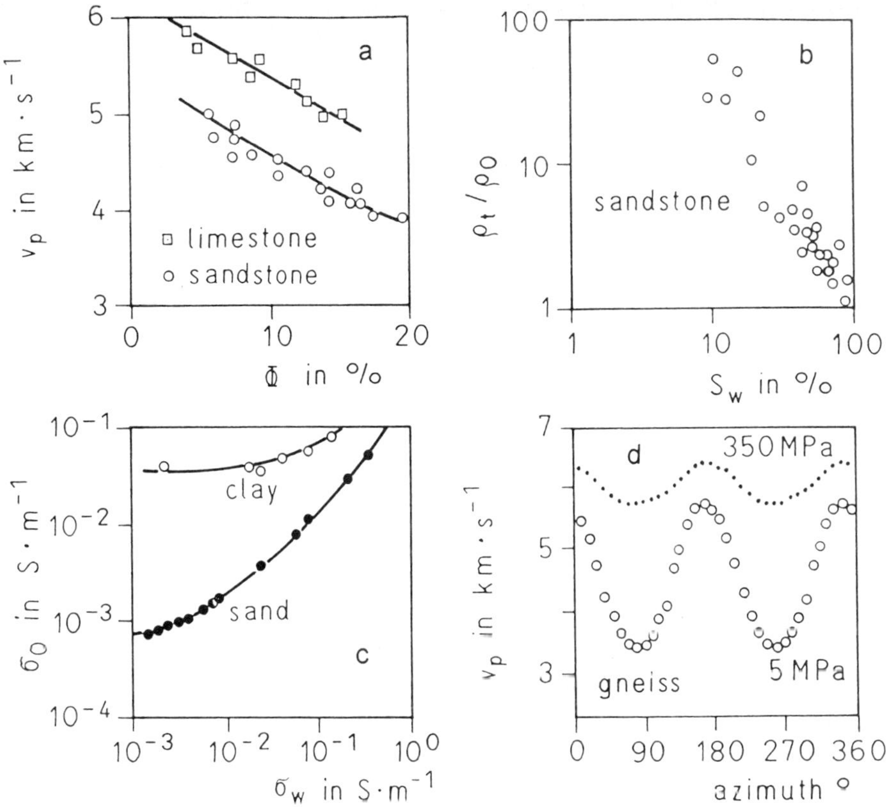

Figure 1.8. Some examples for influences controlling petrophysical parameters
(a) Velocity of longitudinal wave v_p vs. porosity Φ ;
(b) Ratio of the specific electrical resistivity for partial and full saturation ρ_t/ρ_0 vs. water saturation S_w;
(c) Electrical conductivity of the water saturated sediment σ_w vs. conductivity of pore water σ_w illustrating the interface effects (Schön and Börner, 1985);
(d) Dependence of the velocity v_p on the direction of propagation (Zang et al., 1989)

1) The model allows the formulation of essential correlations, which are known from experiments based on a physical concept and mathematical calculation (identity of theory and experiment);

2) The model and the derived relations allow the determination (calculation) of properties without direct experimental determination (additional information);
3) The model relates physical and petrographic properties as a contribution to the integration process in geosciences.

Models in rock physics are related to
➢ a particular group or *type of rocks* (e.g. sedimentary porous rocks)
➢ a particular group or *type of properties* (e.g. elastic properties).

The spectrum of model concepts in rock physics is very broad and extends from solid state physics (relevant to the earth interior) to suspensions (relevant to highly porous marine sediments during the stage of sedimentation). Despite this diversity, we can distinguish four "basic types" of models (see Figure 1.9.):

1) The model consists of parallel elements: Each element represents one component of the rock (mineral, pore content) with its specific properties and volume fraction (such as given by the relative thickness of a schist). The field (e.g. thermal field) or wave (e.g. elastic wave) is directed either parallel or perpendicular to the planes between the schists.

The equations for this model are of a simple linear nature. The model is "volume-oriented" in its philosophy, it is also "volume-true" in the resulting equations. It can be applied to many practical problems and integrated into most complex interpretation systems of well-logging (see chapter 10).

One of the best known relations is the "time-average relation" of Wyllie, Gregory, and Gardner (1956), which allows for the determination of porosity from acoustic logs (see chapter 6). A critical element of this type of model is that it is only applicable without restrictions for scalar properties (e.g. density, nuclear cross sections). A particularly simple model for fractured rocks (Schön, 1973) can be included in this group; it consists of a solid block of matrix material and a "cut" representing the effects of fractures, cracks and other imperfections of bonding between the constituents (see chapter 6).

2) The model - mostly applied to porous rocks - consists of an arrangement of solid particles or pores of a simple geometric form (e.g. spheres, ellipsoides, tubes, channels, spherical or ellipsoidic inclusions). This model is "particle- or pore-geometry oriented" in its philosophy.

The *sphere model*, one type of this group, can be applied to the problem of propagation of elastic waves in porous rocks. Because of its basic principle, it is normally used for unconsolidated rocks (sand) and their elastic/anelastic properties. The mechanical properties being primarily controlled and influenced by the properties of the grain skeleton. Typical examples of this group are the theoretical investigations of Hara (1935), the classic work of Gassmann (1951) and White (1966, 1983). Generally, theories using such models consist of

- a description of the behaviour of two neighbouring spheres (properties at the contact of spheres),
- an assumption about the arrangement of the spheres (internal structure of the model); in most cases it is assumed as a structure similar to a crystal lattice (e.g. cubic or hexagonal structure).

Examination of the pore space (second type of these models) is recommended for properties which are predominantly determined by processes in the pore space. These are the electrical properties (since the conductivity of pore water is often much higher than that of the matrix), and hydraulic permeability (the parameter that describes the motion of the fluid in the pore channels under a pressure gradient). There are simple *tube-models* and various capillary-bundle types of models. Further developments lead to network-models (Schopper, 1966, 1967, 1972) and to the application of percolation- and fractal- concepts (see section 2.3). This direction represents one way of finding direct correlations between the "microscopic internal rock geometry" and the "macroscopic" rock properties.

In an advanced paper on this topic, Morgan et al. (1989) described the microgeometric quantification of sedimentary rocks by computer image analysis: "Many of the parameters in rock physics models are strongly dependent on the details of microgeometry of rocks....The results of the thin section image analysis are compared directly with laboratory measurements of the same cores... Thus far, microgeometric properties such as porosity, fractal dimension, specific surface area, pore and grain-width distributions, pore shape, tortuosity and coordination number, anisotropy and Fourier spectra are being related to laboratory measurements such as porosity, permeability, capillary pressure curves and electrical formation factors"

3) Models based on special types, restrictions or conditions of coupling and interaction of the two (or more) constituents of "the composite rock". Here we find theories with different types, assumptions or requirements of coupling. The broadest spectrum is given for the elastic properties of rocks. An excellent overview on the physical principles is given by White (1983). This group may be called "coupled systems models" in its philosophy.

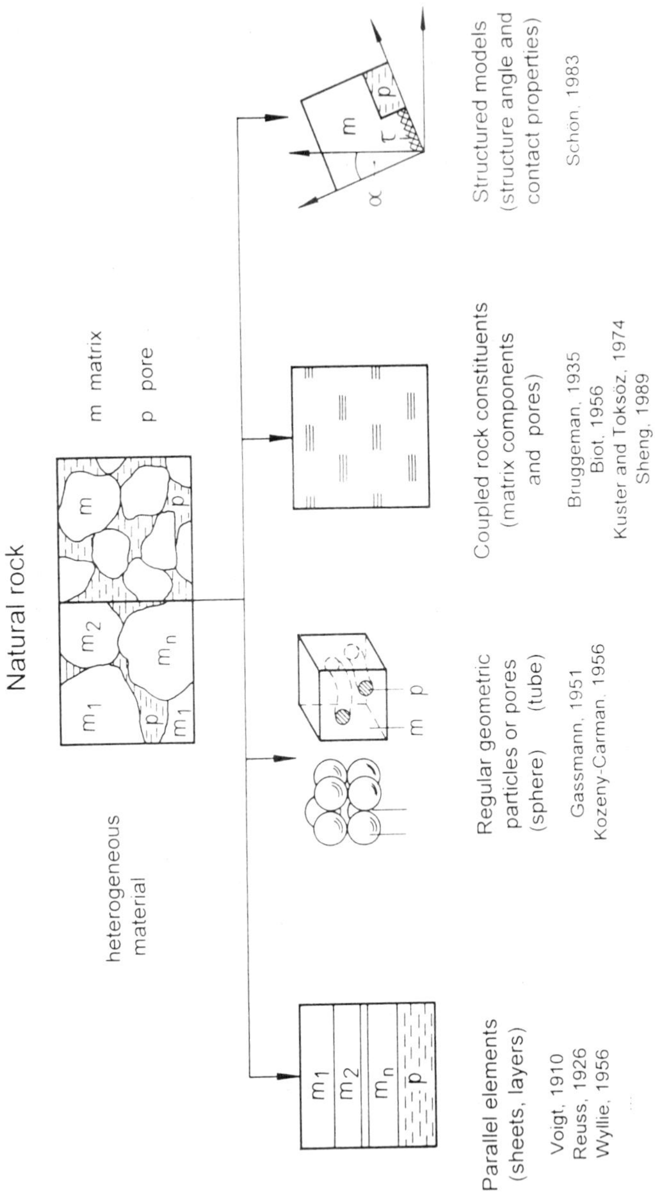

Figure 1.9. The main types of models in rock physics

Bruggeman (1935) computed the average conductivity and dielectric constant of a composite material. In 1937, he published a generalized theory (including elastic properties) for laminated and other compound solids. These ideas were promoted and developed by Helbig (1958, 1995) and Backus (1962) in their theories about transversely isotropic media.

The special importance of porous rocks in their function as reservoirs has generated theories including the idea of a "skeleton" built up by minerals and a pore space filled with fluid. Gassmann (1951) derived his classic theory under the assumption of negligible relative motion between fluid and solid. Relative motion is included in the theoretical concept of Biot (1956).

More general members of this group are the *"effective medium theories"*, which were also pioneered by Bruggeman (1935). Three components are involved - fluid, solid, and cement. For example Sheng (1989) gives a theoretical model and applies it to electrical and elastic properties of sedimentary rocks. Such models and additional considerations (for instance by Korvin, 1982) are also combined with comparisons related to simple or empirical formulae (e.g. Archie-equation, Wyllie-equation).

The self-consistent method is very important as a method concept of this group and is used to calculate properties of rocks containing pores, fractures and cracks (see e.g. Eshelby, 1957; O'Connell and Budiansky, 1974; Kuster and Toksöz, 1974; Yale, 1985).

4) The structured model (Schön, 1974) was developed to calculate different physical properties of porous rocks and to explain several relationships and effects (anisotropic behaviour, structure influence, etc.), which cannot be explained by "volume oriented" models (see chapter 6). One of the objectives of this development is to find a simple model that explains most of the rock physical properties (elastic, electrical, thermal etc.).

Such models are constructed using a simple model element ("microelement") with the properties of the solid, the pore and a "contact" region. The spatial orientation of this "microelement" relating to the "macrosystem" of the measurements is embodied in a "structural angle" as a measure of the internal geometry. This model concept includes a "volume-orientation" by the geometric dimensions of the microelement, a "structure-orientation" by the structural angle, and a "bonding-orientation" by introduction of the properties of the contact region.

A detailed description of some models will be given in the chapters discussing individual properties and correlation between physical properties.

CHAPTER 2

Pore Space Properties: Porosity, Specific Internal Surface, and Permeability

"Rocks have pore-space properties that are very complex. Simple mathematical or experimental models only approximate the pore-space properties of natural rocks". *Jorden and Campbell (1984)*

2.1. Introduction

The geometry of pore space - commonly described by pore- (and pore-throat-) size distribution, connection and orientation - controls reservoir characteristics. The description of vugs, cracks, fractures with their particular geometric shape and spatial orientation can be a very difficult problem.

Three properties are related to the pore space:
1) porosity: the fraction of the entire volume part occupied by pores, cracks etc.,
2) specific internal surface: the magnitude of the surface of pores as related to the pore volume or rock mass. The specific internal surface characterizes the "internal morphology" of pores and controls interface-effects at the boundary grain - pore fluid,
3) permeability: the ability to flow fluid through rock pores.
Porosity and specific internal surface are scalar properties, permeability is a tensor.

There are relationships between these properties, though each is a physical property and is independent of the others. Such relationships are described by additional parameters, for example tortuosity and constrictivity. These interrelationships are determined either by experimental investigations (e.g.

Sen et al., 1990), or by theoretical studies based on percolation and fractal concepts (see Thompson et al., 1987; Morgan et al., 1989; Pape et al., 1982, 1987).

The importance of these three properties results from the fact,

- that the parameters themselves are the fundamental properties for any reservoir characterization, including reservoir dynamics (fluid motion), and of processes (such as contamination and environmental problems).
- permeability seems to be the most important (and hard to determine) property for all reservoir problems - it controls whether the rock can deliver or transmit fluids or not.
- porosity, in particular influences most physical properties (velocity of elastic waves, electrical and thermal conductivity, etc.). It should be noted that the specific surface and the related interface effects also have a strong influence on the electrical, elastic and non-elastic properties.

2.2. Porosity

2.2.1. Definition and terminology

Pores are local enlargements in a pore-space system that provides most of the volume available for fluid storage. Pore throats are the smaller connecting spaces linking pores and providing the more significant restrictions to fluid flow.

Porosity Φ is defined as the ratio of the volume of void or pore space V_p to the total or bulk volume V of the rock (Figure 2.1)

$$\Phi = \frac{V_p}{V} = 1 - \frac{V_m}{V} \qquad (2\text{-}1)$$

V_m is the volume of the solid matrix substance. Porosity is a dimensionless quantity and expressed either as a decimal fraction or as a percentage.

Serra (1984) gives the following short definition: "Porosity is the fraction of the total volume of a rock that is not occupied by the solid constituents." This is a complete definition of "total porosity" (for other types see below).

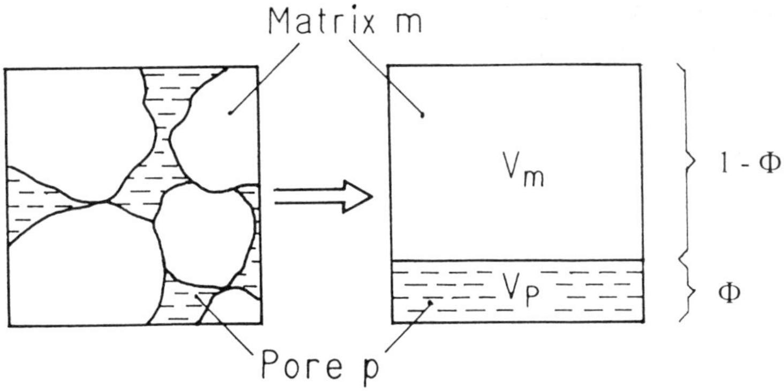

Figure 2.1. Definition of porosity

Porosity is the result of various geological, physical and chemical processes, and is generated during the genesis of the rock as "*primary porosity*" (clastic sedimentation, organogenesis) and/or during the geological history of the rock as "*secondary porosity*" (tectonic processes, chemical processes, dissolution, etc.).

In addition to these two main types, porosity can be classified with respect to the petrographic fabric and the type and degree of interconnection between the individual pores.

According to the petrographic fabric and following Schopper (1982) we can distinguish between:

a) "*intergranular porosity*": The void space between grains, particles or fragments of clastic materials, loosely packed, compacted or even cemented. This is the *primary porosity;*
b) "*intercrystalline or intragranular porosity*": Generated by shrinking or contraction of grains;
c) "*fissure or fracture porosity*": Caused mainly by mechanical, partial chemical action on primarily massive rock (e.g.. limestone) and is a secondary porosity;
d) "*vugular porosity*": Either caused by organisms during genesis, or by chemical action at a later stage; therefore, it can be of both primary or secondary type.

A more detailed description of these types is given by Schopper (1982). Following Choquette and Pray (1970), a classification was published by Serra (1984) using three "basic porosity types":

- fabric selective (inter-particle, intra-particle, inter-crystal, moldic, fenestral, shelter, and growth framework)
- non-fabric selective (fracture, channel, vug, and cavern)
- both fabric selective and non-fabric selective (breccia, boring, burrow, and shrinkage).

This classification clarifies the development of porosity in relation to the history of the rock (Bourbie et al., 1987).

The type and degree of interconnection between the pores is not easily classified since the geometry of pore space is very complex. The pores can be interconnected or partially isolated from each other (i.e., enclosed by the solid matrix). These circumstances result in various definitions being given for porosity of reservoirs. Serra (1984) defined the following porosities:

a) *Total porosity* Φ_{total} is related to all void spaces (pores, channels, fissures, vugs) between the solid components. It is defined by equation (2-1) where V_p is the volume of all empty spaces. The total porosity can have two constituents - primary and secondary porosity:

$$\Phi = \Phi_{total} = \Phi_{primary} + \Phi_{secondary} \tag{2-2}$$

b) *Interconnected porosity* $\Phi_{connected}$ is related only to those spaces which are connected. Pores are considered to be connected when electrical current and fluids can flow between them. Schopper (1982) pointed out that "the term interconnected is often somewhat vague in practice, since interconnection can be gradual, and furthermore the interconnected pore space might contain dead end pores as well." He also gives an extreme example to illustrate the difference between total and connected porosity: A pumice-stone has a total porosity of the order of 50%, but the interconnected porosity is zero because each pore-space is isolated from the others and there are no connecting channels.

c) *Potential porosity* Φ_{pot} is that part of the interconnected porosity in which the diameter of the connecting channels is large enough to permit fluid to flow (greater than 50 μm for oil, only 5 μm for gas). In some cases is $\Phi_{pot} < \Phi_{connected}$ (clays, shales).

d) *Effective porosity* Φ_{eff} is the porosity that is available for free fluids; it excludes all non-connected porosity including the space occupied by the clay-bound water. Effective porosity is a term used specifically in log analysis.

2.2.2. Mean values and ranges of porosity

Porosity of rocks is "very individual" and it is generally difficult to give mean values or mean value ranges for groups or types of rocks (e.g., for sandstones). It is recommended to consider porosity values for different rock types related to region, geological formation and depth.

We find decreasing porosity in the following sequence:

high porous marine sediments
unconsolidated sediments (decrease from clay to silt to sand and gravel)
sandstones
carbonate rocks (limestone - dolomite)
anhydrite
↓ fractured igneous and other initially "dense" rock types.

Schopper (1982) has published data from American, British, German, and Swiss literature and includes over 900 items of porosity (and density). Table 2.1 contains some typical examples from this list expanded by some results from other authors. The main factors which influence (primary) porosity are:

- grain and pore geometrical properties (arrangement and shape of the rock grains, grain size distribution),
- diagenetic processes, amount of cement,
- depth and pressure (which also influences secondary porosity).

The following shows some of the resulting relationships.

2.2.3. The influence of grain geometric properties

The grain geometric properties have the following effect on primary porosity:

a) *grain size*: With small grain sizes, porosity is especially liable to be inversely related to decreasing grain size (Table 2.2 and Figure 2.2). Theoretically, porosity for a given packing is independent of grain size (see section 2.5.). However, a dependence results from the increasing effect of surface forces for fine-grained sediments and the possibility of forming rock-skeletons with higher porosity as in the case of coarse-grained material.

Table 2.1. Some minimal and maximal porosity values for typical rocks; *n* -number of samples (data without signature from Schopper, 1982;)[1] Börner, 1991;)[2] Hübner et al., 1985;)[3] Kulenkampff, 1995; v. Engelhardt, 1960

Rock type, locality, Stratigraphic unit	Source of material, depth	Porosity in % min	max	*n*
Sandstones:				
St.Peter sandstone,Ozark plateau,	outcrop	3.6	14.1	12
Berea sandstone, Ohio, Mississippian	439...458 m	4.7	17.1	8
Bunter sandstone, Great Britain,Triassic	quarry, outcrop	5.8	30.8	18
Buntsandstein, Germany, Triassic	quarry	7.7	26.4	39
Quadersandstein, Germany, Cretaceous	quarry	12.2	26.2	16
Frio Clay (sand), Anahuac field, Texas	2121...2157 m	23.4	37.1	22
Cretaceous s., Elbsandsteingebirge,Germany)[1]	outcrop	18.1	27.6	17
Cretaceous sandstone, Obernkirchen, Germany)[3]	outcrop	18.2	19.5	10
Cretaceous sandstone, Bentheim, Germany)[4]	outcrop	20	28	39
Sandstone, Rotliegend, Altmark, Germany)[1]	3981...4006 m	4.5	9.7	3
Fontainebleau-Sandstone, France)[1]	outcrop	6.8	22.4	3
Limestones:				
Black River Limestone, Wood county, Virginia, Ordovician	2984...3204 m	0.1	1.4	35
Limestone, Midlands/ England; Carbon.	outcrop	2.2	14.9	24
Muschelkalk, Germany,Triassic	quarry	1.2	36.5	19
Solnhofen Limestone, Germany, Jurassic	outcrop	1.2	5.7	23
Shales:				
Shale, undisturbed, Eastern Venezuela,	89...281 m	31.3	35.8	6
Oligocene and Miocene	619...913 m	22.9	28.9	9
	919...1211 m	17.8	25.6	9
	1526...1677 m	12.8	14.6	4
	2362...2437 m	10.3	10.4	2
Unconsolidated sediments:				
Beach sand, Revere Beach, Mass.; Quart.	surface	38.7	44.8	9
Boulder clay, Bridlington/England;	outcrop	23.9	25.5	3
Boulder clay, Northwest Germany)[2]	outcrop	25	40	6
Glacial clay, N-Germany;	outcrop	37	51	4
Loess soil, Idaho; recent age	surface	53.2	69.4	3

b) *grain size distribution* (sorting): With increased sorting, porosity generally increases. In poorly-sorted sediment, the space between the grains with large diameters is filled with smaller grains. (Engelhardt, 1960; Schopper, 1982).

c) *grain shape*: Porosity shows a tendency to increase with the change from spherical or well-rounded grains to angular particles (Table 2.3.). The highest porosity values result from highly anisometric particles, as for example flat mica sheets (Schopper, 1982). The surface force of fine particles (clay) causes structures to become extremely porous ("card-house structures").

Table 2.2. Mean grain diameter, sediment density and porosity of sediment types; continental terrace (shelf and slope); after Hamilton and Bachman (1982)

Sediment type	Mean grain diameter in mm	Density in g/cm^3	Porosity in %
Sand, coarse	0.5285	2.034	38.6
fine	0.1638	1.962	44.5
very fine	0.0988	1.878	48.5
Silty sand	0.0529	1.783	54.2
Sandy silt	0.0340	1.769	54.7
Silt	0.0237	1.740	56.2
Sand-silt-clay	0.0177	1.575	66.3
Clayey silt	0.0071	1.489	71.6
Silty clay	0.0022	1.480	73.0

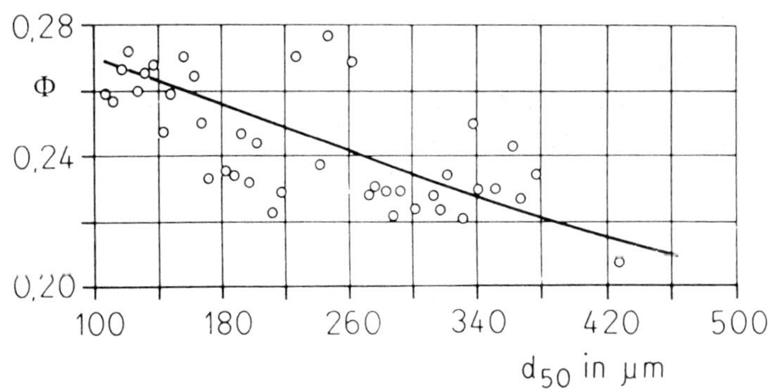

Figure 2.2. Porosity Φ vs. median grain diameter d_{50}; Bentheim Sandstone, Valendis, NW-Germany (v. Engelhardt, 1960)

Table 2.3. Influence of grain shape on porosity (after data from Frazer, 1935; cited in Schopper, 1982 / selected data)

Material (dry)	Porosity in %	
	loose	compressed
Sea sand	38.52	34.78
Dune sand	41.17	37.60
Quartz, ground	48.13	41.20
Mica, ground	93.53	86.62

2.2.4. Influence of diagenetic processes, depth and pressure

Three major mechanisms causing porosity loss and compaction are described by Schneider et al. (1994) related to sedimentary basin modelling (Figure 2.3): The first causes the mechanical rearrangement of the porous medium under the effect of sedimentary loading with two stages (expulsion of free water, expulsion of bound water). The second mechanism is associated with the mechanical deformation (may be elastic, plastic, viscous or even fragile) of the grains making up the skeleton of the porous medium. The third mechanism - the chemical deformation - includes solution and crystallization processes (pressure-dissolution, dissolutions and precipitations due to chemical imbalances).

Schopper (1982) identified the following corresponding stages of the intergranular porosity decrease:

1) *Packing*: The "loosely sedimented particles slide into more stable positions under the pressure of increasing overburden, forming tighter packing of less porosity and more grain-to-grain contacts."
2) *Compaction*: Porosity decreases further "by (mainly mechanical but partially chemical) deformation of grains under even increasing overburden pressure, whereby point contacts between grains gradually change to flat contacts and finally to concave-convex or sutured contacts" (Figure 2.4.).
3) *Cementation*: "Deposition of dissolved material (including alien substances) on free grain surfaces, especially around contacts, thereby still further reducing pore space and also blocking narrow passages".

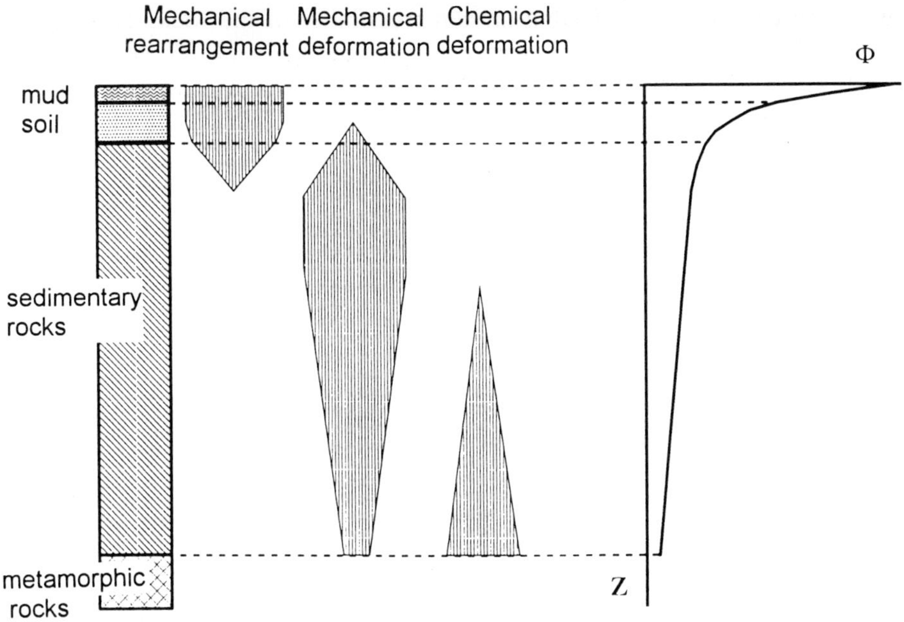

Figure 2.3 The four horizons in a sedimentary basin related to the principal mechanisms causing compaction and porosity loss. The synthetic sedimentary column (left) shows the vertical division as a function of the mechanical behaviour of the sediments. From top to bottom: 1) a liquid zone made up of muds, 2) sediments that behave like soils, 3) sediments *sensu stricto*, 3) metamorphic rocks. The $\Phi(z)$ curve shows that most of the reduction of porosity takes place in the upper two horizons; Schneider et al., 1994

Decrease of porosity primarily results from packing and cementation for sands and sandstones, and from compaction for clays and shales. This reflects a general tendency of decreasing porosity with increasing depth. A typical porosity-depth relationship is illustrated in Figure 2.5. (note the logarithmic scale for porosity).

This nonlinear decrease of porosity with increasing overburden pressure p or burial depth z is described by two main types of relationships:

a) Logarithmic depth - dependence

$$\Phi(z) = \Phi_0 - A_1 \cdot \ln z \tag{2-3}$$

$$\Phi(p) = \Phi_0 - A_2 \cdot \ln p \tag{2-4}$$

where Φ_0 is the initial porosity at depth (or pressure) $z = 0$ ($p = 0$) and A_1, A_2 are empirical factors, depending on compressibility of the rock. This equation was used by Engelhard (1960) and Baldwin and Butler (1985, see also Table 2.4.). There is also a relationship to the density-depth correlation of Polak and Rapoport (1961); see section 3.2.3.

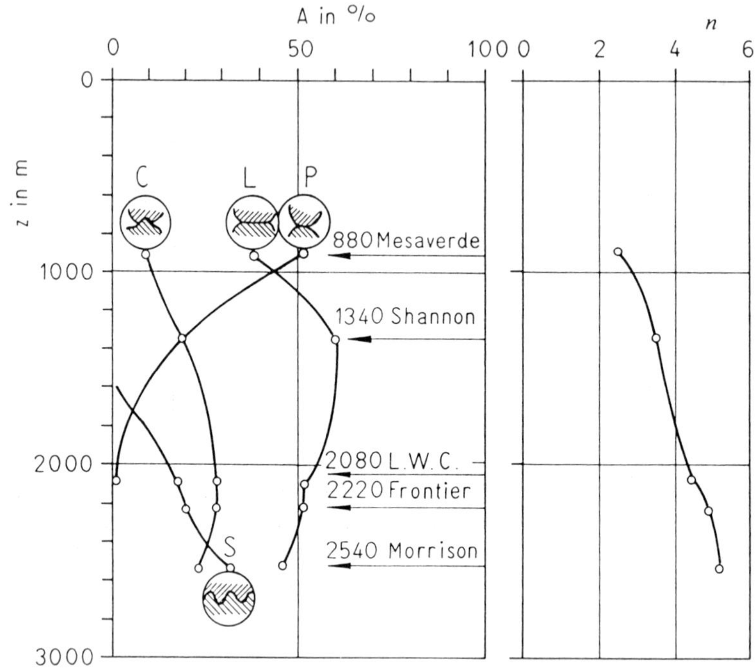

Figure 2.4. Mean number of contacts per grain (in thin sections) n and frequency of occurrence of point contacts (P), long contacts (L), concave-convex (C) and sutured contacts (S) in North American sandstones vs burial depth. After data and a figure from Taylor (1950), cited in Schopper (1982).

b) Exponential depth - dependence

Based on Athy's (1930) porosity-depth relationship for the Paleozoic shales in Oklahoma, Hubbert and Rubey (1959) derived the following formula for the porosity as a function of depth z

$$\Phi(z) = \Phi_0 \cdot \exp(-B_1 \cdot z) \qquad (2\text{-}5)$$

or in terms of pressure p (see also Nagumo, 1965)

$$\Phi(p) = \Phi_0 \cdot \exp(-B_2 \cdot p) \quad (2\text{-}6)$$

where Φ_0 is the initial porosity and the parameters B_1 and B_2 depend on the compressibility of the rock. A detailed discussion of this equation is given by Magara (1985).

Equation 2-6 can be derived from the assumption that the differential decrease of porosity with increasing pressure (or depth) is proportional to the actual porosity value

$$-\frac{d\Phi}{dp} \sim \Phi \quad (2\text{-}7)$$

The resulting differential equation

$$\frac{d\Phi}{dp} = -B_2 \cdot \Phi(p) \quad (2\text{-}8)$$

gives

$$\Phi(p) = \Phi_0 \cdot \exp(-B_2 \cdot p) \quad (2\text{-}9)$$

or

$$\log \Phi(p) = \log(\Phi_0) - B_2 \cdot p \quad (2\text{-}10)$$

Figure 2.5. shows an example for this approximation related to the depth.

Sclater and Christie (1980) used an exponential curve for the description of the depth dependence of porosity of sediments from the Central Graben of the North Sea (see Table 2.4), and Kaelin et al. (1992) for the autochthonous sandstone from the Swiss Molasse basin.

The following empirical relationship (Dortman, 1976), based on this equation was derived for sedimentary rocks of the Russian Platform to a depth of about 3 km:

$$\Phi(z) = \Phi_0 \cdot \exp(-0.45 \cdot z) \quad (2\text{-}11)$$

where the depth z is given in km.

Jelic (1984) has derived a relationship for sandstones (Savska Potolina/formerly Yugoslavia) with an initial porosity of $\Phi_0 = 0.496$

$$\Phi = 0.496 \cdot \exp(-0.556 \cdot z) \quad (2\text{-}12)$$

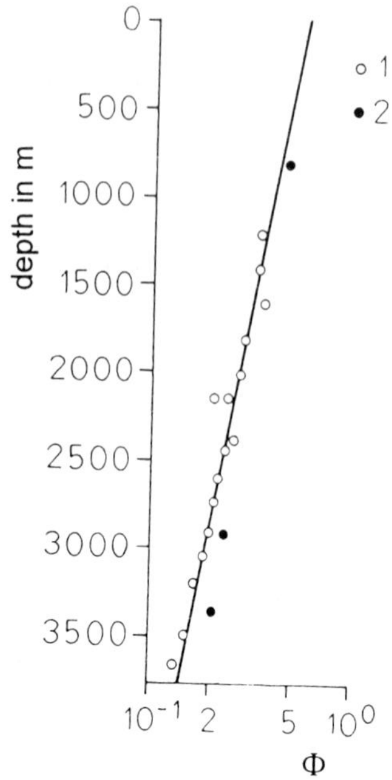

Figure 2.5. Porosity vs. depth z in a semilogarithmic plot; Well Kambara GS-1/Niigata Prefecture; open circles: mudstone; full circles: sandstone; after Nagumo (1965)

Wyble (1958) published a similar relationship

$$\Phi(p) = \Phi_0 \cdot [C + (1 - C) \cdot \exp(-D \cdot p)] \tag{2-13}$$

where C and D empirical values. C decreases with increasing clay content (Voigt, 1979).

Liu and Roaldset (1994) compared various porosity versus depth relationships including a polynomial equation (parabolic model). The equations were used to analyze sediments from the Northern North Sea and are summarized in Table 2.4.. The authors recommended that the parabolic model be used to fit the data (core data and logging data) for the Northern North Sea region.

Furthermore, there are relationships which include the influences of (geological) time t and temperature T as a result of "chemical compaction" (Maxwell and Verall, 1954; Maxwell, 1964; Dobrynin, 1970; Wieseneder, 1976).

Table 2.4. Porosity versus depth relationships for sandstone and shale from the Northern North Sea, after Liu and Roaldset, 1994; z in m, Φ as fraction; References: B - Baldwin and Butler, 1985; S - Sclater and Christie, 1985; L - Liu and Roaldset, 1994

Sediment	Relationship	Ref.
Sandstone	$\Phi = 0.49 \cdot \exp(-2.7 \cdot 10^{-4} \cdot z)$	S, B
Sandstone	$\Phi = 0.728 - 2.719 \cdot 10^{-4} \cdot z + 2.604 \cdot 10^{-8} \cdot z^2$	L
Shale	$\Phi = 0.803 \cdot \exp(-5.1 \cdot 10^{-4} \cdot z)$	S
Shale	$\Phi = 0.803 - 4.3 \cdot 10^{-2} \cdot \ln(z+1) - 5.4 \cdot 10^{-3} \cdot \ln(z+1)^2$	B
Shale	$\Phi = 0.803 - 2.34 \cdot 10^{-4} \cdot z + 2.604 \cdot 10^{-8} \cdot z^2$	L

In all cases, the pressure is the so-called "effective pressure" (i.e., the pressure acting on the rock skeleton causing a compression of the pore space). It originates from the lithostatic (lith) and hydrostatic (hydr) pressure:

$$p_{\text{eff}} = p_{\text{lith}} - \beta \cdot p_{\text{hydr}} \tag{2-14}$$

The factor ß has a mean value of about 0.85 (Schopper, 1982).

It should be noted that for other types of porosity (e.g., secondary porosity, crack porosity) and their dependence on pressure or depth, there is also a relationship following equations (2-4, 2-6) which gives satisfactory approximations.

2.2.5. Porosities of sphere packings

Regular sphere packings - for example cubic and hexagonal packing - are models frequently used for unconsolidated sands (e.g., calculation of elastic properties, see chapter 6.). Figure 2.6 shows two examples.

The calculation of porosity assuming spheres of equal size in a regular arrangement (like a crystal lattice) results for a cubic packing of spheres in a porosity of

$$\Phi_{\text{cubic}} = \frac{V_{\text{pore}}}{V_{\text{cubus}}} = 1 - \frac{V_{\text{sphere}}}{V_{\text{cubus}}} = 1 - \frac{4/3\pi \cdot r^3}{(2 \cdot r)^3} = 1 - \pi/6 = 0.4764 \tag{2-15}$$

Geometry and values for two regular packings are given in Figure 2.6. Figure 2.6 also contains the number of contact points per sphere (coordination number); see also Graton and Frazer (1935).
Obviously, the calculated porosity for such packings with spheres of equal size

- has discrete values only,
- has a range between 0.259 and 0.476,
- is independent of the sphere diameter ("grain size").

These models seem to be of limited validity, but are often used for lack of more realistic models.

Packing type	Solid phase crystal structure	Pore structure	Porosity (%)	Number of contact points per sphere
Simple cubic		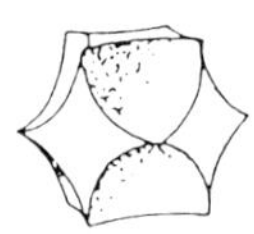 Cubic void	47.6	6
Compact hexagonal or tetrahedral	Separated view 	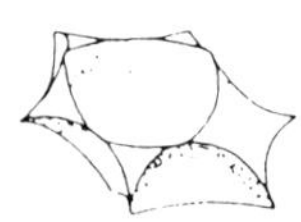 Tetrahedral void	25.9	12

Figure 2.6. Characteristics of some spherical packings (after Bourbie et al., 1987)

2.3. Specific internal surface

Porosity describes the volume of the voids (pores, cracks, fissures, fractures etc.) in relation to the considered rock volume. The specific internal surface describes the surface area of these voids in relation to either

- the total rock volume (S_{tot})
- the pore volume (S_{por})
- the volume of the solid matrix (S_m)
- the mass of the dry rock (S_{ma})

of the considered rock part. The following relations result

$$S_{tot} = \Phi \cdot S_{por} = (1 - \Phi) \cdot S_m \qquad (2\text{-}16)$$

$$S_{ma} = S_m / d_m \qquad (2\text{-}17)$$

where d_m is the density of the solid matrix material. The specific surface S_{por} is identical to the mean capillarity (Schopper, 1982).
The following *units* are used:
- for S_{tot}, S_{por}, and S_m the unit is $m^2/m^3 = m^{-1}$, usually μm^{-1} is preferable,
- for S_{ma} the unit is m^2/g or m^2/kg.

Methods of measuring internal surface areas can be classified into two groups (Schopper, 1982):

1) *Physical methods* based on the surface area affecting the adsorption of gases, liquids, dyes etc. The resolution depends on the size of adsorbed particles (e.g. molecules). The most popular method is the BET-method (after Brunauer, Emmet and Teller). It is based on monomolecular adsorption (or condensation) of nitrogen gas on the internal surface. Details can be found in literature on this subject.

2) *Stereological methods* to determine surface density or component-specific surfaces from section images. The methods are based on stereological principles for estimating geometrical properties of a higher dimensional space (3-dimensional space) by properties in a lower dimensional section (2-dimensional section image) of the considered space (for details see Underwood, 1970; Thompson et al., 1987).

The specific internal surface depends mainly on

- the shape and size of pores,
- the microstructure and morphology of the matrix - pore interface.

The last factor entails a dependence of the measured value on the power of resolution of the applied method. Schopper (1982) has described this feature of the internal surface as follows: "Internal surfaces of natural materials are not smooth, but wavy with many superimposed frequencies; i.e., they contain many superimposed levels of fine structures. Every method of measuring the internal surface invariably applies a certain high frequency choke to the measurement, i.e. smooth surface only is measured. Thus, the measured value always depends on the power of resolution of the method. It can differ by orders of magnitude within the conventional methods and does not even converge towards a limiting value, but theoretically goes beyond all limits with increasing resolution. (This problem is called the "coastline of Britain problem", due to the fact that the length of a coastline as determined from a map depends on the scale)."

Thus, the specific internal surface varies over a wide range of magnitudes. The "micromorphology" of the pore surface has a strong influence on the value of the specific internal surface. As a general tendency, the specific internal surface increases with

- the decrease of the mean pore or grain size,
- the presence of clay particles, carbonate and other (e.g. hydroxidic) minerals on the pore surface and, generally, the growth of various types of surface structures.

Especially for unconsolidated sediments, the specific internal surface related to the matrix volume S_m increases with the decrease of the following properties: mean grain size, sphericity, roundness, and size variance (Schopper, 1982).

A mathematical description of pore geometry based on a fractal "pigeon hole" pore model was developed by Pape, Riepe and Schopper (1982). It allows an analytical description of the fine structure of the pore space and opens the possibility of understanding the effects of interface phenomenona on physical properties, such as hydraulical and electrical properties.

2.4. Permeability

"Over the last two decades, researchers who have been trying to understand transport processes in porous media have begun to recognize that we do not know nearly as much as we once thought we did. Natural porous media are generally not homogeneous, nor are they uniformly random.... Transport in porous media has become a truly multidisciplinary science that requires tools and concepts from many fields." (John H. Cushman, 1990)

2.4.1. Definition and terminology

Permeability describes the property of a porous rock regarding fluid flow through the pore space. It depends on the porosity, on the pore space dimension and geometry, and is generally a tensor.

In 1856, the French engineer Henry Darcy found the fundamental relation for defining laminar flow of a viscous fluid through a porous rock to be

$$u = -\frac{k}{\eta} \cdot \nabla p \qquad (2\text{-}18)$$

where u is the volume flow density or flow rate per cross section area (filtration velocity), p is the fluid pressure, η is the dynamic viscosity of the fluid, k is the rock permeability. Therefore,

$$k = -\eta \cdot \frac{u}{\nabla p} \qquad (2\text{-}19)$$

Equations (2-18) and (2-19) are valid in this form for the isotropic case. They define the permeability of the laminar flow of a viscous fluid.
For anisotropic material permeability must be considered as a tensor; a detailed discussion for laminated rocks is given by Pickup et al. (1994, 1995)

If, in a porous and permeable rock, water (considered as the wetting fluid) and a second fluid (for example oil, non-wetting) flows, the relative permeabilities of the two fluids must be distinguished. The *relative permeability* is defined as the ratio of the effective permeability of the individual fluid to the absolute or intrinsic rock permeability.

Units: Permeability has the dimension of an area (see eq. 2-19). SI-units are therefore m^2 or usually μm^2. In practice, the unit "Darcy" (d) is commonly used. It can be defined as the permeability of a material which permits a volume flow density of 1 cm/s of a fluid with a viscosity of 1 cP (Centi-Poise) under a pressure gradient of 1 atm/cm. The "Millidarcy" md (1 md = 10^{-3} d) is frequently used.

Conversions: 1 d = 0.9869 μm^2 = 0.9869 x 10^{-12} m^2
1 μm^2 = 1.0133 d
In most practical cases, the approximation 1 d $\approx$ 1 μm^2 is sufficient.

In *hydrogeology*, the streaming fluid is always water, and the fluid pressure gradient results from the difference of the water level Δh over a considered distance l. Therefore, in many practical cases equation (2-18) is used in the modified form

$$u = k_f \cdot \frac{\Delta h}{l} \qquad (2\text{-}20)$$

where the hydraulic permeability (or conductivity) coefficient k_f has the dimension of a velocity (mostly given in cm/s). It is defined only for the medium water (i.e. for a constant viscosity and density of the fluid). Under this assumption a relationship between the units results:
a k-value of 1 md corresponds to a k_f-value of about 10^{-6} cm/s
1 md $\approx 10^{-6}$ cm/s = 10^{-8} m/s 1 m/s $\approx 10^5$ d

For an exact determination, it is necessary to take into consideration the following effects and phenomena (see also Klinkenberg, 1941; Geertsma, 1974; Scheidegger, 1960; Schopper, 1982; Debschütz et al., 1989):
a) Streaming *fluid is compressible* (e.g. gas): Correction regarding the expanded Darcy equation must be applied ("normalization of the pressure").
b) The *"Klinkenberg - effect"*: Result of the slip of gas-molecules in small pores. It acts "when the pore dimensions approach the mean free path of the gas molecules " and the resulting "apparent permeability becomes dependent on the mean absolute gas pressure" (Schopper, 1982). The Klinkenberg-factor α allows the correction of the measured permeability values:

$$k_{\text{corr}} = k_{\text{measured}} \cdot \left(1 + \frac{\alpha}{p}\right)^{-1} \qquad (2\text{-}21)$$

c) Above the so-called "critical flow density", the flow is (partly) nonlaminar. In this case, the relation between flow rate and pressure gradient is described by a nonlinear relation. This is described by the *Forchheimer equation*.
d) *Solid-fluid-interface reactions*: The forces at the internal surface can create a "structural viscosity" in the liquid (especially in polar liquids and electrolytes in semipermeable materials) and can have a strong influence on the flow properties at very low flow velocities (interface forces are comparable to the viscous forces).
e) For *two- or polyphase flow*, a relative permeability must be determined depending on the composition of the flowing fluid.

2.4.2. Mean values and ranges of permeability

Corresponding to the classification of different porosity types, we can define

- intergranular permeability
- intragranular permeability
- fracture, crack and fissure permeability
- vugular permeability (Schopper, 1982).

Permeability depends in a very complex way on the properties of the pore space or crack space. The dominant influences are

► porosity,
► pore size and its distribution,
► pore shape, pore surface morphology, specific internal surface,
► arrangement of pores and pore throats (topology of the pore network).

Figure 2.7 gives an overview of the mean permeability ranges for certain rock types. Note that the scale spans thirteen orders of magnitude. The following tendencies are obvious:

◆ permeability increases with increasing porosity (see the increase from unfractured rocks to fractured and porous rocks),
◆ permeability increases with increasing grain size (especially in unconsolidated sediments from shale/clay to gravel),
◆ permeability decreases with compaction and cementation (decrease of porosity and pore-throat radii).

In the following, the influencing parameters are discussed in more detail. This discussion is concentrated on sedimentary rocks, although igneous rocks are also of increasing interest, mostly in connection with problems in geothermal energy sites, waste disposal sites, and hydrogeological research.

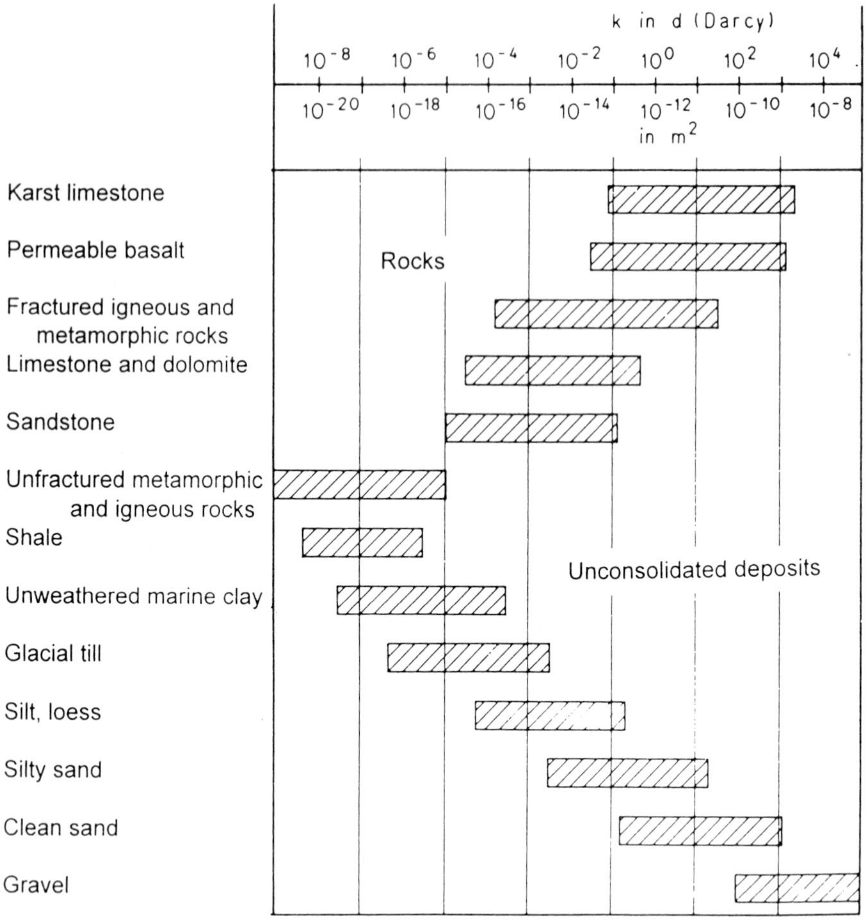

Figure 2.7. Range of permeability values in rocks (drafted after Freeze and Cherry, 1979; cited by Hearst and Nelson, 1985)

2.4.3. The influence of porosity and grain geometric properties on permeability

Porosity and permeability are related to different properties of pore space geometry. This explains the correlation between porosity and permeability, but also the scattering of such correlations indicating strong additional influences. Figures 2.8 and 2.9 are examples of typical permeability vs. porosity plots (a detailed description of the permeability-porosity relationships in sedimentary rocks is given by Nelson, 1994).

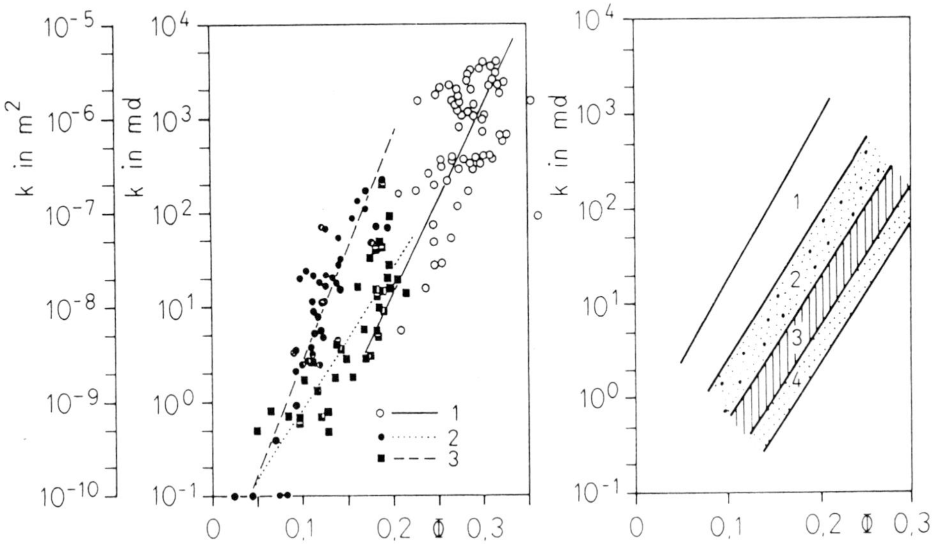

Figure 2.8 Permeability vs. porosity. Left: Core samples from three sandstone reservoirs (1 - Gulf Coast Field; 2 - Colorado Field; 3 - California Field), from Timur (1968);
Right: Core samples from well-sorted sandstones with various clay minerals present in the reservoir rock (1 - clay-free; 2 - kaolinitic; 3 - chloritic; 4 - illitic), from Dresser Atlas (1985)

It is obvious that the increase of permeability with growing porosity is strongly influenced by rock type. The data follow a quasi-linear log(k)-Φ relationship. Figure 2.9 shows this type of plot for two wells. SEM photographs from well 12 showed the presence of clay coating on sand grains, which results in a lower permeability and confirms the tendency observed in Figure 2.8.

Figure 2.10 indicates that the dominant parameter for sediments is grain size. Figure 2.11 shows a logarithmic plot of permeability k vs. grain size d. A regression analysis of these data results in the empirical relation

$$\log k = -2.1007 + 2.221 \cdot \log d \qquad (2\text{-}22)$$

where k is in md and d in μm (Schopper, 1982).

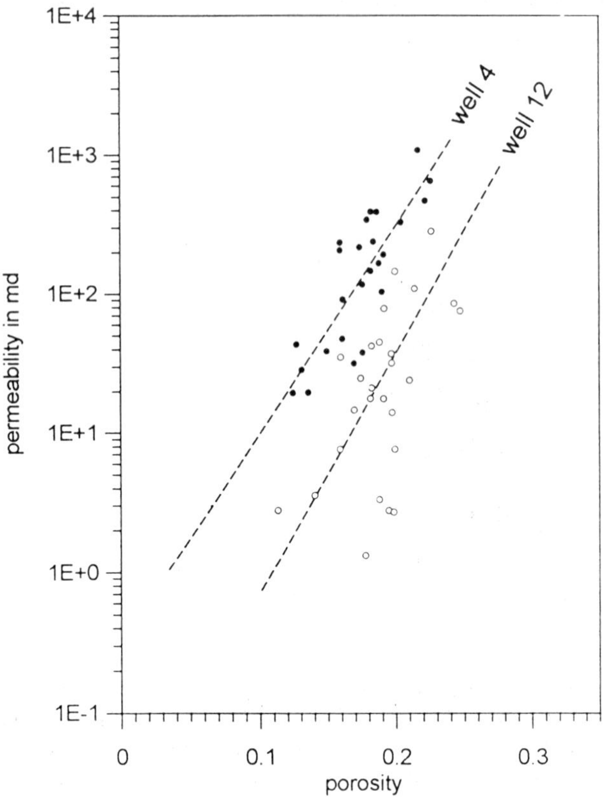

Figure 2.9. Permeability vs. porosity (Rotliegend sandstone); data after Diederix (1982)

The dependence of permeability on grain size (and porosity) for *unconsolidated rocks* is obvious. The correlation between permeability and the square of the grain size is found in the empirical equations of Hazen (1893)

$$k_f = 100 \cdot d_{\mathrm{m}}^2 \tag{2-23}$$

and Terzaghi (1955)

$$k_f = 200 \cdot d_{\mathrm{w}}^2 \cdot \left(\frac{\Phi}{1-\Phi}\right)^2 \tag{2-24}$$

for unconsolidated sand. Contained in equations (2-23) and (2-24) is the permeability coefficient k_f in cm/s and $d_{\mathrm{m}}, d_{\mathrm{w}}$ are median and effective grain diameter values (in mm) derived from the grain size distribution curve.

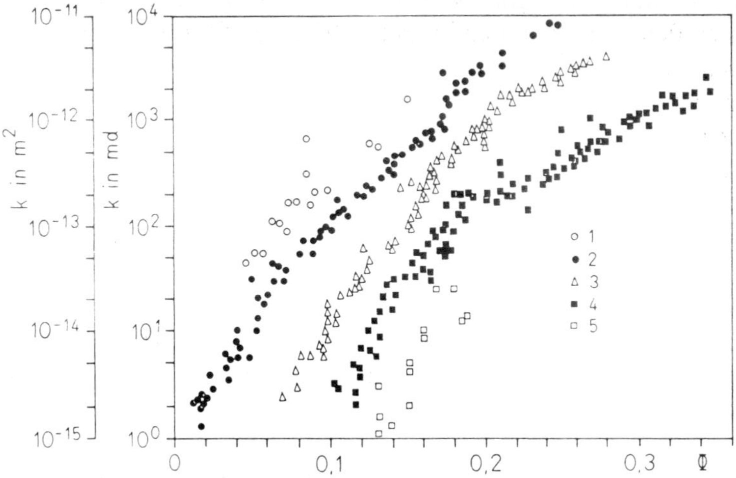

Figure 2.10. Permeability vs. porosity and the influence of grain size of sediments: 1 - coarse- and very coarse-grained; 2 - coarse- and medium coarse grained; 3 - fine-grained; 4 - silty; 5 - clayey; after a figure from Chilingar (1969, see Serra, 1984).

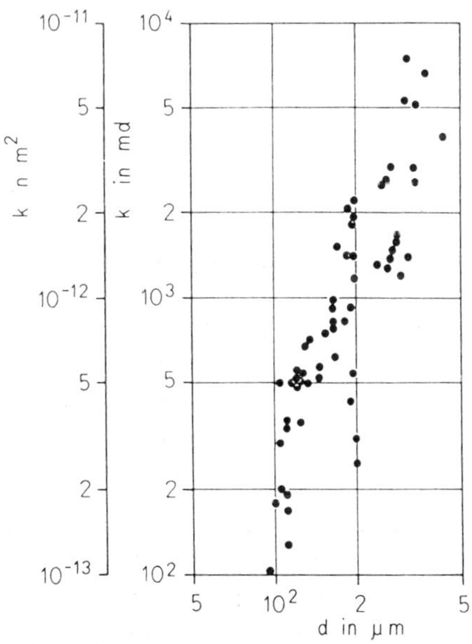

Figure 2.11. Permeability k vs. grain size d (Bentheim Sandstone, Scherhorn oilfield, Germany); from Engelhardt (1960) and Schopper (1982)

The dominant influence of grain size is shown in a hydraulic characterization of sediments by Pettijohn, Potter and Siever, 1972 (see Schopper, 1982):

- *Good aquifers:*
  Clean gravel (10^5 ... 10^4 d)
  Clean sands; mixtures of clean sands and gravel (10^3 ... 10^1 d)
- *Poor aquifers:*
  Very fine sands; silts; mixtures of sand, silt, and clay; glacial till; stratified clays (10^0 .... 10^{-3} d)
- *Impervious :*
  Unweathered clays (10^{-4} ... 10^{-5} d)

This corresponds to magnitudes published by Hölting (1989) for unconsolidated sediments (Table 2.5). The data are given in m/s and were converted for comparison to d.

Table 2.5. Magnitude of permeability for unconsolidated sediments (after Hölting, 1989)

Rock type	k_f in m/s	k in d (converted)
Gravel, clean	$10^{-2}...10^{-1}$	$10^3...10^4$
Sand, coarse	about 10^{-3}	about 10^2
Sand , medium	$10^{-4}...10^{-3}$	$10^1...10^2$
Sand, fine	$10^{-4}...10^{-5}$	$10^1...10^0$
Sand, silty	$10^{-5}...10^{-7}$	$10^{-2}...10^0$
Silt, clayey	$10^{-6}...10^{-9}$	$10^{-4}...10^{-1}$
Clay	$< 10^{-9}$	$< 10^{-4}$

A classification for hydrogeologic investigations is based on this (Hölting, 1989):

very low permeable	$k_f < 10^{-8}$ m/s
low permeable	$k_f = 10^{-8} ...10^{-6}$ m/s
permeable	$k_f = 10^{-6}... 10^{-4}$ m/s
high permeable	$k_f > 10^{-4}$ m/s

Further investigations of sedimentary rocks (see Pettijohn et al., 1972 and Schopper, 1982, Pickup et al., 1994) show the influence of grain size sorting or distribution, cementation, sedimentary structures, etc., on permeability and the anisotropy of permeability (maximum permeability parallel to the plane of bedding or to the mean shape fabric). Nelson (1994) published a systematic discussion of various empirical (and theoretical) equations. Berg (1970) derived a relationship

$$k = 5.1 \cdot 10^{-6} \cdot \Phi^{5.1} \cdot d^2 \cdot \exp(-1.385 \cdot \psi) \qquad \text{(2-25)}$$

where permeability is in d (Darcy), porosity in percent, and median grain diameter d in mm. ψ is a sorting term $\psi = P_{90} - P_{10}$, called the percentile deviation, to account for the spread in grain size distribution; it is expressed in phi units.

Iverson and Satchwell (1989) derived a multi-dimensional correlation between permeability and other petrophysical and petrographic parameters using cores from the Tensleep Sandstone (Wyoming):

$$\frac{k}{d^2} = B_1 + B_2 \cdot 10^{\left(\Phi - s/d^2\right)} + B_3 \cdot s^2 + B_4 \cdot S_k \cdot V_f \cdot 10^{(-s/d)} \qquad \text{(2-26)}$$

where k is in md, d is the mean grain size diameter (in mm), s is the standard deviation from the mean, S_k is the coefficient of skewness, and V_f the weight fraction of fines. The coefficients are empirical (B_1 = 0.05408, B_2 = 0.05714, B_3 = 0.7020, B_4 = - 0.09427).

As a result of these interrelationships between porosity Φ, mean grain (or pore) size d and specific internal surface S, additional correlations exist between S and k (see also section 2.4.5). As a first approximation, such correlations follow a proportionality of

$$k \sim S^{-2} \qquad \text{or} \qquad k \sim S^{-2} \cdot \Phi^{3/2} \qquad \text{(2-27)}$$

Resulting from a study of the relationships between different pore geometric properties, charge density and nuclear magnetic relaxation of some 100 clay-bearing sandstone core samples (mainly from oilfields from various parts of the world), Sen et al. (1990) calculated a regression (written in notation of this book) based on laboratory data:

$$k = 10^{6.59} \cdot \Phi^m \cdot (S_{\text{por}})^{-2.08} \qquad \text{(2-28)}$$

where the regression coefficient is R = 0.90; m is the Archie-exponent (see chapter 9), k is in md and S_{por} in µm^{-1}.

The pore space geometry determines permeability and also the capillary forces; these forces control retention of water in the angles and capillaries between the grains. This so-called "*irreducible water*" cannot be displaced by the forces acting on the fluids in the large pores. The irreducible water saturation $S_{w,irr}$ tends to be larger in low-permeability rocks, where the capillaries are fine (Serra, 1984). On this basis, a derivation of empirical relations between permeability and irreducible water saturation is well-founded (Figure 2.12). Examples are the following relationships for k in md ($\Phi, S_{w,irr}$ in volume fractions):

Tixier (1949) $$k = \left[250 \cdot \frac{\Phi^3}{S_{w,irr}}\right]^2 \qquad (2\text{-}29)$$

Timur (1968) $$k = \left[100 \cdot \frac{\Phi^{2.25}}{S_{w,irr}}\right]^2 \qquad (2\text{-}30)$$

Coates-Dumanoir (1974) $$k = \left[\frac{300}{w^4} \cdot \frac{\Phi^w}{S_{w,irr}^w}\right]^2 \qquad (2\text{-}31)$$

where w is a textural parameter related to the cementation and saturation exponents m and n in Archie's law (see chapter 9), $w \approx m \approx n$ (Schlumberger, 1989).

A more generalized expression, as proposed by Wyllie and Rose (1950; see also Serra, 1972 and Schlumberger, 1989) is:

$$k = a \cdot \frac{\Phi^b}{S_{w,irr}^c} \qquad (2\text{-}32)$$

where a, b, and c are empirically determined. Hearst and Nelson (1985) noted that b is between 3 and 6, and $c = 2$ is used, following Timurs equation (1968). A common relation is derived for the three sandstone types illustrated in Figure 2.8 (left) and shown in Figure 2.12. The relation according to equation (2-32) results with $a = 0.136$, $b = 4.4$, and $c = 2$ if k is in md.

Such correlations are fundamental for permeability calculations utilizing well log measurements. Nelson, 1994 concluded "the best of the log-based predictors rely on good estimates of irreducible water saturations". Section 2.2.5 also shows some relationships based on model calculations.

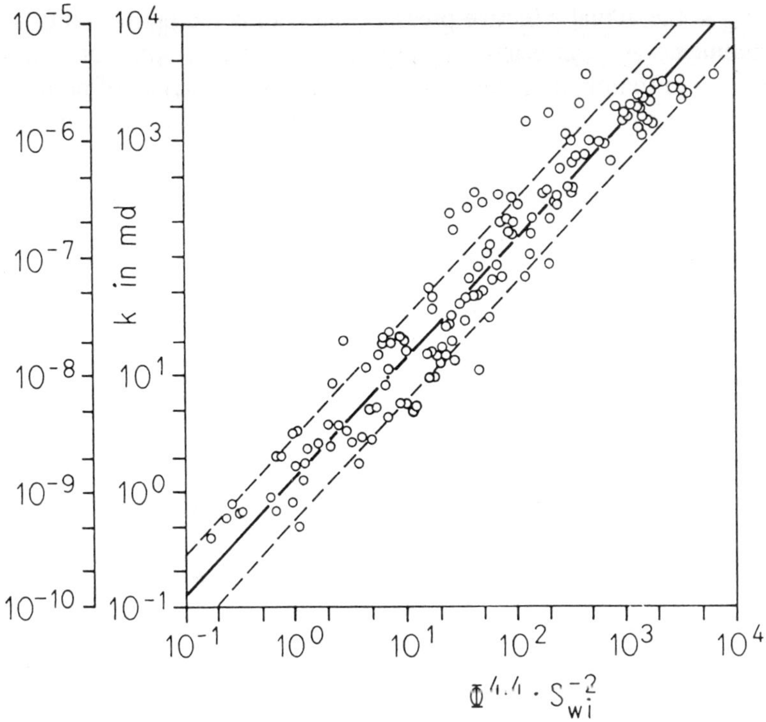

Figure 2.12. Correlation between permeability and porosity and irreducible water saturation for the core samples plotted in Figure 2.7 (left), from Timur (1968).

2.4.4. The influence of depth and pressure on permeability

The reduction of permeability with overburden pressure was described in Fatt and Davis (1952) and Fatt (1953). Increasing pressure reduces the pore space and causes a deformation of the pore channel cross section and its geometry. These effects chiefly result in a permeability decrease with growing pressure. This correlation is nonlinear (see Figure 2.13 and Figure 2.14).

The pressure dependence of permeability can be described to a first approximation by the equation

$$k = k_0 \cdot \exp(-A_k/p_{\text{eff}}) \qquad (2\text{-}33)$$

where p_{eff} is the actual effective pressure, and k_0 is the permeability at $p = 0$. The parameter A_k represents the pressure dependence of permeability and features a deformation modulus (also "compaction coefficient of permeability"). This approximation was also found for some crystalline rocks (Figure 2.14).

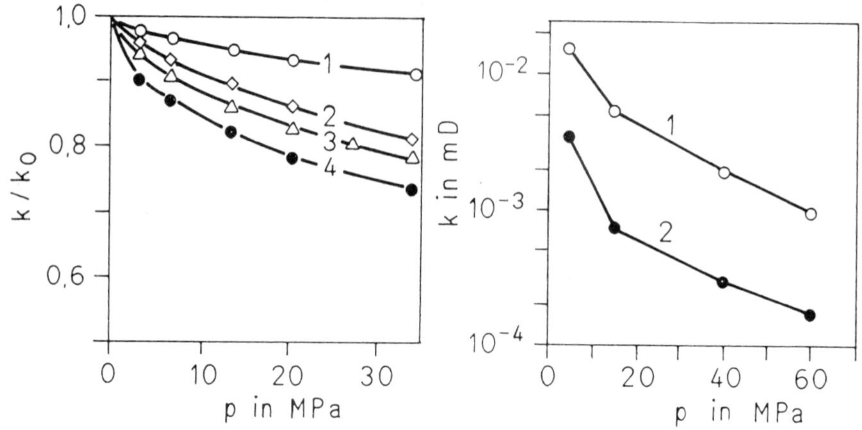

Figure 2.13. Permeability vs. pressure; left - permeability k (normalized with the value at pressure $p = 0$) as a function of the overburden pressure p; sandstone samples (Fatt, 1953)
1 - $\Phi = 15$ % initial permeability 249 md 2 - $\Phi =$ 24 % initial permeability 163 md
3 - $\Phi = 25$ % initial permeability 335 md 4 - $\Phi =$ 22 % initial permeability 110 md
right - permeability k as a function of confining pressure, measured on a gneiss sample in two directions (after Nover et. al., 1989); 1 - axial, i.e. nearly parallel to the plane of foliation 2 - radial, inclined at an angle of 40°.

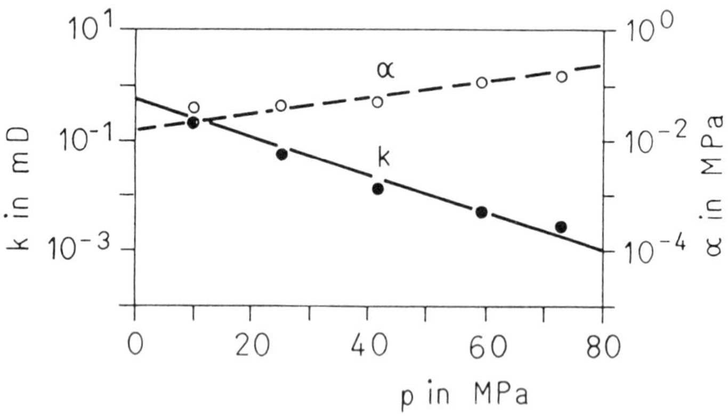

Figure 2.14. Permeability k and Klinkenberg factor α vs. pressure p; Garnet-biotite-gneiss (KTB, depth 708.8 m); after Debschütz et al., (1991).

For cracked or fractured rocks, a nonlinear decrease of permeability with increasing pressure as a result of a "closure process" is obvious, but difficult to specify in detail. For example Yoon and Lee (1993) found a linear correlation between the logarithm of permeability and the logarithm of pressure for low porous granite. Oda et al. (1989) described the change of hydraulic conductivity and crack aperture with depth for a jointed granite when investigating a rock massive for underground oil storage. The permeability decrease with depth z (in m) follows the proportionality

$$k(z) \sim \left(1 - \frac{z}{z+20}\right)^3 \qquad (2\text{-}34)$$

Siegesmund et al., (1993) found at orthogneisses a permeability decrease by about one order in magnitude at hydrostatic pressures between 5 MPa up to 100 MPa and a strong anisotropy. The permeability is low when measured perpendicular to the foliation plane. This is illustrated by the permeability values at 5 MPa for one sample (sample 767; porosity 1.43 %):

$k = 24 \times 10^{-6}$ d (in the foliation in direction of lineation)

$k = 10^{-7}$ d and $k = 1.3 \times 10^{-8}$ d (perpendicular to foliation plane).

A *non-hydrostatic* pressure results in a change of increasing and decreasing permeability depending on the rock anisotropy and the acting stress tensor. Pusch and Weber (1993) found that for crystalline rock samples investigated in a triaxial permeameter cell (radial pressure up to 160 MPa, axial pressure up to 230 MPa) the rock permeability initially decreases with increasing deviatoric stress. After exceeding a certain ratio of radial to axial stress, permeability increases again. This is originated by the overlapped growth of new or the expansion of existing flow paths due to anisotropy of effective stress. The influence of the stress distribution in the near-borehole zone upon the permeability was studied by Buchsteiner et. al. (1993). Fluid production and the corresponding pressure drawdown lead to an increase of the effective stress and unequal reduction of the components of the permeability tensor and, thus, in permeability anisotropy.

Theoretical studies of permeability vs. pressure relationships use different models and assumptions. For example, Walsh and Brace (1984) use the equivalent medium concept with hydraulic radius, tortuosity and a closure law for the pores. Somewhat different with respect to the elastic closure, the model of Seeburger and Nur (1984) is based on a numerical technique. Theories using statistical distributions of crack and pore parameters and their interconnection have been published, e.g., by Gueguen and Dienes (1989) and Gavrilenko and Gueguen (1989).

2.4.5. Relationships between k, Φ and S based on models

The simplest model for describing the streaming process in a porous rock is the pore channel or "capillary tube model" (Figure 2.15). It was derived by Carman (1956). The pore is idealized as a capillary channel with radius r and length l within the considered cube of length L.

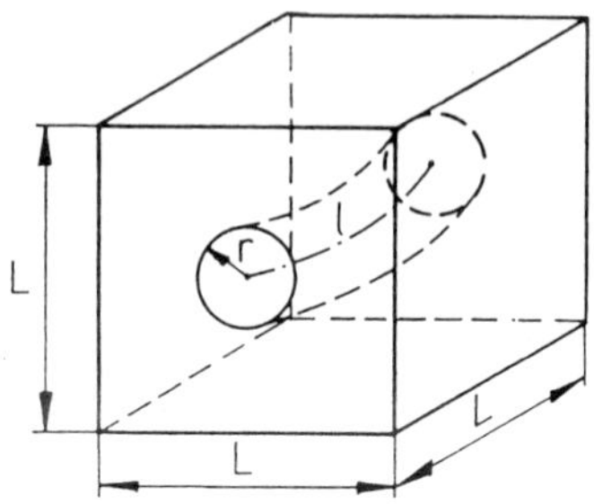

Fig. 2.15. The simple pore channel model

For this model one has

the porosity
$$\Phi = \frac{\pi \cdot r^2 \cdot l}{L^3} = \pi \cdot T \cdot \left(\frac{r}{L}\right)^2 \tag{2-35}$$

the specific surface
$$S_{\mathrm{por}} = \frac{2}{r} \tag{2-36}$$

the tortuosity
$$T = \frac{l}{L} \tag{2-37}$$

The volume flow rate for a tube follows the Hagen-Poiseulle law

$$\frac{\mathrm{d}V}{\mathrm{d}t} = \frac{\pi \cdot r^4}{8\eta} \cdot \frac{\nabla p}{T} \tag{2-38}$$

With the volume flow density $u = \frac{\mathrm{d}V}{\mathrm{d}t} \cdot L^{-2}$, a direct comparison with the Darcy equation (2-18) results in a relation for permeability:

$$k = \frac{\Phi \cdot r^2}{8 \cdot T^2} = \frac{\Phi}{2 \cdot S_{por}^2 \cdot T^2} = \frac{\Phi^3}{2 \cdot S_{tot}^2 \cdot T^2} = \frac{\Phi^3}{2 \cdot (1-\Phi)^2 \cdot S_m^2 \cdot T^2} \quad (2\text{-}39)$$

For cross section types, other than a circle (such as rectangle, triangle, ellipse), the introduction of the Kozeny-Carman shape or cross section shape-factor χ is recommended (see Engelhardt, 1960; Scheidegger, 1960; Brace, 1977; Riepe, 1984; Gueguen et al., 1986; Georgi and Menger, 1994):

$$k = \frac{\Phi \cdot r_{hyd}^2}{4 \cdot T^2 \cdot \chi} = \frac{\Phi}{\chi \cdot S_{por}^2 \cdot T^2} = \frac{\Phi^3}{\chi \cdot S_{tot}^2 \cdot T^2} = \frac{\Phi^3}{\chi \cdot (1-\Phi)^2 \cdot S_m^2 \cdot T^2} \quad (2\text{-}40)$$

Table 2.7 contains some values for χ .

The capillary radius r is replaced by the so-called "hydraulic radius" r_{hyd} for non-circular cross section shape

$$r_{hyd} = 2 \cdot \frac{\text{cross-sectional area normal to flow}}{\text{wetted pore perimeter}} \quad (2\text{-}41)$$

Using this definition of hydraulic radius for a circle results in $r = r_{hyd}$.

Table 2.7. Cross section shape factor

Cross section shape		χ
Circle		2.0
Ellipse, semi-axes a and b	a/b = 2	2.13
	a/b = 10	2.45
	a/b = 50	2.96
Square		1.78
Rectangle, sides a and b	a/b = 2	1.94
	a/b = 10	2.65
	$a/b = \infty$	3.0
Triangle, equilateral		1.67

This set of equations (from 2-38 to 2-41) relates permeability to the other geometric properties of the pore channel based on the simple model and modifications to them.

Note especially

- the dominant influence of the pore radius (compare equation 2-38)
- and of the pore geometry expressed by specific surface properties,
- the influence of porosity.

For *reservoir characterization* some derived parameters are used. Rewriting equation 2-40, results (Georgi and Menger, 1994):

$$k = \frac{1}{\chi \cdot S_{\mathrm{m}}^{2} \cdot T^{2}} \cdot \frac{\Phi^{3}}{(1-\Phi)^{2}} = \frac{1}{C^{2}} \cdot \frac{\Phi^{3}}{(1-\Phi)^{3}} \qquad (2\text{-}42)$$

where

$$C = \sqrt{\chi} \cdot S_{\mathrm{m}} \cdot T \qquad (2\text{-}43)$$

and related to the *"Flow Zone Indicator FZI"* (Amaefule, 1993). The rock-specific parameter *FZI* characterizes the permeability versus porosity correlation for a porous rock. It is constant within the same hydraulic unit, but varies from one unit to another.

In section 9.3.3.2. the formation factor F is introduced. Equation 9-82 gives a relationship between formation factor and porosity and tortuosity for the pore channel model. Assuming that the pathways for fluid flow and electrical current are the same, a relationship between permeability and formation factor results

$$k = \chi^{-1} \cdot \frac{1}{F \cdot S_{\mathrm{por}}} \qquad (2\text{-}44)$$

A plot k versus F with a logarithmic scale results in an empirical modification of the permeability versus formation factor relationship (Worthington, 1975; Katsube and Hume, 1987):

$$k = a \cdot F^{-u} \qquad (2\text{-}45)$$

where a and u are empirical coefficients. Two examples may illustrate the empirical relationship 2-45:

a) Bunter sandstone (Northwest Lancashire, England; Worthington, 1973) with hydraulic k_f in mm/s

$$k_f = 1119 \cdot F^{-5.88} \qquad (2\text{-}46)$$

b) Granite (Manitoba, USA; Katsube and Hume, 1987) with k in m^2

$$k = 2.51 \cdot 10^{7} \cdot F^{-2.22} \tag{2-47}$$

Freund and Nover (1995) collected values of the exponent u for some rock types (Table 2.8).

Table 2.8 Exponent u (equation 2-45) for some rock types; from a data collection of Freund and Nover, 1995; n - number of samples; R - correlation coefficient

Rock type	n	u	R	Reference
Sandstone, with clay bearing	126	3.92	0.82	Sen et al., 1990
samples	48	3.48	0.84	Thompson et al., 1987
Fontainebleau sandstone, clean	10	1.69	0.87	Doyen, 1988; David and Darot, 1989
Westerly granite	6	1.5 ... 2.4		Brace, 1984
Granite, Ontario	29	1.96	0.94	Katsube and Hume, 1987
Granite, Manitoba	24	2.22	0.84	Katsube and Hume, 1987

In the discussions and equations above, the effects of pore size distribution and pore shape distribution, "pore wall morphology", pore connectivity, and interlayer effects are either neglected or idealized. These influences are the subject of more sophisticated models and theories, such as:

- the "*spaghetti - model*" after Cornell and Katz (1953), with a bundle of different tubes in a parallel array;
- the "*network - model*" after Schopper (1972), with a statistical distribution of cross section geometry, pore channel length and an arbitrarily changeable configuration of the pore channel network,
- models based on the *percolation theory* (Kirkpatrik, 1973; Koplik, 1982; Ferrand et al., 1990) including a statistical description of pore or crack geometry (Gueguen et al., 1986; Gueguen and Dienes, 1989),
- models including the interface morphology (topology of the pore space). The "*pigeon hole model*" (Pape et al., 1981, 1985) is based on self similar pore structures. The pore wall has a fractal dimension (compare also with Wong, 1987) in this model type.

The research of Pape et al., (1985) results in a further generalization of the Kozeny-Carman equation and allows the calculation of permeability from porosity, formation factor F (see chapter 9) and a quantity describing the specific internal surface (S_{por}), as in the so-called "PARIS" equation:

$$k = \frac{475.3}{F} \cdot \left(\frac{S_{por}}{q_0}\right)^{-3.1085} \tag{2-48}$$

where the permeability k is in d (Darcy), S_{por} is in μm^{-1} after a BET-determination. The parameter q_0 is a "lithology factor", derived from a comparison between core and log data. The main difference between the equations based on the capillary model after Kozeny-Carman and the equations on the basis of a fractal concept is expressed by the exponent of the surface term S_{por}.

Three ways are described for determination the "key property S_{por}" by other methods (for example logs):

1) *from electrical measurements* based on the correlation between the interface conductivity (see chapter 9) and the specific internal surface (Pape et al., 1984). This establishes a connection to various electrical properties related to grain surface phenomenons like induced polarization, cation exchange capacity etc. (see chapter 9). Based on the laboratory data, Sen et al., (1990) found an equivalent empirical relationship

$$k = 10^{2.65}(\Phi^m/Q_v)^{2.11} \qquad R = 0.88 \tag{2-49}$$

where the exchange cation molarity Q_v is related to the specific internal surface (see section 9.3.3.4).

2) *from NMR (nuclear magnetic resonance) measurements*. According to Timur (1969), S_{por} correlates directly to the reciprocal spin-relaxation time t_1. Because the resolution power of the atomic relaxation process is of the same order of magnitude as the BET-adsorption method, S_{por} can replaced directly by the NMR-parameters (Riepe et al., 1986). Also in this case, Sen et al., (1990) found an empirical equivalent relationship

$$k = 0.794 \cdot (\Phi^m \cdot t_1)^{2.15} \qquad R = 0.94 \tag{2-50}$$

where permeability is in md, porosity as fraction, and spin relaxation time in milliseconds. Note the high value for the regression coefficient R.

The derivation of the relationship between permeability and NMR relaxation time on the basis of the Kozeny-Carman model results (Georgi and Menger, 1994)

$$k = \frac{\varsigma^2}{\chi \cdot T^2} \cdot \Phi \cdot t_1^2 \tag{2-51}$$

where ς is the "killing strength" and for clastics of the order of 0.003 cm/s (Thompson et al., 1989; Georgi and Menger, 1994).

3) *from natural gamma ray spectroscopy* based on the observation that the thorium concentration shows a direct correlation to the specific surface as a result of its adsorption on the mineral surface (see Riepe et al., 1986, Wehr, 1989).

Thompson et al. (1987) have applied the *fractal concept* to quantify the pore geometry for length scales smaller than the grain diameter and have used percolation theory to describe the fluid flow. Wheathcraft et al. (1990) also used as a fractal concept the Sierpinski carpet for the calculation of hydraulic conductivity distribution. These developments and their results confirm the statement by Morgan et al., (1989): "Many of the parameters in rock physics models are strongly dependent on the details of the microgeometry of rocks. One way to access the microgeometry quantitatively is to perform appropriate digital analysis of images of thin-sections of rocks ... Thus far, microgeometric properties such as porosity, fractal dimension, specific surface area, pore and grain-width distributions, pore shape, tortuosity and coordination number, anisotropy and Fourier spectra are being related to laboratory measurements such as porosity, permeability, capillary pressure curves and electrical formation factors." For example, they determined a relationship between permeability, image-derived porosity and specific surface:

$$k_i = \frac{1}{2} \cdot \left(\frac{\Phi_i}{S_i}\right)^2 \cdot \Phi_i^{1.73} = \frac{1}{2} \cdot \Phi_i^{3.73} \cdot S_i^{-2} \tag{2-52}$$

where the subscript i denotes the "image derived properties".
The specific surface as a fractal property depends on the resolution or image magnification, respectively. For the relationship in equation (2-52), the specific surface was determined with a resolution of 20 A° = 10^{-1} nm (as suggested by

Katz and Thompson, 1984) and by a back extrapolation or demagnification on the computer.
The following papers are recommended for further study of the image-analysis of porous materials: Berryman and Blair (1986); Ehrlich et al., (1984); Rink and Schopper, (1978).

Herron (1987) derived a mineralogical model; in place of the pore geometric properties in equation (2-40) he used mineralogical abundances and properties:

$$k = A_f \cdot \left[\frac{\Phi^3}{(1-\Phi)^2}\right] \cdot \exp\left(\sum_i B_i \cdot M_i\right) \tag{2-53}$$

where M_i is the weight fraction of each mineral component and B_i a parameter for each mineral (Table 2.9). Clay minerals with a negative parameter B_i depress permeability. Nelson (1994) analyzed the equation and the effect of mineral composition variation. The coefficient A_f represents the textural maturity of the sediment, described by the maximum feldspar content F_{max} over the interval of interest:

$$A_f = 4.9 + 2 \cdot F_{max} \tag{2-54}$$

Table 2.9 Mineral parameter B_i in Herron's (1987) equation

Mineral	B_i
quartz	0.1
feldspar	1.0
calcite	-2.5
kaolinite	-4.5
illite	-5.5
smectite	-7.5

Other concepts use acoustic or *seismic parameters* for permeability estimation (Bourbie and Zinsner, 1985; Owen, 1993).

CHAPTER 3

Density of Rocks

"Density is a physical property that changes significantly among various rock types owing to differences in mineralogy and porosity. If the distribution of underground rock densities is known, potentially much information can be learned about subsurface geology"
(Olhoeft and Johnson, 1989)

3.1. Definition and units

Density d is defined as the quotient of the mass m and the volume V of a material

$$d = \frac{m}{V} \tag{3-1}$$

Usually the symbol for the density is ρ; but because this symbol is also used for specific electrical resistivity, the symbol d is used for density in this book. The SI unit for density is kg m^{-3}. The following conversions give the relationships to other units:

1 g cm^{-3} = 10^3 kg m^{-3}
1 g cm^{-3} = 62.43 lb ft^{-3} = 8.345 lb/gal (US).

Due to the heterogeneity of rocks, it is necessary to distinguish between different densities which are related to different parts or components of the composite material we call "rock":

- d - bulk density: the mean density of the considered rock volume (including pores etc.); for example: density of sandstone
- d_i - density of an individual rock component i ; for example: density of quartz;

d_m - mean density of the solid matrix material; for example: density of a carbonate matrix (i.e., without pores);

d_p - mean density of the pore-(or fracture-) fluid; for example: density of pore water.

The following relationship exists between these quantities: The density of a composite material (rock) consisting of n components is (see Figure 3.1a) is

$$d = \sum_{i=1}^{n} \left(\frac{V_i}{V}\right) \cdot d_i \tag{3-2}$$

where d_i is the density and V_i is the volume of the component i; the ratio $\left(\frac{V_i}{V}\right)$ is the volume fraction of the component i. Equation (3-2) and Figure 3.1a represent the volumetric rock model, which is valid for density as a scalar property.

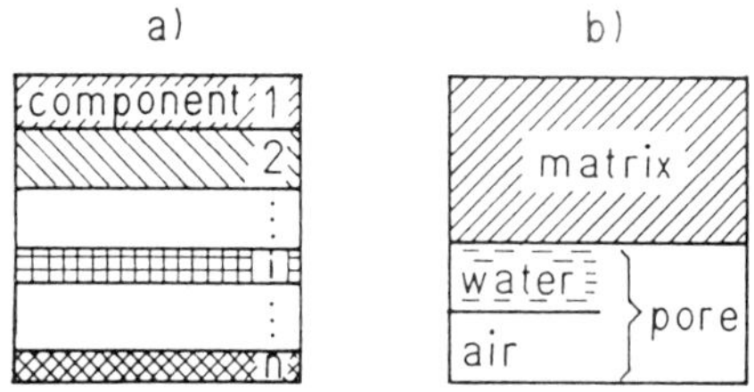

Figure 3.1 Model for density (and other scalar properties) of rocks
a) consisting of n constituents, b) consisting of matrix and two pore constituents.

With a gamma-gamma density log, density is measured based on Compton scattering. This gamma-gamma density d_{GG} (or electron density) is related to the density defined by equation 3-1 as follows:

$$d_{GG} = d \cdot \left(2 \cdot \frac{Z}{A}\right) \tag{3-3}$$

where Z is the atomic number and A the atomic mass (see also chapter 10).

For a porous rock, the volume fraction for the pore space is given by the porosity (eq. 2-1) and can be written as (see Figure 3.2b):

$$d = (1 - \Phi) \cdot d_m + \Phi \cdot d_p \tag{3-4}$$

where Φ is the total porosity.

For the complete volumetric description of a porous rock, a second parameter is needed for the characterization of the pore filling material - the saturation (or degree of saturation). The saturation S_f related to a pore fluid f is defined as the ratio of the volume of this fluid V_f to the whole pore volume V_p

$$S_f = \frac{V_f}{V_p} \tag{3-5}$$

For instance, the water saturation is $S_w = V_w/V_p$. A gas-(subscript g) and water-(subscript w) bearing porous rock has the density

$$d = [1 - \Phi] \cdot d_m + \Phi \cdot [S_w \cdot d_w + (1 - S_w) \cdot d_g] \tag{3-6}$$

3.2. Density of rock constituents

3.2.1. Density of minerals

The density of minerals is controlled by their elemental composition and by internal bonding and structure. Extensive compilations of mineral densities are published by Clark (1966), Dortman (1976), Olhoeft and Johnson (1989), Rösler and Lange (1972), Serra (1984), Wohlenberg (1982).

Densities (see Wohlenberg, 1982 and Table 3.1) range between *low values* of about 1.1 and 1.6 x 10^3 kg/m^3 (e.g. flagstaffite 1.092 x 10^3 kg/m^3, bischofite 1.59 x 10^3 kg/m^3, carnallite 1.604 x 10^3 kg/m^3) and *high values* of about 7 to 14 x 10^3 kg/m^3 (e.g. iron (α-Fe) 7.3 ... 7.87 x 10^3 kg/m^3, lead 11.273 x 10^3 kg/m^3, mercury 13.59 x 10^3 kg/m^3).

The following are ranges for many minerals:
- most abundant rock forming minerals $d = 2.2...3.5 \cdot 10^3$ kg/m^3
- ore minerals $d = 4.0...8.0 \cdot 10^3$ kg/m^3

In composite minerals (series), composition influences density. Table 3.2 demonstrates this with the example of plagioclases as isomorphic mixture of albite and anorthite.

Table 3.1. Mean density values (in 10^3 kg/m^3 = g/cm^3) for selected minerals; based on data from Olhoeft and Johnson (1989)

Mineral	Density	Mineral	Density	Mineral	Density
Actinolite	3.200	Fluorite	3.179	Montmorillonite	2.608
Albite	2.620	Forsterite	3.213	Muscovite	2.831
Almandine	4.318	Galena	7.598	Natrolite	2.245
Andalusite	3.145	Glauconite	2.300	Nepheline	2.623
Anhydrite	2.963	Graphite	2.267	Orthoclase	2.570
Anorthite	2.760	Gypsum	2.305	Phlogopite	2.784
Apatite	3.180	Glaucophane	3.200	Polyhalite	2.780
Aragonite	2.931	Halite	2.163	Pseudobrockite	4.390
Arsenopyrite	6.162	Hedenbergite	3.632	Pyrite	5.011
Augite	3.300	Hematite	5.275	CaAl-Pyroxene	3.360
Barite	4.480	Hornblende	3.080	Pyrrotite	4.610
Beryl	2.641	Illite	2.660	Quartz	2.648
Biotite	2.900	Ilmenite	4.788	Realgar	3.590
Calcite	2.710	Jacobsite	4.990	Rutile	4.245
Cassiterite	6.993	Kainite	2.130	Serpentine	2.600
Chalkopyrite	4.200	Kaolinite	2.594	Siderite	3.944
Chlorite	2.800	Kieserite	2.573	Sillimanite	3.241
Chromite	5.086	Kyanite	3.675	Sphalerite	4.089
Cordierite	2.508	Labradorite	2.710	Spinel	3.583
Danburite	3.000	Langbeinite	2.830	Sylvite	1.987
Diamond	3.515	Leucite	2.469	Talc	2.784
Diaspore	3.378	Maghemite	4.880	Titanomagnetite	4.776
Diopside	3.277	Magnesite	3.010	Tremolite	2.977
Dolomite	2.866	Magnetite	5.200	Trona	2.170
Enstatite	3.209	Malachite	4.031	Wollastonite	2.909
Epidote	3.587	Marcasite	4.870	Wustite	5.722
Fayalite	4.393	Microcline	2.560		

Table 3.2. Influence of mineral composition on density; (data from Dortman, 1976; the density data in the last raw are from the data found in Wohlenberg, 1982)

Type of plagioclase	Albite content in %	Anorthite content in %	Density in 10^3 kg/m^3	Density in 10^3 kg/m^3
Albite	100	0	2.61	2.603 ... 2.688
Oligoclase	80	20	2.64	2.612 ... 2.690
Andesine	60	40	2.67	2.647 ... 2.690
Labradorite	40	60	2.70	2.686 ... 2.718
Bytownite	20	80	2.73	2.75
Anorthite	0	100	2.76	2.703 ... 2.763

3.2.2. Density of pore (and crack) fluids

The density of liquids and gases is controlled by

- chemical composition (including soluble components),
- temperature and pressure (the density generally increases with increasing pressure and decreasing temperature).

For liquid substances, Schlumberger (1972, 1989) gives the following mean density values:

Fresh water	1.000 x 10^3 kg/m^3
Salt water (200,000 ppm)	1.146 x 10^3 kg/m^3
"Oil" , n(CH_2)	0.85 x 10^3 kg/m^3

Some detailed data are given in Table 3.3. The influence of pressure and temperature can be seen, as well as, the influence of the NaCl-concentration in water.

Worthington et al., (1990) published relationships between the concentration, density, and electrical resistivity of NaCl solutions as a part of the "Society of Core Analysts Guidelines for sample preparation". The density of NaCl solutions at 25 °C is

$$d_{25\,\mathrm{C}} = 0.99708 + 4.0785 \cdot 10^{-2} c - 9.5818 \cdot 10^{-4} c^2 + 5.1208 \cdot 10^{-5} c^3 \qquad (3\text{-}7)$$

where c is the solution concentration in mol/l. The value of d in the above equation is accurate to $\pm 10^{-4}\,\mathrm{g/cm^3}$.

Table 3.3. Density of different pore fluids (fluids and gases) d_f in g/cm^3 = 10^3 kg/m^3 after Gearhard, 1978; the table contains the *Z/A* ratio and the gamma-gamma apparent density d_{GG},; see chapter 10.

Substance	d_f	Z/A	d_{GG}
Water (dest.)	1.00	0.5551	1.11
(30 000 ppm NaCl)	1.022	0.5528	1.13
(50 000 ppm NaCl)	1.0365	0.5513	1.143
(100 000 ppm NaCl)	1.073	0.5476	1.18
(200 000 ppm NaCl)	1.146	0.54	1.238
(300 000 ppm NaCl)	1.219	0.5325	1.14
Oil, 10° API, STP	1.00	0.5703	1.14
30°API, STP	0.88		1.00
50° API, STP	0.78		0.85
70°API, STP	0.70	0.5778	0.81
N-pentane C_5H_{12}, STP	0.626	0.5823	0.733
200°F, 7000psi (93°C, 48 MPa)	0.603		0.702
N-hexane C_6H_{14}, STP	0.659	0.58	0.765
200°F, 7000psi (93°C, 48 MPa)	0.628		0.739
Average natural gas, STP	0.0007726	0.5735	0.000886
200°F, 7000psi (93°C, 48 MPa)	0.252		0.289
Methane CH_4, STP	0.000677	0.5703	0.00076
200°F, 7000psi (93°C, 48 MPa)	0.2189		0.2497
Air (dry)	0.001224	0.4997	0.001223

A detailed description of the relationships of fluid density, temperature, pressure, and NaCl concentration (weight fraction) was published by Batzle and Wang (1992). For water and brine (sodium chloride solution) a polynomial was derived:

$$d_w = 1 + 10^{-6} \cdot (-80 \cdot T - 3.3 \cdot T^2 + 0.00175 \cdot T^3 + 489 \cdot p - 2 \cdot T \cdot p$$

$$+0.016 \cdot T^2 \cdot p - 1.3 \cdot 10^{-5} T^3 \cdot p - 0.333 \cdot p^2 - 0.002 \cdot T \cdot p^2) \qquad (3\text{-}8)$$

$$d_{\mathrm{NaCl\ solution}} = d_w + C \cdot [0.668 + 0.44 \cdot C + 10^{-6} \cdot f(p, T, C)] \qquad (3\text{-}9)$$

with

$$f(p,T,C) = 300 \cdot p - 2400 \cdot p \cdot C + T \cdot (80 + 3 \cdot T - 3300 \cdot C - 13 \cdot p + 47 \cdot p \cdot C) \qquad (3\text{-}10)$$

where

d_w	is the density of the (pure) water in g/cm^3 = 10^3 kg/m^3
$d_{\text{NaCl solution}}$	is the density of the NaCl solution in g/cm^3 = 10^3 kg/m^3
T	is the temperature in °C
p	is the pressure in MPa
C	is the weight fraction of NaCl.

For water at 75° F (= 23.9° C) and atmospheric pressure, the relation between NaCl concentration C and density can be shown approximated by (Schlumberger, 1973)

$$d_w = 1 + 0.73\, C \qquad (3\text{-}11)$$

This approximation is in good agreement with equation 3-7 for concentrations of about C= 0.1200...0.1700 or 120,000 ppm...170,000 ppm.

The conversion between C (weight fraction) and c (mol/l) for NaCl solutions is

$$C[\text{ppm}] = \frac{58.443\ \text{g} \cdot 10^3 \cdot c[\text{mol/l}]}{d\left[\text{g/cm}^3\right]} \qquad (3\text{-}12)$$

Natural oils range from light liquids of low carbon number to very heavy tars, bitumen, and kerogen (which may be denser than water). Under room conditions, oil densities can range from under 0.5 g/cm^3 to greater than 1 g/cm^3, with most produced oils in the 0.7 to 0.8 g/cm^3 range (Batzle and Wang, 1992). The American Petroleum Institute oil gravity (API) number is a widely used classification for crude oils. It is defined as

$$\text{API} = \frac{141.5}{d_0} - 131.5 \qquad (3\text{-}13)$$

where d_0 is the oil density measured at 15.6 °C and atmospheric pressure. Batzle and Wang (1992) published polynomial relations for oil to describe the pressure and temperature dependence at a constant composition. The effects of pressure and temperature are largely independent.

The pressure dependence is relatively small and can be described by the polynomial

$$d(p) = d_0 + (0.00277 \cdot p - 1.71 \cdot 10^{-7} p^3)(d_0 - 1.15)^2 + 3.49 \cdot 10^{-4} p \qquad (3\text{-}14)$$

The effect of temperature is larger. Dodson and Standing (1945) have derived one of the most common expressions used to calculate the in-situ density (Batzle and Wang, 1992):

$$d(T) = d(p) \cdot \left[0.972 + 3.81 \cdot 10^{-4} (T + 17.78)^{1.175}\right]^{-1} \qquad (3\text{-}15)$$

where the temperature T is in °C.

Equation 3-2 is valid for mixtures of different substances (e.g. oil and water).

In *frozen rocks* or soils, the material between the matrix components is ice, having a mean density value of $d_{ice} = 0.914 \cdot 10^3$ kg/m^3 (Clark, 1966).

3.3. Density of rocks

3.3.1. Overview

Following equation (3-2), the bulk density of rocks depends on the

- mineral composition,
- content of enclosed pore or fracture space and the filling material.

In *igneous rocks* (and in most metamorphic rocks), the volume content of pores and/or cracks is relatively small and can be ignored.
In *sedimentary rocks*, we find "dense" i.e., pore free chemogenic sediments (e.g. salts), but generally porous rocks with a strong influence of pore volume or porosity and pore filling, respectively, on bulk density.
This distinct difference in properties and behaviour justifies a separate description of igneous and metamorphic rocks, and sedimentary rocks.

3.3.2. The density of igneous and metamorphic rocks

Figure 3.2 shows the mean density ranges for igneous and metamorphic rock types. It illustrates

- a *range of values* for each rock type. In comparison with most sedimentary rocks, this range is small (see section 3.3.3) as a result of the very small influence of the pore- or fracture-volume. The "bandwidth" is generated mainly by the variations of mineralogical composition within the rock type,
- the tendency towards *increasing density from acid to basic rocks.* The correlation between density and quartz-content is demonstrated in Figure 3.3.

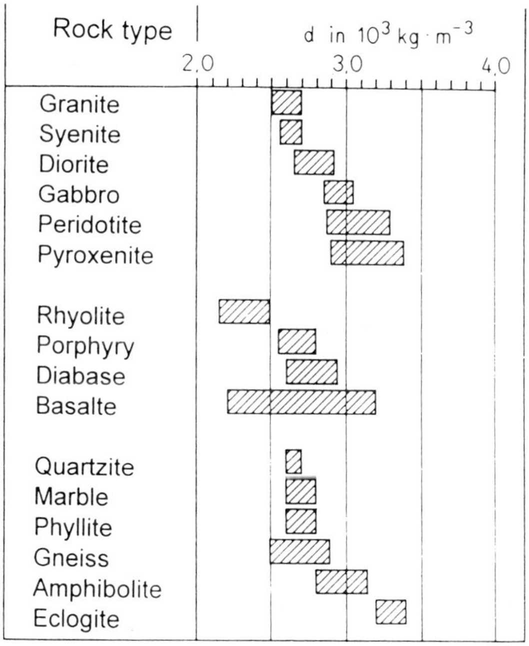

Figure 3.2. Mean density ranges for magmatic and metamorphic rocks

The complete relation between mineral content and density for igneous rocks is given in Figure 3.4.. The upper part shows the range of densities for rock types from granite to peridotite. The lower part shows the mineral compositions. Densities were calculated (circles) at the "border of composition" (vertical lines) using the mineral densities (in parentheses). These calculations are in satisfactory agreement with measured values.

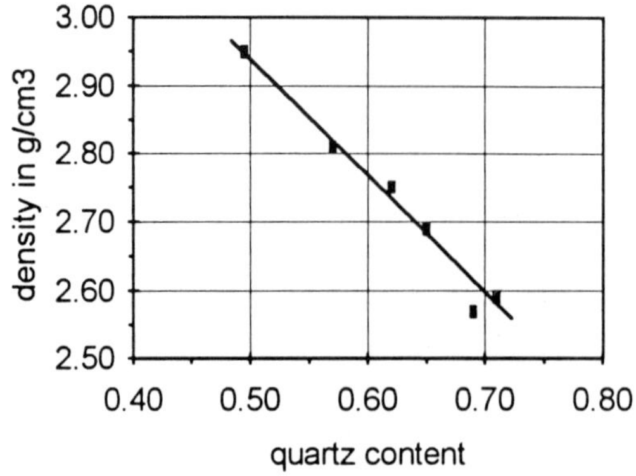

Figure 3.3. Density vs. quartz content for granites and diorites (data from Dortman, 1976)

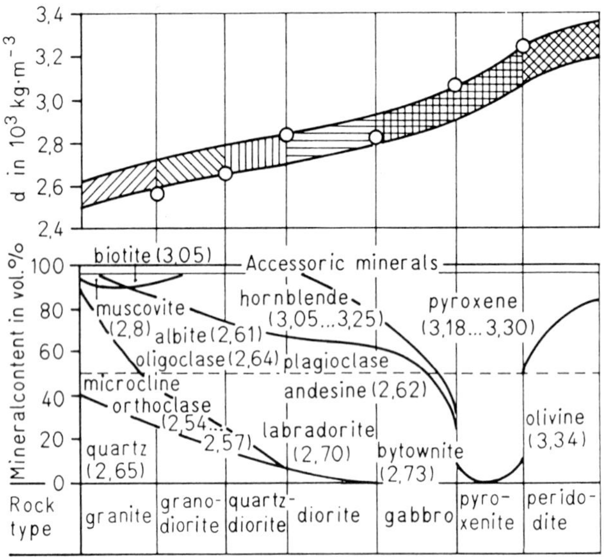

Figure 3.4. The correlation between mineral content and density (in 10^3 kg m^{-3}) of magmatic rocks (after Dortman, 1976);
top: Density range and calculated values (circles),
bottom: Mineral content; the mineral density values used for calculations are in parentheses.

Fractures, cracks, and other defects are the main cause of a (mostly small) dependence of density on pressure. With increasing pressure, such defects are (partially) closed and density increases. This effect is strongest in the initial pressure range and is nonlinear (Figure 3.5). After the closure of the defects, a further density increase is possible, but it corresponds only to mineral compressibility. This increase can be neglected for practical problems.

This behaviour of igneous rocks has frequently lead to the assumption that for depths over about 3...6 km, practically all cracks are closed and the density is nearly constant. However, further research - particularly in superdeep boreholes - show that even at these (and greater) depths, there is fracture/crack-porosity within igneous and metamorphic rocks: At a depth of about 10 km in the KOLA superdeep well, a porosity of 1.6...4.5 % was detected in gneiss and amphibolite (Kozlovskij, 1984). The results of the Continental Deep Drilling Program (KTB - Kontinentales Tiefbohrprogramm) of Germany show intensive fracturing in cores and well logs from the pilot hole (Vorbohrung) up to a depth of 4 km.

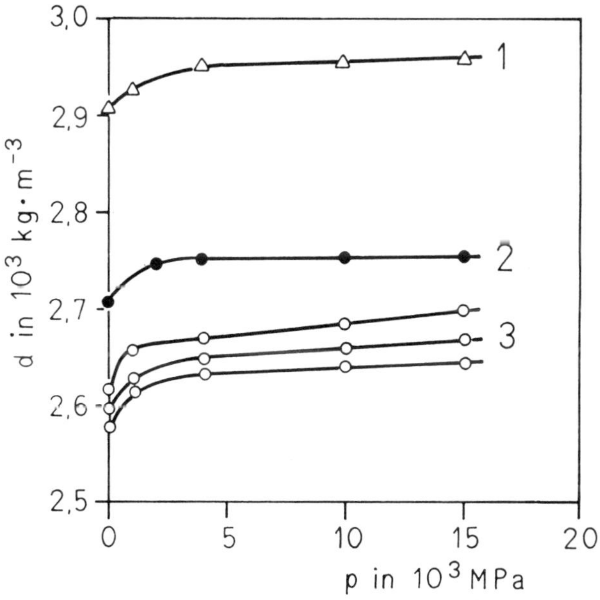

Figure 3.5. Dependence of density on pressure (data after Volarovich et al., 1970); 1 - Gabbro 2 - Diorite 3 - Granite

The density of *metamorphic rocks* is strongly influenced by the composition and density of the initial rock material (the "educt"), the degree of metamorphism, thermodynamic conditions and processes. For example, the "Red gneisses" from the Saxonian Erzgebirge (Germany) show densities of about 2.64×10^3 kg/m^3 corresponding to the granitic educt material, whereas, the "Gray gneisses" with a sedimentary (pelitic, psammitic) educt have an average density of about 2.72 x 10^3 kg/m^3 (Kopf, 1977). This influence is also reflected in the increase of the density of orthogneisses from an acid to a basic educt (Table 3.4 and 3.5).

Table 3.4. Mean densities of orthogneisses (after Kopf, 1977)

Gneiss type	Density in 10^3 kg/m^3
Granite gneiss	2.62
Granodiorite gneiss	2.70
Diorite gneiss	2.80
Gabbro gneiss	3.00
Ultrabasite gneiss	3.35

Table 3.5. Density variations for some gneisses (after Dortman, 1976)

Gneiss-type	Density in 10^3 kg/m^3 mean	Density in 10^3 kg/m^3 range
two-mica	2.60	2.58 ... 2.62
biotitic	2.63	2.60 ... 2.68
amphibolitic	2.78	2.75 ... 2.82
pyroxenitic	2.85	2.80 ... 2.90

A significant change of density can be found with the variation of chemical composition from the addition or removal of individual components or minerals during the metamorphism/metasomatism (e.g. skarnes). Figure 3.6 shows an impressive example of density inhomogeneity over short distances due to the change of mineral composition. In an amphibolite-core from the Kontinentale Tiefbohrung (KTB), a density profile was measured with a gamma-gamma density device for core investigations (Bücker et al., 1990). A collimated gamma-ray crosses the core in four radial directions (azimuths), and the intensity of forward Compton scattering is measured on the core mantle. The measurements were made as a "profile" along the core. A significant minimum for the 135° azimuth between 40 and 50 mm correlates with a quartz-feldspar-mobilizate-band of about 5 mm thickness. Using smaller minima in the other azimuths, an orientation of the band is possible. The density of the visible homogeneous amphibolite varies between 2.950 and 3.050 x 10^3 kg m^{-3} according to the gamma-gamma measurements, whereas, the mean density determined by weight and volume is 2.981 x 10^3 kg m^{-3}. It should be noted that gamma-gamma investigations of cores also detects core disking in its initial stage (see Bücker et al., 1990). The method has been developed for "density tomography".

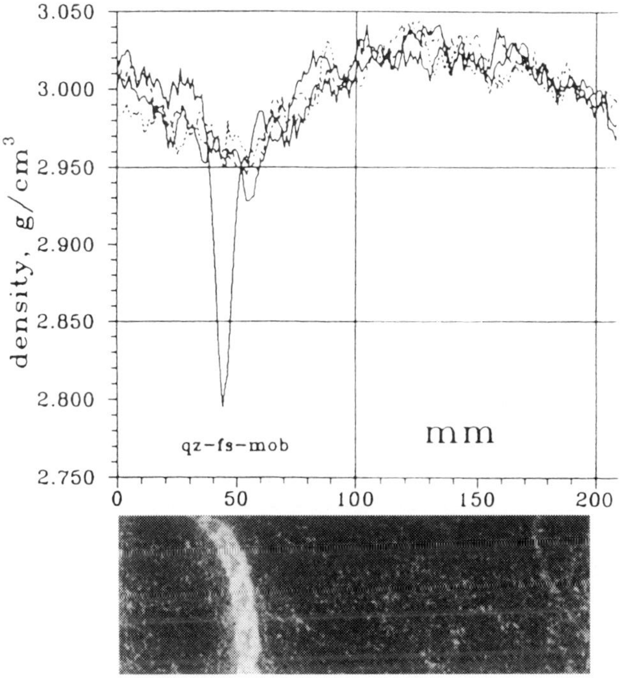

Figure 3.6. Density profiles at the four azimuths along an amphibolite core (Kontinentale Tiefbohrung FRG, Windischeschenbach) from a depth of 3793 m measured with a gamma-gamma-device (upper figure). Photography of the same core (lower figure). The density minimum between 40 and 50 mm correlates with the position of the light band of a quartz-feldspar mobilizate (Bücker et al., 1990).

3.3.3. The density of sedimentary rocks

With regard to density controlling factors, there are two main groups of sedimentary rocks

- pore-free (chemogenic) sediments,
- porous sediments (the most abundant group, the typical sedimentary rock).

Members of the first group (e.g. carbonates, salt) are formed by chemical processes. In their original state, they are practically pore free. But it is possible that later processes (mechanical or chemical) can cause "secondary" porosity (e.g. carbonate porosity; see section 2.2).

The density of *pore free sediments* is determined by their mineral composition; some of them are monomineralic rocks (especially salts). Table 3.6 shows data for this group of pore free sediments. As a result of nearly constant composition, this group is characterized by narrow ranges of density values. A practical conclusion is that for lithological identification we can use gamma-gamma-density measurements in such profiles (e.g. Zechstein-profiles, see also chapter 10).

Table 3.6. Density of pore free sedimentary rocks; after Schlumberger, 1989 (S), Hearst and Nelson, 1985 (H), Gearhart, 1978 (G), Dortman, 1976 (D).

Material	Density in 10^3 kg/m^3			
	S	H	G	D
Dolomite	2.850	2.88	2.85	2.88
Limestone (Calcite)	2.710	2.71	2.71	
Anhydrite	2.960	2.8...3.0	2.8...3.0	2.95
Gypsum	2.320	2.2...2.35	2.2...2.35	2.32
Halite	2.165	2.0...2.2	2.0...2.2	2.17
Sylvite	1.984	1.98	1.98	1.99
Carnallite		1.61	1.61	1.61
Polyhalite		2.78	2.78	2.78
Langbeinite		2.83	2.83	2.83
Kainite		2.12	2.12	2.12

The density of *porous sediments* is controlled by the

- mineralic composition (matrix density),
- porosity,
- composition of pore content (saturation conditions).

As a result of the distinct differences between the mean matrix density range (about 2.5 ... 2.9 x 10^3 kg m^{-3}) and the mean density range of the pore fluid (liquids about 0.8 ... 1.2 x 10^3 kg m^{-3} ; gases < 1 kg m^{-3}), there is a strong correlation between density and porosity. For a given rock type, there is generally a broad range of density values which reflect variations of porosity and saturation.

The description of these correlations is determined by the equation (compare equation 3-4):

$$d = (1 - \Phi) \cdot d_m + \Phi \cdot d_p = d_m - (d_m - d_p) \cdot \Phi \tag{3-16}$$

where d_m is the density of the matrix material

d_p is the density of the pore fluid

Φ is the porosity.

This equation expresses the "contravariant" behaviour of density and porosity: increasing porosity causes decreasing density and vice versa; but the exact correlation is controlled by the densities of matrix and pore fluid.

Figure 3.7 shows mean density ranges for sediments. The lowest density values are for loose, dry soils (see Table 3.7 and 3.8), and marine sediments, with their extremely high porosity. Relatively high density values of more than 2.7 x 10^3 kg m^{-3} can be found in very low porosity carbonate rocks (and sandstones).

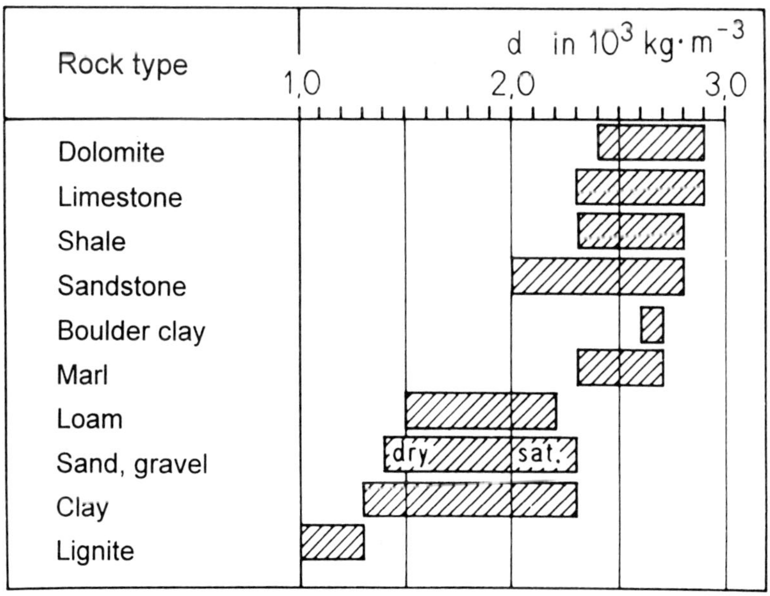

Figure 3.7. Mean density range of sedimentary rocks; data after Wohlenberg, 1982

Table 3.7 Density ranges for soils and unconsolidated sediments; after a compilation of Wohlenberg, 1982

Material	Density in 10^3 kg m^{-3}	Material	Density in 10^3 kg m^{-3}
Soil - dry	1.1 ... 1.3	Clay	2.0 ... 2.7
- wet	1.2 ... 1.7	Boulder clay	2.6 ... 2.7
- stamped, dry	1.6 ... 1.9	Loam	1.5 ... 2.2
- stamped, wet	2.1 ... 2.2	Marl	2.3 ... 2.6
Sand - dry	1.4 ... 2.0	Mould	1.4 ... 1.9
- wet	1.7 ... 2.3		

A special type of sediments are the biogenetic sediments (causto-biolithes). The density increases with the degree of coalification (see Table 3.8).

Table 3.8. Density of biogenetic sediments; after Haalck, 1953 (H), Kopf, 1977 (K), Reeves, 1981 (R), Schlumberger, 1989 (S), Western Atlas, 1992 (W)

Material	Density in 10^3 kg m^{-3}				
	H	K	R	S	W
Peat	1.05	1.03			
Brown coal, lignite	1.10...1.44		1.05...1.25	1.19	1.10
soft		1.20			
hard		1.25			
Bituminous			1.2 ... 1.4	1.24	1.35
Mineral coal (hard coal)		1.35			
Anthracite mineral coal		1.65			
Anthracite	1.30...1.80	1.80	1.5 ... 1.7	1.47	1.60

A second feature of porous sediments is their relatively high compressibility, which results in a dependence of density on pressure or depth (Figure 3.8).

It is obvious that the *density versus depth function* is nonlinear; with increasing depth the relative density change decreases as a result of the decreasing rock compressibility with increasing compaction. There seems to be an asymptotic density value for great depth.

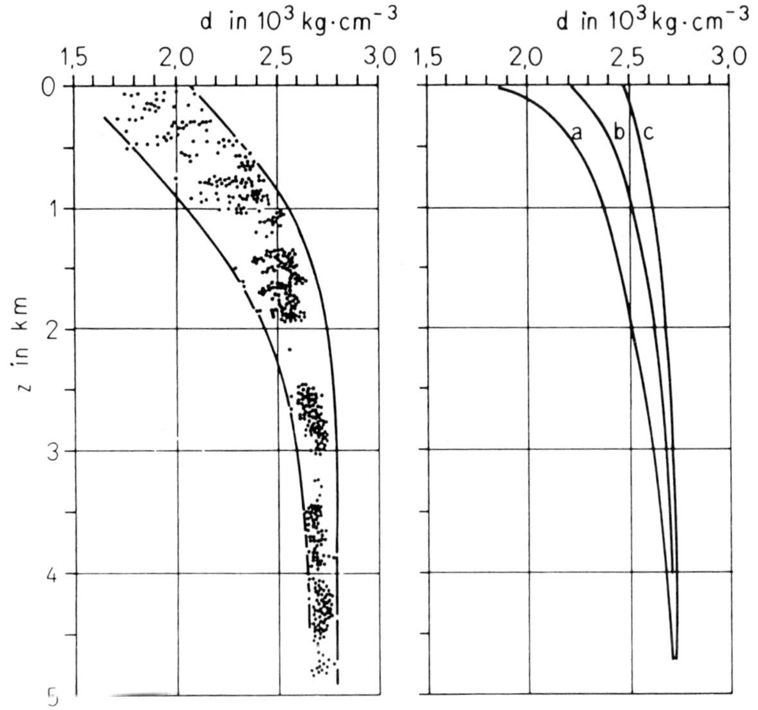

Figure 3.8. Density of sedimentary rocks vs. depth for the North German-Polish Basin (data after Kopf, 1979); left: Sand- and siltstones; right: Mean values for different stratigraphic units (a - Quaternary, Cretaceous, Jurassic; b - Bunter sandstone; c - Permian)

A simple analytical description for the density-depth function was published by Polak and Rapoport (1961):

$$d(z) = d(z_0) + A \cdot \ln\left(\frac{z}{z_0}\right) \qquad (3\text{-}17)$$

where z is the actual depth
z_0 is the upper depth value for the considered layer or section
A is an empirical factor, related to the rock compressibility.

The equation was applied to clay layers and expresses this non-linearity. The equation has no asymptotic value.

Stegena (1964) recommended a relation

$$\begin{aligned} d(z) &= d(z_0) + [d(z_m) - d(z_0)] \cdot [1 - \exp(-B \cdot z)] \\ &= d(z_m) - [d(z_m) - d(z_0)] \cdot \exp(-B \cdot z) \end{aligned} \quad (3\text{-}18)$$

where z_0 is the upper depth value for the considered layer
z_m is the maximum depth of the sediment (maximum depth of burial),
B is an empirical factor, connected with the compressibility.

This equation has an asymptotic value $d(z_m)$ for $z \rightarrow \infty$.

An empirical relation from Jelic (1984) for sandstones and siltstones may be given as an example

$$d(z) = 2.72 - 1.244 \cdot \exp(-0.846 \cdot z) \quad (3\text{-}19)$$

with d in g cm^{-3} and z in km.

The dominant physical process of the density increase is the reduction of the pore space (see also section 2.2.4). The general shape of equation (3-18) can be derived from the porosity-depth relation (2-9) and the density-porosity relation (3-3):

$$d = d_m - (d_m - d_p) \cdot \Phi_0 \cdot \exp(-B^* \cdot z) \quad (3\text{-}20)$$

CHAPTER 4

Magnetic Properties of Rocks

The knowledge of magnetism started in early times with the discovery of "load-stone", which is identical to a natural mass of magnetite.... The magnetic properties of rocks, however, look so complex and so heterogeneous that their systematic description based on the physics of magnetic matter is hardly possible..."
(Takesi Nagata, Preface to the first edition of his book "Rock Magnetism", 1953)

4.1. Physical basis and units

Magnetic properties describe the behaviour of any substance under the influence of a magnetic field. Magnetic phenomena arise from the motion of electrically charged particles within the substance. There are three main groups of materials with regard to magnetic properties:

(1) *Diamagnetic substances:* The electron shells of the substance are complete; under the influence of a magnetic field the electron spins precess and produce a magnetization in the opposite direction to that of the applied field in conformity with Lenz's law.

(2) *Paramagnetic substances:* The electron shells of these substances are incomplete; under the influence of a magnetic field the electron spins precess and produce a magnetic moment that has the same direction as the applied field. Due to electron spin of unpaired electrons, there is a partial alignment of the atomic dipole moments to augment the net magnetization in the direction of the magnetic field.

Thus, these effects exist only in the presence of an applied field and are weak. In both cases, the strength of the induced magnetization M (magnetic dipole moment per unit volume) is directly related to the strength of the applied magnetic field *H:*

$$M = \kappa \cdot H \tag{4-1}$$

where κ is the magnetic susceptibility

$$\kappa = \frac{M}{H} \tag{4-2}$$

In general, the susceptibility is a tensor of rank two. If not otherwise mentioned, the symbol κ means a "mean, quasi isotropic" susceptibility.

Diamagnetic substances, therefore, have negative susceptibilities; the magnitude is for common rock-forming minerals often in the region of -10^{-5} (SI). Paramagnetic substances have positive susceptibilities that extend over a range between 10^{-4} and 10^{-2} (SI) for the common rock-forming minerals (Tarling and Hrouda, 1993).
Diamagnetic susceptibility is independent of temperature. The temperature dependence of paramagnetic susceptibility is given by the Curie law or the Curie-Weiss law.

(3) *Ferro-, antiferro-, and ferrimagnetic substances*: This type of magnetism occurs in substances in which the electron spins have been spontaneously coupled (spontaneous magnetization) in a way that alines the individual spin magnetizations even in the absence of an externally applied magnetic field. The magnetic behaviour is described by the existence and properties of magnetic volume elements termed "magnetic domains" (single-domain, multi-domain). The phenomenon of ferro-, antiferro-, and ferrimagnetism can also be discussed as that of a material which contains ions with an intrinsic magnetic moment. This exists with an exchange interaction between these ions (Bleil and Petersen, 1982).
Such a material is (Figure 4.1):

- ferromagnetic for parallel orientation of neighboring intrinsic moments and a resulting macroscopic external moment,
- antiferromagnetic for equal but antiparallel orientation of the intrinsic moments and, therefore, a zero macroscopic external moment,
- ferrimagnetic for different antiparallel intrinsic moments of different magnitudes and, therefore, a resulting external moment.

When the temperature is higher than

- Curie-temperature T_C for ferro-/ferrimagnetic,
- Ne'el-temperature T_N for antiferromagnetic

the material has paramagnetic properties.

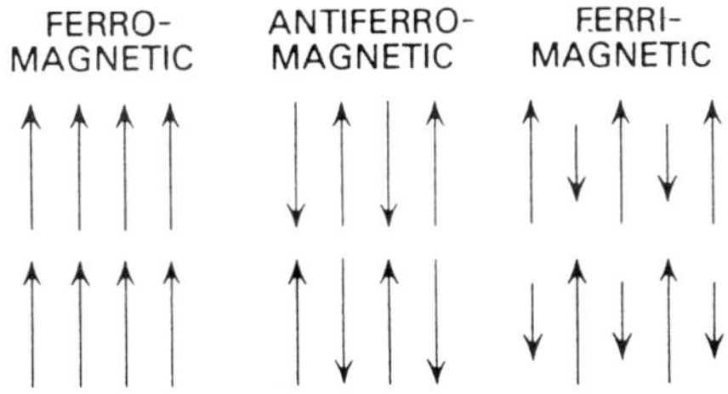

Figure 4.1. Schematic illustration of the orientation of magnetic moments for ferromagnetic, antiferromagnetic, and ferrimagnetic materials.

Ferro- and ferrimagnetic materials show a hysteresis loop for the dependence of the magnetization on the strength of the magnetic field. The magnetization depends on the field strength and the "magnetic history" and shows the phenomenon of remanent magnetization (Figure 4.2). The magnetic susceptibility of ferro- and ferrimagnetic substances is much stronger than that of paramagnetic substances and also depends on the strength of the magnetic field and the magnetic history. Figure 4.2 shows an example of a hysteresis curve for a natural rock.

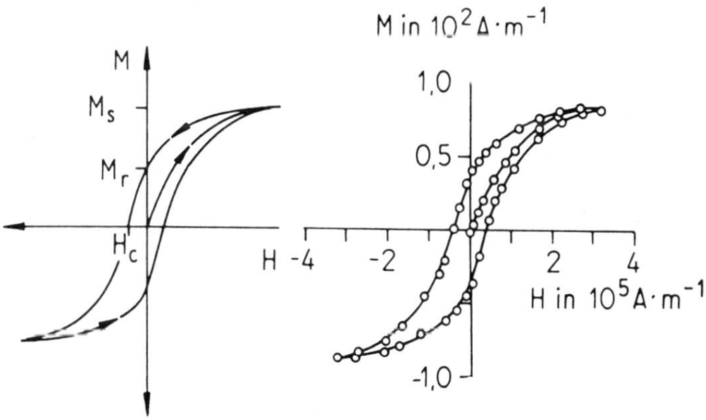

Figure 4.2. The magnetic hysteresis curve: a) schematic, b) hysteresis curve of a volcanic rock; after Nagata (1961), values converted to SI units. M_s- saturation magnetization M_r- remanent magnetization H_c- coercitive field strength

It should be noted that the shape of the hysteresis curve and the magnitude of the parameters vary, depending upon

- the intrinsic properties of the ferrimagnetic components,
- the grain size, and
- the internal stress.

Therefore, information on the domain structure can be derived from the ratio of saturation remanence to saturation magnetization and the ratio of remanence coercitivity to coercive force (Petersen and Bleil, 1982).

Equation 4-1 relates magnetization to the strength of the applied magnetic field; both magnitudes have the unit A/m (Ampere per meter) in SI. The magnetic induction B is defined as the vector sum of the earth's magnetic induction $\mu_0 H$ and the magnetism induced in the rock by the earth's magnetic field $\mu_0 M$

$$B = \mu_0 H + \mu_0 M = \mu_0 (H + M) = \mu_0 (1 + \kappa) H = \mu_0 \mu H \tag{4-3}$$

where μ_o is the magnetic permeability for vacuum;
$\mu_o = 4\,\pi \times 10^{-7}$ V s A^{-1} m^{-1}
μ is the relative magnetic permeability

from which follows a simple relationship between the two material properties μ and κ

$$\mu = 1 + \kappa \tag{4-4}$$

In some cases, the "specific mass susceptibility" is used

$$\kappa_g = \kappa / d \tag{4-5}$$

where d is the density and κ_g is in $m^3\,kg^{-1}$. Consequently, κ is the "specific volumetric susceptibility".

In general for ferro- and ferrimagnetic substances, the magnetization M is the sum of the induced magnetization M_i and the remanent magnetization M_r

$$M = M_i + M_r \tag{4-6}$$

The ratio of the remanent magnetization M_r and induced magnetization M_i is called the "*Koenigsberger Q-ratio*"

$$Q = \frac{M_r}{M_i} \tag{4-7}$$

Table 4.1 shows the units and conversions for the properties used.

Table 4.1. Magnetic units and conversions

Symbol	SI unit	cgs unit	Conversions
H	$A\ m^{-1}$	Oe (Oersted)	$1\ Oe = 10^3/4\ \pi\ A\ m^{-1} = 79.6\ A\ m^{-1}$ $1\ A\ m^{-1} = 4\ \pi\ 10^{-3}\ Oe = 1.257\ 10^{-2}\ Oe$
B	T (Tesla) $1\ T = 1\ V\ s\ m^{-2}$	Gauss	$1\ Gauss = 10^{-4}\ T$ $1\ T = 10^4\ Gauss$
M	$A\ m^{-1}$	Gauss	$1\ Gauss = 10^3\ A\ m^{-1}$ $1\ A\ m^{-1} = 10^{-3}\ Gauss$
κ	dimensionless		$\kappa\ [SI] = 4\ \pi\ \kappa\ [cgs]$ $\kappa\ [cgs] = 1/(4\pi)\ \kappa\ [SI]$
Q	dimensionless		

Fundamental publications and comprehensive reviews about rock magnetism are: Mooney and Bleifuss (1953), Nagata (1961, 1966), Angenheister and Soffel (1972), Stacey and Banerjee (1974), Bleil and Petersen (1982), Petersen and Bleil (1982), Tarling and Hrouda (1993). Literature is available on palaeomagnetism and its applications for geology, geophysics and archaeology (e.g. Tarling, 1983). Recent results concerning aspects of paleomagnetism are published in a Special Issue "Paleomagnetism: Database and experimental techniques" (McClelland et al. (ed.), 1990).

4.2. Magnetic properties of rock constituents

4.2.1. Magnetic properties of minerals

Minerals also can be classified as

- diamagnetic minerals,
- paramagnetic minerals,
- ferromagnetic minerals,
- ferrimagnetic minerals,
- antiferromagnetic minerals.

4.2.1.1. Diamagnetic and paramagnetic minerals

Table 4.2 shows susceptibility values for selected diamagnetic minerals. For more detailed data see Clark (1966), Lindsley et al. (1966), Melnikov et al. (1975), Bleil and Petersen (1982)

Table 4.2. Susceptibility of diamagnetic minerals;
References: B, P - Bleil and Petersen (1982) and Petersen (1985), T, H - Tarling and Hrouda, 1993 (individual values); D, M - Dortman (1976) and Melnikov et al.(1975); data converted into SI; the values in the second column are calculated from the data of the first column and mean density values.

Mineral	κ_g [10^{-8} kg^{-1} m^3]	κ [10^{-6}]	κ [10^{-6}]	κ [10^{-6}]
	B, P	B, P	D,M	T, H
Orthoclase	- 0.58	- 12.5	- 6.3	
Zircon (a-axis)	- 0.21	- 10.0	- 10	
(c-axis)	0.92	43.0		
Quartz	- 0.58	- 12.4	- 16.3	-13.4, -15.4
Anhydrite	- 2.11	- 59.3		
Galena	- 0.44	- 33.4	- 32.7	
Sphalerite	- 0.33	- 13.5	- 60.3	
Aragonite	- 0.52	- 15.2	-17	
Calcite	- 0.48	- 13.0	-14...-40	-13.8
Dolomite				-38.0
Halite	- 0.48	- 10.4	-14	
Apatite	- 0.33	- 10.6		
Diamond	- 0.62	- 21.7		
Fluorite	- 0.79	- 24.0	- 11	

Due to the presence of non-stoichiometric Fe- or Mn-ions, some of these minerals may have paramagnetic properties (Petersen, 1985). Some of the values published by Dortman (1976) are positive and relatively high; so it must be assumed that the samples investigated have impurities (Fe, Ti), which produce superimposed positive components. Bleil and Petersen (1985) remark that in their data "a number of minerals ... are listed in both the table for paramagnetic minerals and the table of diamagnetic minerals." An example is gypsum with the one value (pure mineral) of κ_g = - 1.25 x 10^{-8} m^3 kg^{-1} and the range from 50 x 10^{-8} to 330 x 10^{-8} m^3 kg^{-1} (due to ferrimagnetic impurities).

Table 4.3 shows susceptibility values for paramagnetic minerals.

Table 4.3. Susceptibility of paramagnetic minerals; References: B, P - Bleil and Petersen (1982) and Petersen (1985), T, H - Tarling and Hrouda, 1993 (individual values); M - Melnikov et al., 1975; D - Dortman (1976); data converted into SI; the values in the second column are calculated from the data of the first column and mean density values.

Mineral	κ_g [10^{-8} kg^{-1} m^3] B, P range	mean	κ [10^{-6}] B, P	κ [10^{-6}] D, M	κ [10^{-6}] T, H
Olivine	5...130	29	990	250...5000	
Orthopyroxene	3...92	43	1420	380...5000	
Clinopyroxene	16...77	47	1550	250...4800	
Amphibole	16...100	49	1570	630...5000	
Garnet	13...155	69	2600	13...3000	
Hornblende	6...100				8920
Biotite		52	1500	1250...3000	998, 1180, 1230, 1290
Muscovite	0...25	8	226	4800	122, 165
Chlorite					70, 358, 370, 1550
Tourmaline	2...40	18	560	630...1260	1690
Cordierite	7...42	21	530		
Dolomite			1080	13...1300	

The fairly high values particularly in the data of Dortman (1976) are probably caused by a relatively high content of ferrimagnetic impurities.
The Langevin theory gives relationships between mass susceptibility and temperature, as well as the amounts and magnetic moments of the magnetic ions (e.g. Fe^{2+}, Fe^{3+}, Mn^{2+}).

Because the paramagnetic susceptibility is controlled mainly by the content of Fe^{2+}, Fe^{3+} and (less important) of Mn^{2+} (Bleil and Petersen, 1982), an empirical correlation between susceptibility and Fe-content C_{Fe} (in weight - %) is recommended for practical applications (see Petersen, 1985)

$$\kappa_g = 3.48 \cdot C_{Fe} \qquad (4\text{-}8)$$

where κ_g results in m^3 kg^{-1}. This correlation was derived for the minerals garnet, tourmaline, biotite, muscovite, amphibole, cordierite, and pyroxene with a Fe-content between 0 and 30 weight %.

Dortman (1976) has described the strong correlation between the content of ferrimagnetic materials and the mineral susceptibility, demonstrated in Figure 4.3.

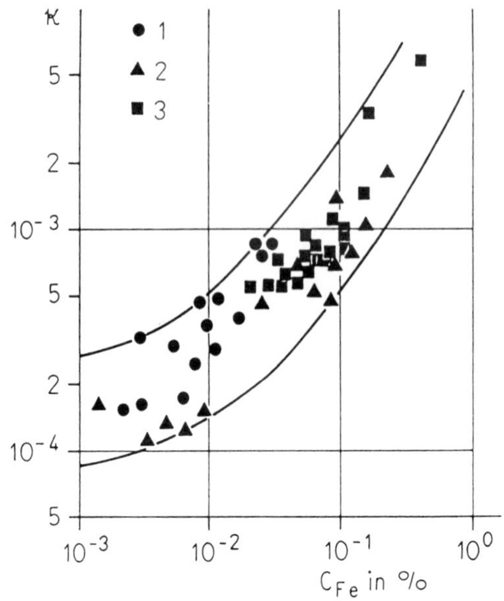

Figure 4.3. Magnetic susceptibility κ vs. content of ferrimagnetica C_{Fe} (in %); after data from Dortman (1976); 1 - biotite; 2 - amphibolite; 3 - pyroxene.

4.2.1.2. Ferro-, antiferro-, and ferrimagnetic minerals

The most important and abundant group of ferrimagnetic minerals in rocks are *iron and iron-titanium oxides*. Iron oxyhydroxides and iron sulphides are significant, but not abundant (Bleil and Petersen, 1982).

Fe-Ti-oxides are the dominant magnetic "substance", particularly in magmatic rocks, being present chiefly in a *ternary system* (see Figure 4.4) i.e.:

FeO (wüstite) - Fe_2O_3 (hematite, maghemite) - TiO_2 (rutile).

It can be seen that this system gives "the most basic knowledge of understanding the ferrimagnetic characteristic of general rocks" (Nagata, 1966). The ternary system contains the chemical composition of

- "simple oxide minerals of interest in rock magnetism" (Nagata, 1961): FeO (wüstite), Fe_3O_4 (magnetite), γ-Fe_2O_3 (maghemite), α-Fe_2O_3 (hematite), $FeTiO_3$ (ilmenite), Fe_2TiO_4 (ulvöspinel), Fe_2TiO_5 (pseudobrookite), and $FeTi_2O_5$(ilmeno-rutile, ferropseudobrookite) and
- four series (solid solution series) of the system: Titanomagnetite, ilmeno-hematite, pseudobrookite, titanomaghemite.

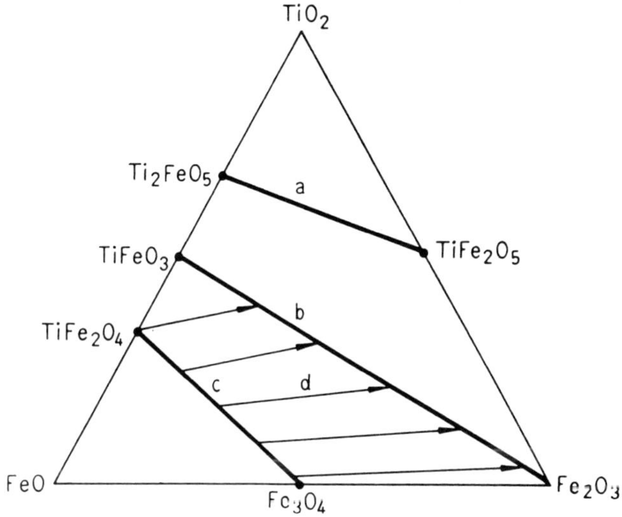

Figure 4.4. The ternary system FeO - Fe_2O_3 - TiO_2 with typical series:(a - pseudobrookite, b - ilmenite-hematite, c - titanomagnetite), the arrows represent the directions of oxidation

The following description contains some of the relevant properties and parameters; for further study, see literature on this subject, eg. Nagata (1961), Stacey and Banerjee (1974), Bleil and Petersen (1982).

➢ *Titanomagnetite series:* Cubic structure/inverse spinel; this series has the end members magnetite and ulvöspinel with the general formula

$$Fe_{3-x}Ti_xO_4 \text{ with } 0 \leq x \leq 1$$

Specific saturation magnetization, initial susceptibility, and the Curie-temperature (see Table 4.4) decrease with increasing x following a relationship (Bleil and Petersen, 1982):

$$T_C = 851 - 580\,x - 150\,x^2 \qquad (4\text{-}9)$$

With respect to the abundance of titanomagnetites, Bleil and Petersen (1982) may be cited: "Titanomagnetites ... are the most common magnetic minerals in igneous rocks... Magnetite occurs in a great variety of igneous, metamorphic, and sedimentary rock types, in certain meteorites, but not in lunar samples. Typically, it is formed in various types of subsolidus reactions. As a carrier of rock magnetism, magnetite is the most abundant and important oxide mineral. Ulvöspinel is a rare natural crystal in terrestrial rocks, almost always intergrown with magnetite. It is frequently observed in lunar samples."

➢ *Ilmenite-hematite-series:* Hexagonal/rhombohedral structure; this series has the end members ilmenite and hematite and the general formula

$$Fe_{2-x}Ti_xO_3 \text{ with } 0 \le x \le 1$$

For the complex relations between the properties and composition refer to Bleil and Petersen (1982). This series yields the following naturally occuring characteristic orientations: Hematite is a carrier of remanent magnetization in sediments (mainly in specular grains and the pigment). In igneous rocks, the primary composition of the series relates to the bulk chemistry of the rock. With decreasing total basicity, the content of ilmenite decreases; subsolidus reactions lead to ilmenite enrichment. This series also occurs in a wide variety of metamorphic rocks.

➢ *Pseudobrookite series:* Orthorhombic structure; this series is defined by the end-members pseudobrookite Fe_2TiO_5 and ferropseudobrookite $FeTi_2O_5$. At room temperature, pseudobrookites are completely paramagnetic (Bleil and Petersen, 1982). Natural occurrence is in igneous and metamorphic rocks.

Table 4.4. Curie and Neel temperatures for some minerals (after Bleil and Petersen, 1982)

Series, group	T_C, T_N in K
Titanomagnetite	
Magnetite	$T_C = 850$
Ulvöspinel	$T_C = 115, 120$
Ilmenite-hematite	
Hematite	$T_N = 950$
Ilmenite	$T_N = 55...68$

➢ *Titanomaghemite series:* Spinel structure; titanomaghemite is generated by oxidation of titanomagnetites at temperatures below 300°C (Petersen, 1985) with the change $Fe^{2+} \rightarrow Fe^{3+}$. At the one end is maghemite and the other members are described by the formula $(Fe,Ti,\delta)_3O_4$, where δ denotes a varying vacancy in the metal ion site of the crystal structure. The magnetic properties are strongly controlled by the composition and are influenced by the "oxidation ratio" $Fe_2O_3/(Fe_2O_3+FeO)$; e.g., the Curie-temperature increases with increasing oxidation ratio. Titanomaghemites are the main magnetic constituents in the basaltic oceanic basement, but they also occur in continental igneous rocks (Bleil and Petersen, 1982).

Pyrrhotite FeS_{1+x} is a representative of iron sulphides (monoclinic and hexagonal), with ferrimagnetic behaviour. Representatives of iron oxyhydroxides are goethite α-FeOOH and lepidocrocite γ-FeOOH (both orthorhombic).

Nagata (1966) analyzed samples of eruptive rocks and showed that more than 90% of the magnetically effective substance are parts of the Titanomagnetite and the Ilmenite-hematite series.

The dependence of susceptibility on magnetic field strength results in the difficulty of giving "representative mean values" for ferri- and ferromagnetic minerals. Thus, the values in the table 4.5 are given for general orientation only.

Table 4.5. Mean values of susceptibility for some ferro- and ferrimagnetic minerals; after Dortman (1976) and Telford et al., (1990)

Mineral	κ (SI) Dortman, 1976 range	κ (SI) Telford et al., 1990 range	mean
Magnetite	8.8...25	1.2...19.2	6
Ulvöspinel	3.8...25		
Maghemite	10^{-5}...10		
Hematite	10^{-4}...10^{-3}	0.5×10^{-3}...3.5×10^{-2}	6.5×10^{-3}
Pyrrhotite	10^{-1}...10^{-2}	10^{-3}...6	1.5
Ilmenite		0.3...3.5	1.8
Chromite	10^{-3}...10^{-2}	3×10^{-3}...1.1×10^{-1}	7×10^{-3}

4.2.2. Magnetic properties of fluids

Most fluids are diamagnetic and have only a very small influence on the magnetic rock properties. For liquids Kobranova (1989) gives the following susceptibility values:

$$\kappa_{water} = -0.9 \cdot 10^{-5} \quad \text{and} \quad \kappa_{oil} = -1.04 \cdot 10^{-5}$$

The mineralization has a small effect, because most salts are also diamagnetic.

Most gas components are also diamagnetic, exept the paramagnetic oxygen. The low value for air therefore is about $\kappa_{air} = 0.04 \cdot 10^{-5}$. For hydrocarbon gases Kobranova (1989) gives susceptibilities of about -10^{-8}.

4.3. Magnetic properties of rocks

4.3.1. Overview

The magnetic properties of rocks are controlled by those mineral constituents which have a magnetic effect. The fraction of these minerals with respect to the total rock volume may be small. Therefore, two consequences result (Carmichael, 1989):

1. Magnetic properties can be quite variable within a rock type, depending on chemical inhomogenity, depositional and/or crystallization, and postformational conditions.
2. The magnetic properties are not necessarily closely predictable by the lithologic rock type (geologic name). This is because the geologic rock name (and the geologic classification, see chapter 1.) is generally given on the basis of the genesis and the gross mineralogy, but a minor fraction of the mineral constituents controls the magnetic properties.

The most abundant minerals in common rocks are paramagnetic or diamagnetic. The magnetic rock properties are controlled by the ferrimagnetic minerals, although their concentration "in major rock types rarely exceeds 10 % vol." (Petersen and Bleil, 1982). Minerals of the Fe-Ti-system (igneous rocks) are dominant; in sedimentary rocks, the Fe-hydroxides are also important. Table 4.6 gives mean values for contents of FeO, Fe_2O_3, and TiO_2 and Figure 4.4 shows schematically the mineral contribution to the susceptibility of a rock.

Susceptibility has a wide range of values for the individual rock types and more or less distinct tendencies and rules as demonstrated in Figure 4.5.. Obviously,

- the susceptibility for each rock type varies with magnitude,
- the susceptibility of magmatic rocks increases from acid to basic rocks,
- the susceptibility of sedimentary rocks increases with increasing clay content.

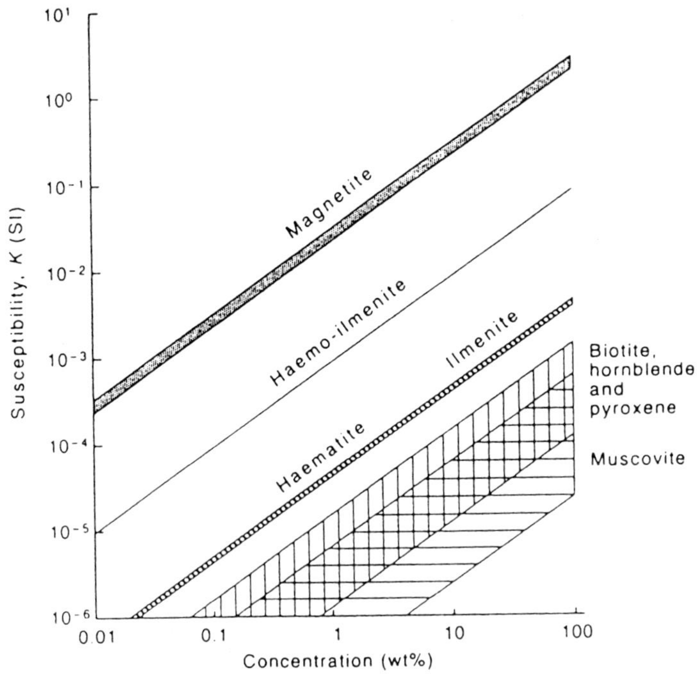

Figure 4.4 The mineral contribution to the susceptibility of a rock; all mineral grains within a rock contribute to its total susceptibility, but their individual influence depends on their intrinsic susceptibility as well as on their concentration (after Tarling and Hrouda, 1993)

Table 4.6. Mean FeO-, Fe_2O_3-, and TiO_2- contents (vol. %) of some rocks (Frölich, 1959)

Rock type		FeO	Fe_2O_3	TiO_2
Intrusive rocks	Granite	1.78	1.57	0.39
	Syenite	3.28	2.74	0.67
	Diorite	4.40	3.16	0.84
	Gabbro	5.95	3.16	0.97
	Dunite	5.54	2.84	0.02
Extrusive rocks	Rhyolite	0.31	1.65	0.48
	Trachyte	2.62	2.64	0.38
	Dacite	1.90	2.38	0.57
	Andesite	3.13	3.33	0.77
	Basalt	6.37	5.38	1.36
Sedimentary rocks	Sandstone	0.30	1.08	0.25
	Shale	2.46	4.03	0.65
	Limestone		0.54	0.06

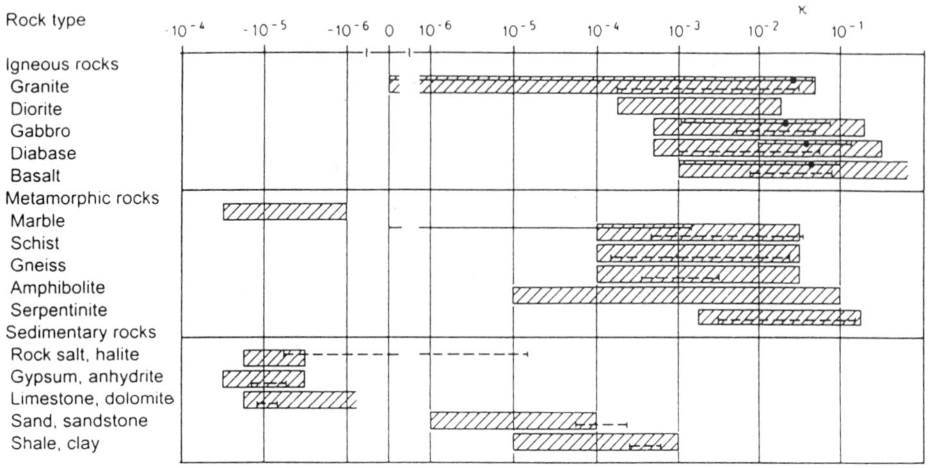

Figure 4.5. Mean range of susceptibility values for different rock types. Box - after data from Melnikov (1975), Dortman (1976), Militzer and Scheibe (1978); line (solid) - after data from Mooney and Bleifuss (1953); line (dotted) - after data from Jakosky (1950)

Using cores from the KTB pilot hole, susceptibility values were plotted for the two main rock types, gneiss and metabasite (see Figure 4.6), which shows significant rock type separation by susceptibility.

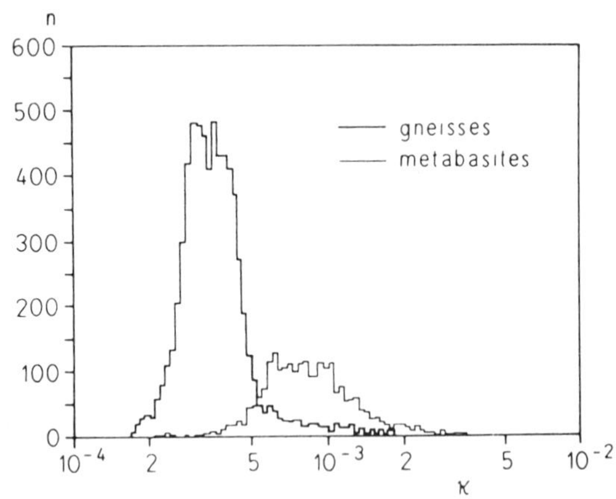

Figure 4.6. Log-normal frequency distribution of magnetic susceptibility; depth 0...3890 m; *n* - number of measured values. Cores taken from the KTB (Kontinentale Tiefbohrung, Germany); after Bücker et al., (1990) and Rauen et al., (1990).

Some additional remarks illustrate the complex behaviour of both magnetic properties and their potential information content:
a) Susceptibility of an igneous rock type is influenced by specific genetic conditions. Figure 4.7. illustrates this by comparing two intrusions of a granite massif in the Saxonian Erzgebirge.

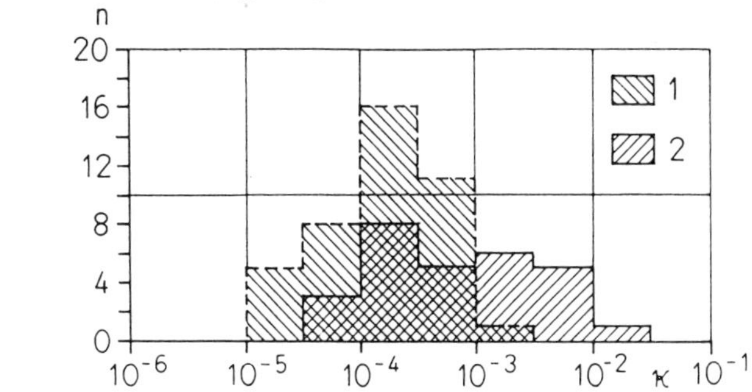

Figure 4.7. Histogram of susceptibility values for the two intrusions of the Kirchberg Granite (Saxony, Germany); after Damm (1988). 1 - younger intrusion 2 - older intrusion

For metamorphic rocks, Dortman (1976) schematically described the different character of rocks of different genesis (para- and ortho-metamorphites), see Figure 4.8..

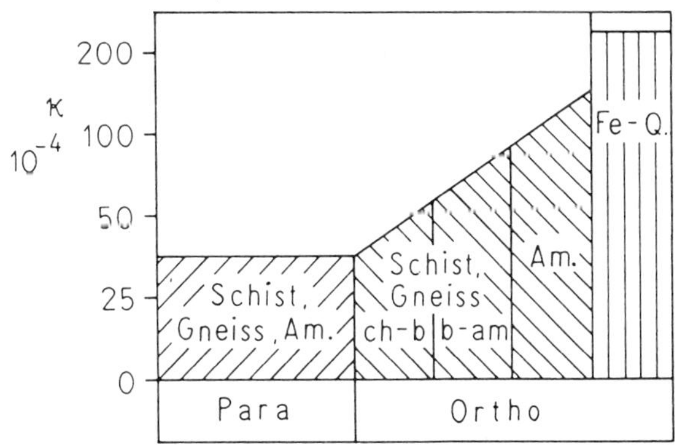

Figure 4.8. Susceptibility of para- and ortho-metamorphites after Dortman (1976); Am - amphibolites; ch-b - chloritic-biotitic; b-am - biotitic- amphibolitic; Fe-Q.- Fe-quartzite.

b) Alteration processes can greatly influence magnetic behaviour. As an example, Henkel and Guzman (1977) reported martization (oxidation of magnetite to hematite) at an outcropping fracture zone with a negative magnetic anomaly. The effect of alteration is demonstrated using a susceptibility κ vs.

Koenigsberger ratio Q plot, which shows decreasing κ and increasing Q with the alteration.
Other studies such as of the Svaneke granites (Sweden) show that the alteration of mafic minerals (hornblende and biotite) into chlorite and magnetite results in increasing susceptibility with an increasing degree of alteration (Platou, 1968). For complex studies (rock classification and diagnostic problems), Henkel (1976) recommended a density-susceptibility plot.

c) Kopf et al., (1981) found a correlation between the susceptibility and clay-mineral content V_{clay} (in volume percentage according X-ray analysis) for clay-, silt-, and sandstones (Cenozoic and Mesozoic sediments from the North-German - Polish basin) and derived an empirical equation (converted to SI units)

$$\kappa = 66.8 \cdot 10^{-6} \cdot V_{clay}^{0.562} \qquad (4\text{-}10)$$

with a regression coefficient of 0.95.

4.3.2. Correlations between susceptibility and content of magnetic substances

The susceptibility of rocks is strongly controlled by the magnetic mineral type and its concentration in the rock. Because "magnetite is the most common and the most magnetic mineral of the iron-titanium oxide series" (Hearst and Nelson, 1985), there is a distinct correlation between rock susceptibility and magnetite content as shown in Figure 4.9 and Figure 4.10..

The correlation for a specific site can be expressed by a relationship of the general form

$$\kappa = a \cdot V_m^b \qquad (4\text{-}11)$$

where V_m is the magnetite volume fraction (mostly in %); a and b are empirical values.
Normally, b ranges between 1.0 and 1.4 (Grant and West, 1965; see Hearst and Nelson, 1985). Table 4.8 contains some values for the empirical parameters in equation (4-11).

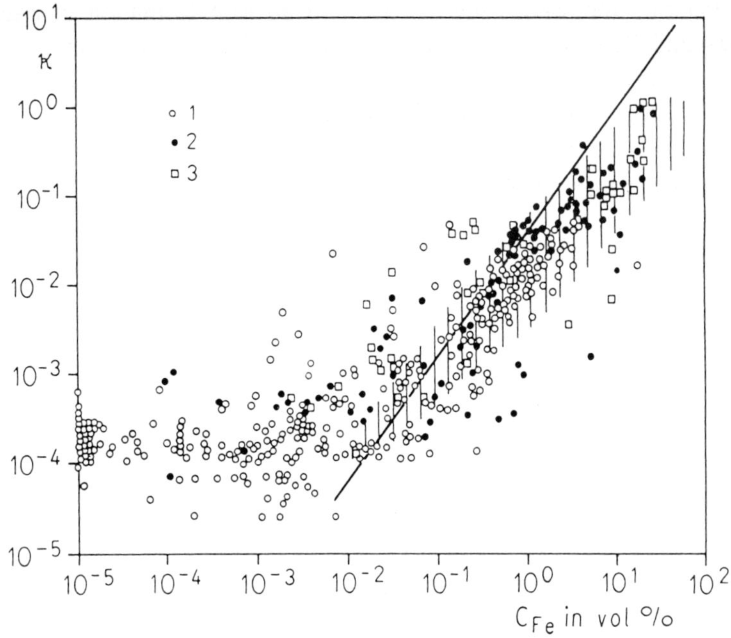

Figure 4.9. Correlation between susceptibility κ and content of the ferromagnetic fraction C_{Fe} (in %); after Dortman (1976); 1 - granite 2 - diorite and gabbro 3 - hyperbasite; area with the signature follows a correlation as given by equation 4-11. Parameters of the curve see Table 4.8

Table 4.8. Empirical parameters of the susceptibility vs magnetite content relation 4-11; all equations converted into SI units; contents are given in vol.%; Reference: M - Mooney and Bleifuss (1953); B - Balsley and Buddington (1958); J - Jahren (1963); D - data (Fe content and susceptibility) after Dortman (1976)

Rock type, magnetite content	*a*	*b*	Ref.
Metamorphic rocks, about 0.01...80 %, Adirondack Mountains, USA	0.0330	1.33	B
Rocks, about 1 ... 80%	0.0140	1.39	J
Basalt, Minnesota	0.0475	1.08	M
Diabase, Minnesota	0.0336	1.14	M
Granite, Minnesota	0.0244	0.47	M
Gabbro, Minnesota	0.0155	0.36	M
All rocks, Minnesota	0.0363	1.01	M
Igneous rocks (Figure 4.9), Russia	0.042	1.38	D

As an example, the values for diabase and an iron formation in Figure 4.10 were analyzed. The lines shown are the best fit using a power relation following equation (4-11)

- for diabase $$\kappa = 0.0336 \cdot V_{\mathrm{m}}^{1.14} \tag{4-12}$$

- for iron formation $$\kappa = 0.0116 \cdot V_{\mathrm{m}}^{1.43} \tag{4-13}$$

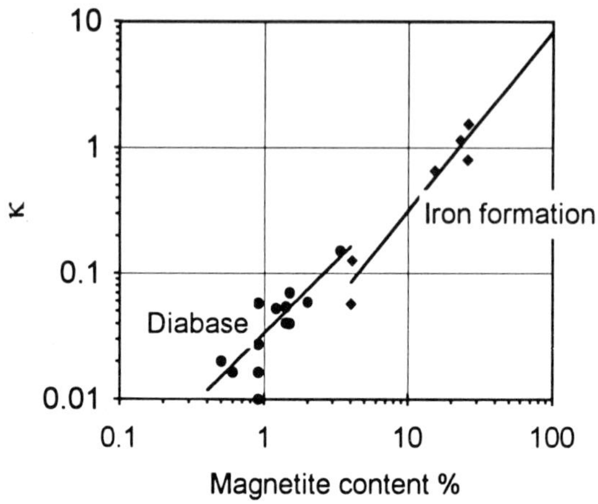

Figure 4.10. Correlation between susceptibility and magnetite content (in vol %) of rocks and ores from Minnesota; data from Mooney and Bleifuss (1953)

Wänstedt (1992) investigated correlations between Fe-content, density and magnetic susceptibility of the Swedish Malmbergit mine. Figure 4.11 shows the results in various plots. There are correlations between Fe-content and susceptibility, but also between Fe-content and density (as a result of the higher density of the ore):

- the density correlation can be described by a linear relationship following equation 3-2
$$d = 33.7 \cdot (\text{Fe-content}) + 2583.5 \tag{4-14}$$

- the susceptibility correlation can be described by the nonlinear relationship following equation 4-11
$$\kappa = 0.0061 \cdot (\text{Fe-content})^{1.71} \tag{4-15}$$

In these equations, density is in kg/m^3 and Fe-content in %.

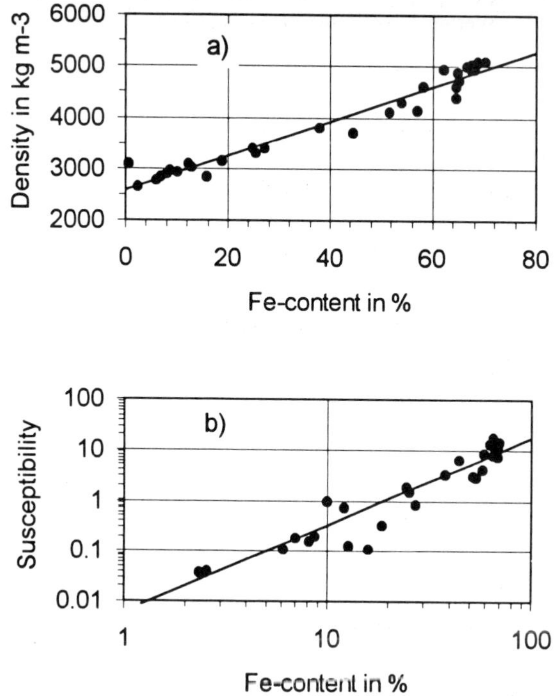

Figure 4.11. Correlation between Fe-content, density *d*, and magnetic susceptibility κ of the Malmberget mine, Sweden; after Wänstedt, 1992; a) Density vs Fe-content, linear scale b) Susceptibility vs. Fe-content, logarithmic scale

Parasnis (1973) commented about the relations (parameters see Table 4.8) of Balsley and Buddington (1958) and Jahren (1963): "Many other relations have also been suggested which make it clear that no universally valid relation between the susceptibility and the Fe_3O_4 content of rocks exist. Furthermore, where a relation does exist, the same susceptibility value may correspond to different Fe_3O_4 contents and vice versa so that a great caution must be exercised in predicting one from the other. It is therefore advisable to directly determine the susceptibilities of rocks and ores within the area of interest and not rely on formulas of the above type."

Table 4.9 contains some examples of data for ores from various deposits.

Table 4.9. Susceptibility values from various ore deposits;
References: D - Dortman (1976); M - Mooney and Bleifuss (1953); N - Neumann (1932), see Parasnis (1973); W - Werner (1945), see Parasnis (1973).

Ore type; locality	κ	Ref.
Hematite; Striberg, Sweden	$(41...98) \times 10^{-4}$	W
Magnetite (magnetite content 86.6...94.6 %); Grängesberg, Sweden	12.5...14	W
Magnetite (magnetite content 31.1...63.3 %); Ställberg, Sweden	3.0...6.15	W
Magnetite-quartzite (Fe content 30...35 %), Kursk, Russia	1...4	D
Magnetite-skarne; Gornaja Shoria, Ural		D
Fe content 25...35 %	0.75...2.5	
Fe content 35...50 %	1.5...5	
Fe content 50...60 %	3...10	
Magnetite; Berggießhübel, Germany	0.071...0.097	N
Titanomagnetite; Kackan/Kusin, Russia		D
Fe content 16...17 %	0.38...1	
Fe content 47...49 %	1.9...4.4	
Magnesioferrite-skarne; Angara, Russia		D
Fe content 33...53 %	1.2...5.6	
Pyrite; Kankberg, Sweden	53×10^{-4}	M
Martite; Kursk magnetic anomalie, Russia	$(12...25) \times 10^{-4}$	D

Dortman (1976) used data from ore deposits in the former USSR to form an empirical relation which differs from equation 4.11. This is

$$\kappa = 0.038 \cdot V_{\mathrm{m}} + 0.0011 \cdot V_{\mathrm{m}}^2 \qquad (4\text{-}16)$$

where V_{m} is the magnetite content in percent. The equation is converted to SI units and is derived for magnetite contents between approximately 10 and nearly 100 percent. Further results of systematic investigations of magnetic properties - including the susceptibility vs. magnetic mineral content for natural and synthetic magnetite of various grain size - have been published by Mauritsch et al. (1987).

A general expression for the "effective" susceptibility of a composite material has been derived by Weinberg (1967)

$$\kappa_{\text{eff}} = \sum (V_{p,d} \cdot \kappa_{p,d}) + \sum V_{f1} \cdot \kappa_{f1} + V_{f2} \cdot \kappa_{f2} \cdot \left(1 + \frac{4}{3} \cdot \pi \cdot \kappa_{f2}\right) \qquad (4\text{-}17)$$

where the first term contains the dia- and paramagnetic contribution, the second the contribution of fine grained ferrimagnetic substances with a content of less than 0.01...0.1%, and the third the contribution of ferrimagnetic substances with a content greater than 0.01...0.1%. The last term contains the "demagnetization effect", indicating the influence of grain size (see next section).

4.3.3. The influence of grain size (and grain shape) on susceptibility

The physical reason for the influence of grain size and grain shape are interaction effects of the grains found within the rock matrix. Susceptibility decreases with decreasing grain size of magnetic minerals in the rock matrix (Figure 4.12).

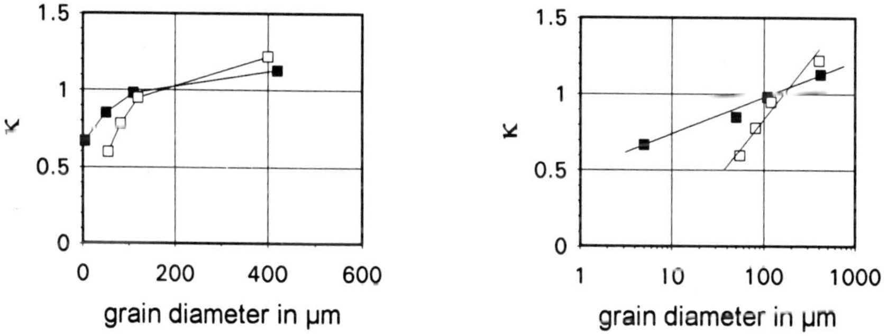

Figure 4.12. Susceptibility κ vs. grain diameter *d* of magnetic particles (magnetite), titaniferous magnetite; ■ data after Nagata (1961) □data after Spravocnik Geofiz.(1966) left: both axes linear; right: *d*-axis logarithmic; best fit lines see text.

The right hand side of the figure shows data in a logarithmic scale for grain diameter with the resulting best fit lines assuming a logarithmic relation, as follows:

$$\kappa = 0.101 \cdot \ln d + 0.502 \qquad (4\text{-}18)$$

(data from Spravocnik Geofiz., 1966)

$$\kappa = 0.277 \cdot \ln d - 0.423 \qquad (4\text{-}19)$$

(data from Nagata, 1961)

where the grain diameter *d* is in μm.

For disseminated ores with larger grain sizes (multi-domain grain size range), susceptibility is influenced by the effect of demagnetization. The resulting susceptibility of the rock is related to the volume content V_m of the magnetic substance, the intrinsic susceptibility κ_{intr}, and the demagnetizing factor of the magnetic grains N:

$$\kappa = \frac{V_m \cdot \kappa_{intr}}{1 + N \cdot \kappa_{intr}} \tag{4-20}$$

The demagnetizing factor is (using SI) $N = 1/3$ for spheres.
Carmichael (1989) has published demagnetizing factors for ellipsoids, cylinders, and rectangular prisms of various dimension ratios; a selection is given in table 4.10.

Table 4.10. Demagnetization factor N for ellipsoids, cylinders, and rectangular prisms (after Carmichael, 1989; converted for SI units). The dimension ratio a/b is
- the ratio of the two diameters in case of the ellipsoid ($a/b < 1$ oblate, $a/b > 1$ prolate ellipsoid)
- the ratio height / diameter in case of the cylinder
- the ratio height / width in case of the rectangular prism.

a/b	Ellipsoid	Cylinder	Prism
0.01	0.9845	0.9650	0.9660
0.1	0.8608	0.7967	0.8051
0.4	0.5882	0.5281	0.5482
1	0.3333	0.3116	0.3333
2	0.1736	0.1819	0.1983
4	0.0754	0.0984	0.1085
10	0.0203	0.04012	0.0457

In Nagata (1961), the empirical value of N was found to be 10...20 % less than that of a sphere, i.e. $N \approx 0.26...0.30$. This corresponds to the values of $N \approx 0.25...0.31$ (converted in SI units) after Puzicha (1977).

Mauritsch et al. (1987) noted that, as the result of a series of systematic experiments, synthetic samples also have a strong correlation between κ and magnetite content (and grain size). A correlation was observed between the saturation magnetisation and the initial magnetic susceptibility. In a logarithmic plot samples with different content of magnetite or hematite lie on straight lines which differ for the separate grain sizes (between 10 and 160 μm). The authors conclude "that not only the content of the magnetite or hematite in the rock, but also the approximate grain size of magnetite could be roughly determined ...".

4.3.4. Influence of rock structure on susceptibility

The influence of the internal rock structure on the magnetic properties is the phenomenon of magnetic anisotropy; the magnetic susceptibility is a tensor of the second rank. It is characterized by three principal susceptibilities $\kappa_1, \kappa_2, \kappa_3$ The susceptibility tensor can be represented by a susceptibility ellipsoid.

A detailed description of the magnetic anisotropy of rocks published Tarling and Hrouda (1993). The following list shows some commonly used parameters for the characterization of the magnetic anisotropy, where κ_1 is the maximum, κ_2 the medium, and κ_3 the minimum susceptibility of the tensor:

Mean susceptibility $$\kappa_{mean} = \frac{\kappa_1 + \kappa_2 + \kappa_3}{3} \qquad (4\text{-}21)$$

Anisotropy degree (P-factor) $$P = \frac{\kappa_1}{\kappa_3} \qquad (4\text{-}22)$$

Foliation (F-factor) $$F = \frac{\kappa_2}{\kappa_3} \qquad (4\text{-}23)$$

Lineation (L-factor) $$L = \frac{\kappa_1}{\kappa_2} \qquad (4\text{-}24)$$

Some of the parameters were used as "normalized" parameters; they are calculated by division with the mean susceptibility, such as

Normalized anisotropy degree $$\overline{P} = \frac{\kappa_1 - \kappa_3}{\kappa_{mean}} \qquad (4\text{-}25)$$

Normalized foliation $$\overline{F} = \frac{\kappa_2 - \kappa_3}{\kappa_{mean}} \qquad (4\text{-}26)$$

Normalized lineation $$\overline{L} = \frac{\kappa_1 - \kappa_2}{\kappa_{mean}} \qquad (4\text{-}27)$$

Petersen and Bleil (1982), state that the anisotropy of susceptibility of *mineral grains* is related to the crystalline anisotropy of the magnetic minerals particularly hematite, ilmenite, and pyrrhotite containing minerals. In this case we must determine the anisotropy of the demagnetization factor (especially for magnetite or titanomagnetite grains with preferred orientation), depending on the dimension ratio of the grains (e.g axes of elliptically assumed grains). In this case N_a, N_b, N_c must be determined.

For *rocks*, the anisotropy relation to petrography usually follows the rule that the long axes of ferrimagnetic minerals are arranged

- along the foliation planes (metamorphic rocks),
- along the bedding plane of detrital sediments,
- also along flow lines of magmatic rocks (Petersen and Bleil, 1982).

An early paper of Ising (1942) shows results of measurements made on *varved clays* from Sweden. There is an anisotropic ellipsoid with the minimum axis perpendicular to the bedding planes. The research of Hamilton et al. (1968) and Hamilton and Rees (1971) show that the magnetic susceptibility anisotropy is related to the orientation of magnetic particles depending on the strength and direction of the depositing current, on the strength and direction of the magnetic field, on the shape of the particles, and on the morphology (dip) of the bedding plane. One important result is the demonstration of the strong influence of gravity and hydrodynamic conditions during sedimentation. Laboratory experiments produced preferential alignment of the short axes of the grains normal to the bedding and the long axes of the grains parallel to direction of the current flow.

Hrouda and Janak (1971) confirm as a result of their investigations on Devonian red *sandstones* from Moravia/Czechien that the basal planes of hematite grains are parallel to the bedding (Figure 4.13a). Schultz-Krutisch and Heller (1985) measured the magnetic susceptibility anisotropy in Upper Buntsandstein (Triassic Plattensandstein) deposits from Bavaria, Germany and found that the anisotropy ellipsoids are "strongly oblate with minimum susceptibility axes normal to sedimentary bedding. The directions of maximum susceptibility axes are consistent with the NNE-NE-trending general sediment transport direction..." (Figure 4.13b). They emphasize that very small differences between the susceptibility values in various directions require a "very sensitive measurement technique" (as the result of comparative measurements they conclude that the most consistent results were obtained with the spinner magnetometer).

An example of susceptibility anisotropy in *magmatic rocks* is shown in Figure 4.13c. Here, the axes of minimum susceptibility correspond to the direction of foliation pole of granodiorite.

Siegesmund et al. (1993) have studied the fabric-controlled anisotropy of KTB core samples. For the gneisses and amphibolites the minimum susceptibility also is oriented perpendicular to foliation. The paragneisses have a degree of anisotropy between 1.2 and 1.3; their maximum susceptibility lies approximately parallel to the lineation within the foliation, for amphibolites this coincidence was not observed. Damm (1988) studied the anisotropy of susceptibility for magmatic and metamorphic rocks (from southeast Germany and some locations in Antarctica) and found a correlation between an increasing mean susceptibility value and an increasing degree of anisotropy.

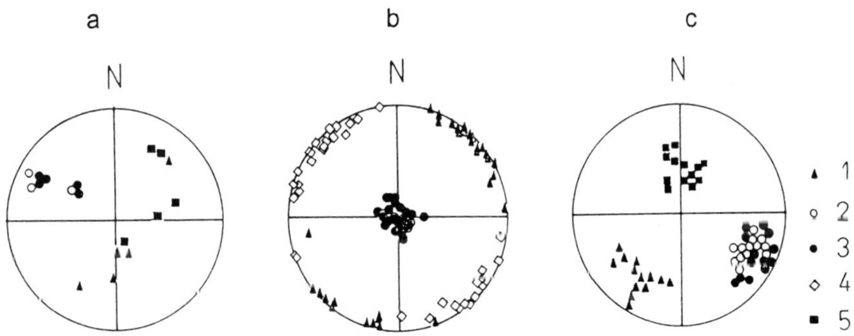

Figure 4.13. Examples of the anisotropy of susceptibility related to the internal rock structure.
1 - direction of maximum susceptibility
2 - direction of bed foliation pole, respectively (lower hemisphere)
3 - direction of minimum susceptibility
4, 5 - direction of intermediate susceptibility

a) Direction of principal susceptibilities, beds and bed poles for a Devonian clastic sedimentary rock from Moravia, Czechien (sample D1); after Hrouda and Janak (1971).
b) Direction of principal susceptibilities for a Triassic sandstone from Bavaria, Germany); the general transport direction (derived from geological investigations) is NNE-NE; after a Figure of Schultze-Krutisch and Heller (1985).
c) Directions of principal susceptibilities and foliation direction for a Variscian granodiorite (Jesenice-massif, Czechien); after Hrouda et al. (1971).

A *mathematical model* to describe and interpret magnetic anisotropy has been developed by Owens (1974). The model allows for calculation of the tensor of rock susceptibility on the basis of a distribution function of the tensor of grain susceptibility, and an angular density function for the grains with regard to their spatial orientation. Hrouda and Schulmann (1990) also describe this theory for the conversion of the magnetic susceptibility tensor into the orientation tensor and emphasize conditions required with respect to their experimental analysis of gneisses: The rock contains one mineral type with a known uniaxial grain anisotropy, the grains of this mineral being of equal size and not interacting magnetically.

4.3.5 Influence of temperature and stress

The influence of temperature on susceptibility is controlled by the characteristic temperature dependence of different minerals (see section 4.1) and their mixing series.

The stress influence is related to the anisotropic behaviour of the susceptibility of rocks. Petersen and Bleil (1982) note:

- Under uniaxial compression, the susceptibility decreases with increasing pressure in a field parallel to the stress direction and increases normal to the stress direction.
- For titanomagnetites the stress sensitivity increases with increasing Ti-content and grain size.

Kapicka (1990) has studied the variation of susceptibility under external uniaxial pressure and under hydrostatic pressure for basaltic rocks and polycrystalline magnetite. In both cases, a decrease of the mean susceptibility with increasing pressure results. The changes under uniaxial stress being larger than those under hydrostatic conditions. The magnitude of decrease depends on the ratio of "stress sensitivity constants" of the various directional susceptibilities. It is remarkable that the stress-induced change is reversible. Lebedev and Poznanskaja (1975) have modeled the behaviour of magnetic properties (susceptibility and remanent magnetization) with increasing depth for rocks from the Ukrainian shield.

The stress and strain influence on the magnetic anisotropy is described by Tarling and Hrouda (1993). Effects were observed in all types of rocks, from unconsolidated soils to igneous rocks. In general, the deformation process is connected with an orientation of magnetic minerals or particles. This process can be described by various models (such as the "line/plane model" in which the magnetic grains do not change their shape, but their orientation is in response to the stress, or the "ductile model" in which the components of the matrix and magnetic particles behave differently as a result of a different viscosity).

4.3.6. Natural remanent magnetization (NRM)

The total magnetization M of any rock is the sum of two vectors (4-6)

- induced magnetization, dependent on the external field, and
- remanent magnetization, independent of the external field.

Figure 4.14 gives as an overview of the mean ranges and magnetization values for some important rock types. The distributions correlate to the abundance of Fe- and Ti-oxides in different rock types (Table 4.6).

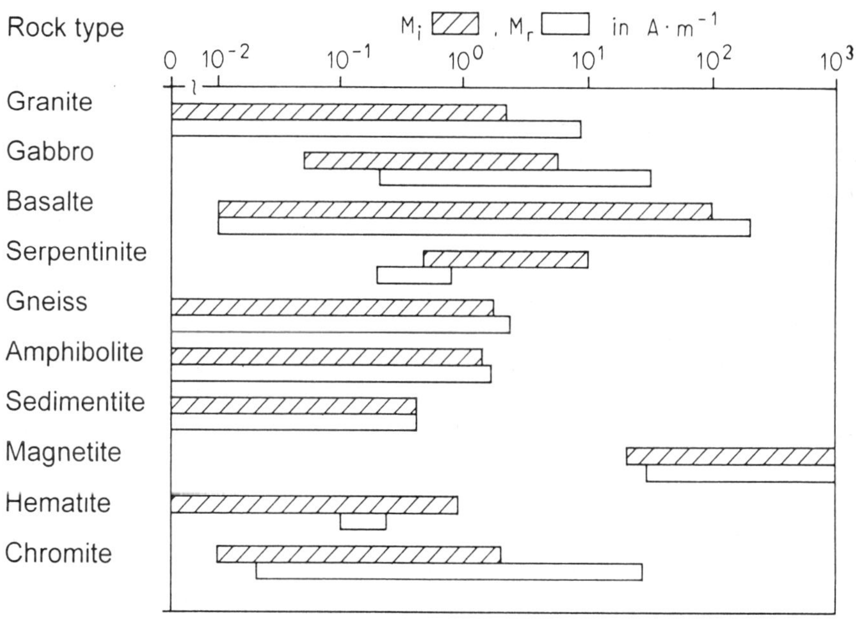

Figure 4.14. Induced M_i (upper column) and remanent M_r (lower column) magnetization for some rock types; after data from Bosum (1981).

The *Koenigsberger-ratio* Q of rocks is the ratio of natural remanent magnetization to the magnetization induced by the earth magnetic field (compare eq. 4-7). Q shows a wide scattering of values for a rock type. Carmichael (1989) gives the following average values

- for typical igneous rocks $Q = 1 \ldots 40$
- for typical sedimentary rocks $Q = 0.02 \ldots 10$.

The natural remanent magnetization (NRM) is the field-independent and irreversible part of the total (natural) magnetization. Ferrimagnetic minerals obtain NMR and there are different types and causes or origins of this phenomenon in rocks, depending on

- the magnetic minerals in the rock matrix, and
- the "physico-chemical history" and related effects.

The following overview contains only a short summary of different types of NMR and their characteristics; more detailed information can be found in literature on this subject (see section 4.1).
The main types of natural remanent magnetization of rocks are:

1) *Thermoremanent magnetization (TRM)*: TRM is the remanence acquired by a rock (containing ferrimagnetic substance) when it is cooled from a temperature above its Curie temperature to a lower (e.g. room) temperature in the presence of a magnetic field. Hunt et al. (1979) remark that the entire TRM is not always acquired at the Curie-temperature, but can also be acquired in discrete temperature intervals (partial TRM). Generally, most of the magmatic and high-temperature metamorphic rocks are characterized by a distinct TRM higher than the induced magnetization, thus, the Koenigsberger ratio is $Q > 1$. TRM acquisition depends on grain size (increasing grain size entails decreasing magnitude). For low fields, TRM is proportional to the applied field, whereas it saturates in stronger fields (Hunt et al., 1979). TRM has an important application in the study of palaeomagnetism.

2) *Chemical remanent magnetization (CRM)*: CRM occurs during the formation of a magnetic mineral (origin and growing process), for example as the result of a chemical reaction or phase transition below its Curie temperature under a magnetic field. CRM is therefore related to processes such as oxidation of magnetite to hematite or maghemite, oxidation of titanomagnetite to titanomaghemite, dehydration of iron hydroxide to hematite, precipitation of ferromagnesian minerals (biotite, hornblende, augite), and recrystallisation of ferrimagnetic minerals below T_C (Hunt et al., 1979; Petersen and Bleil, 1982). Ocean floor basalts are typical examples of CRM (Petersen and Bleil, 1982). Often CRM is connected with DRM (see below) in sediments. Hunt et al. (1979) remark that CRM is "due to the unusually large volumes of hematite in the form of either pigmentation or specularite ... the most probable source of magnetization in red beds."

3) *Detrital or depositional remanent magnetization (DRM)*: DRM originates from the oriented deposition of previously magnetized mineral grains

under the influence of the earth's magnetic field. The magnetic moments of the particles are aligned to the field direction, so that this direction is "conserved" in the sediment. This is a process which depends on the depositional environment (low turbulence) and also on the sediment type (obvious relations are shown by clays). The grains are preferentially oriented by the earth's magnetic field. After deposition, minor changes are possible upon compaction (Post-depositional remanent magnetization PDRM). The DRM can be important in marine sediments, lake sediments, and varved clays (Carmichael, 1989).
DRM is often associated with CRM. Hunt et al. (1979) comment that the original magnetization of all sediments is due to DRM, but in the case of alteration (e.g. red beds) the "CRM component soon dominates that of the DRM."

Further types of remanent magnetization are isothermal remanent magnetization IRM (e.g. produced by lightning), viscous remanent magnetization VRM (time effects), and piezo-remanent magnetization PRM (pressure effects), details are described in the literature (see for example Carmichael, 1989; Petersen and Bleil, 1982).

CHAPTER 5

Natural Radioactivity of Rocks

"One of the best ways to determine clay volume is by using natural gamma ray spectral data ..."
(Walter H. Fertl, 1983)

5.1. Physical basis

Natural radioactivity is the spontaneous decay of the atoms of certain isotopes into other isotopes. This decay is characterized by emission of alpha particles (α - radiation), or beta particles (β - radiation), or an electromagnetic radiation (γ - radiation). For practical use in geophysics the gamma radiation is most important, because the particle radiation (α and β) has a very low penetration length in rocks. The gamma radiation is characterized by wave lengths in the range of 10^{-11} ... 10^{-13} m, equivalent to frequencies between 10^{19} and 10^{21} s^{-1}. The energy of the electromagnetic wave is related to the frequency:

$$E = h \cdot \nu \tag{5-1}$$

where h is the Planck constant (h = 6.6256 x 10^{-34} J s). The energy is usually expressed in electron-volts (eV). The radiation energy is of the order of keV (10^3 eV) to MeV (10^6 eV).

The natural radioactivity of rocks is due to the decay of natural radionuclides found in rocks or minerals. The decay process is described by its probability. A given nuclide has a probability of decay per unit time which is independent of any external conditions (temperature, pressure). The decay process can be characterized either by the *half-life time* $T_{1/2}$ or by the *decay constant* λ. The decay process is described as follows:

$$N(t) = N_0 \cdot \exp(-\lambda \cdot t) \tag{5-2}$$

where $N(t)$ is the actual number of nuclei and N_0 its initial value. After the *half-life time* $T_{1/2}$, is obtained $N(t) = N_0/2$; thus

$$\ln 2 = \lambda \cdot T_{1/2}$$

or

$$T_{1/2} = \frac{\ln 2}{\lambda} \approx \frac{0.693}{\lambda} \tag{5-3}$$

The *activity* of a sample is defined as the number of decays per unit time, i.e.,

$$-\frac{dN}{dt} = \lambda \cdot N_0 \cdot \exp(-\lambda \cdot t) \tag{5-4}$$

so that the activity of a source declines exponentially with time. The unit of activity originally was defined as the number of disintegrations per second occuring in 1 gram of ^{226}Ra. The new definition one Curie (Ci) is 3.7 x 10^{10} disintegrations per second. One disintegration per second is one Bequerel (Bq), thus

$$1 \text{ Ci} = 3.7 \times 10^{10} \text{ Bq}$$

For most purposes the millicurie (1mCi = 10^{-3} Ci) or microcurie (μCi = 10^{-6} Ci) is used. For relative units see section 5.3.

If after the decay of nuclide 1 ("parent") the resulting nuclide 2 ("daughter") is also radioactive, etc., to the n-th nuclide, the condition for the *radioactive equilibrium* is

$$\lambda_1 \cdot N_1 = \lambda_2 \cdot N_2 = \ldots = \lambda_n \cdot N_n \tag{5-5}$$

Three types of natural radionuclides are common in the rocks of the earth's crust:

- the *uranium series:* parent isotopes are ^{238}U and ^{235}U and the end of the decay series are ^{206}Pb and ^{205}Pb, respectively. The half-life time is 4.5 x 10^9 and 7.1 x 10^8 years, respectively.
- the *thorium series:* parent isotope ^{232}Th, end of the series is ^{208}Pb. The half-life time is 1.4 x 10^{10} years.
- the *potassium isotope* ^{40}K: it has an abundance of about 0.012% of natural potassium, and decays to ^{40}Ca emitting beta radiation (89 %). 11 percent of the decays of ^{40}K involve electron capture and ^{40}Ar is the product. This ^{40}Ar is in an non stable excited state; upon returning to its ground-state, it emits the gamma radiation at 1.46 MeV.The half-life time is 1.3 x 10^9 years.

The uranium- and thorium series result in a spectrum with typical energies of the various individual decays. Potassium originates monoenergetic radiation (Figure 5.1).

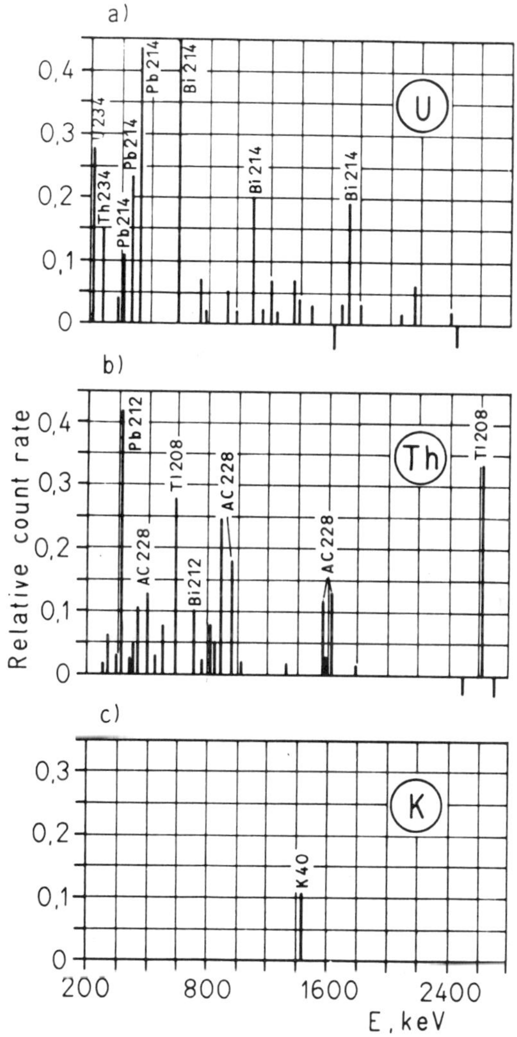

Figure 5.1. Prominent gamma-ray energies from the uranium- and thorium-series and for potassium monoenergetic radiation. The height of the vertical lines is proportional to the relative intensity (after International Atomic Energy Agency, 1976; from Hearst and Nelson, 1985)

For many practical applications (e.g. borehole gamma-spectrometry), energy-windows (or channels) are used in the range of typical energies, for example

1.46 MeV for K
1.76 MeV for U (from bismuth ^{214}Bi)
2.61 MeV for Th (from thallium ^{208}Th).

5.2. Abundance of uranium, thorium, and potassium in minerals

The contents are given generally in ppm for uranium and thorium (1 ppm = 10^{-6} kg U and Th for 1 kg rock mass) and in percent (%) for potassium (1% = 10^{-2} kg K for 1 kg rock mass).

Potassium is a constituent of a number of typical and abundant minerals. It is generally chemically combined

- in clay minerals where it occurs in the clay mineral structures,
- in evaporites where it occurs chemically as salts, such as sylvite, carnallite,
- in rock forming minerals (for example the feldspars: orthoclase, microcline, and micas: biotite, muscovite) where it is chemically combined into the silicate structure.

Table 5.1 gives potassium content of minerals and evaporites. The potassium content in *clays* varies. Illites contain the greatest amount (3.51...8.31 %), while kaolinite has very little or none. Montmorillonite, chlorite, and kaolinite have potassium fixed by adsorption; kaolinite contents sometimes more potassium due to imperfectly degraded feldspars (Schlumberger, 1982). Most clays are mixtures, and therefore shale has a relatively constant potassium content of about 2.7 % (Rider, 1986). The most important potassium-bearing rock-forming minerals are the *feldspars*: Microcline contains approximately 16 %, and orthoclase 14 % (by weight) potassium (Rider, 1986).

The most abundant rock forming minerals contain only small quantities of *thorium and uranium*. Higher concentrations of U occur in uranium minerals (for example, autunite, bequerelite, carnotite, pechblende, uraninite, tyuyamunite with contents up to 76 % U) and uranium-bearing minerals (such as: betafite, chalcolite, fergusonite, pyrochlore, uranotile with contents up to 56 % U). Larger quantities of Th are concentrated in some accessory minerals (see

Table 5.1. Potassium bearing minerals and evaporites. References: S - Schlumberger (1982), R - Rider (1986), W - Western Atlas (1992)

Name	Chemical formula	K content in % weight	Ref.
Evaporite			
Sylvite	KCl	52.44	S
		52.5	R
Langbeinite	$K_2SO_4\,(MgSO_4)_2$	18.84	S
Kainite	$MgSO_4\,KCl\,(H_2O)_3$	15.7	S
Carnallite	$KCl\,MgCl_2(H_2O)_6$	14.07	S
		14.1	R
Polyhalite	$K_2SO_4\,MgSO_4\,(CaSO_4)\,(H_2O)_2$	13.37	S
		12.9	R
Glaserite	$(K, Na)_2SO_4$	24.7	S
Mica and Clay			
Muscovite	$KAl_2\,(AlSi_3O_{10})\,(OH, F)_2$	9.8 (ideal) to 7.9)*	S
		7.9 ... 9.8	W
Biotite	$K\,(Mg, Fe)_3\,(AlSi_3O_{10})\,(OH, F)_2$	6.2 ... 10.1 (av.: 8.5)	S
		6.7 ... 8.3	W
Illite	$K_{1\text{-}1.5}\,Al_4\,(Si_{7\text{-}6.5}\,Al_{1\text{-}1.5})\,O_{20}$	3.5 ... 8.3 (av.: 6.7)	S
	$(OH)_4$	3.5 ... 8.3 (av.: 5.2)	R
		4.5	W
Glauconite	$K_2\,(Mg, Fe)_2\,Al_6\,(Si\,O_{10})_3$	3.2 ... 5.8 (av.: 4.5)	S, R
	$(OH)_{12}$	5.08 ... 5.30	W
Montmorillonite		0 ... 4.9 (av.: 1.6)	S
	K is fixed by adsorption	0.16	W
Chlorite		0 ... 0.35 (av.: 0.1)	S
Kaolinite	K is fixed by adsorption	0 ... 0.6 (av.: 0.35)	S
	K is fixed by adsorption	0.42	W
Feldspars			
Microcline	$KAlSi_3O_8$ triclinic	16 (ideal) to 10.9)*	S
		10.9	W
Orthoclase	$KAlSi_3O_8$ monoclinic	14 (ideal) to 11.8)*	S
		11.8 ... 14.0	W
Feldspathoids			
Leucite	$K\,Al\,(SiO_3)_2$	17.9 (idcal)	S
Nephelite	$(Na, K)\,AlSiO_4$	4 to 8	S

)* corresponds to beginning alteration

Table 5.2. U and Th content of some rock forming and accessory minerals (.... range, [] mean value); References: F - Fertl, 1979, 1984; H - Haack, 1982 (and cited by Haack, 1982); S - Schlumberger, 1982 (and cited by Schlumberger, 1982); Se - Serra, 1984; R - Rider, 1986; W - Western Atlas, 1992.

Mineral	U in ppm	Th in ppm	Ref.
Mica and clay minerals			
Biotite	1 ... 40	0.5 ...50	S
in igneous rocks	0.098...0.29 [0.23]	3.8...4.9 [4.5]	H
in metamorphic rocks	0.022...0.044 [0.03]	0.03	H
Muscovite	2 ...8	10 ...25)*	S
	0.0003...0.05		H
Bauxite	3...30	10...130	F, W
	3...30	8...132 [42]	R
	3...30	10 ...132	S
Glauconite		< 10	S
		2...8	R
Bentonite	1 ...36	4 ..55	S
	10...20	6...50	F; W
Montmorillonite	2...5	14...24	S
Smectite		10...24	R
Kaolinite	1.5...9	6 ..42	S
	1.5...3	6...19	W
Illite	1.5	10...25	S
		6 ..22	R
Chlorite		3 ...5	S, Se
Feldspars			
Plagioclase	0.2...5	0.5...3	S
in basaltes,	0.005...0.025	0.01...0.025	H
in dacites	0.014	0.16...0.64	H
Potassium feldspar	0.2...3	3...7	S
Hornblende	1...30	5...50	S
	0.0005...0.28	about 0.5	H
Accessory minerals			
Allanite	30...700	500...5000	W
Apatite	5...150	20...150	W
Epidote	20...50	50...500	W
Monazite	500...3000	(2.5...20) x 10^4	W
Sphene	100...700	100...600	W
Zircon	300...3000	100...2500	W
Xenotime	500...34000	low	W

)* pure muscovite does not content thorium. But in sedimentary rocks the deposition of muscovite is generally accompanied by deposition of finer heavy minerals wich are thorium and uranium bearing (Schlumberger, 1982).

Table 5.2), which also contain large quantities of U. However, the main portion of Uranium is finely dispersed as submicroscopic, ill-defined microinclusions, loosely bound on grain boundaries, fractures and internal surfaces; it can easily be leached by weak acids (Haack, 1982). This causes the high mobility of uranium under geological processes.

A description of abundance and geological processes of uranium and thorium is given by Serra (1984) and Rider (1986).

Original *uranium* is generally associated with acid igneous rocks (the average is about 4.65 ppm). It forms soluble salts, especially the uranyl ion (UO_2^{2+}) and is it present in this form in rivers and sea water. The salts are unstable and pass easily out of solution.

From the water, uranium can be 'fixed' and pass into sediments in three ways (Serra, 1984):

- chemical precipitation in acid (pH 2.5...4.0) or reducing (pH 0...0.4) environment,
- adsorption by organic matter (plankton, plants, shells, animal skeletons),
- absorption by phosphates.

In general, uranium is not chemically combined with rock constituents like potassium, but is loosely associated with secondary components (Rider, 1986). For the most part, uranium has a very heterogeneous sedimentary distribution, is movable, and can be used as an environmental indicator (see section 5.3.4).

Thorium has its main origin in acid and intermediate rocks. But compared with uranium, it is very stable, and will not generally pass into solution. Thorium and thorium minerals, therefore, occur in sediments as detrial grains. They are typically in stable heavy minerals such as zircon, thorite, monazite, epidote, and sphene (Rider, 1986). Thorium is found in relatively high amounts in bauxite (residual soils), and among clay minerals, it is more abundant in kaolinite (of terrestrial origin) than in glauconites (of marine origin).

As a result of alteration of the source minerals and related processes, *clay minerals* show relatively high concentrations of U, Th, and K.

Clay minerals (and other phyllosilicates) show remarkable differences of their Th/K ratios (Figure 5.2). This property is recommended for "clay mineral identification" on the basis of spectrometric gamma log measurements (see Fertl, 1979, 1983; Schlumberger, 1982; compare section 5.4).

The *total-count gamma-ray logs* have been utilized routinely in well logging and so a standardized calibration of the measuring units is necessary to create comparable logs. The calibration facility is the American Petroleum

Institute (API) device in Houston, Texas. It is briefly described by Hearst and Nelson (1985) as follows: "The API facility is constructed of concrete with an admixture of radium to provide the ^{238}U decay series, monazite ore as a source of thorium, and mica as a source of potassium (Belknap et al., 1959). The concentrations, which are selected to be equivalent to twice those of an average shale in the mid-continent United States, were determined to be equivalent to 4.07 % K, 24.2 ppm Th, and 13.1 ppm U. It was desired that the unit recorded by the log be such that 100 units would correspond to the deflection recorded opposite a shale formation; therefore, 200 API gamma-ray units are defined as the difference in count rate recorded in concrete without the radioactive admixture and in the radioactive concrete.

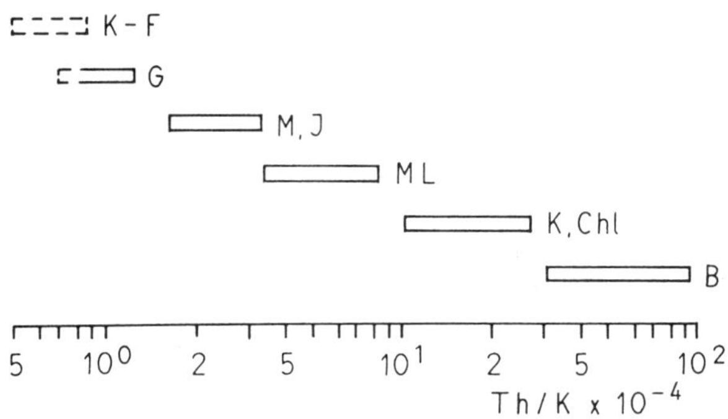

Figure 5.2. Th/K ratio in potassium-rich feldspars (K-F), glauconite (G), muscovite (M), illite (J), mixed layer (ML), kaolinite (K), chlorite (Chl), and bauxite (B) minerals; after Hassan (1976), as quoted by Fertl (1979).

The API standard is adequate for qualitative comparison between logs. It is, however, sometimes convenient to predict tool response to known concentrations of U, Th, and K. For this purpose the count rate recorded by a total count gamma sonde is considered to be linear and additive.
The counts from each type of source can be converted to equivalent uranium concentration:

$$J = k\,(\,a\,\mathrm{K} + \mathrm{U} + b\,\mathrm{Th}\,) \qquad (5\text{-}6)$$

where k is a constant for a given sonde, a is the concentration of U (in ppm) that will give the same total count rate as 1 % K, and b is the concentration of U that will give the same count rate as 1 ppm Th. For an NaI detector of dimensions 150 by 80 mm with a discrimination threshold of 200 keV, the value of *a* is 1.05 and *b* is 0.47 (IAEA, 1976). Some API-values and value ranges for selected rock types are shown in Table 5.3.

Table 5.3. Gamma-ray API values for some minerals; S - data from Schlumberger, 1989; H - data from a Table of Hearst and Nelson (1985) made on the basis of values from Brom and Driedonks (1981), Edmundson and Raymer (1979), Fertl and Frost (1980), Patchett (1975), Reeves (1981), Tixier and Alger (1970)

Rock type	Gamma-ray in API	Ref.
Quartz, Dolomite, Calcite (clean)	0	H; S
Feldspars		
- Plagioclase (Albite, Anorthite)	0	S
- Alcali f. (Orthoclase, Anorthoclase, Microcline)	220	S
Micas (Muscovite, Biotite)	270	S
Shale	80...150	H
Clays		
Kaolinite	80...130	S
Chlorite	180...250	S
Illite	250...300	S
Montmorillonite	150...200	S
Evaporite		
Sylvite	500	H; S
Carnallite	200	H; S
Kainite	225	H
	245	S
Polyhalite	180	H
	200	S
Langbeinite	275	H
	290	S

The old equivalent units "microgram radium-equivalent per ton rock" (μg Ra-equiv./to) corresponds to about 16.5 API units and "microroentgen per hour" (μR/h) to about 7.2 API units (Bradel, 1985).

5.3. The abundance of uranium, thorium, and potassium in rocks - the natural radioactivity of rocks

5.3.1. Overview

The natural radioactivity of rocks is due to the contribution of the three radiation components U, Th, and K. Concentrations of U, Th, and K vary widely in crustal rocks as the result of the mineral composition of the rock type and the rock genesis and environment. Mean values for the crust are shown in Table 5.4. Some more selected values are given in Table 5.5. Detailed data are published by Haack (1982) and Clark (1966). The table also contains the ratios K/Th, K/U, and Th/U and the heat generation rate *A* (compare section 5.4).

Table 5.4. Mean values of the uranium, thorium, and potassium content for the earth' crust after Heier and Rogers, 1963 (H) and Prutkina and Saskin, 1975 (P)

Subject	Uranium ppm	Thorium ppm	Potassium %	Ref.
Earth's crust	2.5	13	2.5	P
Earth's crust	2.1	7.8	2.1	H
Oceanic crust	0.64	2.8	0.87	H
Continental crust	2.8	10	2.6	H

Table 5.5. Mean U, Th, and K contents and heat generation rate *A* of selected parts of the crust (data from Haack, 1982).

Subject	U ppm	Th ppm	K %	K/Th 10^3	K/U 10^3	Th/U	A 10^{-6} Wm^{-3}
Continents as a whole	1.6	6.3	1.64	2.6	10.1	3.9	1.01
Intrusive rocks of the upper crust	3.8	16	2.74	1.7	7.2	4.2	2.38
Canadian shield							
archaic	1.2	9.7	2.21	2.3	18.4	8.1	1.19
- proterocoic	2.2	13.6	2.80	2.1	12.7	6.2	1.78
Baltic and Ukrainian shield, basement of Russian platform	2.8	18	2.56	1.4	9.1	6.2	2.26

In practice, the "*integral or total*" radioactivity is of interest and provides the basis for interpretation.

The integral radioactivity of a rock depends on

- the U, Th, and K content,
- the specific activity of U, Th, and K (this means the specific contribution of the elements with respect to the integral radiation),
- the detector characteristics.

Figure 5.3 shows the general tendency for main rock groups.

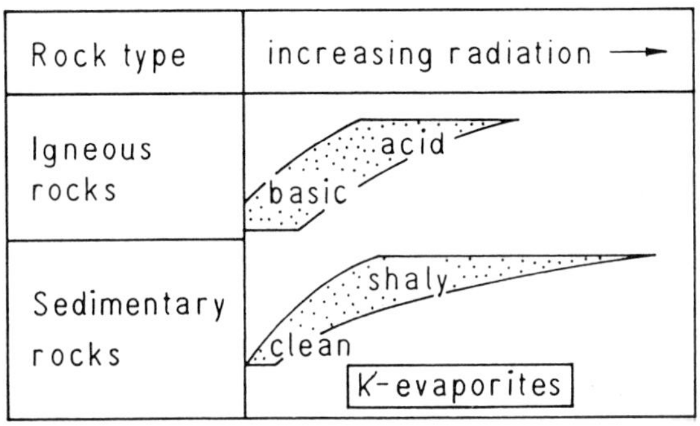

Figure 5.3. General tendencies of variation of natural radioactivity

Two comments should be made here:

1) The radioactivity of igneous rocks increases from basic to acid rocks,
2) The radioactivity of sedimentary rocks increases from "clean" to "shaly" rocks, i.e. with increasing clay content,

5.3.2 Igneous rocks

Within each rock type, a broad scatter of elemental contents and resulting radioactivity is typical (Table 5.6). Generally, we find that concentrations of all three elements in common igneous rocks are distinctly higher (about 10 times) in acidic than in ultrabasic rocks and follow a trend of the SiO_2 content. Exceptions to this are the alkali feldspathoidal rocks, where U and Th concentrations in excess of 100 ppm are not unusual. This rock type, however, represents only a small fraction of crustal igneous rocks (Hearst and Nelson, 1985).

Higher radioactivity of magmatic rocks is mainly related to the presence of accessory uranium- and thorium-bearing minerals. This is partly due to uranium-and thorium-forming minerals and to a minor degree common rock-forming minerals (Kobranova, 1989).

Table 5.6. Mean U, Th, and K content of igneous rocks (after Dortman, 1976; Saporosec, 1978; Fertl, 1979; Kobranova, 1989)

Rock type	U ppm	Th ppm	K %	Th/U
Granite	4...7	15...40	3.4...4.0	3.5...5.6
Granodiorite	2.1	8.3	2.3	4.0
Diorite	1.8	6.0	1.8	3.3
Gabbro, Diabase	0.6	1.8	0.7	3.0
Dunite, Pyroxenite	0.03	0.08	0.15	2.7
Basalts	0.7	2.3	1.0	3.2

For comparison, values and ranges are presented in Table 5.7 from U.S.G.S. standards after Adams and Casparini (1970, see also Schlumberger, 1982).

Nielson et al. (1990) give (based on data from Malan, 1972) for Precambrian granite (47 samples) the following means and ranges:

Uranium	4.70 ppm	(0.3...17.5 ppm)
Thorium	37.8 ppm	(1...253.2 ppm)
Potassium	4.1 wt %	

Contents and content ratios (Th/U, K/U, K/Th) show a distinct dependence and correlation to genetic conditions, for instance (see Haack, 1982):

- granites and granodiorites above active subduction zones have lower U, Th, and K concentrations than granites which are not above subduction zones
- volcanics show various contents and ratios for different geotectonic positions (Table 5.8 and 5.9.). Lava shows different contents and ratios depending on chemical and mineralogical composition (Figure 5.4).

For this reason, research into U, Th and K concentrations and their ratios is used in genetic studies. For example, Simov (1989) discusses these values with respect to the genetic and petrographic character of rocks in the foothills of a granite massif in the Southeastern Alps. The granites (Table 5.10) show relatively high U contents and a distinct difference between the granitic types.

Table 5.7. U, Th, and K content of some igneous rocks;)*from U.S.G.S. standards; after Adams and Casparini (1970, see Schlumberger, 1982)

Rock type	Locality	U ppm	Th ppm	K %
Acidic				
Intrusives	Rhode Island)*	1.32...3.4(1.99)	21.5...26.6(25.2)	3.92...4.8 (4.51)
Granite	Rhode Island)*	1.3...4.7 (4)	6.5...80 (52)	5.06...7.4 (5.48)
	Precambrian	3.2...4.6	14...27	2...6
	- average -	4.35	15.2	4.11
Syenite)*		2500	1338	2.63
Basic Intrusives				
Gabbro		0.84...0.9	2.7...3.85	0.46...0.58
Granodiorite		2.6	9.3...11	2...2.5
Diorite		2.0	8.5	1.1
Ultra Basic				
Dunite		0.01	0.01	0.02
Peridotite		0.01	0.05	0.2
Acid Extrusive		2.5...5	6...15	2...4
Rhyolitc		2...7	9...25	5.7
Trachyte				
Basic Extrusive		0.99	4.6	0.61
Alcali basalt		0.53	1.96	0.61
Plateau basalt	Oregon)*	1.4	3.9	1.4
Alcali olivine basalt		1.2...2.2 (1.73)	5.5...15 (6.81)	1.4...3.23 (1.68)

Table 5.8. Element ratios in volcanics of different geotectonic settings (selected values after Haack, 1982); the data for the Long Valley are calculated from meansafter Nielson et al., (1990)

Volcanics from	K/Th 10^3	K/U 10^3	Th/U
Ocean floor	3.3	12.4	3.7
Ocean islands	3.9	12.5	3.2
Island arcs	2.4	9.0	3.8
Continental margins	4.5	10	2.2
Intracontinental	2.8	8.4	3.0
Italy	1.7	4.8	2.8
Long Valley, California	2.1	6.7	3.2

Table 5.9. Mean U and Th contents and ratios for rhyolitic eruptives from two volcanic provinces in SE-Germany (after Marle, 1978)

Locality	U ppm	Th ppm	Th/U
NW - Saxonia	3.7	18.8	5.1
Elbe-river zone	2.9	24.8	6.0

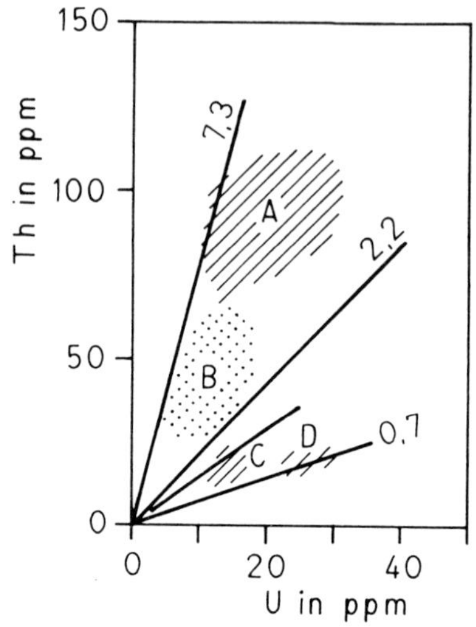

Figure 5.4. Th vs. U content in Italian lavas (Fertl, 1979 after data from Gasparini (1963), Luongo and Rapolla (1964)) Ischian lavas: A - phonolites, alkali trachites; B - trachites, latites Somma-Vesuvia: C - trachitic lavas; D - leucite tephrites and tephritic leucites

Table 5.10. U, Th and K content in granites and associated rocks (Southeastern Alps) after Simov (1989); *n* - number of samples

Rock type	*n*	U ppm	Th ppm	K %	Th/U	U/K 10^{-4}
Coarse grained porphyritic granites	5	10.5	11.3	2.16	1.1	4.9
Medium grained granites	239	6.7	13.5	2.66	2.0	2.5
- biotite granites	165	5.9	13.0	2.76	2.2	2.1
- porphyritic granites	11	8.7	11.5	2.35	1.3	3.7
Fine grained biotite granites	55	8.6	9.4	2.86	1.1	3.0
Migmatites	11	2.4	16.7	3.34	6.9	0.7
Lamprophyres	14	4.4	11.9	1.91	2.4	2.6
Old pegmatites	42	5.5	7.1	2.65	1.3	2.1
Granodiorites	65	6.0	16.0	2.25	2.7	2.7

The highest U content being found in the porphyritic facies. From the low Th/U ratio (1.1), the author concludes a "repeated activation and differentiation of the rock and/or formation of granites through granitization."
Just (1980) used gammaspectrometric results for regional studies of magmatic rocks.

5.3.3 Metamorphic rocks

The uranium, thorium, and potassium content of metamorphic rocks is a result of the original contents of the educt material (igneous or sedimentary) and may have been changed by metamorphic processes. The content is adsorbed and redistributed according to the degree of metamorphic transformation (Rybach and Cermak, 1982).

Figure 5.5 shows values for samples taken from the KTB and the distribution for the two main lithological units.

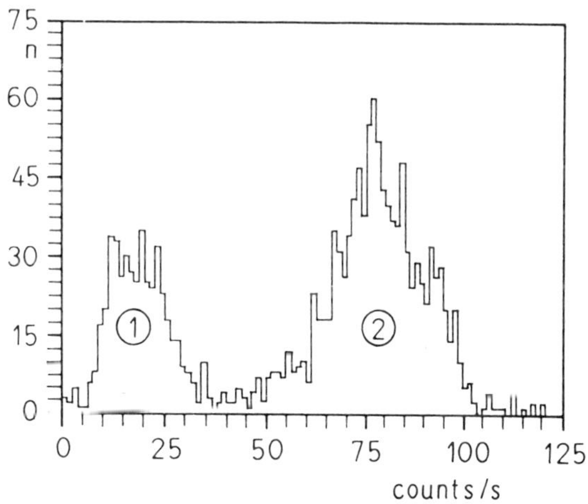

Figure 5.5. Normal frequency distribution of the gamma-ray counting rate of drill cores for the depth interval of 0 ... 3900 m; after Bücker et al. (1990); n - number of samples; 1 - metabasites 2 - gneisses

A comparison of the contents of igneous and metamorphic rocks shows the following general observations:

- Radioactive elements are decreased with increasing metamorphism. Rybach and Cermak (1982) found that the "depletion of U and Th, caused by progressive metamorphism, is most markedly evident in rocks of the granulite facies... U and Th have the tendency towards upward migration in the earth's crust because of reactions due to dehydration (middle level of crust) or because of partial melting near the base of the crust (migmatites) ... K seems to be more or less unaffected by these processes."

- The mean Th/U ratios of the metamorphic rocks deviate from the typical values of many intrusive rocks (3...4). "This is due to unequal losses of K, Th and U during metamorphic processes in which the mobility of U plays a great part (U being easily soluble in the hexavalent form as the uranyl ion).... A reason for the better mobility of U is that, in contrast to Th, it is in part loosely bound on grain boundaries and internal surfaces" (Haack, 1982).

Table 5.11. Ranges and mean U, Th and K content and Th/U ratio for metamorphic rocks; References: D - data after Dortman (1976) for granulite and eclogite facies; B - Bücker et al., (1990); R - after a material published by Rybach and Cermak, (1982) on the basis of data from different authors (P - Puzankov, Bobrov and Duchkov, 1977; Rybach, 1973)

Rock type Locality	U ppm	Th ppm	K %	Th/U	Ref.
Quartzite Swiss Alps	0.4 0.2...0.6	2.2 2.1...2.3	1.06 0.95...1.16	5.5	R
Gneiss KTB, Germany	2.6 ± 1.2	8.2 ± 2.0	2.28 ± 0.17	3.2	B
Gneiss Swiss Alps	4.95 0.9...24	13.1 1.2...25.7	3.11 0.32...4.71	2.64 0.69...8.2	R
Eclogite	0.2	0.4	0.8	2.0	D
Amphibolite	0.7	1.8	0.6	2.6	D
Amphibolite Swiss Alps	1.65 0...7.8	3.00 0.01...13.7	1.23 0.11...2.22	1.82 0.01...3.5	R
Metabasite KTB, Germany	0.6 + 0.5	2.5 + 1.6	1.05 ± 0.16	4.2	B
Plagioclasic mig- matite, Charnockite	0.8	2.3	1.7	2.9	D
Sillimanitic-cordieritic gneiss and schist	1.3	4.2	2.6	3.2	D
Schist Swiss alps	2.14 0.4...3.7	9.73 1.6...17.2	2.23 0.39...4.44	4.6 0.25...2.28	R
Slate Kamchatka	2.0	6.4	1.83	3.2	P

5.3.4. Sedimentary rocks

Table 5.12 contains mean values of the uranium, thorium, and potassium content for selected sedimentary rocks. Generally, we will follow the characterization given by Hearst and Nelson (1985): "In sedimentary rocks, the relative mean abundances are less predictable than in igneous rocks. On the average, K is lower in effective concentrations than U or Th, and Th contributes about the same level of activity as U. As a class, carbonates are the lowest in natural radioactivity of the sedimentary rocks.... Generally, shales will have a higher level of natural radioactivity than other sediments; consequently, the gamma-ray sonde is used to distinguish between shales and other sediments."

The correlation between clay content (respectively shaliness) and radiation of sedimentary rocks is important for the reservoir characterization with respect to

- the distinction between clay and sand layers,
- the determination of the clay content and
- the characterization of " clay types".

These correlations are changed if other radioactive minerals (for example feldspar, mica, glauconite, and accessories like monazite, zircon) are present in clean clastic rocks.

Two kinds of correlations with different analytic expressions (either linear or exponential) are used:

1. Correlations on the basis of the integral gamma intensity/radiation:

Dachnov (1962, 1975), Vendelstein and Larionov (1964), Larionov and Nefedova (1964), Srebredolski and Matcinova (1977), Fertl (1983) and others found correlations between the shale or clay content V_{sh} and the parameter of a normalized natural gamma intensity or "radioactivity index" ΔI_γ (Figure 5.7)

$$\Delta I_\gamma = V_{sh}^a \tag{5-7}$$

where

$$\Delta I_\gamma = \frac{I_\gamma - I_{\gamma,min}}{I_{\gamma,max} - I_{\gamma,min}} \tag{5-8}$$

with I_γ the actual gamma intensity,
$I_{\gamma,min}$ the gamma intensity in clean sand,
$I_{\gamma,max}$ the gamma intensity in clay

a is an empirically determined exponent of the order of about 0.3...0.8 (Saporosec, 1978).

Table 5.12. U, Th and K content of sedimentary rocks, mean values; References: K - Kappelmeyer and Haenel (1974), M - Murray and Adams (1958), W - Wollenberg, Smith and Bailey (1967) all cited in Rybach and Cermak (1982); P - Puzankov, Bobrov, Duchkov (1977); D - Dortman (1976); F - Fertl (1983); WE - Western Atlas (1992)

Rock type, Locality	U ppm	Th ppm	K %	Ref.
Clay, N-Atlantic, Caribbean	2.1	11	2.5	K
Shale	3.7	12.0	2.7	K
	1.0...13.0	2.0...4.7		
Shale (200 samples)	6.0	12.0	2.0	WA
"Common" shales	3.7	12.0	2.7	F; WA
	1.5...5.5	8...18	1.6...4.2	
Oil shale, Colorado	up to 500	1...30	< 4.0	F
Argillite, Russia	4.0	11.5	2.7	D
Aleurolite, Russia	2.9	10.4	1.7	D
Sandstone, Kamchatka	1.40	4.92	1.33	P
Sandstone	0.5	1.7	1.1	F; WA
	0.2...0.6	0.7...2.0	0.7...3.8	
Sandstone	1.80	5.33	1.19	W
Coast range, California	1.23...2.54	3.39...8.67	0.82...1.64	
Graywacke	2.05	7.19	1.30	W
Coast Range, California	1.06...3.15	3.44...12.1	0.53...1.88	
Carbonates	2.2	1.7	0.3	F; WA
	0.1...9.0	0.1...7.0	0...2.0	
Calcite, chalk, limestone, dolomite - all pure -	< 1.0	< 0.5	< 0.1	WA
Limestone (clean)				
Florida	2.0	1.5	< 0.4	WA
Cretaceous trend, Texas	1.5...15	< 2.0	<0.3	WA
Beach sand	0.5	1.9	0.9	M
Southern U.S.	0.31...0.64	1.5...2.2	0.77...1.1	
Beach sand, Gulf Coast, U.S.	0.84	2.8	< 1.2	F; WA
Atlantic Coast sand				
Florida, North Carolina	3.97	11.27	0.37	WA
New Jersey, Massachusetts	0.8	2.07	0.3	WA
Clay, unconsolidated Pleistocene, Germany	1.1...3.8	5.7...10.2	1.9...2.5	
Silt, unconsolidated Tertiary, Germany	1.2...4.3	1.4...9.3	1.3...2.1	

A linear correlation for sediments of the region of Prikumsk (Russia) was recommended by Itenberg (1978) for the clay content range between 0 and 0.4:

$$\Delta I_{\gamma} = 1.39 \cdot V_{sh} + 0.074 \quad (5\text{-}9)$$

Hochstrate (see Hochstrate and Schön, 1983) found a correlation for the sediments of the North German-Polish-Basin (Rotliegendes) giving

$$\Delta I_{\gamma} = 0.733 \cdot \log V_{0.02} - 0.41 \quad (5\text{-}10)$$

where $V_{0.02}$ is the content of fractions with a grain diameter d < 0.02 mm.

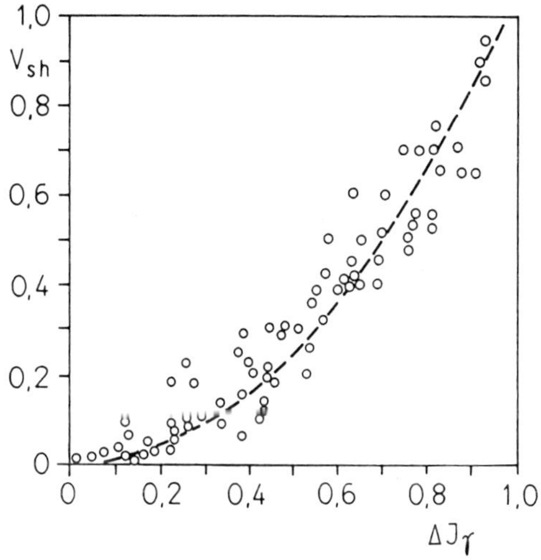

Figure 5.6. The correlation between the normalized integral gamma intensity (radioactivity index) and clay content respectively shaliness V_{sh} for Paleozoic sediments (Dachnov, 1962).

A summary of further relations was published in Fertl (1983) and is given below:

$$V_{sh} = 0.083 \cdot (2^{3.7 \cdot \Delta I_{\gamma}} - 1.0)$$ tertiary rocks (Larionov, 1969) (5-11)

$$V_{sh} = 0.33 \cdot (2^{2 \cdot \Delta I_{\gamma}} - 1.0)$$ older rocks (Larionov, 1969) (5-12)

$$V_{sh} = \frac{\Delta I_{\gamma}}{3.0 - 2.0 \cdot \Delta I_{\gamma}}$$ South Louisiana (5-13)

Miocene and Pliocene (Stieber, 1980)

$$V_{sh} = 1.7 - \left[3.38 - (\Delta I_\gamma + 0.7)^2\right]^{1/2} \quad \text{Clavier et al., 1971} \tag{5-14}$$

Fertl (1983) notes that "in routine well log analysis, several such clay indicators are used comparatively to derive the most likely shaliness value, which then frequently equals the minimum calculated value based on all the indicators available. A constraint inherent to all these techniques is that clay properties in the clastic reservoir and adjacent shales are assumed to be identical - an assumption which is frequently not realistic. Nevertheless, such log-derived shaliness estimates provide useful information as to porosity corrections, empirical permeability transforms and fluid saturation estimates."

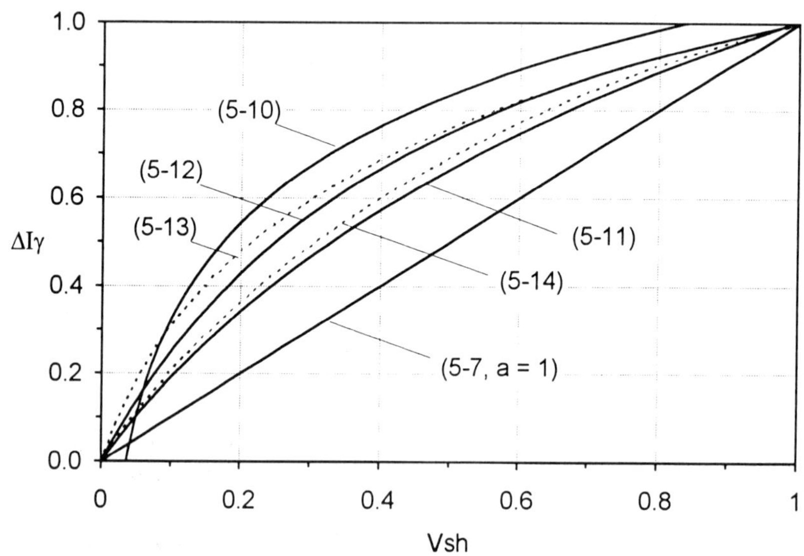

Figure 5.7. Comparison of different V_{sh} vs. ΔI_γ empirical relationships; the numbers indicate the equation numbers

2. Correlations based on spectral components (Th, K) of the gamma intensity: The correlation between the Th content or/and the K content and shaliness should be considered (see Fertl, 1979, 1983), thus excluding the influence of a random associativity of uranium. In analogy to equations (5-11) and (5-12), one gets empiric relationships:

$$V_{sh} = 0.083 \cdot (2^{3.7 \cdot \Delta I_S} - 1) \quad \text{for tertiary clastics} \tag{5-15}$$

$V_{sh} = 0.33 \cdot (2^{2 \cdot \Delta I_s} - 1)$ for highly consolidated and mesozoic rocks (5-16)

where, in this case, ΔI_s is the value analogous to ΔI_γ (equation 5-8) for intensities (or concentrations) of Th or K.

Because the Th and K content varies for different types of clay (Table 5.1, 5.2), the use of only one of these two indicators for clay volume determination can lead to incorrect results. Fertl (1983) recommended a simultaneous use of both by calculating a "product index of thorium and potassium", which is virtually independent of clay type. On the other hand, there are also cases in which the use of Th alone is advantageous. He notes that the thorium curve of a spectral gamma log allows a quantitative clay volume estimate despite the presence of varying amounts of U and K (possibly due to mica, especially muscovite, as for example in Jurassic sandstones in the North Sea). A systematic study of estimation of clay type and volumes from well logs published for example Quirein et al. (1981).

Some typical cases of natural radioctivity of sedimentary rocks are discussed in Schlumberger (1982). Such cases are related to *sandstones*

- Feldspathic sandstones or arcoses: They show some potassium content - dependent on feldspar content in the sand. Thus, they show a very low Th/K ratio (less than 1×10^{-4}) and a some lower matrix density than pure quartz ($2.65 \; 10^3$ kg m^{-3}) due to the low density of feldspar ((2.52...2.53) x 10^3 kg m^{-3}).
- Micaceous sandstones: Micas contain potassium, therefore the potassium content of micaceous sondstones is higher than of pure sandstone; but also the thorium content is higher due to the heavy thorium bearing minerals associated with micas. As a result, the Th/K ratio is close to 2.5×10^{-4}. The density of micaceous sandstones is higher than the density of pure quartz sandstone due to the higher density of mica (2.8...3.1 x 10^3 kg m^{-3}).
- Heavy minerals within sandstones: Heavy minerals like zircon, allanite, monazite, and sphene are thorium and uranium bearing. Therefore in these sandstones, only the thorium and uranium level is high, but the potassium level is very low. Thus, this type of sandstone shows a very high Th/K ratio.

In *carbonate series,* the integral gamma intensity is very often a poor clay indicator, because the measured value is not related to clay content, but to the presence of uranium. Typical cases are

➢ Pure carbonate (chemical origin) which has a thorium and potassium level near zero. If uranium is zero too, this carbonate was precipitated in an oxydizing environment (see below).
➢ If there is a variable uranium content, the carbonate can either have been deposited in a reducing environment, or it corresponds to a carbonate with stylolithes (which contains impurities such as uranium, organic matter, and even clay minerals), or to phosphate-bearing layers.
➢ If thorium and potassium are present with uranium, this indicates clay content of the carbonate (clayey carbonate to marl).
➢ If potassium is present with or without uranium it can corresepond to a carbonate of algal origin or a carbonate with glauconite (Schlumberger, 1982).

The spectral gamma measurement can be important for clay content estimation, clay type and also for a stratigraphic classification of *unconsolidated rocks*. Unconsolidated sediments from two lignite opencast mining districts of Saxonia, Germany show a marked difference between sediments from the Pleistocene and Tertiary (see Figure 5.8).

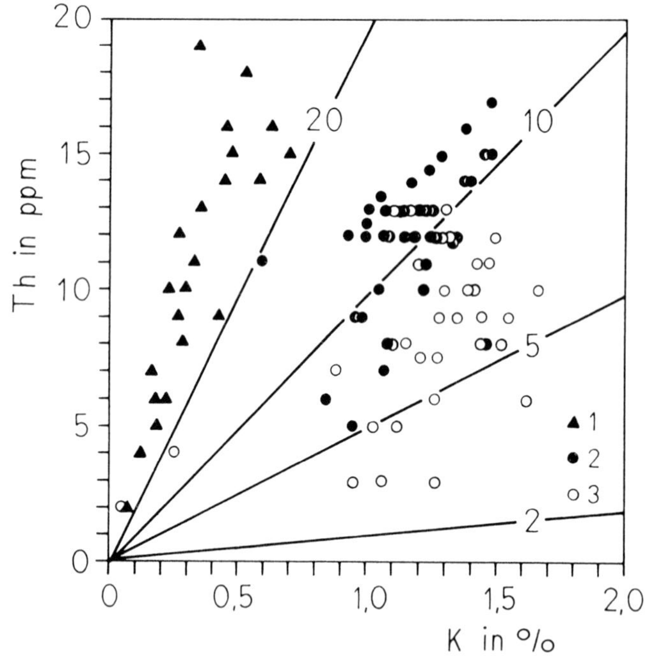

Figure 5.8. Thorium and potassium content for unconsolidated sediments of different stratigraphic units from Saxonia, Germany; the numbers on the line are the Th/K ratio
1 - Pliocene 2 - Miocene (lower) 3 - Miocene (upper)

Further aspects of the *geological significance* of Th/U, U/K and Th/K ratios (e.g. source rock characterization, correlations with the organic carbon content) are summarized and described by Fertl (1979, 1983) and Schlumberger (1982).

The high mobility of uranium and the stable behaviour of thorium results in correlations of the Th/U ratio with the depositional environment as follows:

Th/U > 7 continental, oxidizing environment, weathered soils
Th/U < 7 marine deposits, gray and green shales, graywackes
Th/U < 2 marine black shales, phosphates (Fertl, 1979).

This corresponds to a schematic presentation of the Th content and the Th/U ratio for different sediments after Adams and Weaver (1958; see Serra, 1984).

The behaviour of the radioactive components is also reflected in correlations between thorium and uranium content and sedimentation types and environments, as given by Figure 5.9 after Dortman (1976).

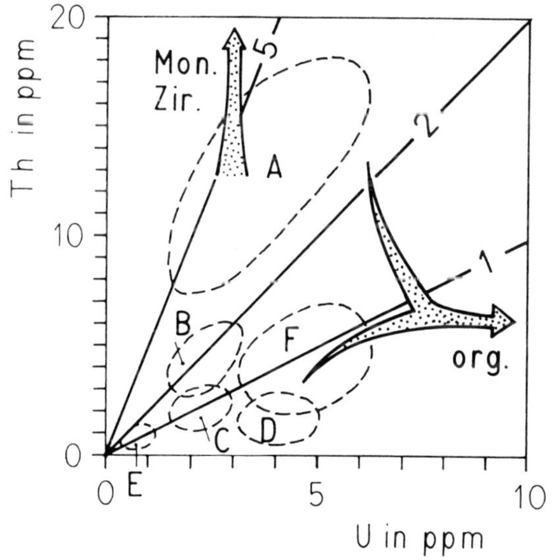

Figure 5.9. Distribution of thorium and uranium in sedimentary rocks, after Dortman (1976). A - sand, clay (terrigeneous); B - quartzitic schists; C - carbonatic rocks; D - dolomite; E - evaporates; F - caustobiolithes; arrow "Mon., Zir." - accessory monazite and zircon; arrow "org." - increasing organic content; numbers on the lines are the Th/U ratio

One example for highly radioactive black-shales, produced in an acid, reducing environment are the North Sea Jurassic shales with the following

averages from a well (well 2/11-1, Norway; data after Gjorlykke, 1975, see Rider, 1986): Uranium = 5.27 ppm ,
Thorium = 10.46 ppm, Th/U = 1.98
Potassium = 1.34 %.

Variations of the natural gamma radioactivity as observed in *oil and gas reservoirs* are related to the uranium concentration in crude oils (Table 5.13) and the alteration during production (see Fertl, 1983). For comparison, uranium in sea water is found at an average concentration of $3 x 10^{-3}$ ppm in the chemical form $UO_2(CO_3)_3^{4-}$ (Fertl, 1983).
Abnormal high gamma values have been detected at the oil/water contact and oil-depleted water flooded strata (Campbell, 1978, Khusnullin, 1973; Doering and Smith, 1974; King and Bradley, 1977; Lehnert and Just, 1979; Fertl, 1983). Toulhoat et al. (1989) studied the retention behaviour of ions by in situ pumping experiments in low permeability formations (aquifers) and found a clear decrease of uranium content as result of the pumping procedure.

Table 5.13. Uranium concentration in several crude oils (from Fertl, 1983)

Location	U in 10^{-3} ppm	Location	U in 10^{-3} ppm
Arkansas	0.5...2.5	Oklahoma	0.32...1.98
Colorado	0.17...0.7	Kansas	0.28...2.6
Montana	0.12	Wyoming	0.24...13.5
New Mexiko	0.54	Lybia	15.0

Table 5.14 summarizes some general tendencies and clues relating chemical properties, transport, and occurrence of K, U, and Th.

5.4. Radioactive heat generation

Heat is generated as a result of the decay of naturally radioactive elements in the earth's crust. The radioactive heat generation A expressed in $\mu W\ m^{-3}$ contributes significantly to terrestrial heat flow. There is another unit, the HGU ("Heat generation unit") with the following conversions:

$$1\ HGU = 0.418\ \ 10^{-6}\ W\ m^{-3} = 10^{-13}\ cal\ s^{-1}\ cm^{-3}$$
$$1\ \mu W\ m^{-3} = 2.39\ HGU = 2.39\ \ 10^{-13}\ cal\ s^{-1}\ cm^{-3}.$$

Table 5.14 Chemical properties, transport, and occurrence of K, U, and Th; Western Atlas, 1992.

	Chemical Properties	Transport	Occurrence
K^{40}	Valence 1+ Soluble	In solution for large distances Feldspars and micaceous materials in suspension · Only a small part of original K transported arrives at the sea · K is extracted from seawater by algae	Detrital sediments · In immature sediments such as feldspars and micas · Clays in general (by absorption) · In micaceous clays structurally chemical sediments · Heaviest concentration in evaporites · In limestone originated from algae
Th^{232}	Valence 4+ Insoluble	Uniquely in suspension · Usually in the silty fraction of shale · Where heavy minerals are more abundant	Only in detrital sediments · In clay minerals by absorption · In heavy minerals (monazite, zircon, rutile) · In volcanic ashes (tuff, tuffaceous sandstone) · In residual sediments that remain after K and U are eliminated by erosion Th has never been detected in pure limestone or dolomite
U^{238}	By far the most soluble Valence 4+ Soluble Valence 6+ Soluble In the complex form UO_2^{2+} depending upon the pH and eH The ion UO^{2+} is combined with sulfates, carbonates, or organic materials UO^{2+} is precipitated with phosphates	Post-depositional transfers are common chiefly in solution Very soluble Very mobile In suspension · Migrates easily during leaching and dissolution of carbonates Concentrates in styolithes in animals that attach to uranium	Detrital sediments · In clay minerals by absorption (high content may indicate high concentration of organic material that in turn may indicate a source bed) · In heavy minerals (zircon) · In volcanic ashes (tuff, tuffaceous sandstone) Chemical sediments · In carbonates (high in organic material) · In phosphates · In organic materials

The mean heat flow at the earth's surface is about 65 mW m^{-2} with the heat flow from the mantle in continental areas being approximately 20 mW m^{-2}. The difference is due to radioactive heat generation in the crustal rocks (Rybach and Cermak, 1982).

The radioactive heat generation of a rock can be calculated as follows (Rybach, 1976 and Rybach and Cermak, 1982):

$$A = d \cdot (9.52 \cdot C_U + 2.56 \cdot C_{Th} + 3.48 \cdot C_K) \cdot 10^{-5} \qquad (5\text{-}16)$$

where d is the density of the rock (in kg m^{-3}), C_U and C_{Th} are the weight content of U and Th (in ppm), and C_K is the weight content of K (in %).

The radioactive heat generated is the sum of the heat generated by the three components U, Th and K. As shown in the foregoing sections, the ratio of these contents vary with rock type. Qualitatively, one has

- ► increasing radioactive heat generation from basic to acid igneous rocks,
- ► increasing radioactive heat generation from clean to shaly sediments.

Rybach's formula (5-16) has been used by many authors. For example, Dorofeyeva (1990) has applied it to study heat generation in the Pre- and Trans-Baikal region (Russia) and has confirmed a correlation between heat generation and density for a rock series from trachites and granites to basalts and gabbros:

$$\ln A = 22.54 - 8.145 \cdot d \qquad (5\text{-}17)$$

where the density d is given in 10^3 kg/m^3 and the heat generation A in µW/m^3.

Such correlations are based on the tendency of increasing heat generation, and decreasing density from basic to acid rocks. The same rules produce a relationship between heat generation and seismic velocity.

Zoth (1989) calculated the heat generation of metamorphic rocks on the basis of data from the Geochemical logging tool (GLT/Schlumberger) which includes an U-,Th-, and K-content determination. The results were compared with core measurements. The mean error for a single value was $\pm$ 0.37 µW/m^3 where a density correction (caliper enlargement) was applied.

CHAPTER 6

Elastic Properties of Rocks

" Yet, despite all the problems with velocity, we are reluctant to give up the hope that somewhere, if only we could find it, there must be a master variable connecting velocity and geology - a master variable meaningful to a geologist, and yet leading directly to the physical property of velocity."
(N.A. Anstey, keynote address at the EAEG/SEG workshop on estimation and practical use of seismic velocities; Cambridge, 1990) Anstey, N.A. (1990) Velocity in thin section, First Break, vol. 9, No. 10.

6.1. Physical basis and units

The theory of elasticity is the fundamental basis for the description of elastic wave propagation. Stress-tensor σ_{ik} and strain-tensor ε_{ik} are connected by the equation of state for a considered medium. For an ideal elastic material the relation between stress tensor and strain tensor (both of rank two) is described by the generalized form of Hooke's law

$$\sigma_{ik} = C_{iklm}\varepsilon_{lm} \tag{6-1}$$

where C_{iklm} represents the elastic stiffness tensor or tensor of elasticity. It is of rank four with 81 components. If the strains are expressed in terms of stresses it results the equation

$$\varepsilon_{ik} = D_{iklm}\sigma_{lm} \tag{6-2}$$

where D_{iklm} is the tensor of the compliances (frequently noted as S_{iklm}).

The existence of an elastic potential (with the independence of the elastic energy from the strain-history) and the symmetry of both stress and strain tensor result in, at most, 21 of the 81 being independent (Helbig, 1992). Material symmetry reduces the number of independent stiffnesses or compliances further (Table 6.1).

Table 6.1. Elastic symmetry properties

Class of symmetry	Number of independent components
triclinic	21
orthorhombic	9
trigonal	6
hexagonal (transversely)	5
cubic	3
isotropic	2

Equation 6-3 is one example the tensor of elasticity for the hexagonal system, where the indices *m* and *n* are obtained from the *ij* and *kl* by the index transformation:

ik or *lm*	11	22	33	23, 32	13,31	12, 21
m or *n*	1	2	3	4	5	6

$$C_{mn} = \begin{pmatrix} c_{11} & c_{11}-2c_{66} & c_{13} & \cdot & \cdot & \cdot \\ c_{11}-2c_{66} & c_{11} & c_{13} & \cdot & \cdot & \cdot \\ c_{13} & c_{13} & c_{33} & \cdot & \cdot & \cdot \\ \cdot & \cdot & \cdot & c_{55} & \cdot & \cdot \\ \cdot & \cdot & \cdot & \cdot & c_{55} & \cdot \\ \cdot & \cdot & \cdot & \cdot & \cdot & c_{66} \end{pmatrix} \tag{6-3}$$

In an *isotropic material* there are only two independent constants; they describe the elastic behaviour completely. In this case the stress-strain relation has the simple form

$$\sigma_{ik} = \lambda \vartheta_{ikmm} \varepsilon_{mm} + 2\mu\varepsilon_{ik} \tag{6-4}$$

where λ, μ are Lamè constants, and ϑ_{iklm} is the Kronecker symbol.

Besides the Lamè constants the following parameters are used, particularly in technical problems:

E modulus of elasticity (Young's modulus)

σ Poisson's ratio.

To describe the elastic properties of *gases and fluids* one normally uses k (modulus of compression), or $c = 1/k$ (compressibility).

Conversion of these quantities are possible using Table 6.2.

Table 6.2. Elastic constants for isotropic solid materials

	Known			
Desired	λ, μ	M, μ	k, μ	E, σ
M	$\lambda + 2\mu$	M	$\frac{3k + 4\mu}{3}$	$\frac{(1-\sigma) \cdot E}{(1+\sigma)(1-2\sigma)}$
μ	μ	μ	μ	$\frac{E}{2 \cdot (1+\sigma)}$
k	$\frac{3\lambda + 2\mu}{3}$	$\frac{3M - 4\mu}{3}$	k	$\frac{E}{3(1-2\sigma)}$
E	$\frac{3\lambda + 2\mu}{\lambda + \mu} \cdot \mu$	$\frac{3M - 4\mu}{M - \mu} \cdot \mu$	$\frac{9\mu k}{\mu + 3k}$	E
σ	$\frac{\lambda}{2(\lambda + \mu)}$	$\frac{M - 2\mu}{2(M - \mu)}$	$\frac{3k - 2\mu}{2(3k + \mu)}$	σ
λ	λ	$M - 2\mu$	$\frac{3k - 2\mu}{3}$	$\frac{\sigma \cdot E}{(1-2\sigma)(1+\sigma)}$

The velocities of the two body waves are related to these parameters and to density by the following relations:

- velocity of the *longitudinal wave* (compressional wave)

$$v_{\mathrm{p}} = \left(\frac{\lambda + 2\mu}{d}\right)^{1/2} = \left(\frac{E}{d} \cdot \frac{1-\sigma}{(1-2\sigma) \cdot (1+\sigma)}\right)^{1/2} = \left(\frac{M}{d}\right)^{1/2} \tag{6-5}$$

where d is the density in this text and M is the "plane wave modulus" (after White, 1983):

$$M = \frac{1-\sigma}{(1-2\sigma)(1+\sigma)} \cdot E \tag{6-6}$$

- velocity of the *transverse wave* (shear wave)

$$v_s = \left(\frac{\mu}{d}\right)^{1/2} = \left(\frac{E}{d} \cdot \frac{1}{2 \cdot (1+\sigma)}\right)^{1/2} \tag{6-7}$$

In laboratory investigations the *rod-wave velocity* is also measured. This type of wave exists only when special relations hold between wavelength and the sample dimensions (diameter < wavelength; sample length > wavelength)

$$v_r = \left(\frac{E}{d}\right)^{1/2} \tag{6-8}$$

Poisson's ratio and velocity quotients are related as follows:

$$\frac{v_p}{v_s} = \left(2 \cdot \frac{1-\sigma}{1-2\sigma}\right)^{1/2} \tag{6-9}$$

$$\sigma = \frac{\theta - 2}{2(\theta - 1)} \tag{6-10}$$

where $\theta = (v_p/v_s)^2$.

In an elastically *anisotropic material,* wave velocities are a function of the direction of propagation. Pure compressional waves and pure shear waves exist only under particular conditions or directions. In transversely isotropic media with vertical axis of symmetry, a so-called "anisotropy ratio" can be defined for a simple description as follows

$$A = \frac{v_{\text{vertical}}}{v_{\text{horizontal}}} = \frac{v_{\perp}}{v_{//}} \tag{6-11}$$

The theory of wave propagation in transversely isotropic media has been given in detail by Rudzki (1898), Riznicenko (1949), Postma (1955), Helbig (1958, 1994, 1995), Cesnokov (1977), Gassmann (1964), and White (1965, 1983), Thomsen (1986).
A discussion of fundamental aspects of anisotropy of wave propagation is given by Helbig (1993, 1995).

Winterstein (1990) published an overview of the terminology with respect to velocity anisotropy.

Two observations are important for the description of the stress-strain behaviour of rocks:

1) The elastic moduli (or the components of the elasticity tensor) depend on stress. Therefore, the stress-strain-relations are *nonlinear*.
2) Rocks are *not ideally elastic* materials. Their reaction to stress depends also on the velocity of deformation and the "history" of deformation. These deviations from Hooke's law result in
   - the phenomenon of energy absorption (see chapter 7),
   - the discrepancy between "statically" and "dynamically" determined moduli (see chapter 10).

For the elastic properties, the following *units* are used:
The SI unit for the *elastic moduli* (Young's modulus, shear modulus, plane wave modulus, Lamè constants, modulus of compression) is the Pascal (Pa)

$1\ \text{Pa} = 1\ \text{N m}^{-2} = 1\ \text{kg m}^{-1}\ \text{s}^{-2}$

Most commonly used is the Gigapascal (GPa) or Megapascal (MPa)

$1\ \text{GPa} = 10^9\ \text{Pa}$ $\qquad$ $1\ \text{MPa} = 10^6\ \text{Pa}$

To convert to kp cm^{-2}, bar, and psi, use the following conversions:

$1\ \text{kp cm}^{-2} = 9.8067\ 10^4\ \text{Pa} \approx 0.1\ \text{MPa}$

$1\ \text{Pa} = 1.0197\ 10^{-5}\ \text{kp cm}^{-2}$ $\qquad$ $1\ \text{MPA} \approx 10\ \text{kp cm}^{-2}$

$1\ \text{bar} = 0.1\ \text{MPa}$

$1\ \text{psi} = 6.894\ 10^3\ \text{Pa} = 6.894\ \text{kPa} = 0.006894\ \text{MPa}$

$1\ \text{Pa} = 1.45038\ 10^{-4}\ \text{psi}$ $\qquad$ $1\ \text{MPa} = 145.038\ \text{psi}$

Poisson's ratio is dimensionless.

The SI unit for *velocity* is m s^{-1}. It results if the moduli units used are in Pa and the density in kg m^{-3}. Frequently, the unit km s^{-1} ($1\ \text{km s}^{-1} = 10^3\ \text{m s}^{-1}$) is applied in seismics. The conversions to the unit ft s^{-1} are:

$1\ \text{ft s}^{-1} = 0.3048\ \text{m s}^{-1}$ $\qquad$ $1\ \text{m s}^{-1} = 3.2808\ \text{ft s}^{-1}$.

6.2. Elastic properties of the rock-forming minerals and of the fluid rock constituents

6.2.1. Elastic properties of rock-forming minerals

The elastic properties of rock-forming minerals are controlled by the properties of their chemical elements, by the bonding of the constituents and by the structural properties of the minerals. Additionally, they are influenced by pressure and temperature.

Depending on crystal-class, a certain number of independent elements of the elasticity tensor are necessary for a complete characterization of the elastic behaviour. The fundamental requirements for a systematic description of the elastic properties of minerals were published by Voigt as early as 1910. Extended collections of data have been published by Clark (1966), Alexandrov et al. (1966), Belikov et al. (1970), Gebrande (1982) and Carmichael (1989). Table 6.3 gives the complete set of elastic constants for four minerals.

Table 6.3. Examples for the elastic constants c_{mn} (in GPa) for minerals of trigonal (quartz, calcite, dolomite) and orthorhombic (olivine) symmetry. After data collections of Belikov, 1970 (B) and Gebrande, 1982 (G)

Index *mn*	α-quartz		calcite		dolomite	olivine forsterite
11	86.6	86.8	144	144	205	328.4
22						199.8
33	106.1	105.9	84.0	83.1	113	235.3
44	57.8	58.3	33.5	32.7	39.8	65.9
55						81.2
66						80.9
12	6.7	7.1	53.9	57.1	71.0	63.9
13	12.6	11.9	51.1	53.4	57.4	68.8
14	-17.8	-18.1	-20.5	-20.5	-19.5	
15					13.7	
23						73.8
Ref.	G	B	G	B	G	G

The elasticity constants of plagioclase are given in Table 6.4 as a function of the anorthite content. The influence of the *mineral composition* on the elasticity properties is also obvious for feldspars and other minerals in Table 6.5.

Table 6.4. Dependence of the elastic constants c_{mn} (in GPa) of plagioclase on the content of anorthite (in %) and mineral density d (in 10^3 kg/m^3); after Rysova (1964)

Index	Anorthite content %	9	29	56
mn	Density	2.61	2.64	2.69
11		74.9	84.5	98.9
22		137.5	150.5	172.0
33		128.9	132.5	141.4
44		17.2	18.5	19.9
55		30.3	31.4	34.1
66		31.1	34.3	37.6
12		36.3	41.7	52.1
13		37.6	40.9	44.1
23		32.6	33.0	36.6
15		-9.1	-8.7	-8.1
25		-10.4	-6.9	-5.1
35		-19.1	-18.5	-19.1
46		-1.3	-1.1	-1.9

To characterize the elastic properties of rocks, moduli are frequently used, which in principle are only for an isotropic material. If we assume that in natural rock the direction of the crystallographic axes are distributed statistically, we can define for such a quasi-isotropic medium the so-called *"effective" or "macroscopic" moduli* for the minerals (e.g. Young's modulus, shear modulus, compressibility, Poisson's ratio). These moduli are determined experimentally or calculated from the complete set of single-crystal values. Calculations are based on theoretical relations developed by Voigt (1910), Reuss (1929), Bruggeman (1936), Hill (1952), Hashin (1962), Hashin and Shtrikman (1962) and also described by Simmons (1967) and Gebrande (1982).

The *Voigt- and Reuss moduli* (see section 6.4.2.) arise from the assumption of homogeneous strain and homogeneous stress, respectively, throughout the aggregate being considered. These represents the upper and lower bounds of the effective moduli. Hill (1952) recommended the use of their arithmetic mean as a better approximation ("*Voigt-Reuss-Hill-averages* VRH").

Table 6.5. shows a collection of some mean values, including data for longitudinal and transverse wave velocities. In the range $d < 4 \times 10^3$ kg m^{-3}, the general tendency of increasing velocity (or elastic moduli) with increasing *density* is obvious. The acidic constituents (quartz) generally have lower densities and also lower values of velocities or moduli than the basic constituents (e.g.. olivine). In the range $d > 4 \times 10^3$ kg m^{-3} the reversed tendency is observed.

Table 6.5. Elastic properties for selected rock forming minerals.
Data from a collection of Gebrande, 1982 (signature G, values of the first and second line) and Alexandrov et al., 1966 (signature A, values of the third and fourth line);
d - density in kg m^3; m_A - mean atomic mass;
k_R, k_V, E_R, E_V, μ_R, μ_V - compression (bulk) modulus, Young's modulus and shear modulus calculated after Reuss and Voigt, respectively, in GPa;
v_p, v_s - velocity of the compressional and shear wave (in m/s), calculated as Voigt-Reuss-Hill-mean value in Ref. G and experimentally determined in Ref. A;
σ - Poisson ratio, calculated as Voigt-Reuss-Hill-mean value in Ref.G

System Mineral	Ref.	d m_A	k_R k_V	E_R E_V	μ_R μ_V	v_p v_s	σ
cubic							
Ca-fluoride	G	3180	86		41	6700	0.28
	G	26.03	86		44	3660	
	A	3150	87.5	111.3	43.2	6800	
	A	87.5	117.9		46.2	3760	
Almandine	G	4318	180.1		94.7	8420	0.28
	G	2488	180.1		94.7	4680	
	A	4183	176.3	241.8	95.1	8510	
	A	176.4	241.8		95.1	5250	
Magnetite	G	5180	161.0		91.1	7390	0.26
	G	33.08	161.0		91.5	4200	
	A	5170	161.7	230.3	91.2	7410	
	A		161.7	231.4	91.7	4200	
Halite	G	2160	25.7		14.3	4560	0.26
	G	29.22	25.7		14.6	2590	
Pyrite	G	5010	143		125	7920	0.16
	G	39.99	143		131	5060	
	A	4930	143.9	286.8	122.8	7910	
			144.0	297.3	128.7	5030	
Spinel	G	3581	196		98	9750	0.27
	G	20.32	196		118	5500	
hexagonal							
Apatite	G	3218	76		47	6680	0.26
	G		86		40	3830	
	A	3180	60.5	112.4	47.2	6460	
	A		78.6	114.6	45.6	3800	

System	Ref.	d	k_R	E_R	μ_R	v_p	σ
Mineral		m_A	k_V	E_V	μ_V	v_s	
Biotite	G	3050	42		13	5350	0.27
	G	23.27	60		42	3000	
	A	2750	41.2	33.8	12.4	5130	
	A		59.7	102.3	42.4	2980	
Muskovite	G	2790	43		22	5810	0.25
	G	20.96	62		41	3370	
	A	2790	42.9	56.8	22.2	5810	
	A	61.5		100.6	41.0	3360	
ß-quartz	G	2533	56.1		40.9	6635	0.20
	G	20.03	56.4		42.0	4043	
trigonal							
Calcite	G	2712	70		27	6540	0.31
	G	20.02	76		37	3430	
	A	2712	71.4	68.8	25.7	6660	
	A		77.7	92.6	35.6	3390	
Dolomite	G	2860	89		40	7370	0.29
	G	18.44	99		52	4000	
α-quartz	G	2650	37.6		41.0	6050	0.08
	G	20.03	38.1		47.6	4090	
	A	2650	38.0	91.8	41.5	6030	
	A		38.6	101.2	47.6	4110	
Hematite	G	5240	97		92	6510	0.14
	G	31.94	98		92	4220	
	A	5100	98.0	209.2	91.4	6700	
	A		98.4	214.4	94.3	4320	
orthorhombic							
Anhydrite	G	2970	52		23	5620	0.27
	G	22.69	58		36	3140	
Barite	G	4500	54.7		22.8	4350	0.32
	G		55.1		24.6	2250	
	A		52.9	54.2	20.4	4340	
	A		53.0	62.8	24.1	2260	
Olivine	G	3224	128.5		79.4	8570	0.24
Forsterite	G	20.10	130.7		82.7	5015	
	A	3324	128.7	197.4	79.3	8400	
	A		133.7	205.4	82.4	5160	

System Mineral	Ref.	d m_A	k_R k_V	E_R E_V	μ_R μ_V	v_p v_s	σ
monocline							
Augite	G	3420	89		56	7100	0.25
	G		101		60	4120	
Diallage	G	3300	81		59	7030	0.21
	G		85		62	4270	
	A	3307	81.2	141.3	58.4	7010	
	A		84.9	148.8	61.6	4260	
Diopside	G	3310	105		62	7700	0.26
	G		117		66	4380	
	A	3280	105.3	154.4	61.5	7800	
	A		117.6	166.4	65.8	4390	
Hornblende	G	3124	84		41	6810	0.29
	G		90		45	3720	
	G	3153	91		41	7030	
	G		98		51	3820	
Feldspars							
Or 78.5	G	2506	45		23	5930	0.28
Ab19.4 An2.1	G		63		32	3260	
Or 66.6	G	2540	43		20	5600	0.29
Ab 28.6	G		54		27	3050	
Albite	G	2630	49		28	5940	0.28
	G	20.17	61		31	3290	
	A	2605	53.0	69.0	26.9	6060	
	A		61.6	80.5	31.4	3350	
Ab76	G	2640	58		28	6220	0.29
An24			65		32	3400	
Ab47	G	2690	69		32	6610	0.29
An53	G		75		37	3570	
Anorthite	G	2760	79		37	7050	0.29
	G		89		43	3800	
Orthoclase	A	2570	40.3	55.8	22.0	5690	
	A		56.0	78.3	30.9	3260	
Oligoclase	A	2640	57.9	72.8	28.2	6240	
	A		65.2	84.3	32.8	3390	
Labradorite	A	2680	66.8	80.1	31.2	6550	
	A		73.9	92.7	35.9	3540	
Microline	A	2560	44.9	58.1	22.6	6000	
	A		62.7	81.8	31.9	3260	

Birch (1960, 1961) based on experimental data for minerals and rocks (see section 6.3.2.2), suggested that the velocity of compressional waves depends primarily upon the mean atomic mass (mean atomic weight) m_A and the density *d*. He derived a linear relationship between seismic velocity and density for oxides with a mean atomic mass $m_A \approx 21$. The equation from Anderson and Nafe (1965) gives a relationship between the bulk modulus, mean atomic mass and density. Also, Simmons (1964) and Duffy and Anderson (1989) emphasized the correlation between longitudinal wave velocity, mean atomic mass and density for different minerals and rocks (see also section 6.2.1). An examination of these equations with experimental data published Simmons and England (1969). They concluded that the relations "... are useful for the geophysical interpretation of the Earth and other planets, there are many exceptions and none of the relations appears to have general validity."

The increase of v_p with increasing density was described by Dortman and Magid (1969) in the form of an empirical relationship

$$v_p = 5.75 \cdot \exp[0.5 \cdot (d - 2.6) + 0.2 \cdot (20 - m_A)] \qquad (6\text{-}12)$$

where the density *d* is given in g cm^{-3} or 10^3 kg m^{-3} and the velocity v_p in km/s. This equation is recommended for most rock forming minerals for densities below about 4 x 10^3 kg m^{-3}.

The elastic properties of minerals are pressure dependent. Extensive data collections from experimental investigations have been published by Clark (1966), Belikov et al. (1970), Dortman (1976), and Gebrande (1982).

6.2.2. Elastic properties of pore fluids

The pore, crack and fissure space of rocks contain gases (air, natural gas), fluids (water, oil, contaminants) or mixtures of these. Generally, the elastic moduli of the gaseous and fluid constituents are *lower than the moduli of the minerals* (matrix material).

The compressibility (or the modulus of compression k_f) are commonly used to characterize the elastic properties of fluids and gases. The compressional wave velocity is given by the following relation

$$v_{p,f} = (k_f / d_f)^{1/2} \qquad (6\text{-}13)$$

where the index f denotes the pore fluid. There are no transverse or shear waves in gases or fluids because the shear modulus is zero ($\mu_f = 0$).

6.2.2.1. Elastic properties of gases

The propagation of elastic waves in a gas can be considered as an adiabatic process. Thus, the sonic or compressional wave velocity is

$$v_{p,gas} = \left(\frac{k_{ad}}{d}\right)^{1/2} = \left(\kappa \cdot \frac{p}{d}\right)^{1/2} \tag{6-14}$$

where k_{ad} is the adiabatic compression modulus, d is the gas density, p is the gas pressure, and κ is the ratio of the specific heat at constant pressure to that at constant volume, i.e., $\kappa = c_p/c_V$.

Table 6.6. contains data of the sonic velocity (at 273° K), the mean temperature coefficient and the ratio of specific heat for various gases.

Table 6.6. Sonic velocity (at 273° K) in m s^{-1}, mean temperature coefficient in m s^{-1} K^{-1} and ratio of the specific heat for different gases after Bergmann (1954)

Gas	v_p	$\Delta v_p/\Delta T$	κ
He	965	0.8	1.66
CO_2	259	0.4	1.30
CO	338	0.6	1.40
Air	331	0.59	1.402
CH_4	430		1.31
SO_2	213	0.47	1.29
N_2	334	0.6	1.40

Ebert (1976) has published data for air and noted that the pressure influence is relatively small (a pressure increase of 50 atm ≈ 5 MPa results in a velocity increase by a factor of only 1.022).
Some of his data and the corresponding mean values after Gearhart (1985) - converted into SI units - are summarized in Table 6.7.

Table 6.7. Compressional wave velocities of gases in m/s after Ebert (1976), Gearhart, (1978), Western Atlas (1992)

Gas	v_p	Remarks
Air	263	$T = 173$ °K
	306.5	$T = 223$° K
	331.8	$T = 273$ °K
	355.3	$T = 313$ °K
	387.2	$T = 373$ °K
Methane	487.70	$p = 15$ psi (= 0.103 MPa)
Ethane	307.80	$T = 10$ °C = 283 °K, $d = 1.25$ kg m^{-3}
Carbondioxide	259.10	$d = 1.9776$ kg m^{-3}

6.2.2.2. Elastic properties of fluids

Compressibility and sonic (compressional) wave velocity of fluids depend on the chemical composition, on pressure and temperature. Table 6.8 shows mean values for water, mud, and some oils.

Many authors have investigated the influence of *salinity, pressure and temperature* on the elastic properties of water; different handbooks (e.g. Serra, 1984) contain nomograms for practical use. The knowledge of such relations is important for porosity calculations made from acoustic-log measurements and for investigations in marine sediments.

As a result of experimental investigations Del Grosso, 1952 (see Bulatova et al., 1970) developed an empirical relation for the sonic velocity of *water* (in m s^{-1}):

$$v_{water} = 1410 + 4.21\,T - 0.037\,T^2 + 0.114\,C + 0.18\,p \qquad (6\text{-}15)$$

where T is the temperature in degrees Celsius, C is the concentration in %, and p the pressure in kp cm^{-2} (1 kp cm^{-2} ≈ 0.1 MPa).

Batzle and Wang (1992) derived on the basis of data from Chen et al. (1978) a relationship between the velocity of a *brine solution* v_B, the velocity of pure water v_W, the temperature T (in °C), the pressure p (in MPa), and the brine concentration C (in weight fraction of sodium chloride):

$$v_B = v_W + C \cdot (1170 - 9.6 \cdot T + 0.055 \cdot T^2 - 8.5 \cdot 10^{-5} \cdot T^3 + 2.6p - 0.0029 \cdot pT$$
$$-0.0476 \cdot p^2) + C^{1.5} \cdot (780 - 10 \cdot p + 0.16 \cdot p^2) - 820 \cdot C^2 \qquad (6\text{-}16)$$

Table 6.8. Sonic velocity (in m s^{-1}) for various fluids after Bergmann (1954), Bulatova et al., (1970), Tagiev and Mustafaev (1975), Gregory (1990), Serra (1984), Ellis, (1987)

Fluid	v_p	Remarks
Water	1497	distilled, $T = 298°$ K
	1585	NaCl-Conc.= 0.10 % (1.0×10^3 ppm); p= 0.103 MPa
	1638	NaCl-Conc. = 0.15 % (1.5×10^3 ppm); p= 0.103 MPa
	1689	NaCl-Conc = 0.20 % (2.0×10^3 ppm); p= 0.103 MPa
Drilling mud	1.830	mean value (Ellis, 1987)
Kerosene	1420	p = 0.103 MPa
Oil	1280	p = 0.103 MPa
Petroleum	1.295	T = 307° K; d = 825 kg m^{-3}
Paraffin oil	1420	T = 306.5° K; d = 835 kg m^{-3}
Gasoline oil	1250	T = 307° K; d = 803 kg m^{-3}
Oil/natural	1035...1370, 1280	
Mesa crude	1362	d = 877 kg m^{-3}
Leona crude	1397	d = 911 kg m^{-3}
Pentane C_5H_{12}	1020	T = 293° K
Heptane C_7H_{16}	1150	T = 293 K
Octane C_8H_{18}	1192	T = 293° K
Nonane C_9H_{20}	1230	T = 293° K
Decane $C_{10}H_{22}$	1090	T = 293° K

For water and water-based *mud*, Podio and Gregory, 1990 investigated the velocity and attenuation as a function of the mud density up to 1.68×10^3 kg m^{-3} in the frequency range from 200 to 600 kHz (see Figure 6.1). The resulting relationship for the compressional wave velocity (in m/s) is

$$v_p = 1463.2 + 99.336\, d_m - 70.420\, d_m^2 \qquad (6\text{-}17)$$

where d_m is the mud density in 10^3 kg m^{-3}. The velocity increases about linearly with fluid pressure as a result of increasing compressional modulus.

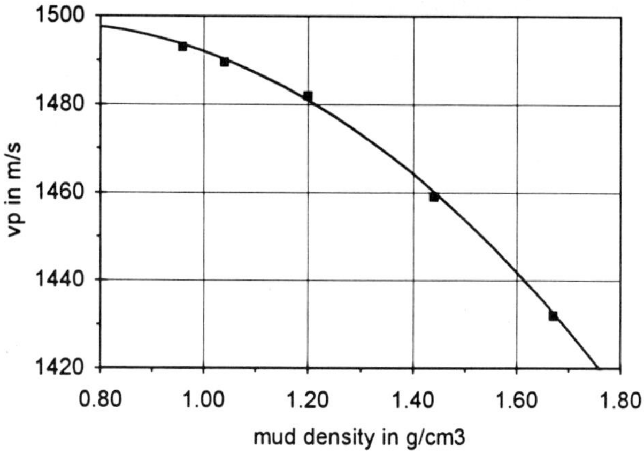

Figure 6.1. Effect of mud density on compressional wave velocity of water; after Podio and Gregory (1990)

For oil Batzle and Wang (1992) published the relation

$$v_{\text{oil}} = 2096 \cdot \left(\frac{d_0}{2.6} - 1\right)^{1/2} - 3.7 \cdot T + 4.64 \cdot p + 0.0115 \cdot \left[4.12\left(\frac{1.08}{d_0} - 1\right)^{1/2} - 1\right] \cdot p \cdot T \tag{6-18}$$

or in terms of API oil gravity

$$v_{\text{oil}} = 15450 \cdot (77.1 + \text{API})^{-1/2} - 3.7 \cdot T + 4.64 \cdot p + 0.0115 \cdot \left(0.36\text{API}^{1/2} - 1\right) \cdot p \cdot T \tag{6-19}$$

Nur (1989) shows for pure hydrocarbons (alkanes and alkenes) a decrease of compressional wave velocity with increasing temperature (between 22 and 110 °C) and decreasing carbon number or increasing molecular weight m (see also Table 6.8):

$$v_{\text{p}}(T) = v(T_0) - \left(0.306 - \frac{7.6}{m}\right) \cdot (T - T_0) \tag{6-20}$$

where $v\ (T_0)$ is the velocity at a reference temperature T_0.

6.2.2.3. Elastic properties of gas-fluid mixtures

If the pores or cracks contain both gas and fluid ("*partial saturation*"), then the effect of this medium on the elastic behaviour of the rock depends on

a) the elastic properties and density of the two (or more) pore constituents,
b) the volume fractions of the constituents,
c) the distribution or arrangement of the constituents in the pore or fracture space and the effects of boundary forces.

The influence of the effects listed under c) is of a very complex nature; any exact analysis would have to include the geometry of the pore space as well as the wettability properties. Domenico (1976, 1977) published two formulae (which include the factors listed under a) and b)) for the "effective compressibility" of a *gas-water mixture*:

$$c_{\text{eff}} = S_{\text{water}} \cdot c_{\text{water}} + (1 - S_{\text{water}}) \cdot c_{\text{gas}} \tag{6-21}$$

$$\frac{1}{c_{\text{eff}}} = \frac{S_{\text{water}}}{c_{\text{water}}} + \frac{1 - S_{\text{water}}}{c_{\text{gas}}} \tag{6-22}$$

where S_{water} is the degree of water saturation, c_{water} the water compressibility and c_{gas} the gas compressibility. Figure 6.2 shows the calculated dependence c_{eff} versus S_{water}.

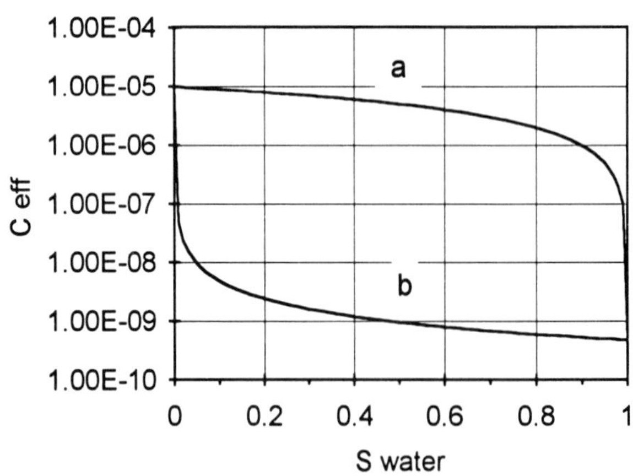

Figure 6.2. Effective compressibility c_{eff} for a gas-water mixture as a function of the water fraction S_{water} (water saturation) calculated after equation 6-21 (curve a) and equation 6-22 (curve b); $c_{\text{water}} = 0.0478\ 10^{-8}\ \text{Pa}^{-1}$ $(= 0.33\ 10^{-5}\ \text{psi}^{-1})$ $c_{\text{gas}} = 0.0098\ 10^{-3}\ \text{Pa}^{-1}$ $(= 0.068\ \text{psi}^{-1})$

Domenico's formulae represent the upper and lower bounds of compressibility. The values calculated from (6-21) and (6-22) differ by more than a factor of 10^3 in the saturation range $0.1 < S_{water} < 0.9$. This, and the fact that c) is neglected, limits the general applicability for practical purposes.

6.3. The velocity of elastic waves in rocks - results of experimental investigations

6.3.1. Overview

The elastic properties and the velocity of elastic waves in rocks are controlled primarily by the elastic properties of the rock forming minerals and constituents, their fractional volume, their contact, cementation or bonding properties, pressure and temperature. Figure 6.3. shows the range of the velocities for groups and types of some commonly occurring rocks.

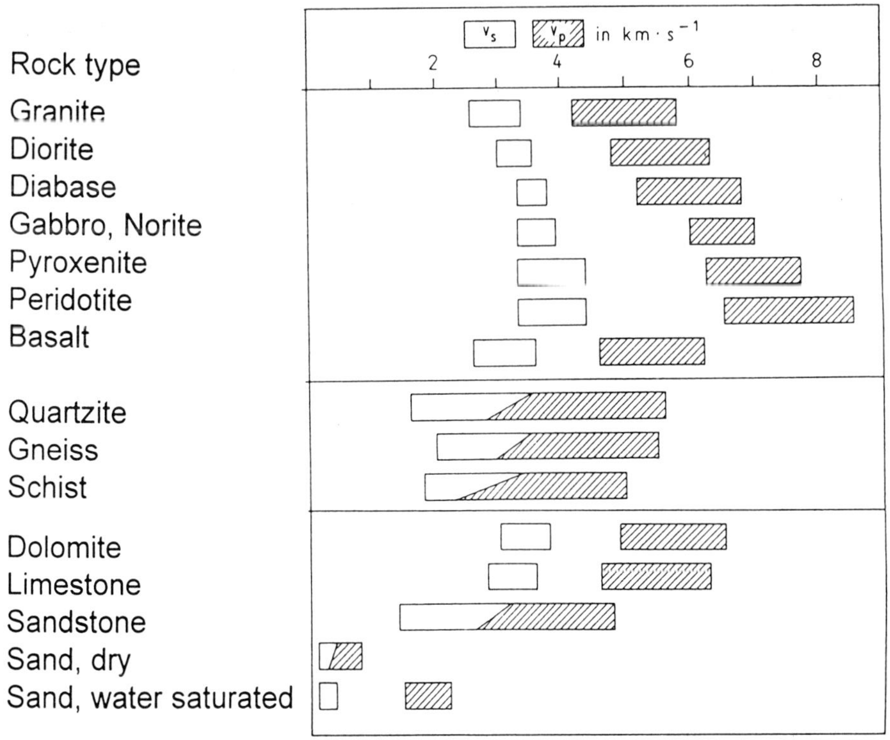

Figure 6.3. Mean-value range of longitudinal and transverse wave velocities

6.3.2. The velocity of elastic waves in magmatic and metamorphic rocks

6.3.2.1. General remarks

Table 6.9. contains a collection of velocity values determined by several authors for samples of magmatic and metamorphic rocks. There is a broad range of values within each rock type (see also Figure 6.3). The variations are caused by the influence of

- differences of the rock composition,
- effects of cracks, fractures and pores,
- anisotropic effects (especially for metamorphic rocks),
- temperature and pressure.

Figure 6.4 shows the distribution of compressional velocities for granites, gabbros and peridodites/pyroxenes at different pressures. Generally, an increasing pressure leads to an increase in velocity. This effect can be observed most clearly in the case of the granites in this figure, and is due to the closing of microcracks. Dense, i.e. non-fractured, rocks show a very small increase. This is due to the low pressure dependence of the elastic properties of the minerals.

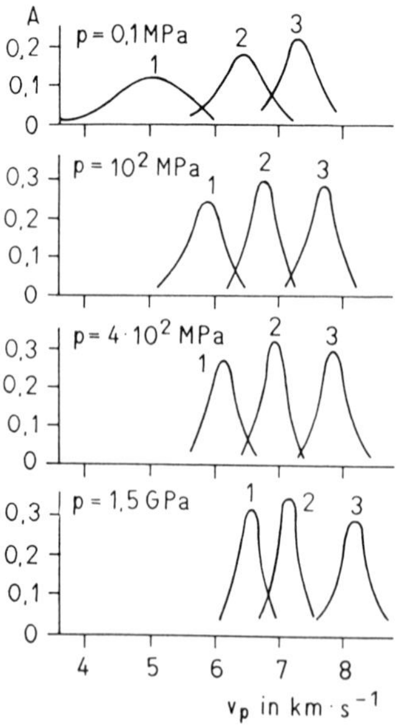

Figure 6.4. Distribution of velocity values at different pressures: (1) granite, (2) gabbro, and (3) peridodite/pyroxene. After Volarovic and Bajuk (in Volarovic et al., 1977)

Table 6.9. Ranges and mean values of velocities of magmatic and metamorphic rocks
References: D - Dortman (1976), different regions of Russia, the range is defined as min...max value; G - Gebrande (1982), data collection from results of different authors, the range is defined by the 80% confidence limit, *n* is the number of data; K - Kopf (1977), rocks from the basement, Northeast Germany

Rock type	*n*	v_p in km/s range	v_p mean	v_s in km/s range	v_s mean	*d* in kg/m³	Ref.
Granite	31	4.18...5.85	5.01	2.61...3.30	2.98	2705	G
		3.70...6.00	5.20	2.70...2.85	2.75	2610	D
Syenite							
- nepheline		5.80...6.45	6.15	2.60...3.45	3.22	2823	D
- amphibole			5.60				K
- pyroxene			5.90				K
Diorite	6	4.83...6.37	5.60	2.96...3.35	3.22	2823	G
		5.60...6.10					D
Gabbro	75	5.97...6.87	6.42				G
			6.30				K
Gabbro, Norite	10	5.99...6.93	6.46	3.31...3.69	3.50	2991	G
		6.10...7.40	6.75	2.70...3.65	3.35	2970	D
Pyroxenite	15	6.23...7.79	7.01	3.35...4.30	3.83	3180	G
Peridotite	16	6.58...8.37	7.48				G
Andesite, Trachite	15	3.97...6.13	5.05	2.44...3.50	2.97	2636	G
Basalt							
- dry, low	29	4.67...6.15	5.41	2.61...3.51	3.06	2840	G
- alcalic	25	4.56...5.77	5.17	2.37...3.07		2726	G
- oceanic	189	4.86...6.09	5.48			2785	G
	65			2.83...3.37	3.10	2785	G
Diabase	17	5.24...6.81	6.30			2880	G
	6			3.38...3.72	3.55	2950	G
Gneiss		5.10...5.50					K
		5.00...6.35		2.55...3.55			D
- parallel	62	3.13...4.90	4.02	2.01...2.91	2.46	2668	G
- perpendicular	56	1.84...3.66	2.75	1.58...2.52	2.05	2668	G
Schist - parallel	62	4.12...6.32	5.22	2.68...3.74	2.71	2741	G
- perpend.	56	2.76...5.44	4.08	1.95...3.47	2.71	2761	G
Quartzite	10	2.46...5.56	4.01	1.79...3.33	2.56	2636	G
Serpentinite	23	4.25...6.23	5.24	2.07...3.49	2.78	2620	G
Granulite	11	5.36...6.50	5.93	3.10...3.57	3.34	2932	G
		5.30...6.10	5.75	2.68...3.10	2.75	2830	D
Eclogite	20	6.53...8.01	7.27	3.56...4.56	4.06	3410	G

6.3.2.2. The dependence of elastic wave velocity on density and mineral composition

In igneous rocks, the velocity of elastic waves is controlled by the mineral composition. This fact is illustrated by the correlation between the longitudinal wave velocity and the SiO_2 content of igneous rocks. Quartz is characterized by a relatively low velocity. Thus, acidic rocks have lower velocities than basic rocks (Figure 6.5).

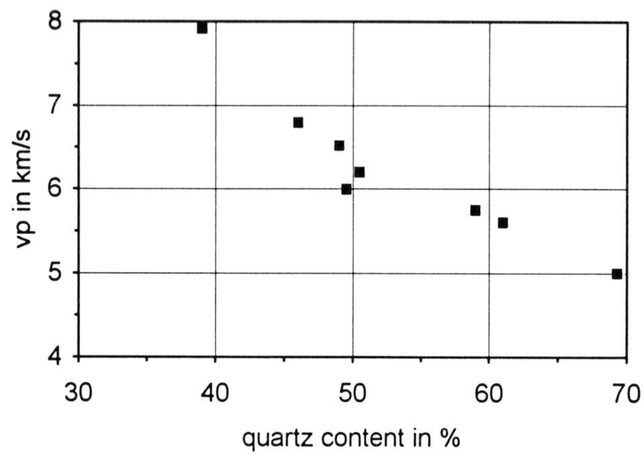

Figure 6.5. Longitudinal wave velocity vs. quartz content of magmatic rocks

Acid constituents have both lower densities (see chapter 3) and lower velocities than basic constituents. From this observation follows as the general tendency, a *velocity-density correlation*. This velocity-density correlation has been the subject of many investigations. Figure 6.6. shows a graph for magmatic and metamorphic rocks from different locations in Russia, published by Dortman (1976); the derived equation is discussed later (see eq. 6-37).

Figure 6.7. is the result of a study summarizing the velocity and density values, determined by samples from the KTB pilot borehole; the figure contains longitudinal and transverse wave velocities and also shows the main rock types.

Many authors have developed empirical relations to describe the correlation between velocity v_p and density d; linear regression analysis is frequently used.

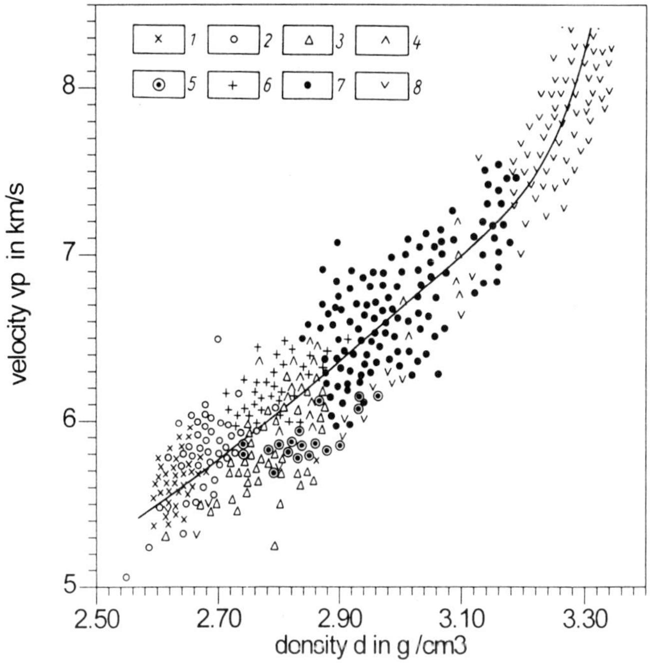

Figure 6.6. Correlation between longitudinal wave velocity and density for magmatic and metamorphic rocks; after Dortman (1976)
1 - granite, 2 - gneiss (biotitic, amphibolitic), 3 - gneiss (garnet-biotitic), 4 - amphibole, gneiss (amphibolitic), 5 - granulite, 6 - diorite, 7 - gabbro-norite, 8 - ultrabasite;
curve - empirical relation, see equation 6-37

Birch (1961) has given the fundamental relation with the two empirical parameters a and b

$$v_p = a + b \cdot d \tag{6-23}$$

He found for magmatic rocks the relation

$$v_p = 2.76\, d - 0.98 \tag{6-24}$$

His samples were from the North American continent and India. In this and in the following empirical relations to eq. 6-37, the density d is given in 10^3 kg m^{-3} (= g cm^{-3}) and the velocity v_p in km/s.

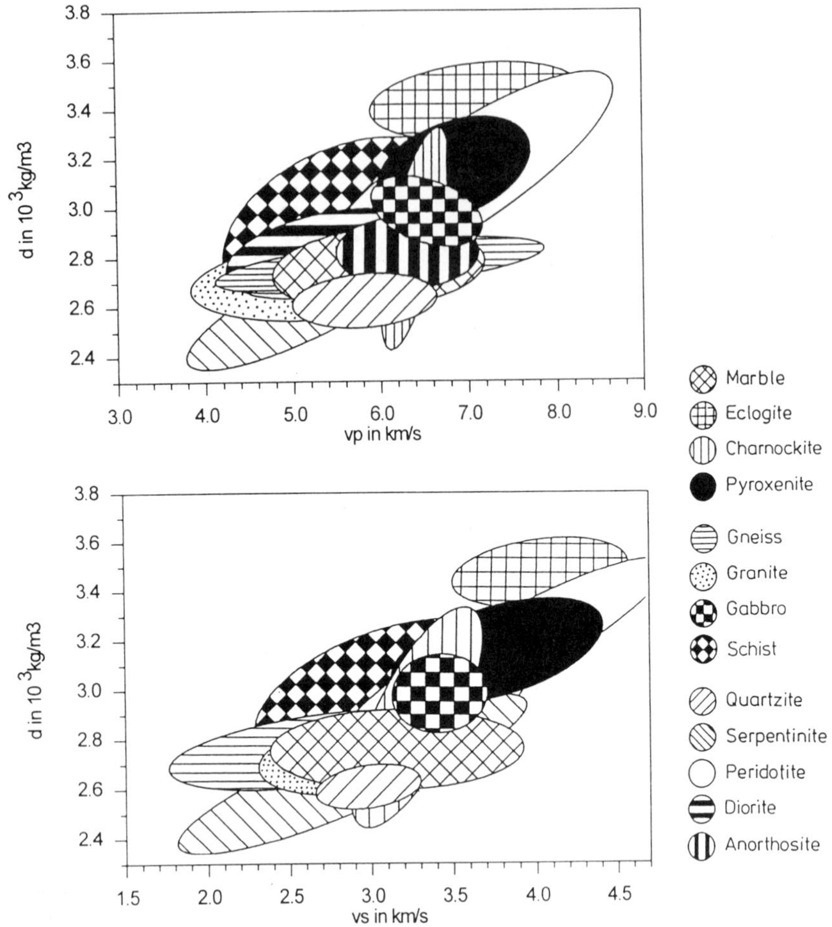

Figure 6.7. Bivariate normal distributions for longitudinal (v_p) and transverse wave velocity (v_s) and density (*d*) data of different rocks from the KTB pilot borehole. The ellipses include approximately 80% of all data. The plot was drawn using the database DABACOS (after Büttgenbach et al., 1990).

The direct correlation between velocity and density is explained by the fact that a variation in the mineral composition of most rocks affects both properties in the same direction.

Birch's relation (6-23) was applied for example by Volarovich and Bajuk (1977). From measurements on magmatic rocks from various regions of the former USSR, they obtained the relationship

$$v_p = 2.67\ d - 1.08 \qquad (6\text{-}25)$$

with a correlation coefficient of R = 0.88. In particular for rocks from Kasakhstan, detailed investigations at different pressures led to the relations given in Table 6.10.

Table 6.10. Empirically derived relations between longitudinal wave velocity v_p (in km s^{-1}) and density d (in 10^3 kg m^{-3}) with the regression coefficient R for rocks from Kasakhstan; after Volarovich and Bajuk, 1977

Pressure	Relation type *A*		Relation type *B*	
in MPa	formula	R	formula	R
0.1	$v_p = 3.25\ d_0 - 3.46$	0.85	$v_p = 3.45\ d(p) - 3.40$	0.82
100	$v_p = 2.72\ d_0 - 1.24$	0.86	$v_p = 1.86\ d(p) - 1.20$	0.86
400	$v_p = 2.54\ d_0 - 0.52$	0.86	$v_p = 1.85\ d(p) + 1.43$	0.88
1 000	$v_p = 2.44\ d_0 - 0.08$	0.85	$v_p = 1.93\ d(p) + 1.34$	0.90

The relation type *A* relates the velocity $v_p(p)$ at pressure p and the density d_0 at atmospheric pressure; the relation type *B* relates velocity $v_p(p)$ and density $d(p)$ at the same pressure p. At high pressures, there is a stronger correlation of the relations *B* than of the relations *A*.

For Precambrian rocks (Fennoscandia) Rybach and Buntebarth (1984) derived comparable relationships.

Marle (1978) and Kopf (1977, 1980) have also used the Birch relation (6-23) to analyze data from magmatic rock samples obtained from the eastern part of Germany (basement). They determined the following relationships for conditions under atmospheric pressure:

- for plutonic rocks: granites, diorites, gabbros

$$v_p = 3.10\ d - 2.98 \qquad (6\text{-}26)$$

- for volcanic rocks: porphyrites, keratophyrites, diabases, and basalt

$$v_p = 2.30\ d - 0.91 \qquad (6\text{-}27)$$

For the conversion of longitudinal wave velocities to densities, Henkel et al. (1990) also used a linear relation. It was derived for mantle rocks in the density range between (2.5 ... 3.5) 10^3 kg m^{-3}

$$d = \frac{v_P + 1.0 \pm 0.4}{2.61} \tag{6-28}$$

where the velocity is given in km s^{-1} and the density results in 10^3 kg m^{-3}. Immediately, we get the relation corresponding to (6-23):

$$v_P = 2.61 \cdot d - 1.0 \pm 0.4 \tag{6-29}$$

A systematic study has been published by Gebrande, 1982. He derived empirical relations (regressions) for longitudinal and transverse wave velocities and their correlation to density (the results are summarized in Table 6.11).

Table 6.11. Correlation between elastic wave velocities and density; after Gebrande (1982)

Rock type	Empirical linear relations	
plutonic	$v_P = 4.36 \cdot d - 6.73 \pm 0.03$	$v_S = 1.66 \cdot d - 1.48 \pm 0.06$
volcanic	$v_P = 2.81 \cdot d - 2.37 \pm 0.18$	$v_S = 1.46 \cdot d - 1.02 \pm 0.22$
metamorphic	$v_P = 4.41 \cdot d - 6.93 \pm 0.37$	$v_S = 1.70 \cdot d - 1.62 \pm 0.22$

Birch (1961) has shown for silicates and oxides the dependence of the parameter a in equation (6-23) on the *mean atomic mass* m_A. For rocks, he gives the equation

$$v_P = 2.76\, d - 0.98 + 0.7\, (21 - m_A) \tag{6-30}$$

The mean atomic mass characterizes the composition of the rock as an additional parameter which influences the velocity.

Gebrande (1982) presented an interesting comparison of regression analysis both with and without the influence of mean atomic mass. The analysis of experimental values was made as a one-dimensional (v vs. d) and a two-dimensional (d vs. d, m_A) regression with the results as given in Table 6.12. for the plutonic and metamorphic rocks at different pressures. The comparison shows that for the rocks investigated, the velocities depend on density more strongly than on mean atomic mass. Shear wave velocities are almost independent of m_A within the usual range of variation of m_A (Gebrande, 1982).

increases in v_p and v_s". A theoretical explanation is given with the Kuster-Toksöz-model (see chapter 6.4.).

The *difference* between the velocities in dry and water-saturated state increases with increasing porosity and decreases with increasing pressure. The influence generally is much stronger for P-waves (due to the distinct difference of the compressibility of fluids and gases) than for S-waves (where only the change of the density and boundary effects play a role). Figure 6.11. illustrates this behaviour.

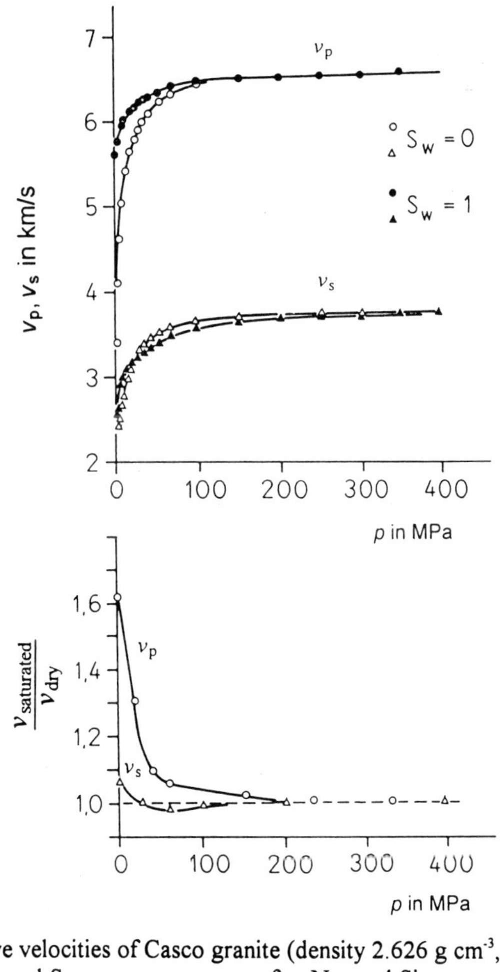

Figure 6.11. Wave velocities of Casco granite (density 2.626 g cm^{-3}, total porosity 0.7 %)
a) Velocity of P- and S-wave vs. pressure; after Nur and Simmons, 1969; open symbols: dry rock, closed symbols: water saturated rock
b) Velocity ratio $v_{saturated}/v_{dry}$ vs. pressure, calculated using values from Figure a

6.3.2.4. The dependence of elastic wave velocity on pressure and temperature

The study of elastic (and other physical) properties of minerals and rocks under the temperatures and pressures found in the earth's crust is a special field of rock physics, involving extensive research programs and highly sophisticated experimental and theoretical work. In this book, these investigations are presented fairly briefly and concentrate mainly on changes in physical properties under the conditions of the *upper* crust.

The variation of the physical properties of rocks and minerals under the influence of pressure and temperature is of fundamental importance for the interpretation of geophysical data corresponding to different depth ranges. The generalized relation for the *variation of velocity with depth z* is (Lebedev, 1975):

$$\frac{\mathrm{d}v}{\mathrm{d}z} = \left(\frac{\delta v}{\delta p}\right)_T \cdot \frac{\mathrm{d}p}{\mathrm{d}z} + \left(\frac{\delta v}{\delta T}\right)_p \cdot \frac{\mathrm{d}T}{\mathrm{d}z} \qquad \text{(6-40)}$$

where $\left(\frac{\delta v}{\delta p}\right)_T$ is the change of velocity with pressure (isotherm),

$\left(\frac{\delta v}{\delta T}\right)_p$ is the change of velocity with temperature (isobar),

$\frac{\mathrm{d}p}{\mathrm{d}z}, \frac{\mathrm{d}T}{\mathrm{d}z}$ are the vertical pressure and temperature gradients .

The pressure and temperature dependence are discussed separately below.

The *dependence of velocity on pressure* shows two dominant features (see also figures 6.12 to 6.15):

1) the relation is *non-linear*; in the higher pressure range the velocity increase is smaller than in the lower pressure range,
2) the velocity change during the loading-unloading cycle is *partially irreversible* (so-called "velocity hysteresis", see Figure 6.13). This fact is one expression for the non ideal elasticity behaviour of natural rocks.

Figures 6.12 and 6.13 show the general pressure dependence and the phenomenon of anisotropy.

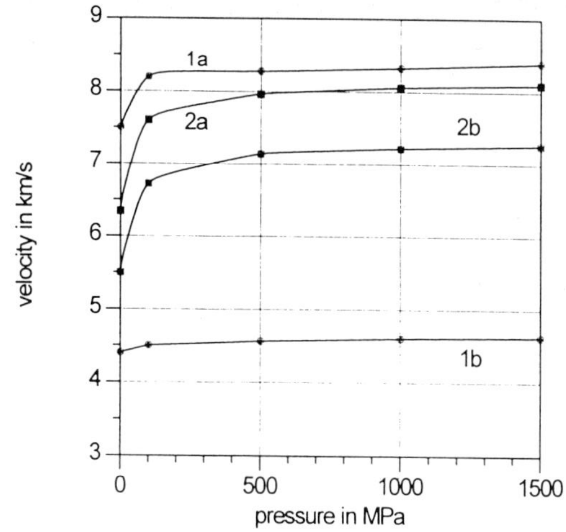

Figure 6.12. Longitudinal and transverse wave velocities as a function of hydrostatic pressure; data after Bajuk and Levitova, 1990;
1 - Peridotite (Kola peninsula): curve 1a - longitudinal wave; curve 1b - transverse wave
2 - Olivinite (Siberia): curve 2a - longitudinal wave, direction [100]; curve 2b - longitudinal wave, direction [010].

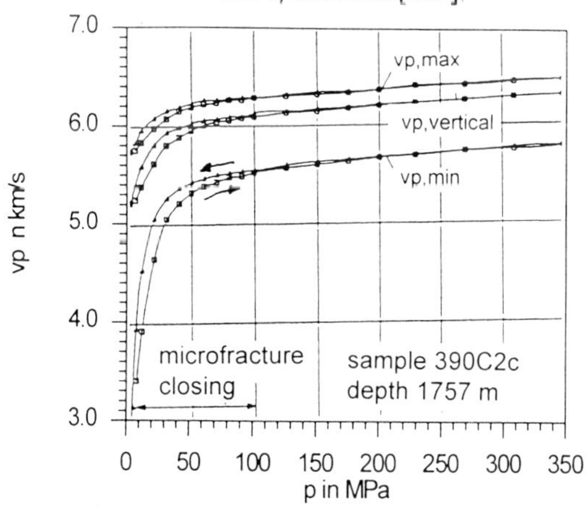

Figure 6.13. Longitudinal wave velocities as a function of hydrostatic pressure (loading and unloading), measured at a gneiss core-sample (390C2c, KTB Borehole, depth: 1757 m) in different directions; after Zang et al. (1989),

$v_{p\,max}$ - maximal velocity value in radial direction of the core
$v_{p\,min}$ - minimal velocity value in radial direction of the core
$v_{p\,vertical}$ - velocity value in axial direction of the core.

The curves obtained for longitudinal and transverse waves appear to be quite similar (Figure 6.12. and 6.14.). Significant differences are visible in a plot of the ratio v_p/v_s vs. pressure (Figure 6.14 b). It is noteworthy that the ratio of the velocities measured in different directions (anisotropy) is also influenced by pressure (see section 6.3.4.).

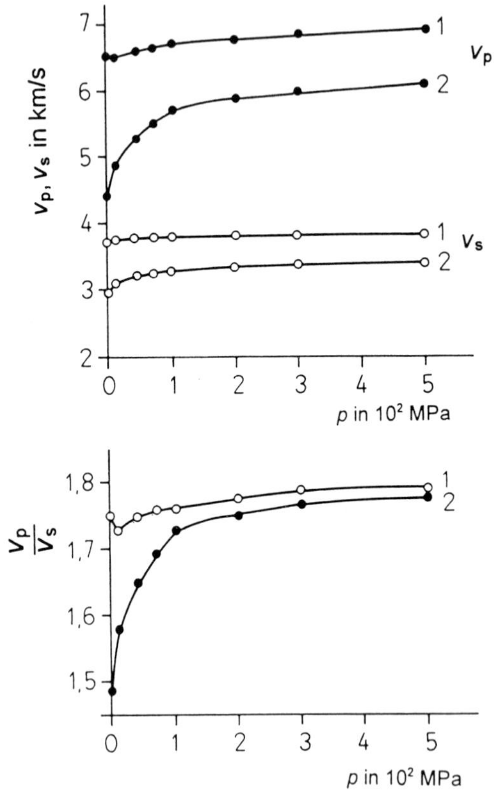

Figure 6.14. Dependence of the longitudinal and transverse wave velocity on the pressure; data after Simmons and Brace, 1965: 1- diabase (Maryland), 2 - granite (Westerly)
a.) velocities vs. pressure b.) ratio v_p/v_s vs. pressure

All examples confirm the distinct increase of velocity with increasing pressure, particularly in the lower pressure range. In this range, the velocity gradients of the individual rocks and samples clearly differ.

The strong correlation of these velocity changes and the structural-textural microproperties on the one hand with the mineralogical composition on the other hand is explained by Lebedev, Sapoval and Korchin (1974) as follows:

- The marked increase of velocity under the influence of pressure in the lower pressure range is due primarily to the closure of the large and penny-shaped pores and microjoints. This closure improves the contact between rock-forming minerals.
- At higher pressures the compaction of the aggregate is nearly complete. Therefore, the further velocity increase is due to the change of elastic properties of the mineral substance (influence of mineralogical composition on the behaviour under pressure).

The influence of *cracks* upon the velocity, the pressure dependence and the influence of pore fluid is demonstrated by Figure 6.15.

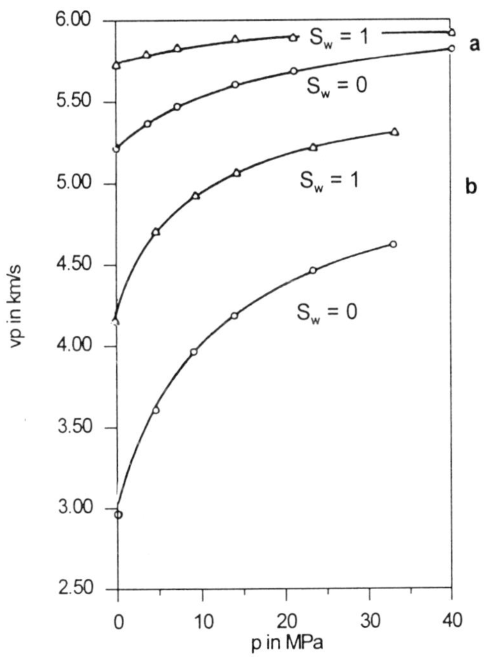

Figure 6.15. Longitudinal wave velocities as a function of uniaxial pressure, granite (California); after King and Paulsson, 1981: a - intact samples b - microcracked samples

The correlation between the character of the velocity-pressure relationship and processes at microcracks, joints, fractures and pores has motivated various investigations with the aim of determining a direct *correlation between velocity changes and the crack parameters*, for example by Simmons et al. (1975),

Lebedev et al. (1974), Zang et al. (1989), Kern (1990) and also theoretical studies (see chapter 6.4.). As an example of phenomenological considerations, the following is from Lebedev and Korchin (1974):
Table 6.13 lists mean values for the longitudinal wave velocity and porosity as a function of pressure. Measurements were made on the samples of three types of granite as shown in Figure 6.10. The variation of velocity and porosity is distinctly different for the three types. The highest pressure effect on both quantities is found for the coarse-grained rock, the lowest for the fine-grained rock. The authors explain this in the following manner: "It is suggested that with growing pressure, microjoints and penny-shaped pores are the first to close. Therefore, penny-shaped porosity may be thought of as prevailing in coarse granite varieties, whereas fine-grained varieties are generally characterized by three-dimensional porosity. Consequently, the dimensions of the rock forming minerals may also serve as a criterion for determining the nature of porosity in a rock."

Table 6.13. Longitudinal wave velocity (in km s^{-1}) and porosity (in %) as function of pressure (in MPa) for granite of different grain size; after Lebedev et al., 1974

Pressure	fine		medium		coarse	
p in MPa	v_p	Φ	v_p	Φ	v_p	Φ
0.1	5.80	0.71	5.68	0.72	5.41	0.75
10	5.88	0.47	5.90	0.37	5.82	0.33
25	5.94	0.33	5.98	0.24	6.00	0.20
50	5.98	0.23	6.05	0.14	6.15	0.10
75	6.00	0.19	6.08	0.10	6.20	0.06
100	6.02	0.15	6.09	0.08	6.23	0.04
125	6.03	0.14	6.11	0.06	6.255	0.03
150	6.03	0.13	6.12	0.05	6.27	0.01
190	6.04	0.11	6.13	0.03	6.28	0
220	6.05	0.10	6.14	0.02		
260	6.05	0.09				

For further discussions of this correlation, we use a plot (Figure 6.16.) with the relative *change of velocity*

$$\frac{\Delta v}{v_0} = \frac{v_p - v_0}{v_0} \tag{6-41}$$

and the relative *change of porosity*

$$\frac{\Delta\Phi}{\Phi_0} = \frac{\Phi_0 - \Phi_p}{\Phi_0} \tag{6-42}$$

where the subscript 0 denotes the atmospheric pressure and the subscript p the actual pressure. The plot shows clearly the *influence of grain-size* and pressure (curve-parameter) on the feature of the velocity vs. pressure relation.

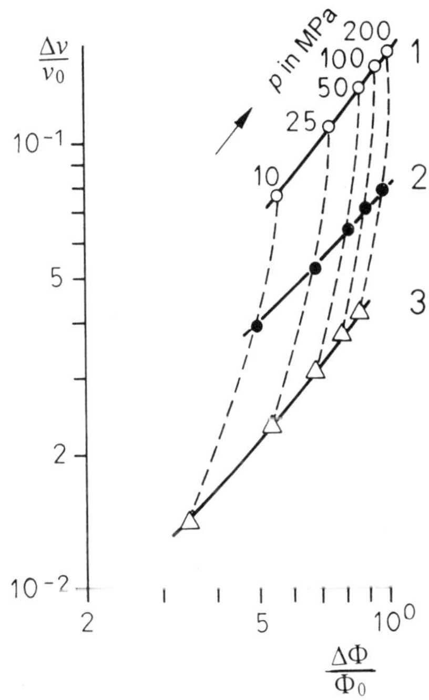

Figure 6.16. Relative increase of velocity vs. relative decrease of porosity resulting from increasing pressure; data from Table 6.13.
1 - coarse grained granite 2 - medium grained granite 3 - fine grained granite

A comparable presentation for the correlation between porosity change and velocity change (Figure 6.17) is used by Simmons, Todd and Baldridge, 1975. They correlate the "normalized compressional velocity $v_{0.1}/v_{100}$ (in the original paper $v_p(0)/v_p(10)$) with the crack porosity Φ_c for terrestrial and lunar rock groups. The velocities are defined as follows

$v_{0.1}$ velocity at pressure $p = 1$ bar $= 0.1$ MPa
v_{1000} velocity at pressure $p = 10$ kbar $= 10^3$ MPa.

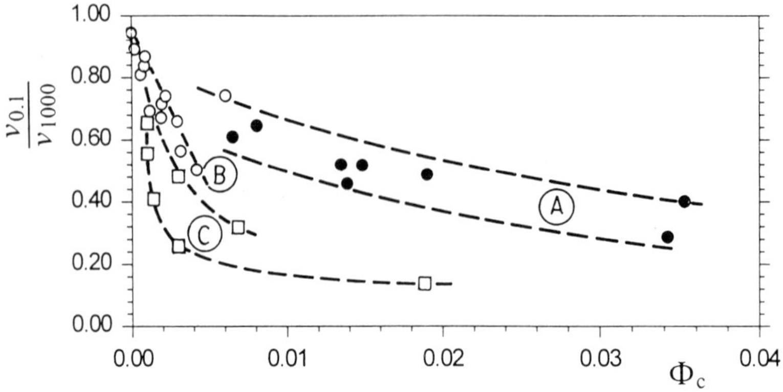

Figure 6.17. Normalized compressional wave velocity as a function of crack porosity; after Simmons, Todd and Baldridge (1975)
A - shock-metamorphosed rocks (granitic rocks / Nördlinger Ries crater - Germany, terrestrial breccias, lunar rocks / meteorite impact);
B - terrestrial igneous rocks and limestones (tectonically loaded)
C - thermally cycled rocks (Westerly granite, Fairfax diabase).

In addition to the correlation between the normalized velocity and crack porosity the influence of the *crack-forming process* is also clearly visible. Three types of such processes, with different effects on the elasticity behaviour, can be seen

- range A: Terrestrial and lunar rocks, shock wave impact
- range B: Terrestrial rocks under tectonic load
- range C: Thermally cycled terrestrial rocks.

Simmons, Todd and Baldridge (1975) conclude that "the cause of cracking in rocks is important in determining the nature of the resulting cracks and that a linear relation between velocity and crack porosity does not hold in general". The paper also includes a discussion of crack porosity modeling using the aspect ratio in form of the Nur-Simmons relation.

Similar results can be derived by an analysis of the entire experimental data determined for granites of the Ukrainian Shield obtained by Lebedev et al. (1974). Figure 6.18. shows a significant differentiation between the fine-, medium and coarse-grained granites.

Differences between Figure 6.17. and Figure 6.18. are caused by the use of different types of porosity (Figure 6.17 crack porosity, Figure 6.18 total porosity), and by differences in the definition of the "normalizing velocity".

This example confirms the existence of a strong correlation between grain size (or structural-textural properties) and the types of the resulting cracks. In summary, we can state that the structural-textural properties and the "thermodynamic history" of the rock controls its velocity-porosity relationship.

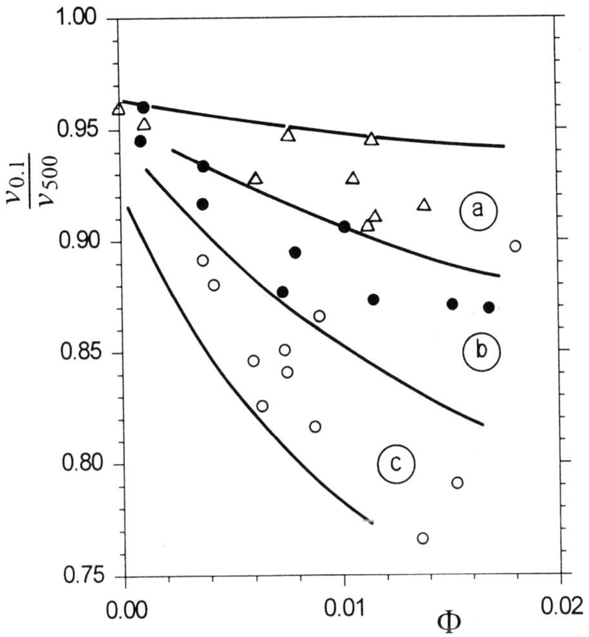

Figure 6.18. Ratio of the longitudinal wave velocities measured at a pressure of 0.1 MPa and 500 MPa vs. total porosity , Lebedev et al., 1974
a - fine-grained granite b - medium-grained granite c - coarse-grained granite

The strong influence of cracks upon velocity (and also on the attenuation) is used in many practical applications to characterize jointed, cracked and fractured rock masses with respect to their mechanical, hydraulic and reservoir-technical behaviour (e.g., see King et al., 1986; compare Figure 6.15 and chapter 10.).

In magmatic and metamorphic rocks, increasing *temperature* generally leads to decreasing velocities. Pressure and temperature have opposite effects on wave velocities, but in contrast to the pressure dependence we find in the lower temperature range (to about 100 ... 150 °C) only small changes in velocity; higher temperatures have a stronger influence on velocity (Figure 6.19).

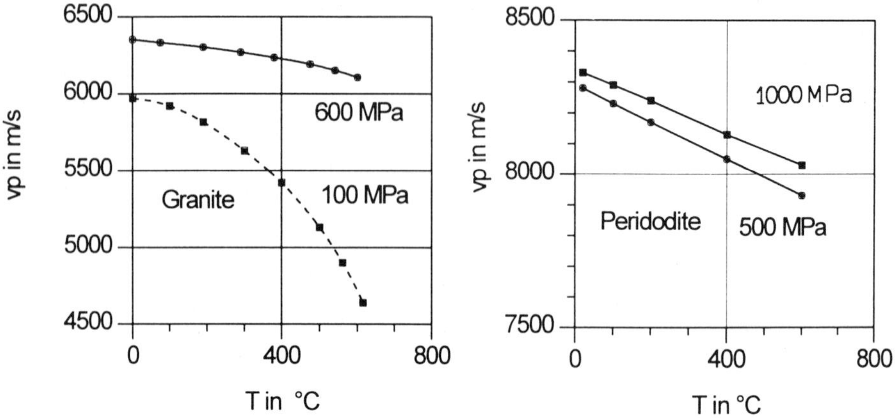

Figure 6.19. Temperature dependence of longitudinal wave velocities at different confining pressures; a.) granite, after Kern, 1990 b.) peridodite, data after Bajuk and Levitova, 1990.

The main causes of the velocity change are:

a) temperature dependence of the elastic properties of the rock-forming minerals and phase change of minerals,
b) temperature dependence of the elastic properties of the pore constituents and change of pore constituents (e.g. pore water) from a liquid to a gaseous state,
c) changes in the contact conditions at grains, crack boundaries, etc., resulting from variations of the interface effects or resulting from different thermal expansion properties of rock-forming minerals.

For temperatures up to about 150 °C, the change of fluid properties (b) dominates, but at higher temperatures the causes noted under point a) are dominant. Contact conditions and microfracturing in all range are influenced in a complex manner by temperature, pressure and rock properties.

For the determination of the *velocity-depth function* of a given region, the knowledge of the vertical gradients of pressure and temperature is necessary in addition to information about the velocity-pressure and velocity-temperature relations (compare equation 6-40).

Velocity-depth relations, determined in this manner, are shown in Figure 6.20 for rocks from the Ukrainian Shield. The values increase rapidly down to depths of 2...3 km and reach a maximum at about 4 ... 7 km (here the strong pressure dependence dominates). With further increase of depth the velocities decrease noticeably down to 20 km (experimental limit). In this second depth-range the reducing effect of temperature dependence dominates.

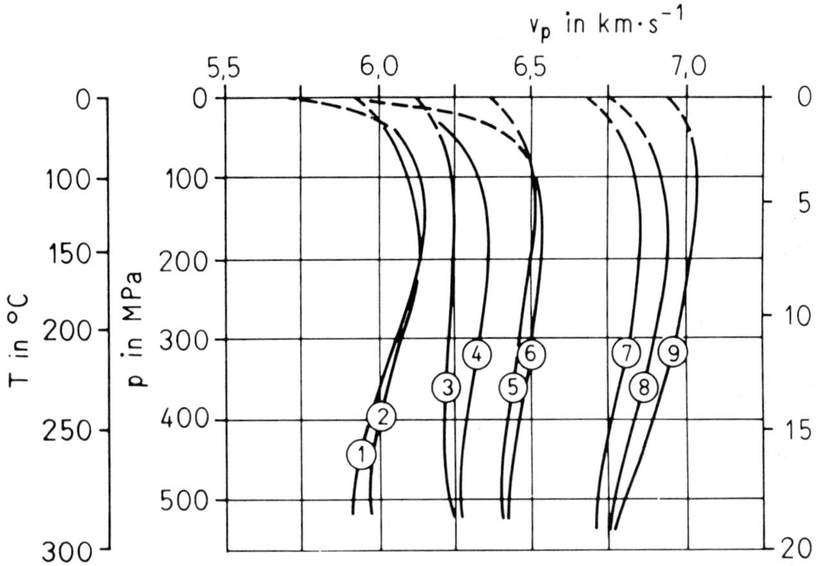

Figure 6.20. Longitudinal wave velocity vs. depth as a result of the combined effect of pressure and temperature dependence of velocity for the central part of the Ukrainian Shield; after Lebedev, 1975

1 - gneiss	2 - migmatite	3 - gabbro-monzonite
4 - granite	5 - charnockite	6 - rapakiwi-granite
7 - gabbro-norite	8 - gabbro-amphibolite	9 - labradorite.

6.3.3. The velocity of elastic waves in sedimentary rocks

"One of the key factors needed for the successful use of seismic wave velocities in reservoir development, characterisation, and recovery is a fuller understanding of what seismic waves can tell about the state of reservoir rocks and the fluids contained in their pore space." (Amos M. Nur and Zhijing Wang, 1988)

6.3.3.1. General remarks

For a study of the physical properties of sedimentary rocks, we recommend the use of the physical classification given in chapter 1. This is important

- in view on the wide variety of rock-types (dense, monomineralic salts... fractured and porous consolidated rocks... unconsolidated sediments),
- particularly for the study of seismic properties with their complex dependence on various rock features.

Table 6.14 gives an overview of the average of velocities of the compressional and shear wave. The sedimentary porous rocks generally show *lower velocities* and a *broader range* for an individual rock type than igneous and metamorphic rocks. Both features are mainly due to the influence of the pore contents with their low elastic parameters.

Table 6.14 gives an indication of the dominant parameters which influence velocity. These are:

- mineralogical composition of the rock matrix,
- consolidation and cementation of the rock matrix,
- porosity, pore shape and pore content,
- pressure and temperature.

Additionally, fracturing can be of influence, particularly in carbonate rocks. Structure and texture lead to different types of anisotropy in the elastic properties.

In the following sections we will study the individual influences. Due to the distinct differences in their physical features and behaviour, we will consider consolidated and unconsolidated sedimentary rocks separately.

Table 6.14 Compressional (v_p) and shear (v_s) wave velocity of sedimentary rocks; values are partly converted and rounded, n - number of individual data; Φ - porosity, numbers in brackets are densities in kg/m^3.
References: B - Bourbie et al. (1987), BH - Baermann and Hübner (1984), E - Ellis (1987), G - Gebrande (1982) - data collection from various authors; K - Kopf (1977), S - Serra (1984), W - Western Atlas (1992)

Rock type	n	v_p in km/s range	mean	v_s in km/s range	mean	Ref.
Sandstone	21	2.80...4.46	3.63			G
	14			1.59...2.93	2.26	G
		3.05...5.94	5.34			S
- porous		2.00...350		0.80...1.80		B
Φ = 0.2..0.05		3.50...5.03		1.83...2.90		E
- semi-consolidated			5.49			W
- consolidated			5.79			W
Shale		2.13...5.18				S, E
		1.83...4.88				W
Silt- and (parallel)	13	3.91...5.55	4.73			G
claystone (perpend.)	13	2.83...5.37	4.10			G
Marl (2688)	8	4.78...6.04	5.41			G
(2350)		2.00...3.00		0.74...1.50		B
Limestone dense			6.40		3.38	E, W
- tight		5.72...6.40	5.86			S
Φ = 0.20...0.05		3.96...5.64		2.13...2.90		E
(2658)	26	4.89...6.37	5.63	2.54...3.70	3.12	G
(2550)		3.50...6.00		2.00...3.30		B
Dolomite (2700)		3.50...6.50		1.90...3.60		B
(2794)	10	4.92...6.50	5.71	2.90...3.74	3.32	G
Φ – 0.20...0.05		4.57...6.10		2.29...3.35		E
Anhydrite			6.09		3.47	E
Gypsum		5.75...5.81	5.80			S
Salt, Halite			4.57			S, W
Polyhalite			5.30			S
Sylvite			4.11			S
Carnallite			3.66			S
Anthracite		2.54...3.39				S
Bituminous coal		1.80...3.05				S
Lignite		1.00...2.18				S
Brown coal, Germany		0.5 ...1.8				K
Uncons. sediments		0.55...1.95		0.15...0.4		BH
Sand, unconsolidated		0.4 ...2.1		0.15...0.6		K

6.3.3.2. The velocity of elastic waves in consolidated sedimentary rocks

Dependence of velocity on matrix composition and porosity

The mineralogical *composition* of a sedimentary rock has a strong influence upon its velocity. This is shown by its effect (depending on the fractional composition) on

- the "effective" elastic moduli of the rock,
- the grain to grain-bonding, cementation and contact conditions in general.

For certain rock types, *porosity* is often the dominant influence.

Evaporates (e.g., halite, sylvite, gypsum, polyhalite etc.) are, for most practical purposes, *pore-free*. Therefore, the velocities in these rocks depend only weakly on pressure or depth. These small velocity variations are caused by changes of intergranular contact conditions. Mean values of these rocks are given in Table 6.14. We note that well defined values for these rocks are useful indicators for lithologic interpretation, e.g., of acousticlog data.

For duo- or polymineralic composition, the relevant "mixing rules" equations give a sufficient approximation (see section 6.4.3). Resulting from such rules (arithmetic or geometric mean), we can explain, for example, the variations of the velocity of anhydrite, determined by Kopf (1977):

anhydrite with 50 % limestone	v_p = 5600 m/s
anhydrite with 50 % dolomite	v_p = 5900 m/s
anhydrite with 50 % gypsum	v_p = 5400 m/s
anhydrite with 50 % halite	v_p = 4900 m/s.

Similarly, we can interpret the velocity-decrease of dolomites with increasing calcite-content.

The elastic properties of *porous* clastic and carbonate rocks are mainly controlled by porosity and matrix composition. The matrix composition also influences the contact conditions, cementation, grain-bonding etc.. The velocity dependence on porosity (Figure 6.21) is stronger for smaller variations of matrix composition. Summarizing the porosity effect, Hearst and Nelson (1985) formulated that "the most obvious effect of porosity on propagation of stress waves is reduction of velocity."

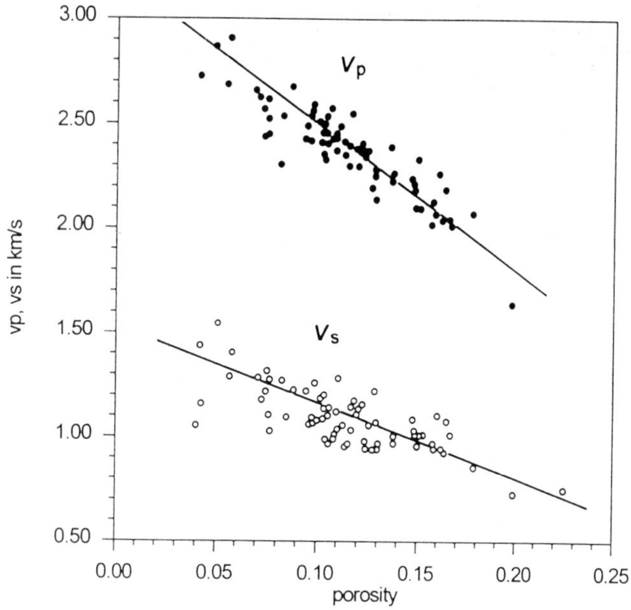

Figure 6.21. Longitudinal and transverse wave velocity versus porosity for water saturated sandstone (Rotliegenes, Northern Germany); pressure 14 MPa.

From the complex influence of matrix properties and porosity the famous *"time-average equation"* was derived by Wyllie, Gregory and Gardner in 1956 (see section 6.4.3):

$$\frac{1}{v_p} = \frac{1-\Phi}{v_m} + \frac{\Phi}{v_{fl}} \quad (6\text{-}43)$$

where

v_p is the compressional wave velocity in the fluid saturated porous rock,
v_m is the compressional wave velocity of the rock matrix (matrix velocity),
v_{fl} is the velocity of the pore fluid.

When v_p is measured and v_m and v_{fl} are known or assumed, the only unknown parameter is the porosity Φ. In section 6.4.3 a detailed discussion of the time average equation and "relatives" is given.

Another type of empirical equation is given for example, by the relation obtained by Raymer et al. (1980) for consolidated rocks

$$v_p = (1-\Phi)^2 v_m + \Phi v_{fl} \quad (6\text{-}44)$$

and by Nafe and Drake (1963) and Gardner et al. (1974) as a relationship between longitudinal wave velocity (in m/s) and density (in g/cm³ = 10³ kg/m³) for saturated sedimentary rocks

$$d = 0.31 \cdot v_p^{0.25} \qquad \text{or} \qquad v_p = 108 \cdot d^4 \tag{6-45}$$

Clay within the rock produces some additional problems: Clay minerals significantly reduce velocities and elasticity moduli. The *effect of clay* with respect to the elastic behaviour has four aspects:

1. The velocity of clay is lower than that of quartz or carbonate matrix material. Thus, there generally is a decrease of velocity with increasing clay content.
2. The deformability or compressibility of clay is higher than quartz or carbonate matrix material. This leads to a growing pressure dependence of the velocity with increasing clay content.
3. The physical properties of clay depend strongly on water content. Thus, porous rocks containing clay show a high and very complex dependence of elastic properties on saturation.
4. These effects of clay are dependent on the distribution and configuration of the clay colloids within the rock frame. If the detrial clay lie between the grains precluding quartz-quartz contact, the elastic properties of the frame are strongly controlled by the clay properties. If the clay is authigenic and resides in the pores of a quartz-grain frame only (clay content below 40 %), the density will be most affected. The frame moduli will be lowered if authigenic clay, rather than quartz overgrowth, forms the cement between the asperities (Murphy et al., 1993).

The first two aspects are demonstrated by Figure 6.22 for a series of carbonate sedimentary rocks (North German - Polish Basin): The velocity of the longitudinal wave v_0 under surface conditions shows a clear decrease with increasing clay content V_{sh}; the ratio v_1/v_0 of the velocities at 1 km depth and at the surface (as a measure of the pressure sensitivity) distinctly increases with increasing clay content.

There are many *empirical relationships* which describe the velocity as a function of porosity and of clay content. A linear correlation of the longitudinal wave velocity with porosity and clay content was found by Tosaya and Nur (1982) for sandstones and shales with a clay fraction up to 0.72.

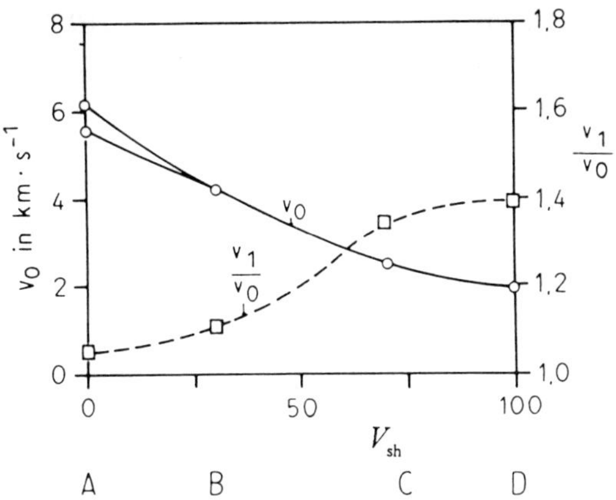

Figure 6.22. Dependence of the longitudinal wave velocity v_o and the ratio v_1/v_o on the clay content of carbonate rocks V_{sh} (data after Kopf, 1977)
v_o - velocity under conditions of 0 km depth v_1 - velocity under conditions of 1 km depth
A - limestone, dolomite (clean) C - marl
B - marl-limestone D - shale

Results of systematic investigations of the clay influence are published in a comprehensive study by Han, Nur and Morgan (1986). Compressional and shear wave velocities were measured as functions of pressure in 80 sandstone samples with varying *clay content* (volume fraction ranging from 0 to 50 percent) *and porosity* (range from 2 to 30 percent). Figure 6.23 shows the velocity-porosity crossplots for longitudinal and transverse waves.
The authors applied, as a first trial, a relation with the form of the time-average relation

$$\Delta t = v^{-1} = B_0 + B_1 \Phi \tag{6-46}$$

and also a linear relation with respect to the velocity

$$v = A_0 - A_1 \Phi \tag{6-47}$$

where Δt is the transit time, v the velocity, and A_0, A_1, B_0, B_1 are empirical values. Using least-squares regression analyses, the following equations with correlation coefficient R and relative *rms* deviation (with 68.3 % confidence) result:

$$\Delta t_p = v_p^{-1} = 0.194 + 0.328 \cdot \Phi \qquad R = 0.844 \quad rms = 6.6\ \% \tag{6-48}$$

$$\Delta t_s = v_s^{-1} = 0.322 + 0.628 \cdot \Phi \qquad R = 0.750 \quad rms = 10{,}3\ \% \qquad (6\text{-}49)$$

$$v_p = 5.02 - 5.63 \cdot \Phi \qquad R = 0.840 \quad rms = 6.8\ \% \qquad (6\text{-}50)$$

$$v_s = 3.03 - 3.78 \cdot \Phi \qquad R = 0.754 \quad rms = 10.8\ \% \qquad (6\text{-}51)$$

where the velocities are given in km/s (Δt in s/km in this case) and the porosity as a decimal fraction.

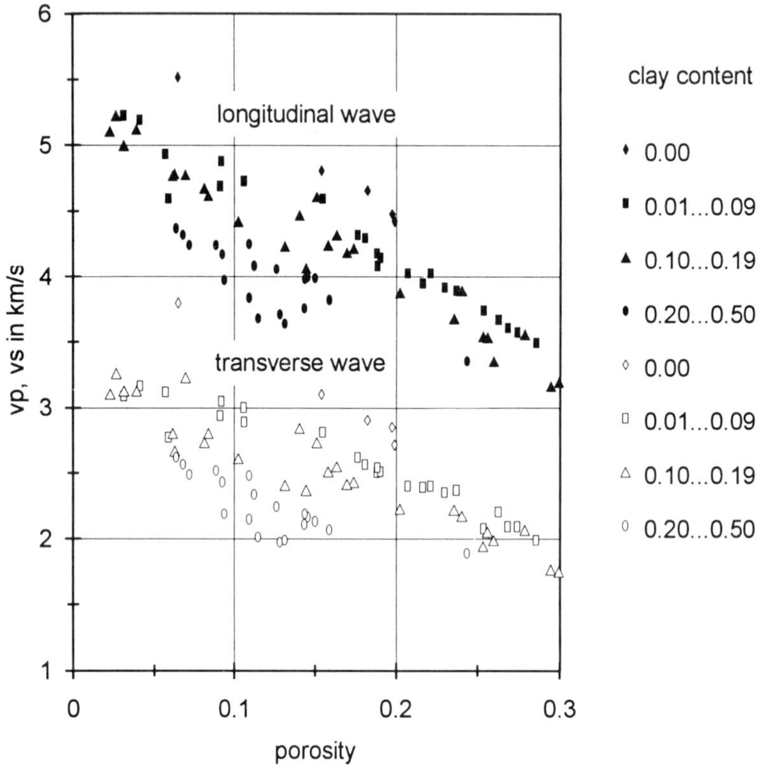

Figure 6.23 Longitudinal and transverse wave velocities versus porosity for 75 sandstone samples at a confining pressure 40 MPa and a pore pressure 1.0 MPa; data after Han et al. (1986).

As a result of further analysis, the authors found that the deviations between measured and predicted values clearly depend upon the clay content. They developed expanded forms of the two equations using a linear clay term (C is the clay content):

$$\Delta t = v^{-1} = B_0 + B_1 \cdot \Phi + B_2 \cdot C \quad (6\text{-}52)$$

$$v = A_0 - A_1 \cdot \Phi - A_2 \cdot C \quad (6\text{-}53)$$

This analysis leads to the result that both velocities (and also travel times) "... were found to correlate linearly with porosity and clay content in shaly sandstones". At the given pressures (confining pressure 40 MPa, pore pressure 1.0 MPa), the best least-squares fits to the velocity data are

$$v_p = 5.59 - 6.93 \cdot \Phi - 2.18 \cdot C \qquad R = 0.985 \quad rms = 2.1\ \% \quad (6\text{-}54)$$

$$v_s = 3.52 - 4.91 \cdot \Phi - 1.89 \cdot C \qquad R = 0.959 \quad rms = 4.3\ \% \quad (6\text{-}55)$$

The velocities of clean sandstones are significantly higher than those predicted by relationships given above (about 7 percent for v_p and 11 percent for v_s), "which indicates that a very small amount of clay (1 or a few percent of volume fraction) significantly reduces the elastic moduli of sandstones" (Han et al., 1986). The influence of clay is also visible at the reduced matrix velocity (longitudinal wave) of 5.59 km/s in comparison with the value for quartz aggregates of 6.05 km/s. Han et al. (1986) formulated that the clay amount "can significantly soften the sandstone matrix." .
The corresponding equations for the travel times are:

$$\Delta t_p = v_p^{-1} = 0.163 + 0.399 \cdot \Phi + 0.119 \cdot C \qquad R = 0.972 \quad rms = 2.8\ \% \quad (6\text{-}56)$$

$$\Delta t_s = v_s^{-1} = 0.242 + 0.812 \cdot \Phi + 0.307 \cdot C \qquad R = 0.945 \quad rms = 5.1\ \% \quad (6\text{-}57)$$

Marion and Jizba (1992) have investigated North sea shaly sand reservoir rocks. They applied the regression equation (6-53) and found the following parameters for the reservoir at 35 MPa:

Pore content	Wave type	A_0	A_1	A_2
Gas	P-wave	4.82	5.02	0.597
	S-wave	3.26	3.03	0.892
Brine	P-wave	5.46	6.29	1.10
	S-wave	3.32	3.62	0.952

We note, that the order of porosity effect on P- and S-wave velocities is comparable, that the S-wave velocities are relative insensitive to saturation, and that the effect of clay is most important for the S wave velocities (Marion and Jizba, 1992).

Three remarks may be added to these considerations:

1) The coefficients in the linear velocity-porosity relations are dependent on pressure; equations with a non linear pressure dependent term are derived by Eberhart-Phillips et al. (1989). This fact is discussed more in detail on page 189.
2) The analysis of both velocities (v_p, v_s) allows the v_p/v_s− ratio and its dependence on clay content, porosity and pressure to be considered (this is discussed in section 6.3.4.3. and 6.4.8.).
3) Parameters derived from compressional and shear wave velocity or travel time values are also used for porosity determination, especially in seismic exploration. An example is the linear relation between the travel time difference and porosity after Mehta and Verma (1991)

$$\Delta t_s - \Delta t_p = (\Delta t_{m,s} - \Delta t_{m,p}) + \Phi \cdot (B_s - B_p) \tag{6-58}$$

where

$\Delta t_s, \Delta t_p$	are the measured travel times for S- and P-wave,
$\Delta t_{m,s}, \Delta t_{m,p}$	are the matrix travel times, and
B_s, B_p	are empirical parameters, dependent on lithology.

This equation was derived for limestones and sandstones. A similar example is the relationship between the seismic impedance and porosity (e.g., Budny, 1991; Owen, 1993).

The influence of the pore content and saturation upon velocities

Nur and Wang (1988) wrote in their preface for the chapter "Effects of Saturation and Pore Fluid Properties" in the SEG-Geophysics reprint series No.10 that the velocity decrease resulting from the increased overall density "sometimes are offset by other effects related to porosity and pore structure, interfacial surface phenomena between the rock and its pore fluid, and viscous relaxation behavior of the pore fluid".

Variation of the pore filling substance (fluids, gases, mixtures and suspensions) influences the elastic wave velocities as a result of different effects:

- effects of changing elastic properties and density of the pores as a part of the whole system "rock";
- effects of changing particle-contact conditions (e.g., solution or swelling of cement);
- effects of interactions of gas-fluid-solid phases (e.g., capillary tensions) and solid-fluid-interface effects.

The complex nature of these different effects and their dependence on porosity, pore structure and physio-chemical properties may be the reason for the difficulty in determining any general description and explanation. Therefore, as a first consideration the influence of different but unmixed pore contents (air, water, kerosene) will be discussed. Secondly, the variable pore content composition (air-water mixtures) will be considered.
Figure 6.24. illustrates the influence of different pore fluids upon the P- and S-wave velocities in a velocity vs. pressure plot.
The results show three typical features:

1) the velocity vs. pressure relation is non-linear,
2) the increase of longitudinal wave velocity from air- to kerosene- to water-filling corresponds to the velocities of these substances (Table 6.7 and 6.8.). The longitudinal wave velocities of the porous rock reflects the influence of the elastic properties of its pore fluid:

$$v_{p,air} < v_{p,kerosene} < v_{p,water}$$

3) the shear modulus of the porous rock may be considered as unaffected by different pore fluids; therefore, the transverse wave velocity is affected almost only by a density effect of the different pore components. This results in the sequence of transverse wave velocities which are opposite to that of the longitudinal wave velocities because

$$d_{air} < d_{kerosene} < d_{water}$$

In all these considerations, it is presumed that the pore fluid has no shear resistance. The effect of heavy oil having a high viscosity was investigated by Wang and Nur (1988); the oil creates much higher velocities than the water saturation).

King (1966) found a satisfactory agreement of the experimental results at Boise sandstone (Figure 6.24) with Biot's theory (see chapter 6.4.). Nur and Wang (1988) wrote as a comment to this: "Agreement of the velocity results with Biot's theory is dependent on the microstructure (e.g. , crack content) and porosity of the rock, the viscosity of the pore fluid, and the interactions between the rock and its pore fluid. King's results show that at high effective pressures and in the absence of the interactions between the saturant and the detrital material, the measured velocities are found to approach the behaviour predicted by the Biot theory."

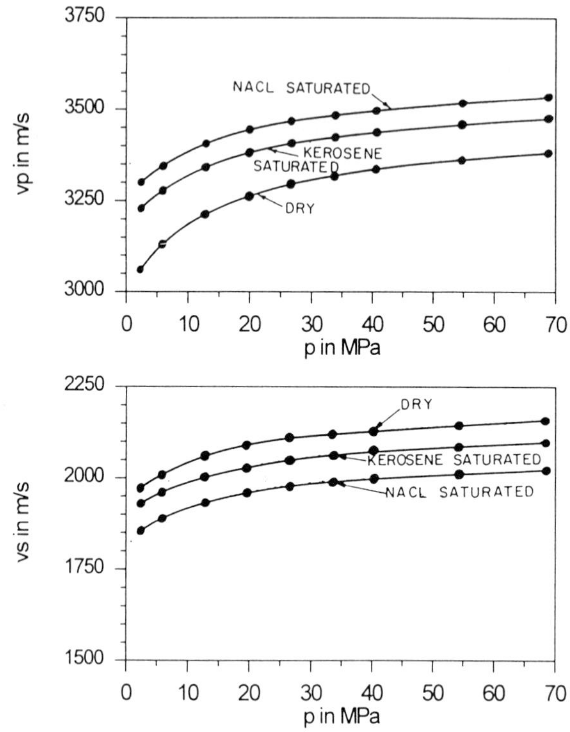

Figure 6.24. Longitudinal and transverse wave velocities of Boise Sandstone as a function of pressure at different pore fluids; after King (1966), converted units.

In a second experimental series of Bandera Sandstone, King (1966) found a somewhat different behaviour: the longitudinal wave velocity has lower values for water saturation than for kerosene saturation (and the transverse wave velocity for water saturated sandstone was extremely low). "The probable reasons for this behavior are that there is a loss of rigidity of the appreciable interstitial clay mineral content of Bandera Sandstone when it is saturated with an aqueous solution, and a reduction in free surface energy of the grains due to saturation with a polar liquid." (King, 1966).

Figure 6.25. gives a further example, demonstrating the influence of matrix properties upon the effect of different pore fluids on longitudinal wave velocity:

- ► clean limestone generally shows an increase of the velocity (excepting the low pressure range) strongly following the sequence of the velocities of pore fluid; in the clean limestone the change of contacts is obviously negligible.
- ► sandstones with a clay cement and shale do not show the same sequence as the pore fluids. In particular the velocity of the water-saturated rock is less than that of the dry rock. Corresponding to the remark on page 179 this is the effect of a "softening of the matrix" by water-clay-interactions (as, for example swelling). Kerosene in our example has no such an effect on the rock skeleton - its modulus of compression is higher than that of air and, thus, the velocity is higher.

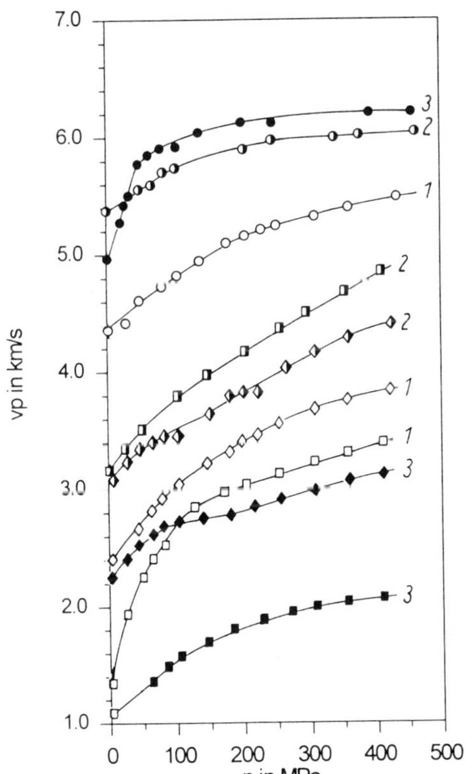

Figure 6.25. Longitudinal wave velocity of various sedimentary rocks as a function of pressure for different pore fluids; after Volarovic et al. (1968).

• - limestone	♦ - sandstone/clay-cemented	■ - shale
1 - dry rock	2 - kerosene saturated rock	3 - water saturated rock

If the *pore fluid is composed* of two (or more) components (e.g., partial water saturation), then the interactions between the individual pore constituents and between pore constituents and solid skeleton (e.g., capillary tension, surface-boundary effects) act additionally. The different influencing factors are acting on the elastic properties of the porous rock in a difficult and complex manner, which results in very different types of velocity-saturation functions. The experimental results of Gregory (1976), measured at 20 different rock samples with porosities in the range between 4 and 41 percent, illustrate this diversity. Figure 6.26. and Figure 6.27. are examples demonstrating the influence of saturation conditions, rock properties (particularly pore-space properties) and pressure.

Kopf (1977), Bacri and Salin (1986), Frisillo and Thomsen (1991), and Cadoret et al. (1992) have investigated the *influence of "saturation technique"* (drainage, imbition, drying) upon the character of the velocity vs. saturation function and found a distinct difference between the observed velocity vs. saturation dependencies.
This confirms the important role of the *spatial distribution* and the feature of interactions *of the fluid components* on the elasticity behaviour. Cadoret et al. (1993) investigated limestone samples and controlled the microscopic fluid distribution using X-ray tomography. For two different saturation methods (drying, progressive depressurization) the experiments show:

- a different fluid distribution: The images of depressurized samples show a great homogeneity of saturation; in the drying experiment, at high saturation levels clusters dominate,
- a clear dependence of the compressional wave velocity on saturation method: At high levels of saturation v_p exhibits a large increase with increasing S_w in the drying experiment; in contrast, v_p remains relatively constant with increasing S_w using depressurization method;
- shear wave velocity is nearly independent on saturation method.

The observation emphasizes the necessity of a careful approach in interpreting elastic wave velocity in term of fluid saturation when the microscopic fluid distribution in the porous media is not perfectly known (Cadoret et al., 1993).
The analytical or theoretical description of the velocity vs. saturation behaviour is one of the most difficult problems in petrophysics.

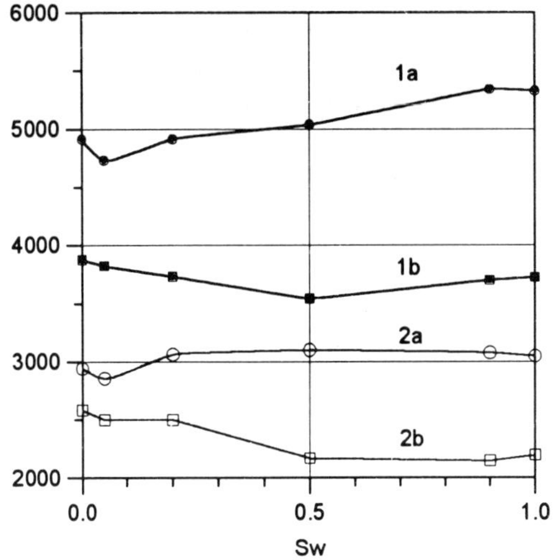

Figure 6.26. Longitudinal (1) and transverse (2) wave velocities as a function of water saturation degree for two samples of different porosity at a pressure of 5000 psi (= 34.48 MPa); after Gregory (1976), converted units.

1a, 2a - Travis Peak sandstone (sample CS-1A), porosity 4.45 percent
1b, 2b - Berea sandstone (sample 107-13), porosity 19.1 percent

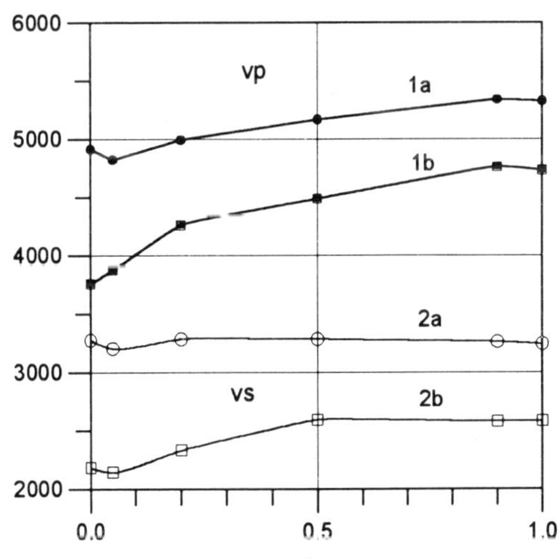

Figure 6.27. Longitudinal (1) and transverse (2) wave velocities as a function of water saturation for the Travis Peak sandstone at two pressures; after Gregory (1976), converted units.

1b, 2b - pressure 1000 psi (= 6.895 MPa) 1a, 2a - pressure 10000 psi (= 68.95 MPa)

Dependence of velocity on depth, pressure, and temperature

Faust (1957) has developed an empirical relationship between the velocity of longitudinal wave v_p (in m/s) and depth z (in meters) from an analysis of field measurements (500 data from shales and sands):

$$v_p = L \cdot (A \cdot z)^{1/6} \tag{6-59}$$

where L is a lithological factor (46.6 for Faust's samples, after Bourbie et al., 1987), and A is the geological age (in years) of the formation.

It is remarkable for all further considerations that this early formula contains a non-linear relation (power-law) between velocity and depth. The non-linearity was demonstrated in some examples of the foregoing sections of this chapter (e.g., Figures 6.24., 6.25.). A compilation of published results shows the dominant features:

- a non-linear dependence of velocity on depth (the gradient decreases with depth),
- the especially strong dependence of velocity on depth for rocks containing clay and shale,
- the very small changes of velocity for dense, pore-free rocks (e.g., anhydrite, salt).

For constant lithology and pore-water composition, the velocity-depth relation reflects the dependence of velocity on pressure and temperature. Both influences are considered in the following section.

Figure 6.28. demonstrates some fundamental influence of pressure on velocity for sedimentary rocks:

► the pressure dependence increases with increasing deformability of the rock skeleton (i.e. with increasing porosity),

► the velocity-pressure relationship is non-linear,

► inelastic deformation results in the phenomenon of an "elastic hysteresis",

► with increasing pressure longitudinal and transverse wave velocities generally increase, but at different rates. This results in a pressure dependence of the velocity ratio and Poisson's ratio (see section 6.3.4).

The increase of the velocity with increasing pressure in sedimentary rocks is caused by

- the decrease of porosity,
- the improvement of contact conditions.

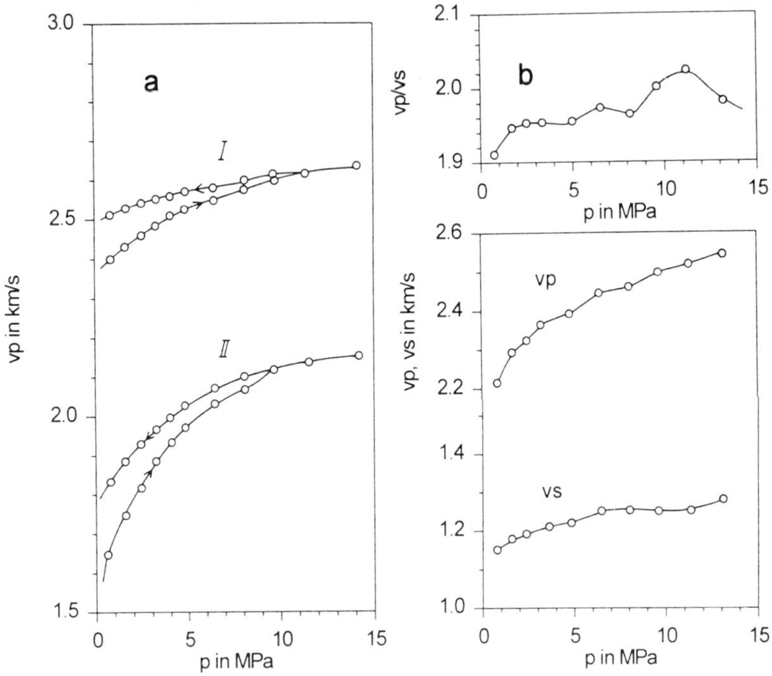

Figure 6.28. Dependence of elastic wave velocities of sandstones (Mesozoic; N-German basin) on the effective pressure; Schön, 1983
a) longitudinal wave velocity at porosity 6.3 (I) and 19 percent (II);.
b) longitudinal and transverse wave velocities and velocity ratio at porosity 12 %.

Both effects related to changes of the properties of the rock-skeleton (the change of the elastic properties of the minerals themselves can be neglected). Therefore, the pressure acting on the rock-skeleton ("*effective pressure*") is the critical parameter. If a part of the total pressure acts on the pore content (pore pressure p_{pore}), then the effective pressure is given by the difference:

$$p_{eff} = \Delta p = p - p_{pore} \tag{6-60}$$

Extremely high pore pressures are found in non-drainable systems as for example in certain "high pressure zones" or zones of pore-water overpressure (soil mechanics implications).

Figure 6.29 emphasizes that the velocity change is a result of the effective pressure: at constant effective pressure there is a small change of velocity only in the first part of the curve.

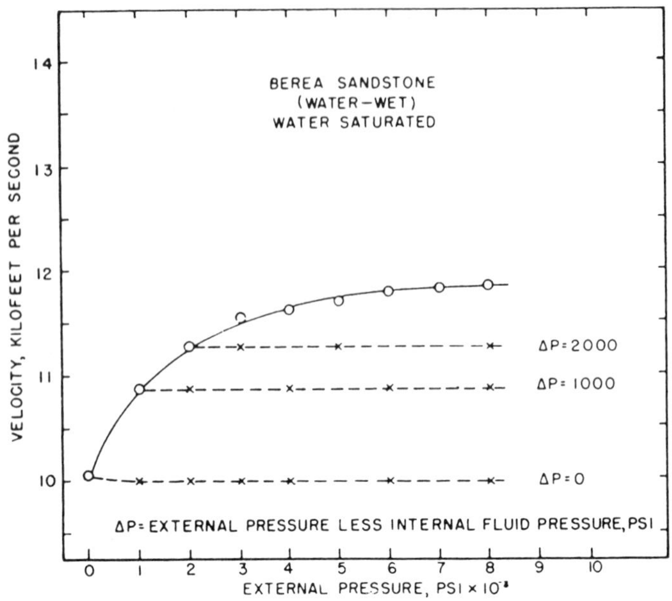

Figure 6.29. Longitudinal wave velocity of a sandstone core under various differential pressures; from the classic paper of Wyllie et al. (1958) - units not converted: pressure in psi (100 psi = 6.895 MPa); velocity in kilofeet/second (1 kft/sec = 304.8 m/s)

Hearst and Nelson (1985) comment on this behaviour: "At small constant differential pressure, the velocity increases with external pressure at first and then becomes approximately constant. At large constant differential pressure, the velocity is constant for all external pressures". This was borne out by experiments of Avcan et. al., (1968), Dobrynin (1963), Desai et. al., (1969) and Todd and Simmons (1972) that resulted in a modification of equation 6-60:

$$p_{\text{eff}} = p - n \cdot p_{\text{pore}} \tag{6-61}$$

The factor n depends on the volume-compressibility of the matrix components. The value is given by Dobrynin (1963) with 0.83 and by Kopf (1977) with 0.63... 1.2 depending on the depth. Nur and Wang (1988) notes that n is found close to 1 for both P-wave and S-wave, although significantly smaller values are common in low porosity rocks.

In addition, we can assume that *high pore pressures* affect the contact conditions and by this the rock skeleton properties. Yale (1985) notes in an overview paper - referring to results of Zoback (1975), Walls and Nur (1979),

Christenson and Wang (1985) - that "elastic properties of clays are affected by pore pressure".

For an *analytical description* of the non-linear velocity vs. pressure relationship, power laws or exponential laws are most commonly used. The power law is preferred for unconsolidated sediments (see section 6.3.4.). It may be noted that for both (power law and exponential law) a theoretical explanation based on a model is possible (see chapter 6.4).

An extended study of the combined *influence of effective pressure, porosity, and clay content* of sandstones using a multivariate analysis has been published by Eberhart-Phillips et al. (1989). For data 64 sandstone samples with porosity ranging between 4.6 % and 29.4 % and a clay content ranging between 0 % and 40 % , the best fit is:

$$v_p = 5.77 - 6.94 \cdot \Phi - 1.73 \cdot C^{1/2} + 0.446 \cdot [p_{\text{eff}} - \exp(16.7 \cdot p_{\text{eff}})] \qquad (6\text{-}62)$$

with an *rms* error 0.105 km/s

$$v_s = 3.70 - 4.94 \cdot \Phi - 1.57 \cdot C^{1/2} + 0.360 \cdot [p_{\text{eff}} - \exp(16.7 \cdot p_{\text{eff}})] \qquad (6\text{-}63)$$

with an *rms* error 0.099 km/s,

where velocities are in km/s, porosity Φ and clay content C as fractions and the pressure in kbar (1 kbar = 100 MPa). The exponent 1/2 of the clay term expresses the strong effect of a small amount of C on the velocity.
The relationships are of the more general form with empirical parameters a, b, c, e, and f:

$$v_{\dots} = v_0 - a \cdot \Phi - b \cdot C^g + e \cdot \left[\left(\frac{p_{\text{eff}}}{p_0}\right) - \exp\left(f \cdot \frac{p_{\text{eff}}}{p}\right)\right] \qquad (6\text{-}64)$$

The effect of *temperature* in the range from 0 °C to about 300 °C is "usually small compared to the effects of porosity and saturation" (Hearst and Nelson, 1985). This is observed for water saturated sediments; the general tendency is a small decrease of the rock velocity with increasing temperature. *Frozen rocks* are discussed in the following section.

More sensitive are *oil bearing* rocks. Wang and Nur (1988), Nur (1989) and Tosaya et al., 1985 (see Bourbie et al., 1987) have published results of laboratory compressional and shear wave velocity measurements on samples of oil sands and sandstones as a function of temperature. Heating from about

10 °C to 150 °C results in a tremendous decrease of the velocity for these samples, such as:

- from 3.4 to 2.8 km/s (Kern River oil sand)
- from 3.4 to 2.2 km/s (Venezuelan oil sand).

6.3.3.3. The velocity of elastic waves in unconsolidated sediments

General remarks

We can distinguish two main groups of unconsolidated sediments (see also chapter 1)

- ➢ non-cohesive rocks (e.g. sand, gravel)
- ➢ cohesive rocks (e.g. clay, loam).

The differences in the features of these two groups are based on various physical conditions at the contacts of the rock particles. In the first group, the conditions are controlled by friction effects and in the second group physio-chemical phenomenona are dominant.

Experimental investigations are chiefly directed at properties of natural non-cohesive sediments, in some cases artificial sediments (glass-spheres) are used. For a given sediment the velocity depends mainly on

- ► porosity,
- ► pressure,
- ► water saturation.

Different sediment types are characterized by different velocity values and velocity dependencies of controlling parameters. These are controlled by differences in the mineralogical composition, grain-size distribution and grain shape, etc..

The dependence of elastic wave velocity on porosity, density and clay content

Similarly to consolidated sedimentary rocks, there exists a significant correlation between velocity and *porosity* for unconsolidated sediments. Velocity values in unconsolidated sediments are distinctly lower than in consolidated sediments. The longitudinal wave velocity, particularly shows a clear difference for the dry (about 200...500 m/s) and water saturated (about 1600...2000 m/s) sediment.

Figure 6.30 shows a collection of mean curves of the velocity vs. porosity correlation for dry sediments and for water saturated marine sediments. Note the strong dependence on porosity in both cases and the clear difference in the absolute values. It is remarkable that the velocity in saturated sediment is comparable to that of water velocity or higher.

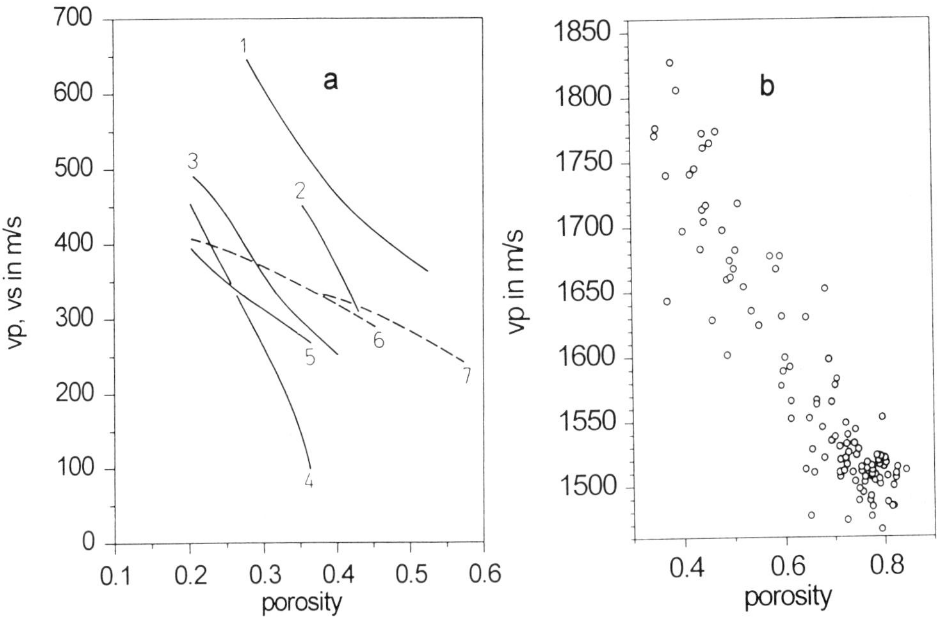

Figure 6.30. Velocity versus porosity for unconsolidated sediments
a) Mean curves for the dependence of elastic wave velocities on porosity of dry unconsolidated sediments; (Schön, 1964, 1969, 1983)

Curve number	Sediment	Wave type
1	quartz sand, pressure 50 kPa	longitudinal
2	glass spheres, pressure 10 kPa	longitudinal
3	sand, loamy, pressure 50 kPa	longitudinal
4, 5	rock fill (depth 0...3 m, Ohra-dam, Germany)	longitudinal
6	sand, rounded	transverse
7	sand, angular	transverse

b) Correlation between longitudinal wave velocity and porosity of water saturated marine sediments (Hamilton, 1971)

For a known grain density, and especially in the case of a nearly constant mineralogical sediment composition, the velocity-porosity relation can be transformed into a velocity-density relationship. On this basis, the empirical relationships are discussed below. In the simplest case, linear relationships are used:

$$v_p = a_1 + a_2 \cdot d \tag{6-65}$$

$$v_p = b_1 - b_2 \cdot \Phi \tag{6-66}$$

where d is the density, Φ the porosity, and a_1, a_2, b_1, b_2 are empirically determined parameters. Note that

1) both equations are equivalent with

$$b_1 = a_1 + a_2 \cdot d_s \tag{6-67}$$

$$b_2 = a_2 \cdot (d_s - d_{fl}) \tag{6-68}$$

where d_s is the density of the solid matrix material, and d_{fl} is the density of the pore fluid,

2) the form of equation (6-63) is identical with that of equation (6-44) for consolidated sediments,
3) there is a formal agreement with the structure of Birch's formula (6-19), although Birch's formula is based on a different physical background (mineralic composition and its influence on velocity and density).

Quadratic regressions are applied in some papers. The following equations are only examples of a large number of empirical relations. The units and symbols are expressed as follows: velocity v_p in m/s, porosity Φ as a fraction, density d in g/cm^3, mean grain size D in mm (or in phi-units), and clay content C as fraction. Hamilton (1970, 1971, 1978, 1982) and Hamilton and Bachman (1982) have published an extended collection of experimental (including granulometric and petrographic) data and derived empirical equations expressing the correlation between *velocity, porosity, density, clay content, and grain size*. For example, Hamilton and Bachman (1982) list for marine sediments correlations for three sediment types (after transformation of units):

continental terrace / shelf and slope:

$$v_p = 2502.0 - 2345 \cdot \Phi + 140 \cdot \Phi^2 \tag{6-69}$$

$$v_p = 2330.4 - 1257.0 \cdot D + 487.7 \cdot D^2 \tag{6-70}$$

abyssal hill / turbidite:

$$v_p = 1564.6 - 59.7 \cdot \Phi \qquad (6\text{-}71)$$

$$v_p = 1591.5 - 63.4 \cdot D \qquad (6\text{-}72)$$

abyssal hill / pelagic:

$$v_p = 1410.6 + 117.7 \cdot \Phi \qquad (6\text{-}73)$$

$$v_p = 1476.7 + 29.7 \cdot D \qquad (6\text{-}74)$$

The opposite tendency, seen in the velocity vs. porosity relationship for abyssal hill pelagic material (porosity = 0.8...0.9), is the typical behaviour of suspensions in the high porosity range (increasing velocity with decreasing particle concentration). This behaviour is described by Wood's equation (see chapter 6.4.5.).

Morgan (1969) has derived a set of linear and quadratic regressions for water-saturated sediments from Lake Erie. Figure 6.31 shows the velocity-density, velocity-porosity, velocity-grain-size plots, and the porosity-mean diameter plot, derived from the published data. As an example of the derived relations the following may be given (with the correlation coefficient R):

- linear relation between porosity and velocity ($R = 0.84367$)

$$v_p = 1917 - 566 \cdot \Phi \qquad (6\text{-}75)$$

- quadratic relation between velocity and porosity ($R = 0.87358$)

$$v_p = 2380 - 2197 \cdot \Phi + 1333 \cdot \Phi^2 \qquad (6\text{-}76)$$

- quadratic relation including grain size in phi-units ($R = 0.88202$)

$$v_p = 2452 - 2269 \cdot \Phi + 1428 \cdot \Phi^2 - 9 \cdot D \qquad (6\text{-}77)$$

Deviating from the above-mentioned general form of the velocity-porosity correlation, Hamdi and Smith (1981) used an exponential relation for the representation of experimental data; instead of porosity, they used the void ratio ε which is related to the porosity as

$$\varepsilon = \Phi/(1 - \Phi) \qquad (6\text{-}78)$$

The relation derived for marine sediments from continental shelf (at uniaxial pressure 96.5 kPa) is

$$v_p = 1.6929 \cdot \varepsilon^{-0.1993} \qquad (6\text{-}79)$$

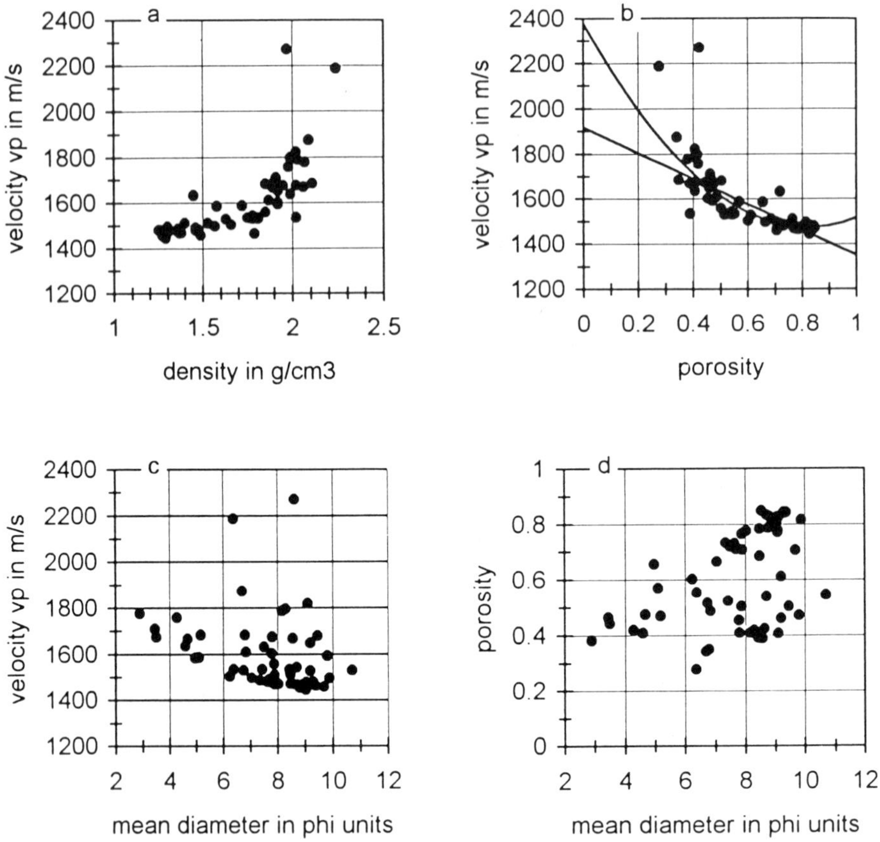

Figure 6.31. Plots of the velocity vs. density (a), vs. porosity (b), vs. median grain diameter (c), and porosity vs. median grain diameter (d) for sediments of Lake Erie; after data from Morgan (1969). The curves in Figure b are equation 6-75 and 6-76

Results of measurements on *synthetic sediments* (glass spheres) are published by Schön (1969). A velocity-porosity graph is shown in Figure 6.61 in connection with theoretical investigations. The graph demonstrates the decrease of longitudinal and transverse wave velocities for vertical and horizontal propagation with increasing porosity in the dry state.

With increasing *clay content*, the velocity generally decreases in unconsolidated sediments. This is the result of the low stiffness of the clay-water aggregates in the sediments. Figure 6.32 shows an example where the data are reasonably well described by a linear relation.

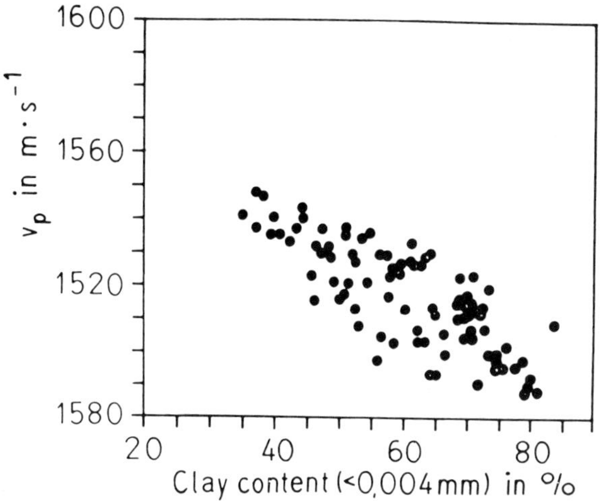

Figure 6.32. Longitudinal wave velocity of water-saturated marine sediments as a function of clay content (clay with particle size > 0.004 mm); after Hamilton, 1970.

The *transverse wave velocity* also decreases with increasing clay content; differences of the clay effect upon the two wave velocities result in a dependence of the v_p/v_s - *ratio* on the clay content (see section 6.3.4.).

Systematic investigations on *boulder clay* by Baermann and Hübner (1984) and Hübner et al. (1985) show the effect of porosity, clay and calcium carbonate content on velocities (Figure 6.33): increasing porosity and clay content result in a decrease of longitudinal wave velocity, whereas, an increasing carbonate content leads to a stabilization of the sediment skeleton and, therefore, to increasing stiffness, and thus to an increasing shear wave velocity.

Dependence of elastic wave velocity on water content

A comparison of Figure 6.30a and Figure 6.30b shows the distinct influence of the elastic properties of the pore fluid, especially upon the *longitudinal wave* velocity. Not only is the velocity itself influenced, but also the character of pressure dependence on the longitudinal wave velocity is controlled by the type of pore fluid (gas or liquid); more about this in the next section of this chapter.

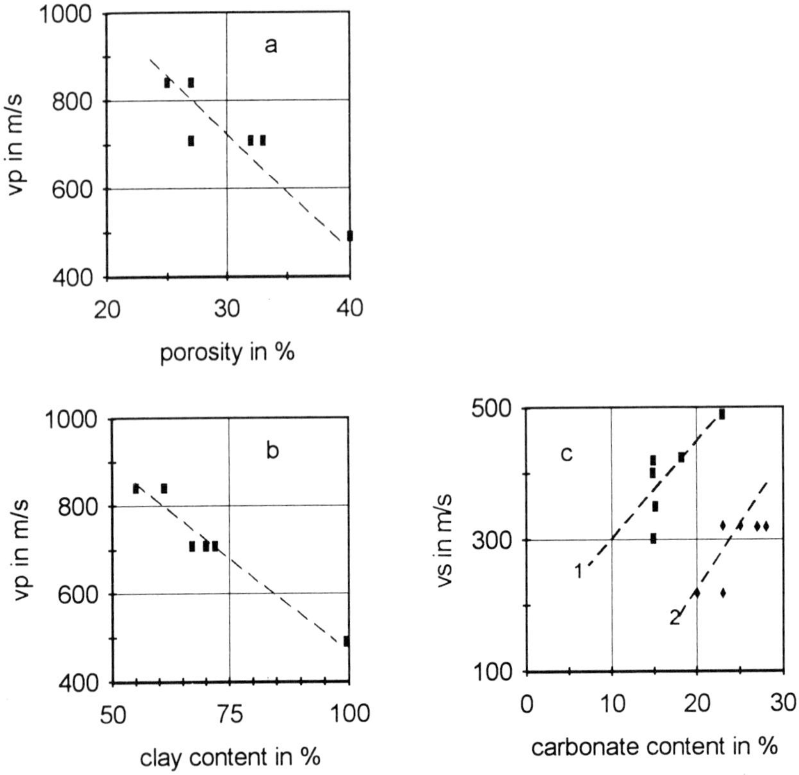

Figure 6.33 Correlation between elastic wave velocities and porosity, clay content and calcium carbonate content for boulder clay (Hübner et al., 1985)

The effects of the pore fluid (particularly the water content) upon the velocity of the consolidated sediment have been described in section 6.3.3.2.2. For unconsolidated sediments these effects are similar. But it seems that for unconsolidated sediments with their "sensitive" *contact* of particles the

- grain boundary effects and
- influence of capillary tension

are of greater importance than for consolidated sediments with their more stable skeleton. In soil mechanics, these effects are well known as 'cohesion' and 'apparent cohesion' (see, e.g., Kezdi, 1976). Studies showing the dependence of velocity on water content have been published, e.g. by Elliott and Wiley (1975), Hübner et al. (1985) and - connected with a detailed analysis of theoretical aspects - by Domenico (1977).

Figure 6.34a illustrates some typical features of the behaviour of velocities on partial saturation:

► the *longitudinal wave velocity* shows only very small changes with saturation increasing to about 80...85%; then the velocity *abruptly increases* to values higher than the velocity in the pore fluid water. Or: starting with the water saturated stage the longitudinal wave velocity drops significantly with the presence of a small amount of gas in the pore fluid (Murphy et al., 1993).

► the *transverse wave velocity* shows only a *small decrease* with increasing water content

Domenico (1977) gave the following comments on his results: "The latter show that compressional-wave velocity decreases gradually with increasing brine saturation, becoming minimum in the vicinity of 90 percent brine saturation and then increasing abruptly and substantially to the highest value at full brine saturation.... Measured shear-wave velocities generally decrease only slightly with increasing brine saturation ...".

The phenomenon can be explained by:

- the *compressibility* of the pore filling mixture (gas and water) is determined up to a high fraction of water (about 80 ...90 % of the pore volume, or S_w = 0.8 ... 0.9) by the very high compressibility of the gas ($c_{gas}/c_{water} \approx 2 \times 10^4$). With full saturation the compression modulus 'jumps' to the value of the water. The non uniform distribution of water and air in the pores of different size and geometry results in a small zone of high velocities about from 90..95 % to 100 % water-saturation.
- the *shear modulus* of the sediment is not influenced by the pore filling if its shear modulus is zero.
- the small decrease of velocities (for longitudinal wave to about 90 percent water saturation, for shear waves over the whole saturation range) are caused by the increase of *density* with increasing water-saturation.

Figure 6.34a shows experimental results from "synthetic" sediments (glass spheres). Both plots in Figure 6.34 confirm the two features described previously. Additionally, there is a small "maximum" of the velocity value at low saturation; this small maximum disappears with increasing differential pressure and increasing grain size (Schön, 1969, 1983). The origin of this effect may be a capillary tension at the contacts (it increases with increasing surface curvature, i.e. decreasing sphere radius) and acts as an additional pressure on the skeleton. This phenomenon is the same as the apparent cohesion of *partial saturated* (fine grained) sands in soil mechanics ... and is one of our first petrophysical childhood experiences from modeling in the sandbox with wet sand.

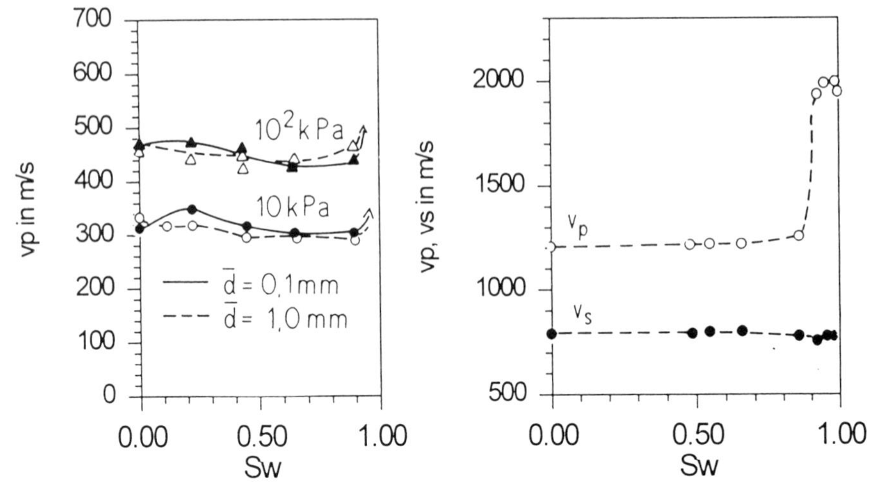

Figure 6.34. Elastic wave velocity as a function of the water saturation
a) longitudinal wave velocity of a glass sphere sediment with two mean diameters under two different effective pressures. Note that only the saturation below about $S_w = 0.95$ is plotted; above 0.95 the velocity "jumps" of values of about 1.7 km/s (Schön, 1983)
b) longitudinal and transverse wave velocity of Ottawa sand at a differential pressure of 1500 psi (= 10.34 MPa), data after Domenico (1977).

Frozen sediments

Frozen unconsolidated sediments, for example under permafrost, show a behaviour more comparable to consolidated sediments; the ice between grains acts like a cement and thus the velocity of longitudinal and transverse wave are distinctly higher than in the non frozen state. Some data from Canadian Arctic permafrost samples shows Table 6.15.

Velocity changes with freezing are discussed in Pandit and King (1978) and Scott et al. (1982). Figure 3.35 shows some examples with the different effect for consolidated and unconsolidated sediments; note that the ice formation in the sediment with its pore size distribution and water salinity is not exactly at 0 °C.

Table 6.15 Seismic wave velocities of permafrost samples from the Mackenzie River Valley and Canadian Arctic islands (selected data from Zimmerman and King, 1986)

Sediment	Porosity	T (°C)	v_p (km/s)	v_s (km/s)	v_p/v_s
Sand	0.36	- 5	4.18	2.42	1.73
	0.36	-15	4.21	2.43	1.73
	0.40	- 5	3.88	2.14	1.81
Silt	0.33	- 5	3.57	1.90	1.88
	0.33	-15	3.86	2.15	1.80
	0.37	- 5	3.38	1.78	1.90
Clay/silt	0.33	- 5	2.92	1.52	1.92
	0.33	-15	3.34	1.74	1.92
	0.44	- 5	2.57	1.15	2.23
Clay	0.37	- 5	2.27	0.91	2.44
	0.37	-15	2.64	1.31	2.02
	0.43	- 5	2.27	0.98	2.32

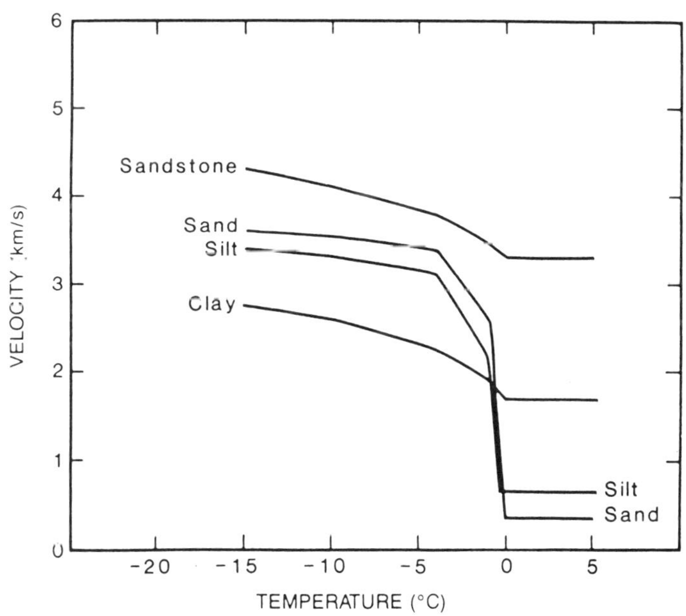

Figure 6.35 Dependence of longitudinal wave velocity on temperature (after measurements of Aptikaev, 1964; see Scott et al., 1982))

Dependence of elastic wave velocity on pressure

Figure 6.36 shows typical characteristics for the velocity-pressure relationship of non-cohesive sediments (when used alone the term "pressure" refers to the "effective pressure"):

- ► increasing velocity with increasing pressure (non-linear relationship),
- ► distinct velocity differences between dry and water-saturated sediment for longitudinal waves,
- ► only small velocity differences between dry and water-saturated sediment for transverse waves.

The dominant physical cause for increasing velocity with increasing pressure is the *change of grain-to-grain contact* conditions. The decrease of porosity alone cannot explain the velocity variation (e.g., during the experiment of Figure 6.37 the porosity decreases only from 38,2 % to 36,7 %, i.e. by 1.5 %).

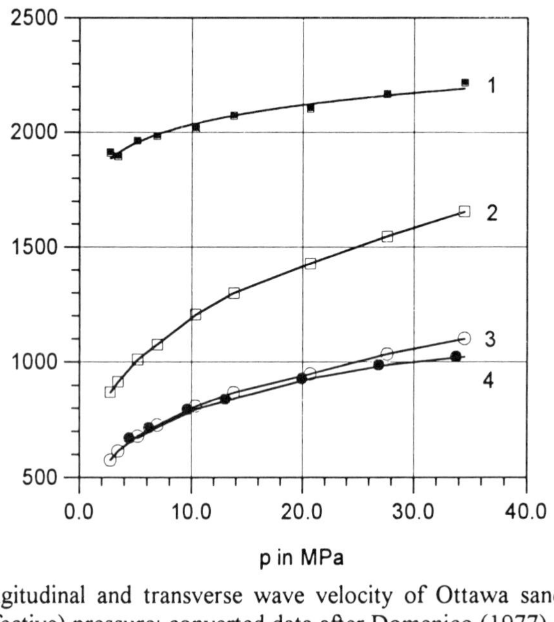

Figure 6.36. Longitudinal and transverse wave velocity of Ottawa sand as a function of differential (or effective) pressure; converted data after Domenico (1977).
1 - longitudinal wave, water-saturated 2 - longitudinal wave, dry
3 - transverse wave, dry 4 - transverse wave, water-saturated

A bi-logarithmic presentation of velocity vs. pressure results in nearly straight lines. Figure 6.37 shows experimental results for glass spheres. The combined presentation of Figure 6.38 illustrate the general validity of the statement. A straight line in a bi-logarithmic plot leads to a relation of the type

$$v = v_0 \cdot \left(\frac{p}{p_0}\right)^m \tag{6-80}$$

where v is the longitudinal or transverse wave velocity, v_0 its value at a pressure p_0, and p the variable effective pressure. v_0, p_0, and the exponent m must determined empirically. They are controlled by individual sediment properties.

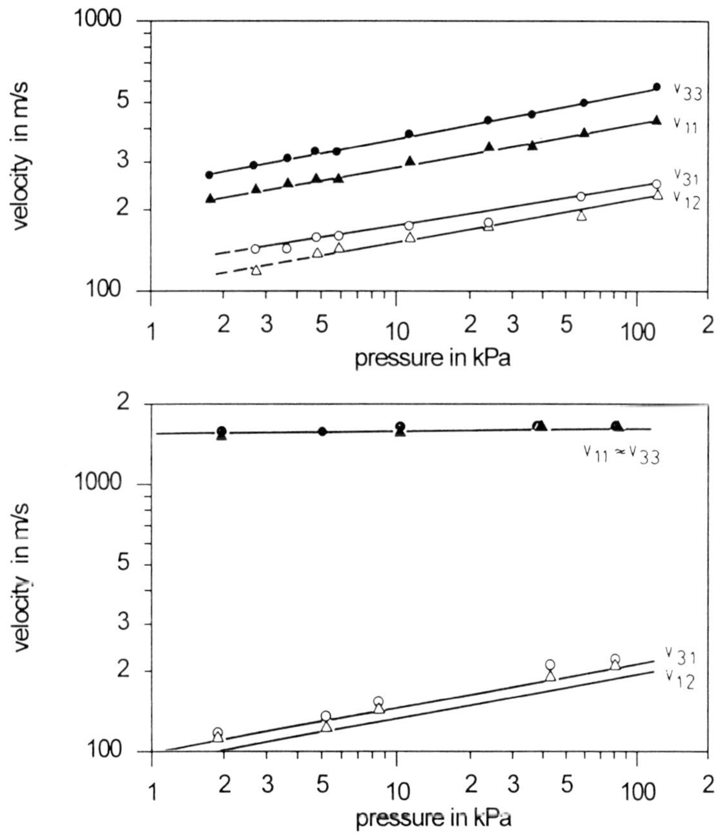

Figure 6.37. Bi-logarithmic presentation velocity vs. pressure; glass spheres (diameter 1.3 mm); a) dry sediment b) water-saturated sediment

v11: P-wave, horizontal propagation v33: P-wave, vertical propagation

v12: S-wave, horizontal propagation, horizontal polarization

v31: S-wave, vertical propagation, horizontal polarization

Table 6.16. contains some data of the exponent m. For longitudinal waves in *dry* sediments and for transverse waves in *dry and water-saturated* sediments, the exponent m lies in the range $1/6 \leq m \leq 1/4$.

Table 6.16. Exponent m in the velocity vs. pressure relation; References: D- Domenico (1977), H - Hunter et al. (1961), S - Schön (1969, 1983), Z - Zareva (1956)

sediment	longitudinal wave		transverse wave		Ref.
	dry	saturated	dry	saturated	
Sand	1/6...1/5				H
Sand	1/6				Z
Ottawa sand	1/4	1/17	1/4	1/4.4	D
Heller sand	1/5...1/4	1/18...1/14	1/5...1/4	1/5...1/4	S
Glass spheres	1/4.1	1/20	1/4.3	1/4.2	D

This relationship is due to the stress-strain properties of the *contact region* (compare with theoretical considerations in section 6.4.4). Therefore, different *grain shapes* result in different contact conditions and thus in different m exponents. Figure 6.39 shows that m is lower for angular grains (e.g., broken rocks) and is higher for well rounded grains. The longitudinal wave in *water saturated* sediments results $m \approx 1/20...1/14$. This is caused by the dominant influence of the compressional modulus of water, which controls the range and the character of the dependence.

Hübner et al. (1985) have achieved comparable results for sand and boulder clay: the effective pressure has a strong influence on the grain-boundary contacts; the longitudinal wave velocity in dry and partially saturated sediments, and the transverse wave velocity in dry, partially- and fully-saturated sediments follow a relationship of the type of equation (6-80). In saturated sediments, the longitudinal wave velocity has a fairly constant value for the low pressure conditions of near-surface layers.

One can summarize:

- ► the longitudinal wave velocity in the dry state is determined by the grain frame properties; the sediment has a "*frame determined*" (or skeleton-determined) behaviour,
- ► the longitudinal wave velocity in the water saturated state is determined by pore material properties; it has a "*pore-determined*" behaviour,
- ► the transverse wave velocity is determined, in both states, by the rock skeleton (frame) properties and has generally a "*frame-determined*" behaviour.

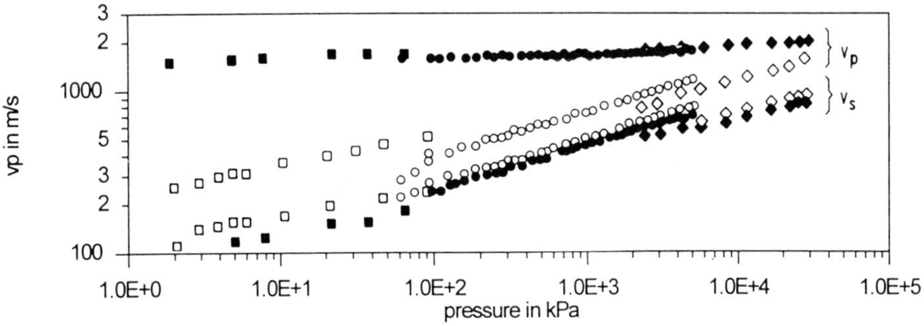

Figure 6.38. Combined presentation of the velocity vs. pressure dependence;

pressure range: 10^0 to 10^2 kPa - glass spheres

pressure range: 7×10^1 to 5×10^3 kPa - sand (Darss/East Sea)

pressure range: 2.7×10^3 to 35×10^4 kPa - Ottawa sand (Domenico, 1977)

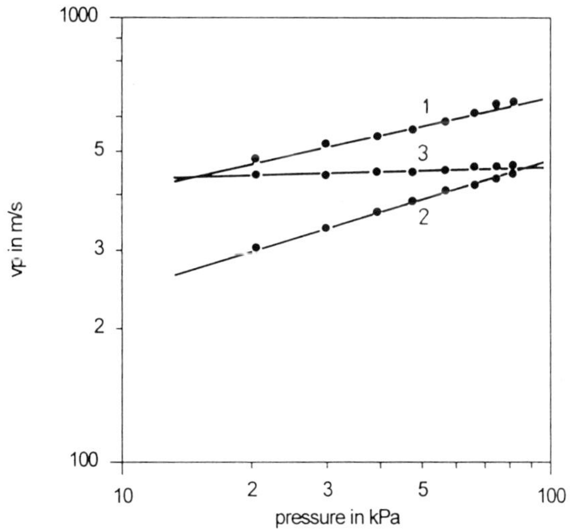

Figure 6.39. Longitudinal wave velocity as a function of pressure for dry sediments with different grain shape

1 - Glass spheres : mean diameter 1.3 mm; $\Phi = 0.40$; $m = 1/5$

2 - Sand: mean diameter 1.4 mm; $\Phi = 0.42$; $m = 1/3.6$

3 - Quartzite, angular: mean diameter 4.5 mm; $\Phi = 0.47$; $m = 1/40$

The *velocity-depth relationships* reflect the pressure dependency. Table 6.17 shows some results, published by Hamilton (1976). It contains also some data for cohesive sediments.

Table 6.17. Velocity-depth-relations for water-saturated marine sediments after Hamilton, 1976 (velocity in m/s; depth z in m)

sediment	depth range (m)	longitudinal wave	transverse wave
Fine sand	0...20	$v_p = 1681\, z^{0.013}$	$v_s = 120\, z^{0.28}$
Coarse sand	0...20	$v_p = 1869\, z^{0.016}$	$v_s = 127\, z^{0.31}$
Silt, clay	0...36		$v_s = 116 + 4.65\, z$
	36...120		$v_s = 237 + 1.28\, z$
	120...650		$v_s = 322 + 0.58\, z$

The velocity-pressure or velocity-depth dependence of both cohesive and non-cohesive sediments could be described by an expression of the form

$$v = v_1 + v_2 \cdot \left(\frac{z}{z_0}\right)^m \qquad (6\text{-}81)$$

with $v_1 \to 0$ $m \leq 1/3$ for non-cohesive sediments
$v_1 \geq v_2$ $m \approx 1$ for cohesive sediments.

The pressure independent velocity term v_1 for cohesive sediments has the same physical origin as the cohesion in Coulomb's law (see section 10.3.4).

The remarks of Domenico (1977) are important for the comparison of velocity-pressure and velocity-depth dependence: "The increase in compressional-wave velocity ... with implied depth of burial in a sedimentary basin (the depth being proportional to differential pressure) is at a much lower rate than that of in-situ sand reservoir velocity measurements. This is attributable to rapidly increasing cementation and corresponding reduction in porosity with depth occuring in natural sand reservoirs."

6.3.4. Some remarks about velocity ratios - the v_p/v_s-ratio and elastic anisotropy

"The orientation of in-situ stresses, the relative magnitude of these stresses, and estimates of permeability and fluid conductivity can be derived from seismic anisotropy. Rock properties, including lithology and porosity, can be obtained from comparative traveltimes of P and S waves. Fluid properties may be determined from comparative P and S traveltime and attenuation measurements." (T. L. Davis, 1995)

6.3.4.1. General remarks

Most of experimental data and dependencies described are related to longitudinal waves. In the last few years - as a result of technological advance - more transverse wave velocity data have become available in seismic exploration and in well logging. The transverse wave velocity itself and the *velocity ratio* v_p/v_s now seem to be useful parameters for rock characterization. Han, Nur and Morgan (1986) note that "the combined use of the velocity and velocity ratio [provide] a useful tool for reliable lithology discrimination."

A second type of velocity ratio is of growing interest and importance - the *anisotropy* of elastic waves. Fundamental papers by Bruggeman (1937), Helbig (1956, 1983), Backus (1962), Postma (1955), White (1983), Thomsen (1986), Schoenberg (1994) et al. discuss the layer-induced anisotropy, its consequences for the interpretation of seismic data and the possibility to determine the internal structure of the rocks. Observationally, "shear wave splitting" is a sensitive indicator of elastic anisotropy and, thus, has a special place in rock physics (see for, example Crampin and Lovell, 1991). Experimental studies, for example by Lo et al. (1986), and theoretical considerations (e.g., Schön, 1983) show a connection between both types of velocity ratios, which is controlled by the internal structure and bonding properties of the rock (see also section 6.4.8.).

Although comments about v_p/v_s-ratio and anisotropy have been given in the foregoing chapters, it may be useful to discuss some general results at this point.

6.3.4.2. The v_p / v_s ratio

Figure 6.40 summarizes data of the v_p/v_s ratio as a function of the transverse wave velocity after Molotova and Wassiljew (1960) and illustrates a general tendency: With decreasing transverse wave velocity, the v_p/v_s ratio increases. This indicates decreasing mechanical strength and compactness, as well as increasing porosity and fracturing. Obviously, this implies an increasing difference between the two elastic wave velocities.

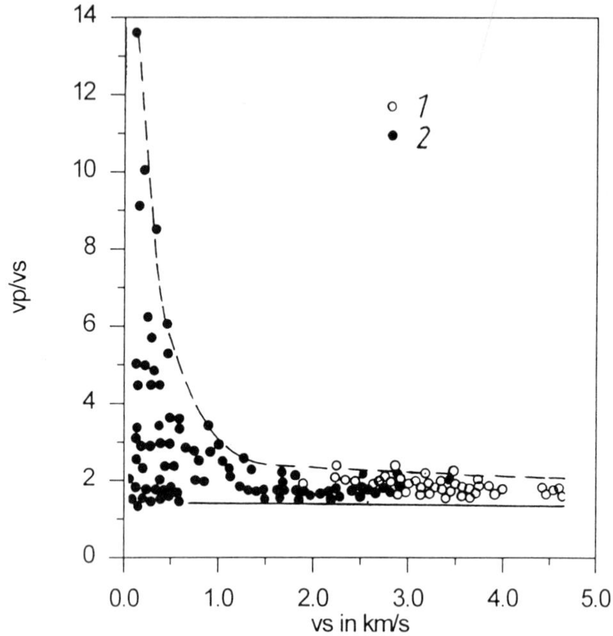

Figure 6.40. The ratio of longitudinal and transverse wave velocity as a function of the transverse wave velocity; after Molotova and Wassiliev (1960).
1 - magmatic and metamorphic rocks 2- sedimentary rocks

Examples of the dependence on different pressures of the two waves and thus a pressure dependent v_p/v_s ratio were given, e.g., by Figure 6.12 and more distinctly in Figure 6.14 for igneous rocks. The different behaviour of the two waves under the influence of the tectonic stress field and the fracture mechanisms may be the cause of observed anomalies in the v_p/v_s ratio in connection with earthquakes (e.g., Aggarwal et al. 1973; see also chapter 10.).

The following references give a small selection from the extensive literature about the v_p/v_s ratio: Pickett, 1963; Gregory, 1977; Eastwood and Castagna,

1983; Domenico, 1984; Castagna et al., 1985; Wilkens et al., 1984; Stümpel et. al. 1984; Tatham, 1985; Han et al., 1986; Hornby and Murphy, 1987; Wang, 1990; Mavko and Jizba, 1990; Marion and Nur, 1990; Tatham and McCormack, 1991; Bunch and Dromgoole, 1995. These papers contain many experimental data and empirical relations. They contain information about the v_p/v_s ratio with respect to

- lithology, facies,
- pore content, especially the presence of gas,
- mechanical properties.

In the following section, we look at some typical relations that concentrate on the first two aspects and on sedimentary rocks. After consolidated rocks are considered, unconsolidated terrestrial and marine sediments are discussed. As an introduction Table 6.18 illustrates the effect of pore fluid and fracturing.

Table 6.18 Mean v_p/v_s values, Cretaceous Fruitland formation (coal bed); Cedar Hill Field, New Mexico (Blott, 1993)

Saturation	Fracturing	v_p/v_s
Air saturated	little or no fracturing	1.81
	considerable fracturing	1.91
Brine saturated	little or no fracturing	1.98
	considerable fracturing	2.40

The distinct difference of ratios for *gas- and water-saturated* (or oil-bearing) porous rocks has an important consequence for the practice of hydrocarbon exploration: The v_p/v_s ratio value is much lower (10...20 % after Tatham, 1982) for gas saturation than for liquid saturation. The physical cause is the strong influence of water (or oil) on the bulk modulus (and longitudinal wave) and the negligible influence on the shear modulus (and transverse wave) - see section 6.3.3.2..

The classic paper by Pickett (1963) recommended the v_p/v_s ratio as a *lithology indicator*. He used the transit times Δt_P and Δt_S, which are the reciprocal velocities (see Figure 6.41).

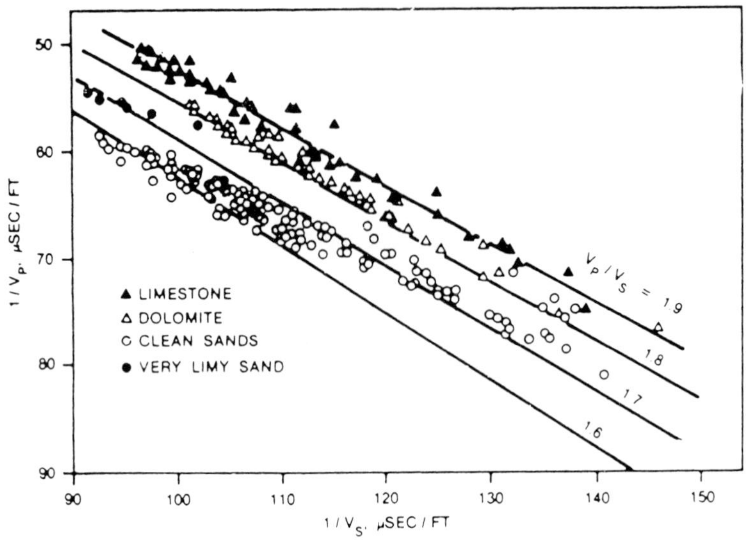

Figure 6.41. Results of laboratory measurements of longitudinal and transversal wave velocities for three sedimentary rock types ("Pickett diagram"); after Pickett (1963), Figure taken from Tatham (1985). Lines of constant slope represent constant velocity ratio.

Such a linear relation between longitudinal and transverse velocities is often approximately valid for one type of sedimentary rocks. Table 6.19 contains some typical ranges for the v_p/v_s ratio in addition to Figure 6.41.

Table 6.19 Ranges of the v_p/v_s ratio for typical sedimentary rocks after Pickett, 1963 and Domenico, 1984

Sediment	v_p/v_s (Pickett)	v_p/v_s (Domenico)
Sandstone	1.65 ... 1.75	1.59 ... 1.76
Calcareous sandstone		1.67 ... 1.76
Dolomite	1.75 ...1.85	1.78 ... 1.84
Limestone	1.85 ...1.95	1.84 ... 1.99
Shale		1.70 ... 3.00

For *mudrocks* (water-saturated clastic silicate sedimentary rocks, composed primarily of clay- and silt-sized particles), Castagna, Batzle and Eastwood (1985) found a relationship from in-situ sonic and seismic field measurements:

$$v_p = 1.16\, v_s + 1.36 \qquad (6\text{-}82)$$

where the velocities are in km/s. A comparable relation is published by Han, Nur and Morgan (1986) based on 75 sandstone samples (porosity = 0.02...0.30; clay content = 0....0.5):

$$v_p = 1.26\, v_s + 1.07 \qquad (6\text{-}83)$$

with a correlation coefficient of 0.97.

The influence of matrix composition upon the v_p/v_s ratio is discussed in a study by Wilkens, Simmons and Caruso (1984). For siliceous *limestones* with varying proportions of carbonate and silica, the values are measured at a pressure of up to 2 kbar (= 0.2 MPa). Figure 6.42 shows the velocities v_p and v_s, and the v_p/v_s ratio as function of the carbonate content: v_p increases and v_s decreases with growing carbonate content. This corresponds to the values for single crystals of quartz and calcite:

$v_{p,calcite} = 6.53$ km/s $\qquad v_{p,quartz} = 6.05$ km/s

$v_{s,calcite} = 3.36$ km/s $\qquad v_{s,quartz} = 4.09$ km/s.

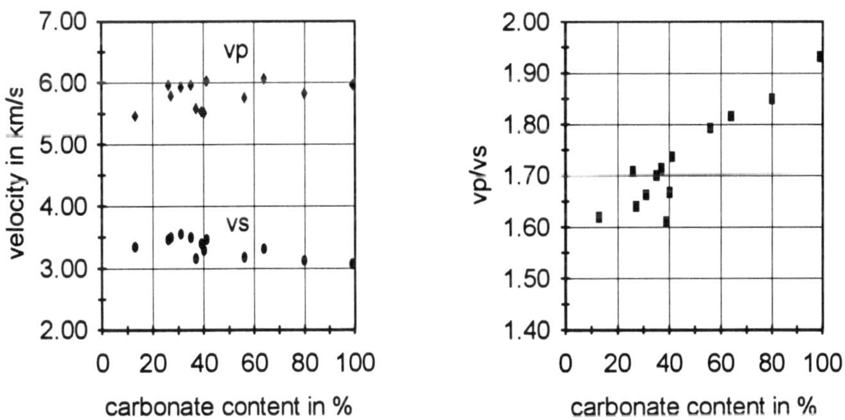

Figure 6.42. v_p and v_s, and the ratio v_p/v_s of siliceous limestone for a pressure of 1 kbar (= 100 MPa) as function of the carbonate content (in wt %); dry rock. Data after Wilkens et al. (1984).

It is remarkable that the correlation between the *velocity ratio* and carbonate content is better than that between individual velocities and carbonate content. Thus, v_p/v_s is a reasonable estimator of rock composition (Wilkens et al. 1984).

For *sandstones*, Han, Nur and Morgan (1986) investigated the effect of *porosity and clay* content on velocities (see section 6.3.3.2) and the velocity ratio. It is important that both - porosity and clay content - are analyzed for their influence on v_p/v_s. Figure 6.43 shows the velocity ratio versus clay content of the sandstone types with different porosities. The porosity was determined for each sample, separately.

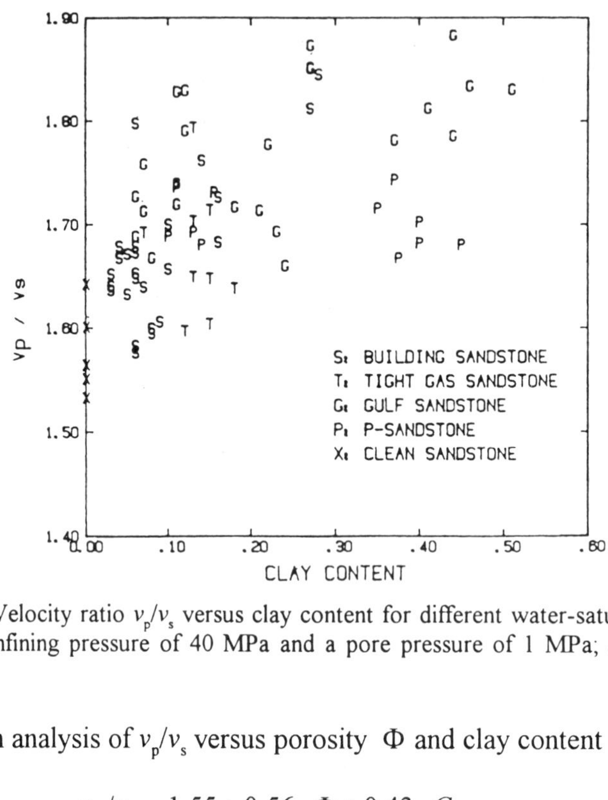

Figure 6.43. Velocity ratio v_p/v_s versus clay content for different water-saturated sandstone types at a confining pressure of 40 MPa and a pore pressure of 1 MPa; after Han et al., 1986.

A regression analysis of v_p/v_s versus porosity Φ and clay content C yields

$$v_p/v_s = 1.55 + 0.56 \cdot \Phi + 0.43 \cdot C \tag{6-84}$$

for the water saturated samples (correlation coefficient 0.7).

The physical interpretation of this increase of the ratio with increasing porosity and clay content is given by the authors as follows: "Generally the effects of porosity and clay content on shear velocity v_s are larger than on the compressional velocity v_p. Thus, a sample with high porosity and clay content tends to have a high v_p/v_s ratio." Because the interaction between clay minerals also influences elastic moduli with differing response of shear and bulk moduli we find a difference between the ratio for dry and water-saturated sediments. The difference increases with the porosity and clay content. Han, Nur and Morgan (1986) developed the following regression

$$(v_p/v_s)_{saturated} - (v_p/v_s)_{dry} = 0.018 + 0.36 \cdot \Phi + 0.47 \cdot C_c \qquad (6\text{-}85)$$

The combined effect of porosity and clay content was investigated also by Castagna et al. (1985), Tosaya and Nur (1982) and Hornby and Murphy (1987). The last-named authors note that the clay affects the moduli, velocities and the ratios only, when the "clay is an integral part of the rock frame".

The clay content affects also the velocity ratio v_p/v_s of *unconsolidated sediments*. The ratio decreases with increasing transverse wave velocity (v_{SH}), particularly for boulder clay and saturated sand, and somewhat less for dry and wet sand (Baermann and Hübner, 1984). Increasing transverse wave velocity reflects an increase of compaction and decrease of porosity. Figure 6.44 shows a v_p/v_{SH} ratio vs. v_{SH} plot. Since

- ➢ the shear wave velocity depends on the skeleton properties and is unaffected by the pore fluid properties, and
- ➢ the compressional wave velocity depends on the skeleton properties (expressed by the shear wave velocity) and the pore fluid properties, a simple relation results

$$v_p = a + b \cdot v_{SH} \qquad (6\text{-}86)$$

or

$$v_p/v_{SH} = (a/v_{SH}) + b \qquad (6\text{-}87)$$

The empirical parameter a expresses the contribution of the pore fluid to the compressional properties of the rock. For high compaction (extremely high values of v_{SH}), the ratio v_p/v_{SH} tends towards the upper limit of about 1.4, and thus follows

$$v_p/v_{SH} = (a/v_{SH}) + 1.4 \qquad (6\text{-}87)$$

The curves in Figure 6.44 are calculated for different values of the parameters a. It is obvious that this parameter is much smaller for dry or wet sand than for a water-saturated sediment. The tendency of this dependency also corresponds to theoretical curves for a structured rock model (section 6.4.8, Figure 6.73).

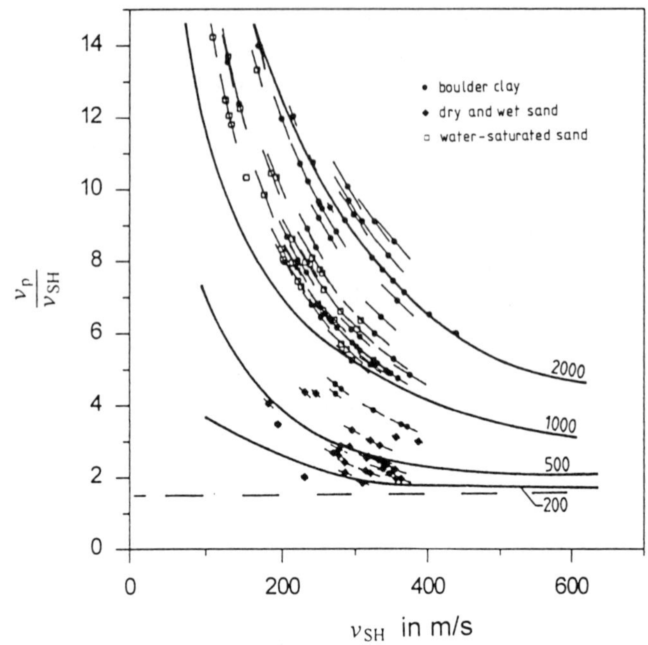

Figure 6.44. Relation between the ratio v_p/v_{SH} and the velocity v_{SH} for unconsolidated sediments (velocity values normalized to a depth of 4 m) after Baermann and Hübner (1984); Curves are calculated with equation (6-87) with different parameters *a*.

The *effect of compaction* (or depth) on the velocity ratio is very clear for unconsolidated marine sediments. Hamilton (1979) has derived the velocity-depth relationship for compressional and shear waves in marine sediments and the resulting depth dependence of the velocity ratio. The individual equations are given with Table 6.17. For example, the relation for fine sand is

$$v_p/v_s = 14 \cdot z^{-0.27} \tag{6-88}$$

The decrease of the ratio with increasing depth is caused mainly by the increase of the transverse wave velocity.

6.3.4.3. Anisotropy of elastic wave velocities

"Anisotropy of wave propagation means primarily that the velocity depends on direction. We have grown accustomed to assuming that most media are isotropic, but this assumption has its root often more in the ensuing convenience than in reasoning or observed fact. ... the majority of solids, and certainly most geologic media, have preferred directions.
For geologic media these preferred directions are either inherited from their mode of emplacement or dictated by external conditions". *Helbig, K. (1993).*

In elastically anisotropic rocks the velocity of wave propagation depends on the direction of propagation. Obvious anisotropy exists in sedimentary rocks with layering (bedding) and schists.
Crampin and Lovell (1991) list five possible sources of seismic anisotropy:

(1) aligned crystals;
(2) direct stress-induced anisotropy;
(3) lithologic anisotropy (for example, aligned grains);
(4) structural anisotropy (for example, fine layering), and
(5) stress aligned crack-induced anisotropy.

A *complete description* of the elastic anisotropy of a rock requires more than the standard measurements in two perpendicular directions. Techniques with polyhedral (Giesel, 1963) or with spherical samples (Pros, 1977; Jahns et al., 1994) give a more detailed picture and allow the study of the correlation between anisotropic properties and the internal rock structure. Results were presented as stereographic projections with isolines for velocity. The example in Figure 6.45 demonstrates the complicated distribution and the pressure effect for granodiorite; note that the principal orientation is virtually unaffected by high hydrostatic pressure.

In many cases, *transverse isotropy* (or hexagonal symmetry) is assumed. In this symmetry, the elastic properties are equal in the x-y plane (plane of stratification or schistosity planes), but vary with the angle from the vertical z-axis. In transversely isotropic rocks, elastic properties are governed by five elastic constants (see Table 6.1).

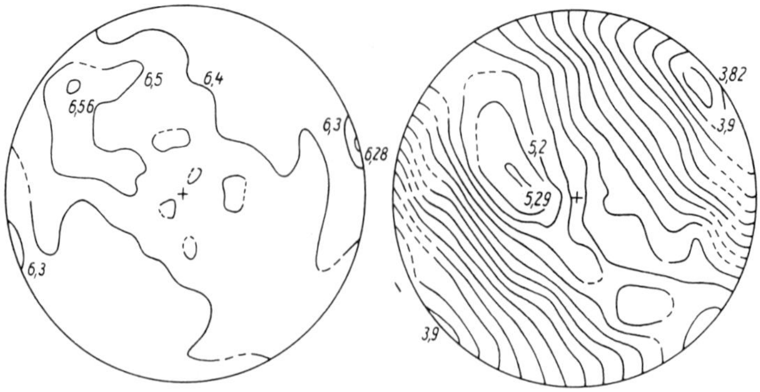

Figure 6.45. Stereographic presentation of anisotropy investigations using spherical samples (granodiorite) after Pros (1977); the isolines are given for longitudinal wave velocity in km/s
a) pressure 300 MPa (anisotropy coefficient = 0.044)
b) pressure 0.1 MPa (anisotropy coefficient = 0.36).

Table 6.20 gives the computed elastic constants for a relatively homogeneous paragneiss sequence of the KTB-pilot hole. It summarizes field and laboratory measurements and various equations for calculation. As elastic property the ratio of the c_{ik} term and density (= velocity^2) was used. The sample has a distinct foliation and weak lineation. A comparison of result b) and c) shows that the sample is not perfect transverse isotropic. There is a large difference between VSP- and laboratory data (the main reasons are: sampling of crack-free cores, different pressure and saturation conditions, Jahns et al., 1994).

Frequently, the anisotropy is *described simply*, using either the maximum and minimum measured values (v_{max}, v_{min}) or the values parallel and perpendicular to the stratification, foliation or main crack orientation ($v_{//}, v_{\perp}$). We apply the following definitions:

- anisotropy coefficient $A_v = \frac{v_{max} - v_{min}}{v_{min}}$ (6-90)

- anisotropy ratio $A_v^* = \frac{v_{max}}{v_{min}} = A_v + 1$ (6-91)

Table 6.20 Elastic constants computed for a paragneiss (KTB-pilot hole), divided by density in 10^6 $(m/s)^2$; Jahns et al., 1994.

a) from VSP data set, segment 2.0 ... 3.2 km depth, (Rabbel, 1992) using equations given by Christensen and Crosson (1968)	38.44	10.16	14.20	0	0	0
		38.44	14.20	0	0	0
			37.21	0	0	0
				10.89	0	0
					10.89	0
						14.36
b) from laboratory data set (sample from depth 3145 m) assuming a hexagonal type of symmetry (Christensen and Crosson, 1968)	42.48	12.84	13.40	0	0	0
		42.48	13.40	0	0	0
			32.59	0	0	0
				10.89	0	0
					10.89	0
						14.82
c) from laboratory data set (sample from depth 3145 m) without assuming any rock symmetry using equations given by Klima, 1973	41.27	11.09	14.00	0.01	0.16	0.12
		43.73	14.14	0.07	0.01	-0.26
			33.28	0.87	0.55	-0.01
				10.14	-0.60	0.04
					10.98	0.11
						15.41

Table 6.21 contains a collection of values from published data. A comprehensive study with much anisotropy data under atmospheric and high pressures was published by Bajuk et al. (1982). There are two significant tendencies

- ► among the magmatic and metamorphic rocks on this list, gneiss and schist have the highest values,
- ► for sedimentary rocks, the anisotropy increases from sandstones to siltstones and shale.

In the following, only some selected aspects will be discussed, particularly the combined influence of stress and internal crack geometry and the phenomenon of acoustic bifrigerence (or shear wave splitting).

Nur and Simmons (1969) presented a study of *stress-induced velocity anisotropy* in rocks connected with results of laboratory measurements of velocities in a granite rock sample under uniaxial stress. They summarized their concept: "The effect of an applied nonhydrostatic stress is to close cracks in some directions and leave cracks open in other directions. The stress necessary to close a penny-shaped crack is proportional to its aspect ratio (Walsh, 1965);

the narrower cracks close at lower pressure. The effective elastic constants of a solid with cracks are determined by the distribution of orientations of open cracks. If cracks are closed in some directions but open in others, rocks that are intrinsically isotropic show a directional dependence of the effective elastic constants and are, in general, anisotropic. The elastic properties of a rock that initially have a random crack distribution could possibly remain isotropic under hydrostatic pressure, attain axial symmetry under uniaxial stress, and attain orthorhombic symmetry under three different principal stresses." The results of their experiments on Barre granite show that the compressional and two shear wave velocities are dependent on stress and direction related to the closure mechanism of cracks.

Table 6.21. Anisotropy parameters of elastic wave velocity under atmospheric conditions
References: B - Bajuk et al., 1982; D - Dortman, 1976; H - calculated using data from Hurtig, 1966, C - Cesnokov, 1977; Ba - Babuska, 1977; J - Jituboh et al.,1991 (value range or mean value; individual values in brackets [])

Rock type	A_v	Wave	Ref.
Granite, biotitic	[0.11]	P	D
, gneissic	[0.39]	P	D
Gabbro	..0.04	P	B
Diabase	...0.03	P	B
Biotite-gneiss/Saxony	0.08...0.51	P	H
Gneiss	0.02...0.32	P	C
	0.12...0.38	S	C
Gneiss, biotitic/ Kola well	0.04...0.19	P	B
	0.04...0.19	S	B
Schist	0.13...0.48	P	Ba
Marble	0.05...0.23	P	C
Limestone	...0.11	P	B
Sandstone	...0.06	P	B
Sandstone/England, Carboniferous	[0.09]	P	J
	[0.05]	S	J
Sandstone/ Germany, Mesozoic	0.12	P	H
Sandstone/ Germany, Paleozoic	0.20	P	H
Siltstone/ England, Carboniferous	[0.21]	P	J
	[0.12]	S	J
Siltstone/ Germany, Paleozoic	0.27	P	H
Shale/Germany, Jurassic	0.48	P	H
Claystone/ England, Carboniferous	0.12...0.31	P	J
	0.20...0.49	S	J

The *influence of cracks and pressure* on anisotropy is also shown in Figure 6.46. It demonstrates that the effect does not depend on temperature. Due to the closure of cracks, there is a strong decrease of anisotropy in the pressure range up to about 200 MPa. In the higher pressure range, the anisotropy can be considered as the *intrinsic* anisotropy of the nearly crack-free rock.

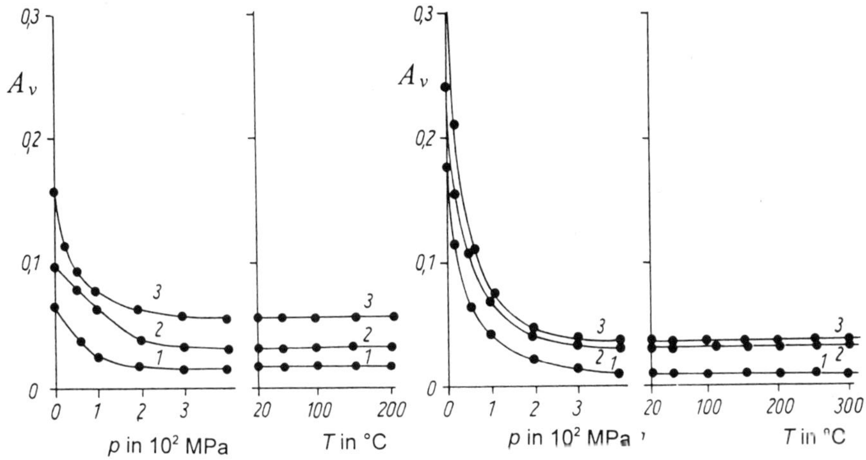

Figure 6.46. Anisotropy coefficient versus hydrostatic pressure and temperature for longitudinal (left) and transverse (right) wave velocities; after data from Bajuk and Tedev (1978); 1 - schist 2, 3 - gneisses.

Comparable results are published by Gebrande (1982) who considers the influence of water saturation on the anisotropy of cracked rocks: the longitudinal wave velocity increases and the anisotropy of longitudinal wave velocity decreases when cracks are water saturated. Transverse wave velocities and anisotropy of transverse velocity are almost independent of water saturation.

Illustrative studies about the influence of stress field, crack distribution and orientation and the intrinsic texture (orientation of fabric elements) on the elastic anisotropy of magmatic and metamorphic rocks have been published in connection with the KTB investigations. An example is Figure 6.47. The figure shows the velocity of longitudinal waves for a paragneiss sample from a depth of 1757 m in different horizontal directions under hydrostatic pressure of 5 MPa, 20 MPa, 60 MPa, and 350 MPa. The anisotropy *decreases with increasing hydrostatic pressure*. Zang et al. (1989) have used the data to "separate crack-caused and textural anisotropy".

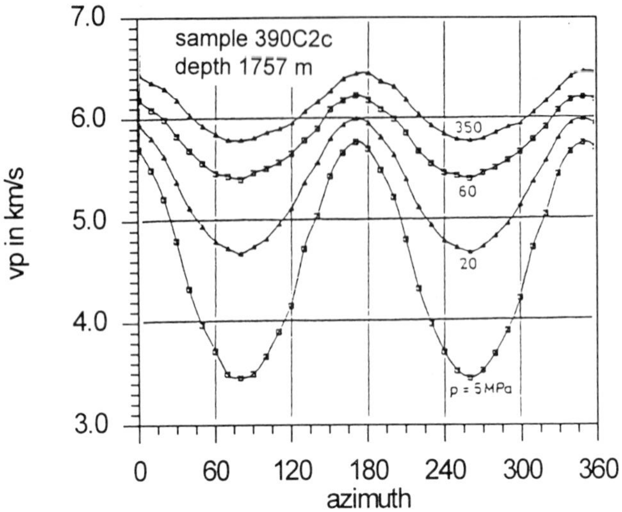

Figure 6.47. Horizontal transmission of longitudinal waves through a gneiss sample (KTB-pilot hole, sample No. 390C2c) for four selected pressure levels; after Zang et al., 1989

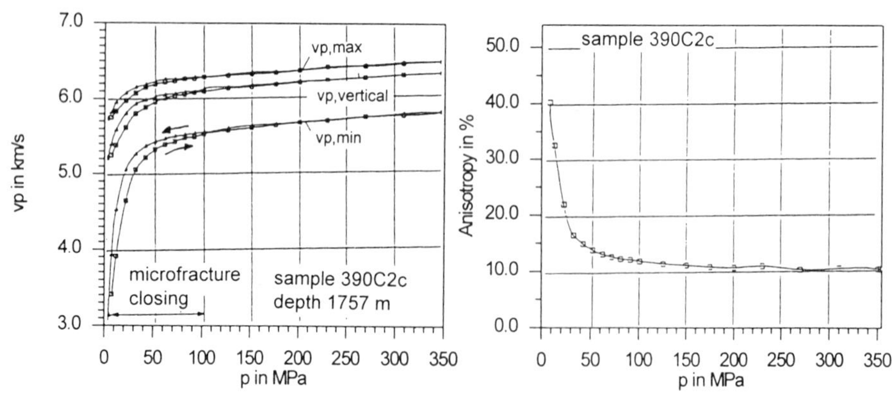

Figure 6.48. Pressure dependence of longitudinal wave velocity and anisotropy, determined on a gneiss sample (KTB-pilot hole, sample No.390C2c); after Zang et al. (1989)

a) pressure dependence of the maximal and minimal horizontal velocity and the vertical velocity value. The strongly increasing velocities below 100 MPa are due to microcrack closure;

b) pressure dependence of the horizontal v_p-anisotropy. It shows a 40 % total anisotropy, of which 30 % is caused by cracks.

For this purpose the extreme values of the horizontal velocities (maximum and minimum) at all pressure steps were plotted in a velocity vs. pressure diagram (Figure 6.48 - left is identical to Figure 6.13) and from this the so-called "horizontal v_p-anisotropy" was calculated:

$$Anisotropy \text{ in } \% = \frac{v_{max} - v_{min}}{v_{min}} \cdot 100\% = A_v \cdot 100\% \qquad (6\text{-}92)$$

The result is plotted in Figure 6.48 (right) as function of pressure. "From this curve a 30 % crack anisotropy and 10 % textural anisotropy may be deduced, because crack anisotropy disappears with increasing pressure and the remaining anisotropy at 350 MPa is restricted to a textural one." (Zang et al., 1989). Using the same method other samples were also investigated (Table 6.22).

Table 6.22. Crack and textural anisotropy of samples from the KTB pilot borehole; after Zang et al. (1989)

Rock type	*n*	Anisotropy in %		
		total	crack	textural
Amphibolite	4	10	7.8	2.2
Gneiss	9	33	23	10
Lamprophyre	2	≈ 4.8	≈ 4	≈ 0.8
Marble	1	7	≈ 7	0

Siegesmund and Vollbrecht (1991) have studied the correlation between texture, microcrack fabrics, and seismic anisotropy. Figure 6.49 shows the pressure dependence of P- and S-waves in the three orthogonal directions, and the anisotropy coefficient (in %) for P- and S-wave of an amphibolithe sample. The same sample was used to show the influence of texture and microcracks on shear wave splitting (Figure 6.51).

The study of the shear wave anisotropy has created a special field of seismic research: the investigation and application of the phenomenon of *shear wave splitting* (acoustic bifrigerence). During the last decade many results were published about the physical background, laboratory and field techniques, and applications. The state-of-the-art is summarized in the Proceedings of the Fourth International Workshop on Seismic Anisotropy (Edinburgh, 1990), in a special issue of the Geophysical Journal International.

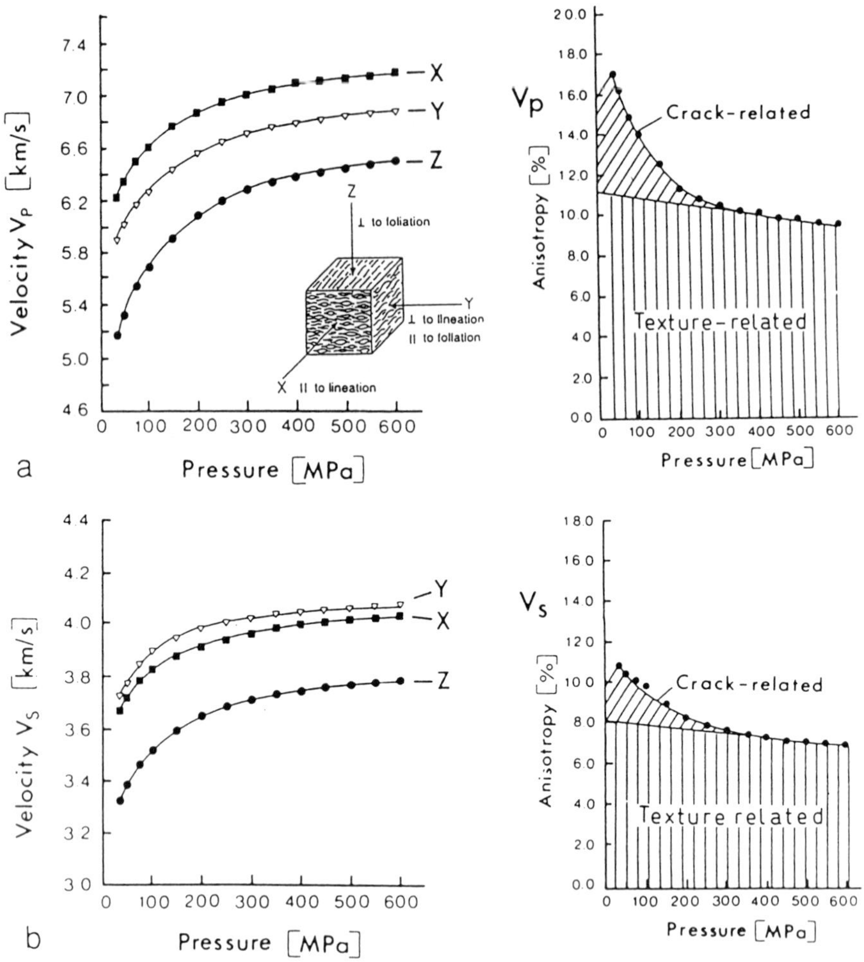

Figure 6.49 Directional dependence of P- and S-wave velocities in the three orthogonal directions as a function of pressure (left) and derived texture- and crack-related anisotropy of P- and S-wave velocities (right); Clinopyroxene amphibolite (Western Alps, Italy), after Siegesmund and Vollbrecht (1991)

The phenomenon of shear wave splitting is schematically illustrated in Figure 6.50. The parallel fractures (or stress-aligned cracks, microcracks, and preferentially oriented pore space) result in two different shear moduli and two different shear wave velocities for the polarization parallel and perpendicular to the fracture plane, repectively. A shear wave traveling through the rock, therefore, will split into two waves with orthogonal polarization: the wave with higher velocity is polarized parallel to the fracture strike and the wave with lower velocity is polarized perpendicular to the fracture strike. The phenomenon is similar to optical bifrigerence.

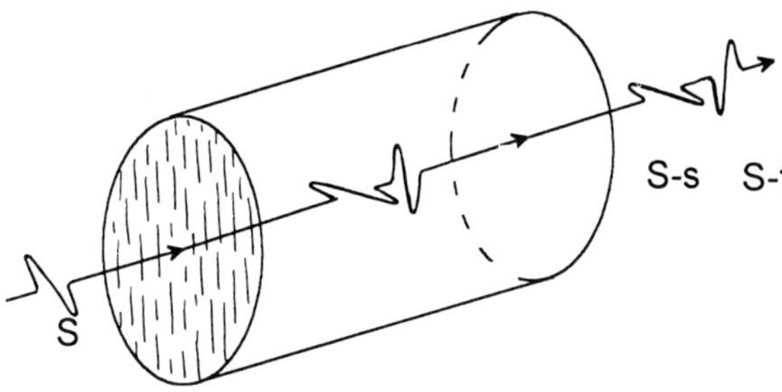

Fig.6.50. Schematic illustration of the shear wave splitting in a fractured rock. The wave propagating along the fractures is splitted into a fast wave polarized parallel and a slow wave polarized perpendicular to the fracture strike.

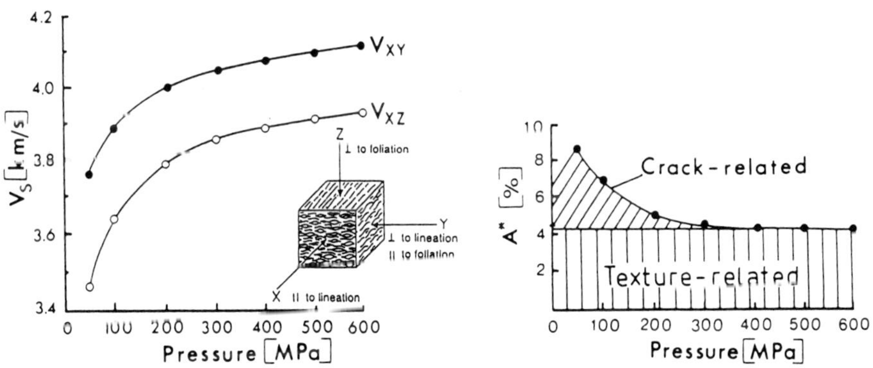

Figure 6.51 Shear-wave splitting along the X-direction (S-waves are polarized to the XY- and XZ-plane) versus pressure at room temperature and respective shear-wave splitting anisotropy related to texture and crack fabrics; Clinopyroxene amphibolite (Western Alps, Italy), after Siegesmund and Vollbrecht (1991)

Crampin and Lovell (1991) list the following potential *applications* of shear wave splitting:

- understanding the geometry of fluid-filled inclusions,
- applications to hydrocarbon production and other industrial applications (e.g. , locating sub-surface fractures, estimating the orientation of maximum compressional stress and of hydraulic fractures),

► more speculative applications (such as monitoring stress before earthquakes and rockbursts.

In addition to the applications described in the Proceedings some further aspects are:

- the application in connection with hot-dry rock experiments (e.g., Crampin, 1988, Crampin and Booth, 1989),
- the prediction of fracture direction - including a brief discussion of the applied techniques (Yale and Sprunt, 1989),
- observations of shear-wave splitting from mining induced stress-failures (Graham and Crampin, 1991),
- correlation between shear wave bifrigerence and directional permeability in fractured rocks (Xu and King, 1989).
- variation of reflection coefficients with crack strike and density in anisotropic media (Li and Crampin, 1992)
- shear-wave splitting studies for tectonic problems (Mjelde, 1991).

For *sediments*, White et al. (1983) have analysed the elastic behaviour of Pierre shale based on seismic field measurements. Seismic velocities measured under different directions are used to determine the anisotropy coefficients for longitudinal and transverse waves that the shale is transversely isotropic. Using the authors data, the anisotropy ratios $A_p^* = v_{p,//}/v_{p,\perp}$ and $A_s^* = v_{s,//}/v_{s,\perp}$ are calculated. The plot of the squares of both ratios shows a correlation (Figure 6.52) with the relationship as a first approximation (see also section 6.4):

$$A_s^{*2} = 1.5 \cdot A_p^{*2} - 0.5 \qquad (6\text{-}93)$$

For marine sediments Bachman (1983) notes the influence of burial depth on anisotropy. Bachman (1969, 1983) gives an analysis of experimental investigations made on marine sediments with longitudinal wave velocities in the range of about $v_p = 1.5...5$ km/s. The relationship between velocities in the vertical ($v_{p,v}$) and horizontal direction ($v_{p,h}$) follows a linear regression

$$v_{p,v} = a + b\, v_{p,h} \qquad (6\text{-}94)$$

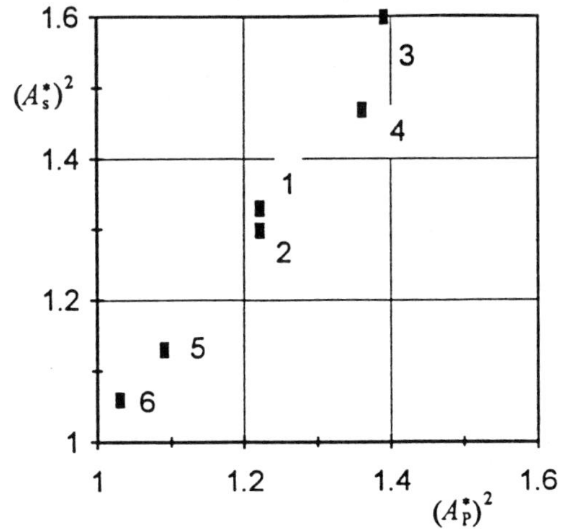

Figure 6.52. Correlation between the squares of the two anisotropy ratios $(A_s^*)^2$ vs. $(A_P^*)^2$ (based on results by White et al., 1983); numbers indicate the depth: 1 - 137 m; 2 - 168 m; 3 - 198 m; 4 - 229 m; 5 - 259 m; 6 - 290 m

where a and b are determined empirically. The values in Table 6.23 show that a increases with decreasing b. The two empirically determined parameters are related by

$$a = 1.53 - 1.52\,b \qquad (6\text{-}95)$$

This leads to a modification of equation (6-94) as a first approximation for the described material

$$v_{p,v} = 1.53 + b \cdot (v_{p,h} - 1.52) \qquad (6\text{-}96)$$

Table 6.23. Empirical parameters and regression coefficient of equation (6-94); after Bachmann (1983); velocities and parameter a in km/s); r is the regression coefficient.

Sediment type	a	b	r
Calcareous	0.174	0.887	0.98
Silt, clay	0.393	0.740	0.93
Siliceous	0.222	0.855	0.92
Marl	0.259	0.832	0.96
Sand	0.248	0.854	0.96

6.4. Theories

"It is a problem of considerable geophysical interest and of some mathematical difficulty to determine the elastic properties of the composite from the separate properties of the inclusions and the matrix material." (Hudson and Knopoff, 1989)

6.4.1. Overview

The equations for the elastic wave velocities (equations 6-5 and 6-7) are derived on the basis of the classic theory of continuum mechanics. This presupposes a homogeneous material. The velocities are related to density and elastic moduli, which have unique values for a homogeneous medium.

Rocks are in general *inhomogeneous*. An inhomogeneous material can be replaced by an "effective" or "equivalent" homogeneous material, if the dimensions of the constituents (e.g., grains, pores, fractures, laminae) are small compared with the shortest wavelength of significance (macroscopic homogeneity). Hudson and Knopoff (1989) formulate: "A material with small-scale variations in properties, with embedded inclusions of a different material or permeated with cracks, acts mechanically as a homogeneous medium if the excitation wavelengths are large compared with the scale of the variations in structure or the size of inclusions".

The effective or macroscopic properties are the results of the elastic properties of the rock components, the rock structure and various interconnections (see chapter 6.3).

Theories of the elastic behaviour of rocks are directed

- under a *theoretical aspect* at a physical-mathematical description of the elastic properties of the compound medium in terms of the material properties, volume fractions and geometry of the constituents, depending on environmental conditions (e.g., pressure and temperature).
- under a more *practical aspect* at an applicable basis for a better understanding and interpretation of seismic data.

How can we derive such relations theoretically? We need to begin with a simplified model of a natural, heterogeneous, structured rock that can be described in mathematical form. Such a model is necessarily an idealization that emphasizes the relevant aspects. The selection of the "relevant aspects and influences" depends on the purpose of the study. The fundamental types were discussed in chapter 1.4.

Hudson and Knopoff (1989) distinguish two types of theories
(1) those concerned with finding upper and lower *bounds* within which the values of the elastic parameters of the compound medium must lie, and
(2) those concerned with the calculation of the elastic *parameters* of the compound medium.

The early work on the properties of composite materials of Voigt (1910) and Reuss (1927) - as described in section 6.4.2 in more detail - involves both ideas because they
◆ result in an exact solution for compound media that conform to their models on the basis of a simple physical model, and
◆ result in lower and upper bounds for the compound media not conforming to either model.
We can sketch a classification of main types of theories based on the corresponding model concept (Figure 6.53).

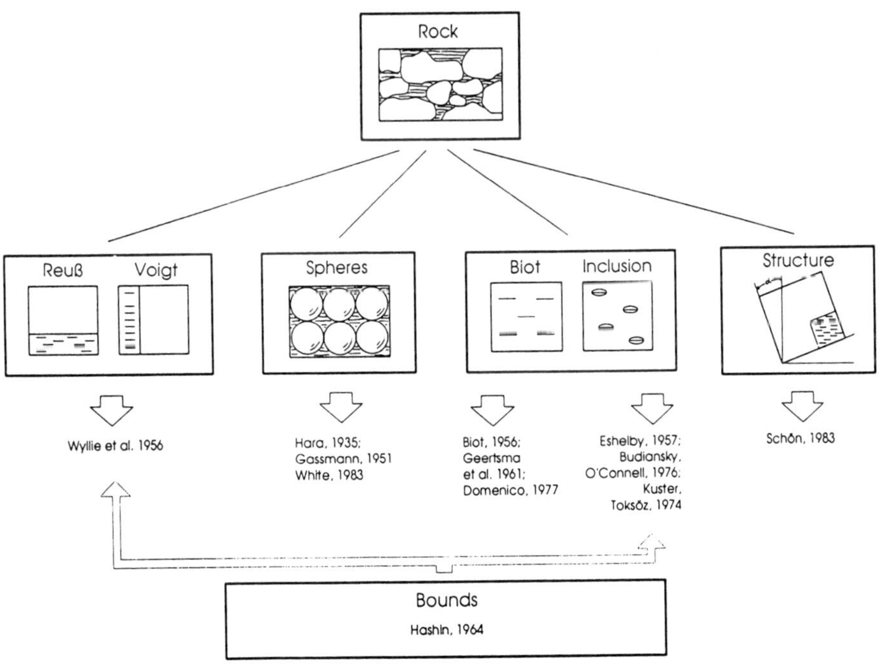

Figure 6.53. Classification of theories based on the model used and its basic concept

This is a "travel guide" for consideration of the following particular groups, but only some representations of theories have been described. The description will concentrate on

- the assumptions and the model concept,
- the fundamental idea of mathematical derivation,
- the principal results.

6.4.2. The fundamental spatial averaging models of Voigt (1910) and Reuss (1927) - Theories based on sheet models

Voigt (1910) and Reuss (1926) developed averaging schemes to estimate the elastic constants of monomineral aggregates. If no preferred orientation of mineral grains and their crystallographic axes exists, the aggregate may behave isotropically.
Under assumption of *homogeneous strain* throughout the aggregate the Voigt moduli (subscript V) are

$$k_{\mathrm{V}} = \tfrac{1}{3} \cdot (A + 2B) \tag{6-97}$$

$$\mu_{\mathrm{V}} = \tfrac{1}{5} \cdot (A - B + 3C) \tag{6-98}$$

where $k_{\mathrm{V}}, \mu_{\mathrm{V}}$ are related to the components of the *stiffness tensor* of the mineral

$$3A = C_{11} + C_{22} + C_{33} \tag{6-99}$$

$$3B = C_{23} + C_{31} + C_{12} \tag{6-100}$$

$$3C = C_{44} + C_{55} + C_{66} \tag{6-101}$$

Under assumption of *homogeneous stress* throughout the aggregate, the Reuss moduli (subscript R) are

$$k_{\mathrm{R}} = \frac{1}{3(a + 2b)} \tag{6-102}$$

$$\mu_{\mathrm{R}} = \frac{5}{4(a - b) + 3c} \tag{6-103}$$

where k_R, μ are related to the components of the *compliance tensor* of the mineral

$$3a = S_{11} + S_{22} + S_{33} \tag{6-104}$$

$$3b = S_{23} + S_{31} + S_{12} \tag{6-105}$$

$$3c = S_{44} + S_{55} + S_{66} \tag{6-106}$$

For *application* on rocks, the fundamental model assumption is based on a separation of the individual rock components (minerals, pore filling material) and its arrangement such as a *sequence of sheets*. The relative thickness of the individual sheet is proportional to the volume fraction of the considered component; the property of the sheet is that of the corresponding component. Also (see Figure 6.54) a state of a macroscopic homogeneity is assumed for this model with

- homogeneous strain (Voigt moduli) or
- homogeneous stress (Reuss moduli).

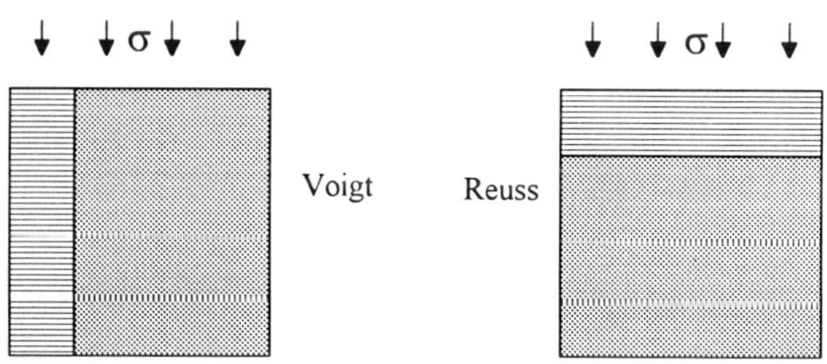

Figure 6.54. The two fundamental models of Voigt (1910) and Reuss (1927)

The n individual components are assumed to be isotropic; their elastic properties are described by k_i and μ_i, where the index i means the component i.

The two sets of fundamental equations for the compression or bulk modulus and for the shear modulus are

$$k_V = \sum_{i=1}^{n} k_i \cdot V_i \tag{6-107}$$

$$\mu_V = \sum_{i=1}^{n} \mu_i \cdot V_i \tag{6-108}$$

$$k_R^{-1} = \sum_{i=1}^{n} k_i^{-1} \cdot V_i \tag{6-109}$$

$$\mu_R^{-1} = \sum_{i=1}^{n} \mu_i^{-1} \cdot V_i \tag{6-110}$$

where V_i is the volume fraction of the component i related to the total volume.

When the effective material is isotropic, the bulk moduli and shear moduli given by these equations form *boundaries* on the corresponding correct values. The Voigt relationship leads to the upper (maximum), and the Reuss relationship to the lower (minimum) bound (see Figure 6.55). The real (measured) values are situated between this two bounds.

To limit the predicted value range, two methods can be used

(1) calculation of mean values (e.g., Hill mean value),

(2) application of further constraints or variational principles (Hashin bounds, see section 6.4.7).

Hill (1952) proposed the arithmetic mean of the two estimates for each modulus (called also Voigt-Reuss-Hill values or VRH-values, a term first used by Chung, 1963, see also Watt et al., 1976)

$$k_{VRH} = \frac{1}{2} \cdot (k_R + k_V) \tag{6-111}$$

$$\mu_{VRH} = \frac{1}{2} \cdot (\mu_R + \mu_V) \tag{6-112}$$

Also the geometric mean is applied (e.g., Kumazava, 1969)

$$k_{geom} = (k_R \cdot k_V)^{1/2} \tag{6-113}$$

$$\mu_{geom} = (\mu_R \cdot \mu_V)^{1/2} \tag{6-114}$$

A modification of Voigt's model has given Alexandrov and Eisenberg (1966), using the logarithm of the bulk and shear modulus

$$(\ln\ k_A) = \sum_{i=1}^{n} V_i \cdot (\ln k_i) \tag{6-115}$$

$$(\ln\ \mu_A) = \sum_{i=1}^{n} V_i \cdot (\ln \mu_i) \qquad (6\text{-}116)$$

Figure 6.55 shows the example of a two component material with the dependence of the material modulus related to the modulus of the component 1 (k/k_1) on the volume fraction V_2. The ratio of the bulk moduli is assumed to be $k_1 : k_2 = 1 : 0.2$ for this example. The curves for the Voigt and Reuss model give the *boundaries*. In the space between the boundaries, there are the curves for the VRH mean and the geometric mean; there is only a small difference between these two curves.

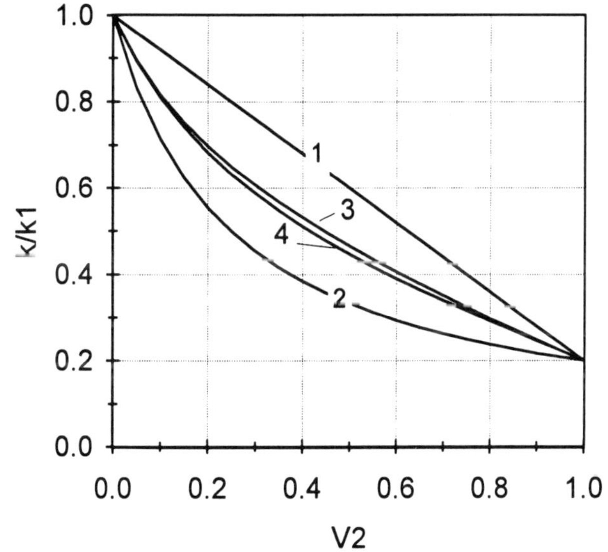

Figure 6.55. Dependence of the normalized bulk modulus k/k_1 on the volume fraction of the second fraction V_2; the ratio of the bulk moduli is assumed as $k_1 : k_2 = 1 : 0.2$
1 - Voigt 2 - Reuss 3 - Hill 4 - Geometric mean

The modification (6-115, 6-116) is recommended for the magmatic rocks by Belikov, Alexandrov and Rysova (1970). Table 6.24 contains a comparison between measured and calculated bulk moduli using equations (6-111) and (6-115).

Crosson and Lin (1971) extended the Voigt and Reuss theory to prediction of anisotropic aggregates and applied this to dunites (Addie and Twin Sisters dunite) using petrofabric data and single-crystal constants. The agreement between the predicted and observed velocities was within the observational error for compressional waves when using the Voigt model. The best agreement was at pressures above 2 kbar (= 0.2 GPa); this may be caused by the decreasing influence of defects like microcracks at higher pressures.

Table 6.24. Comparison of measured (k_{exp}) and calculated (k_{VRH}, k_A) bulk moduli values (in 10^9 Pa = GPa); Φ is the porosity in %; after Belikov, Alexandrov and Rysova (1970)

Rock type	k_{exp}	k_{VRH}	k_A	Φ
Granite	49.1	49.0	49.1	0.50
Granite	54.6	52.3	51.0	0.46
Granodiorite	60.4	57.3	60.8	1.00
Gabbro-diorite	75.3	86.9	86.5	0.91
Gabbro-diorite	81.5	84.3	84.6	0.46
Pyroxenite	94.8	94.2	99.2	0.38

Nur (1991) applied the classic Voigt and Reuss concept to porous sands and modified the porosity by introducing a "*critical porosity*" and achieving a satisfactory agreement with experimental data; the critical porosity seems to be related to the initial porosity of the sediment.

The basic conceptual model and the principle of the averaging process are sources for *modifications* and further developments, such as

- ➢ application to the seismic travel time (time-average relation),
- ➢ development of the theory for laminated materials, which are both described in the next section.

6.4.3. Further theories based on a spatial averaging or sheet models

This section considers the time-average relationship (and their "relatives") and laminated solids. Additionally, a description of a simple model for fractured rocks is considered, although this model assumption is somewhat different.

6.4.3.1. The time-average relationship and their "relatives"

"The work of Wyllie and others was the breakthrough necessary to put acoustic logging into a central role in log interpretation. ... It gave them a way to relate it to porosity of the rock." (Darwin V. Ellis, 1987)

Wyllie et al., (1956) proposed an empirical relationship between interval transit time (reciprocal velocity) and porosity based on numerous laboratory experiments and the application of the sheet model to porous rocks (see Figure 6.56). The derivation is based on the addition of the partial transit times of the compressional wave for a pure matrix (such as silica) and pure pore filling material (such as water). The portion for the two parts in a model cube with the length equals 1 are (1-Φ) for the matrix and Φ for the pore fluid. The addition of the two transit times results in the so-called "time-average relationship"

$$\Delta t = (1 - \Phi) \cdot \Delta t_{\mathrm{m}} + \Phi \cdot \Delta t_{\mathrm{f}} \qquad (6\text{-}117)$$

where

Δt is the transit travel time of the rock,

Δt is the interval transit travel time of the matrix material,

Δt is the interval transit travel time of the pore fluid,

Φ is the porosity.

The unit of the interval transit travel time is μs/m with the conversions:

1 μs/m = 0.305 μs/ft 1 μs/ft = 3.28 μs/m.

It is the reciprocal of velocity and, therefore, the velocity terms yield the equation

$$\frac{1}{v} = \frac{1 - \Phi}{v_{\mathrm{m}}} + \frac{\Phi}{v_{\mathrm{f}}} \qquad (6\text{-}118)$$

From equation 6-117, there results a linear relation between rock interval travel time and porosity of

$$\Phi = \frac{\Delta t - \Delta t_{\mathrm{m}}}{\Delta t_{\mathrm{f}} - \Delta t_{\mathrm{m}}} \qquad (6\text{-}119)$$

The decisive prerequisite for the application of the time-average relationship is the choice of *representative interval travel times* for matrix and pore content.

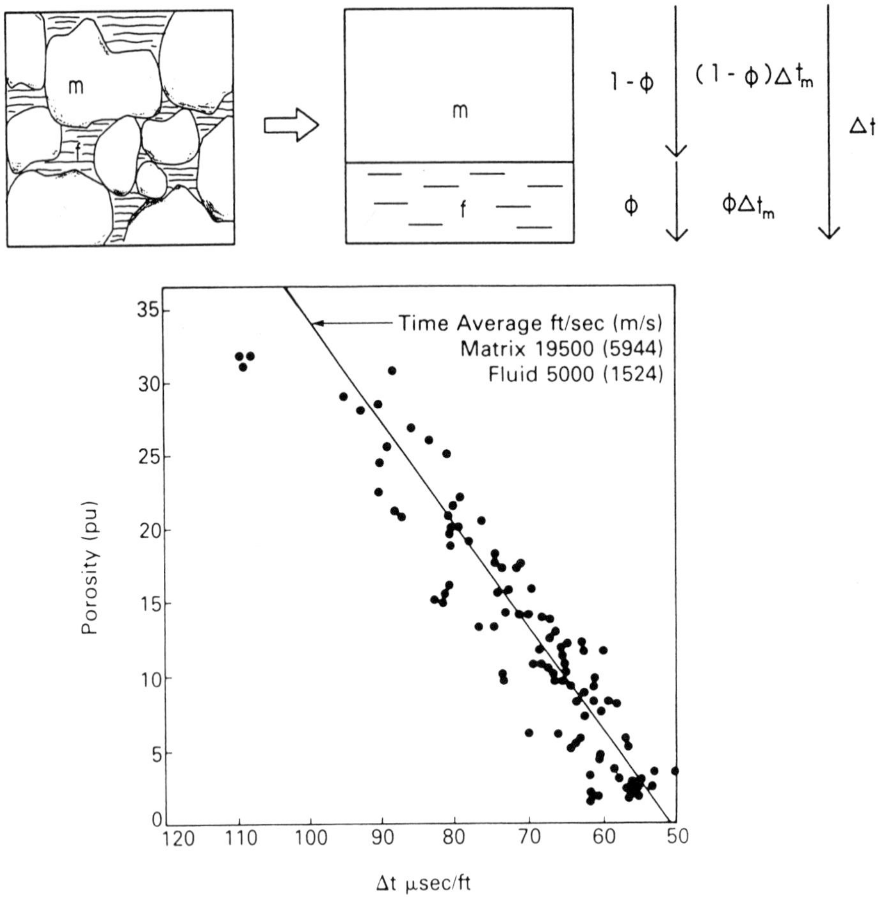

Figure 6.56. Time average relation: Model concept and correlation between compressional transit time and porosity (sandstone measurements), the straight line represents the time average relation (from Wyllie et al.'s original publication, reprinted by Ellis, 1987)

Table 6.25 contains mean values for some matrix and pore filling materials.

The time-average relationship is *applied* in most cases either for a "single" porosity determination at known matrix travel time or as one equation in a set of equations for matrix-porosity estimation (chapter 10.). Agreement between calculated and observed porosities are particularly good for the cases of

- correct individually determined matrix travel times,
- compacted rocks or rocks below a given depth (e.g., 2000 m),
- water saturated conditions (or low gas content).

Table 6.25. Mean values for interval transit travel times of matrix, pore filling, concrete and casing; References: B - Bulatova et al. (1970); I -Itenberg (1978); L - Lehnert et al. (1970), Lehnert (1970), S - Schlumberger (1989); Se - Serra (1984), the reference Se contains mean values and as range the extreme values; W - Western Atlas (1992);

Material	Δt in μs/m	Δt in μs/ft	Ref.
Sandstone			
general	176.5 ... 328 (187)	53.8 ... 100 (57.0)	Se
	167.3 ...182	51.0 ... 55.5	S
consolidated	172.5	52.6	W
	170	51.8	I
semi-consolidated	182.4	55.6	W
	189	51.5	B
unconsolidated	192.9	58.8	W
	400	122	I
Limestone	156.2	47.6	W
	142.7 ... 156.2	43.5 ... 47.6	S
	149.3 ... 155.8 (152.6)	43.5 ... 47.6 (46.5)	Se
Dolomite	142.7	43.5	W
	143.5	43.7	L
	142.7	43.5	S
Anhydrite	131 ... 148 (144 4)	40.0 ... 45.0 (44.0)	Sc
	164.0	50.0	W, L
	164.0	50.0	S
Shale	197 ... 558	60 ... 170	Se
Water (pure)	715	218	W
(10 % NaCl)	862	208	W
(20 % NaCl)	620	189	W
Concrete	273 ... 312 (312)	83.3 ... 95.1 (95.0)	Se
Casing (steel)	187.3	57.1	Se

Empirical *modifications* are recommended and applied for increased confidence in results obtained between porosities determined by sonic logs and laboratory core data (e.g. Raymer et al., 1980).

In *uncompacted* formations, the Wyllie equation yields porosities that are too high. The comparable situation exists for *shaly formations*. A correction factor $c \cdot \Delta t_{sh}$ takes the effect of compaction or pressure and temperature into account and the equation can be rewritten as (see e.g., Serra, 1984):

$$\Phi_{\text{corr}} = \frac{\Delta t - \Delta t_{\text{m}}}{\Delta t_{\text{f}} - \Delta t_{\text{m}}} \cdot \frac{100}{c \cdot \Delta t_{\text{sh}}} \tag{6-120}$$

As "relatives of the time average relation", we will consider

- linear relationships between travel time (as well as velocity) and the rock constituents (matrix, clay, pore fluid), such as the relationships given by Han et al., 1986 (see section 6.3.3.2.1.) and Pickett, 1963,
- non-linear relationships between travel time or velocity and porosity; see Raymer et al., 1980 (see eq. 6-43) and of Chapman, 1981 (see note by Elphick, 1993) and Raiga-Clemenceau et al., 1988:

$$\Delta t = F_{\text{ac}} \cdot \Delta t_{\text{m}} = (1-\Phi)^{x} \cdot \Delta t_{\text{m}} \tag{6-121}$$

where F_{ac} is the "acoustic formation factor" and x an exponent, depending on rock type (1.60 for silica, 1.76 for calcite, 2.00 for dolomite).

If we look for a *"common source relation"* for all equations that describe the velocity-porosity or interval travel time-porosity correlation we can use a generalized "mixing rule" published by Korvin (1982). This gives the relation between a property g of the composite and the properties of the components g_1, g_2 and their volume fractions V_1, V_2

$$g = g_1 \cdot V_1^a + g_2 \cdot V_2^a \tag{6-122}$$

with an exponent a.
In the case of a porous rock, this results in

$$g = g_{\text{m}} \cdot (1-\Phi)^a + g_{\text{f}} \cdot \Phi^a \tag{6-123}$$

Empirically (and according to the relations of Raiga-Clemenceau et al., 1986 and Raymer et al., 1980), we introduce, as a modification, two exponents (a, b) for the two volume fractions which results in a generalized "*source relation*"

$$g = g_{\text{m}} \cdot (1-\Phi)^a + g_{\text{f}} \cdot \Phi^b \tag{6-124}$$

All the equations in this section are special cases of this general relationship. The special exponents and parameters are shown in Figure 6.57 in form of a "*common family tree*".

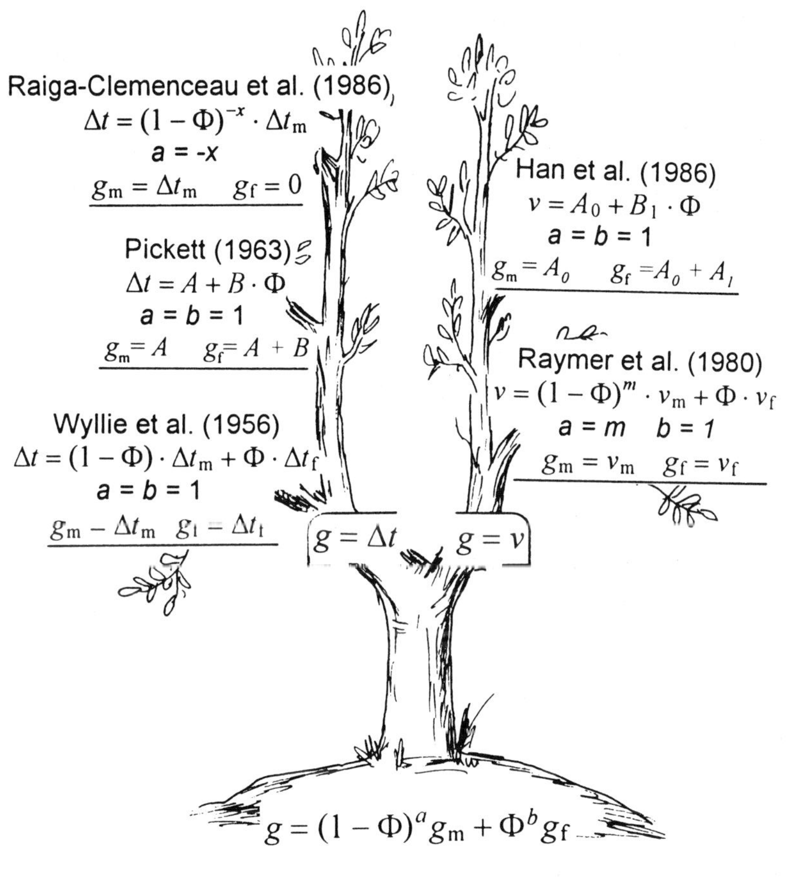

Figure 6.57. The "common family tree" of simple travel time- and velocity- porosity relations

6.4.3.2. Laminated solids

Figure 6.58 shows a model for a finely laminated rock representing a laminated clay or a gneiss. For *long wavelengths* (compared with the dimension of the individual layers) averaging can obtain the effective properties of the medium. The stratified medium consists of individual homogeneous and isotropic layers. For the effective properties (long wave averaging), *transverse isotropy* results and the elastic behaviour is described by the five elastic constants discussed in chapter 6.1.

The first theoretical investigations into the effective properties of such composites are made by Rayleigh in 1892 (see Helbig, 1958). Bruggemann (1937) solved the problem of the elastic properties of a laminated material consisting of two components. A systematic derivation based on the Lamé constants was published by Postma (1955) and Helbig (1958). The theory and model was extended for any number of components (Helbig, 1958; Backus, 1962).

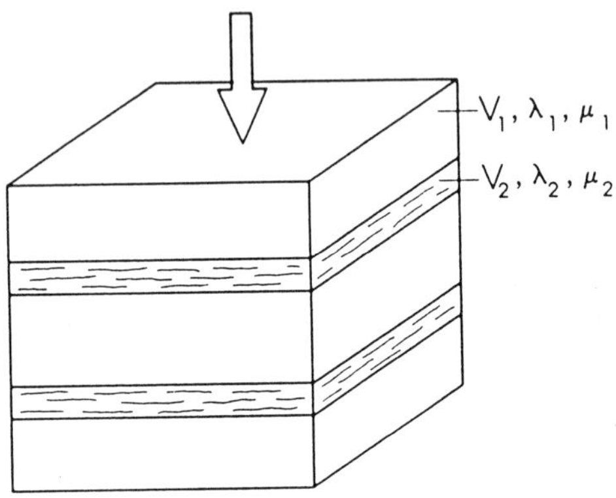

Figure 6.58. Elementary volume of a laminated solid consisting of two components

We follow an elegant derivation by Helbig (personal communication, 1992 and Helbig, 1995): It starts with the stress strain relation for the transverse isotropic medium

$$\begin{pmatrix} \sigma_1 \\ \sigma_2 \\ \sigma_3 \\ \sigma_4 \\ \sigma_5 \\ \sigma_6 \end{pmatrix} = \begin{pmatrix} c_{11} & c_{11}-2c_{66} & c_{13} & 0 & 0 & 0 \\ c_{11}-2c_{66} & c_{11} & c_{13} & 0 & 0 & 0 \\ c_{13} & c_{13} & c_{33} & 0 & 0 & 0 \\ 0 & 0 & 0 & c_{55} & 0 & 0 \\ 0 & 0 & 0 & 0 & c_{55} & 0 \\ 0 & 0 & 0 & 0 & 0 & c_{66} \end{pmatrix} \begin{pmatrix} \varepsilon_1 \\ \varepsilon_2 \\ \varepsilon_3 \\ \varepsilon_4 \\ \varepsilon_5 \\ \varepsilon_6 \end{pmatrix} \tag{6-125}$$

The laminated model consists of a (periodic) *stack of isotropic layers*. It is transversely isotropic with the vertical (3) axis if the wavelength is sufficiently larger than the characteristic thickness of the stack. The 1-2-plane is parallel to the boundaries between the stack members.

The components of the stress and strain tensors fall into two classes:

- class of constant or "*slowly varying components*" (varying with the order of the wavelength): $\varepsilon_1, \varepsilon_2, \varepsilon_6, \sigma_3, \sigma_4, \sigma_5$. Continuity of stresses and strains across the boundaries requires that these are constant components.
- class of variable or "*rapidly fluctuating components*" (changing with the individual layer thickness): $\varepsilon_3, \varepsilon_4, \varepsilon_5, \sigma_1, \sigma_2, \sigma_6$.

Helbig (1992, 1995) expresses the "rapidly fluctuating components" in terms of "slowly varying components" :

$$\begin{pmatrix} \sigma_1 \\ \sigma_2 \\ \varepsilon_3 \\ \varepsilon_4 \\ \varepsilon_5 \\ \sigma_6 \end{pmatrix} = \begin{pmatrix} c_{11} - \frac{c_{13}^2}{c_{33}} & c_{11} - 2c_{66} - \frac{c_{13}^2}{c_{33}} & \frac{c_{13}}{c_{33}} & 0 & 0 & 0 \\ c_{11} - 2c_{66} - \frac{c_{13}^2}{c_{33}} & c_{11} - \frac{c_{13}^2}{c_{33}} & \frac{c_{13}}{c_{33}} & 0 & 0 & 0 \\ -\frac{c_{13}}{c_{33}} & -\frac{c_{13}}{c_{33}} & \frac{1}{c_{33}} & 0 & 0 & 0 \\ 0 & 0 & 0 & \frac{1}{c_{55}} & 0 & 0 \\ 0 & 0 & 0 & 0 & \frac{1}{c_{55}} & 0 \\ 0 & 0 & 0 & 0 & 0 & c_{66} \end{pmatrix} \begin{pmatrix} \varepsilon_1 \\ \varepsilon_2 \\ \sigma_3 \\ \sigma_4 \\ \sigma_5 \\ \varepsilon_6 \end{pmatrix} \quad (6\text{-}126)$$

Each of the right hand strain and stress components are of the "constant class" - they are for a given stress-strain state spatially constant. This results directly in *averaging relationships* to obtain the compound stiffnesses for a medium consisting of transversely isotropic laminae with vertical symmetry axis (x_3):

$$c_{11} - \frac{c_{13}^2}{c_{33}} = \left\langle c_{11}^{(i)} - \frac{c_{13}^{(i)2}}{c_{33}} \right\rangle \qquad \frac{c_{13}}{c_{33}} = \left\langle \frac{c_{13}^{(i)}}{c_{33}^{(i)}} \right\rangle$$

$$\frac{1}{c_{33}} = \left\langle \frac{1}{c_{33}^{(i)}} \right\rangle \qquad \frac{1}{c_{55}} = \left\langle \frac{1}{c_{55}^{(i)}} \right\rangle \qquad c_{66} = \left\langle c_{66}^{(i)} \right\rangle$$

or

$$c_{11} = \left\langle c_{11}^{(i)} - \frac{\left(c_{13}^{(i)}\right)^2}{c_{33}^{(i)}} \right\rangle + \frac{\left\langle c_{13}^{(i)}/c_{33}^{(i)} \right\rangle^2}{\left\langle 1/c_{33}^{(i)} \right\rangle}$$

$$c_{13} = \frac{\left\langle c_{13}^{(i)}/c_{33}^{(i)} \right\rangle}{\left\langle 1/c_{33}^{(i)} \right\rangle} \qquad c_{33} = \frac{1}{\left\langle 1/c_{33}^{(i)} \right\rangle}$$

$$c_{55} = \frac{1}{\left\langle 1/c_{55}^{(i)} \right\rangle} \qquad c_{66} = \left\langle c_{66}^{(i)} \right\rangle \tag{6-127}$$

The parentheses $\langle\ \rangle$ indicate weighted averages of the expressions inside, i.e.

$$\langle c_{\dots}^{(i)} \rangle = \frac{\sum_{i=1}^{n} h^{(i)} c_{\dots}^{(i)}}{\sum_{i=1}^{n} h^{(i)}} \tag{6-128}$$

where *(i)* indicates the layer, $c_{\dots}^{(i)}$ its tensor component and $h^{(i)}$ its layer thickness.

The special case of a medium consisting of two components with Lame constants $\lambda_1, \mu_1, \lambda_2,$ and volume fractions V_1 and V_2 gives expressions for the velocities and the stiffnesses in the main directions (for detailed derivation see Helbig, 1958):

$$v_{p//} = \left(\frac{c_{11}}{d}\right)^{1/2} \qquad v_{p\perp} = \left(\frac{c_{33}}{d}\right)^{1/2} \tag{6-129}$$

$$v_{s//} = v_{SH} = \left(\frac{c_{66}}{d}\right)^{1/2} \qquad v_{s\perp} = v_{SV} = \left(\frac{c_{55}}{d}\right)^{1/2} \tag{6-130}$$

with

$$c_{33} = \frac{1}{V_1/(\lambda_1 + 2\mu_1) + V_2/(\lambda_2 + 2\mu_2)} \tag{6-131}$$

$$c_{55} = \frac{1}{V_1/\mu_1 + V_2/\mu_2} \tag{6-132}$$

$$c_{66} = V_1\mu_1 + V_2\mu_2 \tag{6-133}$$

$$c_{11} = c_{33}\left[1 + 4V_1V_2\frac{(\mu_1 - \mu_2)(\lambda_1 - \lambda_2 + \mu_1 - \mu_2)}{(\lambda_1 + 2\mu_1)(\lambda_2 + 2\mu_2)}\right] \tag{6-134}$$

Examples of the *application* of layered models are: Postma (1955) studied the case of a laminated medium consisting of alternating sandstone and limestone. A study of the Pierre shale properties was published by White et al. (1983); some of the experimental results are presented by Figure 6.52 and discussed in section 6.3.5.3.

Schoenberg (1994) solved the *converse problem*: to find a set of isotropic layers equivalent to a given transversely isotropic media. The decomposition is studied for two isotropic constituent layers (uniquely determination) and for three isotropic constituent layers (in this case two have the same Poisson's ratio and thickness fraction).

Shapiro et al. (1994) investigated the angle-dependent wave transmissivity in a random, multilayered medium and derived a "*frequency dependent anisotropy* (generalized O'Doherty-Anstey formula for waves in finely layered media).

6.4.3.3. A simple model for fractured rocks

Fractures, cracks and other defects of the solid mineralic substance change the elastic properties dramatically, where the elastic wave *velocities decrease* and a strong velocity *dependence on pressure* results. The porosity as the ratio of the volume of the defects (crack) to the total volume cannot express these effects - microcracks with only a small porosity can significantly change the velocity. Another parameter is needed to describe the physical effects of these defects.

A phenomenological description using a simple model for such a fractured rock is demonstrated below. Starting with a cube of solid material we assume that the *effect of all defects* (fractures, defects at grain boundaries, intragranular defects, etc.) is described by a parameter D (Figure 6.59).

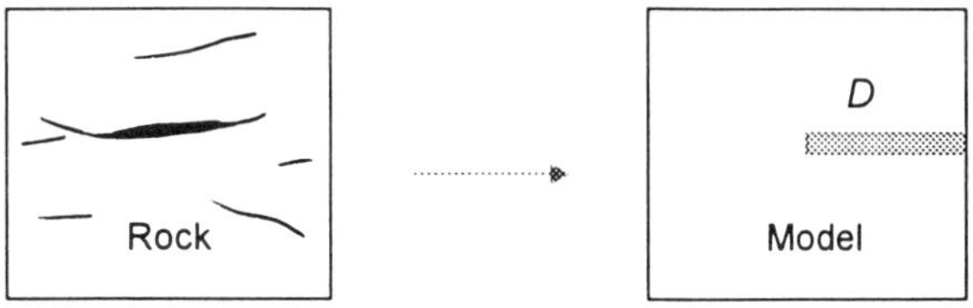

Figure 6.59. A simple model for a rock with internal defects (cracks, fractures etc.) expressed by the parameter D

This parameter is like a "cut" of the relative depth D. Neglecting terms of higher order and any effects of the pore fluid, we can describe the reduction of the elastic properties for the example of the plane wave modulus by the simple relation:

$$\overline{M} = M_{\mathrm{m}} \cdot (1 - D) \qquad (6\text{-}135)$$

where $\overline{M}$ is the resulting rock model modulus and M_{m} is the modulus for the (defectless) solid matrix material. This relation is based on the assumption that only the "non-cut" part of the rock cross-section represents the material rock stiffness. Thus, the parameter D represents the effects of fractures, cracks, and other defects upon the elastic wave velocity. If we neglect the (small) influence of the defects on density, the velocity is

$$v_{\mathrm{p}} = v_{\mathrm{m}} \cdot (1 - D)^{1/2} \qquad (6\text{-}136)$$

The *form* of this equation is similar the approximation in Eshelby's equations (6-192 and 6-193) in section 6.4.6.1. These equations give a more specific and detailed expression for the fracture effect based on a model.

Increasing pressure p results in closure of fractures - or more generally - in a decrease of defects expressed by $D(p)$. For this dependency, we can assume that the decrease of D with increasing p is proportional to the magnitude of the defects, thus

$$-\frac{\mathrm{d}D}{\mathrm{d}p} \sim D \qquad (6\text{-}137)$$

Introducing a proportionality factor which depends on the deformation properties, a differential equation results with the solution

$$D(p) = D_0 \cdot \exp\left(-\frac{p}{p^*}\right) \qquad (6\text{-}138)$$

D_0 is the initial value of the defect parameter at the pressure $p = 0$ and p^* is a reference pressure. The relationship between velocity, pressure and degree of initial defects is now

$$v_{\mathrm{p}} = v_{\mathrm{m}} \cdot \left[1 - D_0 \cdot \exp\left(-\frac{p}{p^*}\right)\right]^{1/2} \qquad (6\text{-}139)$$

The exponential character of the velocity vs. pressure relation, including the closure process, suggests that a " half closure pressure" $p_{1/2}$ should be introduced. It can be defined as that pressure which creates a decrease of the defect parameter to its half initial value, i.e.

$$D = \frac{D_0}{2} = D_0 \cdot \exp\left(-\frac{p_{1/2}}{p^*}\right) \tag{6-140}$$

thus the half closure pressure and its relation to the reference pressure is:

$$p_{1/2} = -(\ln 0.5) \cdot p^* = 0.693 \cdot p^* \tag{6-141}$$

Figure 6.60 shows the graphical representation of equation 6-139. The abscissa is the exponent (p/p^*) and the ordinate the normalized velocity (v_p/v_m). The initial value D_0 is the curve parameter. The intercept is the initial normalized velocity at zero pressure. The analysis can be made graphically (using the curves of Figure 6.60a as master curves) or numerically.

Two examples show the application of this simple model for a description of the defect controlled velocity-pressure functions:

- Figure 6.60b shows the results of an analysis of experimental data, measured on terrestrial and lunar rock specimens. In particular, the lunar basalt shows intensive internal disturbance (probably created by numerous impacts) and a high value of the parameter D_0, but a very low porosity (its density is that of a nearly pore-free rock).
The corresponding equations are:

Dunite	$v_p = 7.5 \cdot [1 - 0.2 \cdot \exp(-0.012 \cdot p)]^{1/2}$
Gneiss	$v_p = 6.0 \cdot [1 - 0.5 \cdot \exp(-0.0087 \cdot p)]^{1/2}$
Basalt	$v_p = 7.6 \cdot [1 - 0.8 \cdot \exp(-0.0092 \cdot p)]^{1/2}$
Breccia	$v_p = 4.4 \cdot [1 - 0.9 \cdot \exp(-0.0097 \cdot p)]^{1/2}$ (6-142)

- The analysis of rock specimens of the KTB sample (Figure 6.13) was impossible with only one exponential term. But a *series of two terms*, gives reasonable results (v_p in m/s, p in MPa)

$$v_{p,max} = 6500 \cdot [1 - 0.12 \cdot \exp(-0.063 \cdot p) - 0.12 \cdot \exp(-0.005 \cdot p)]^{1/2}$$
$$v_{p,min} = 5900 \cdot [1 - 0.60 \cdot \exp(-0.050 \cdot p) - 0.20 \cdot \exp(-0.005 \cdot p)]^{1/2} \tag{6-143}$$

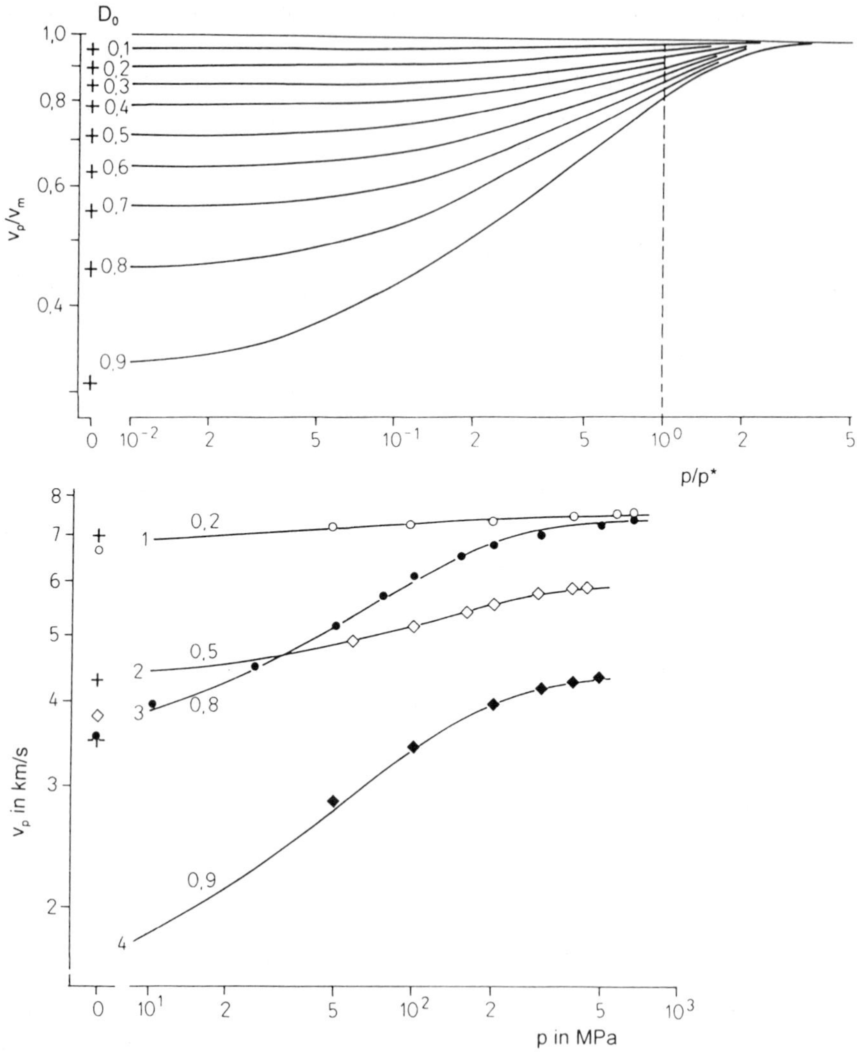

Figure 6.60. P-wave velocity as a function of pressure under the influence of internal defects
a) dependence of normalized velocity v_p/v_m on the normalized pressure p/p^* for different parameters D_0, calculated after equation (6-139);
b) analyzed experimental data (data after Wang et al., 1971); v_p in km/s, pressure in MPa;
1 - dunite: $d = 3.31 \times 10^3$ kg/m^3; $D_0 = 0.2$; $p_{1/2} = 58$ MPa
2 - gneiss (terrestrial): $d = 2.72 \times 10^3$ kg/m^3; $D_0 = 0.5$; $p_{1/2} = 80$ MPa
3 - basalt: $d = 3.33 \times 10^3$ kg/m^3; $D_0 = 0.8$; $p_{1/2} = 75$ MPa
4 - breccia (lunar): $d = 2.35 \times 10^3$ kg/m^3; $D_0 = 0.9$; $p_{1/2} = 71$ MPa

This can be interpreted as follows:

- ► there are two defect systems with different parameters D_0, p^* (or $p_{1/2}$)
- ► the first dominates at low pressure and the second at higher pressure.

The first defect system may be correlated with the sampling process and the expansion of the material, and the second with the in-situ stress and material conditions. We note that the second system has the same closure half pressure for the maximal and minimal velocity.

Three remarks may be added:

1.) The form of equation (6-135) is in agreement with the equation (6-184) for elliptical cracks derived by Eshelby (1957) and discussed in section 6.4.6; the derivation here is greatly simplified.

2.) The same model is also applied to other properties (e.g., thermal conductivity, strength) and for the interrelationships between them (see chapters 8 and 10).

3.) The model concept can be integrated into structural models, as described in section 6.4.8; the defect then has a variable orientation with respect to the direction of wave propagation.

6.4.1. Theories based on packing of spheres

The concept of packing of spheres is natural for *unconsolidated rocks*. In most theories it is assumed that

- grains are spherical,
- grains have single diameter, and
- centres of the spheres occupy a crystal lattice system (e.g. cubic, hexagonal).

These assumptions determine the *possibilities and limitations* of each theory. The derivations are based on

a) the *stress-strain relations for the contact* of spherical particles. The fundamentals of stress-strain behaviour due to normal forces at the point of contact were derived 1881 by G. Hertz (see Love, 1944, Landau and Lifshitz, 1965, White, 1983). Cattaneo (1938) and Mindlin (1949) also introduced tangential forces; a summary discussion is given by Deresiewics (1958).

b) the *sphere pack model.* After the early paper of Hara (1935) assuming the cubic packing of spheres as a model to describe the behaviour of a carbon-granule microphone, Gassmann (1951) published his classic paper about the elasticity of a hexagonal packing of spheres. Further studies are by White and Sengbush (1953) and Brandt (1955) assuming various packings.

In the following, the principle is described by *cubic packing*. Fig. 6.61 shows two spheres of a radius R in contact.

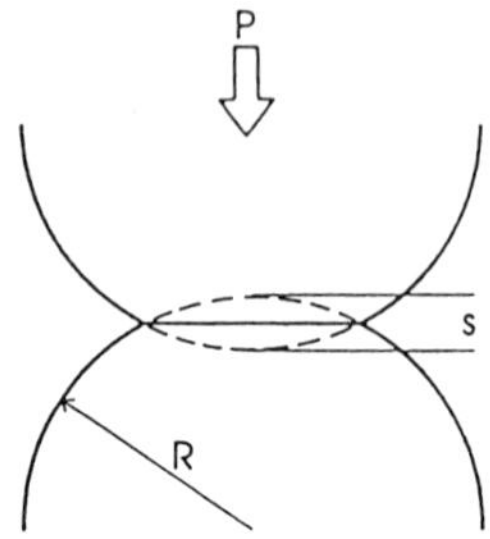

Figure 6.61 The sphere model and the behaviour of contact of two spheres under normal loading (Hertz's theory)

The force P results in a deformation of the two spheres described by displacement s between the centres of two neighbouring spheres $s = 2 \cdot (\Delta R)$. The relation between deformation and force is (after Hertz):

$$s = 2\Delta R = \left[\frac{9 \cdot (1 - \sigma_s^2)^2 P^2}{2 \cdot E_s^2 \cdot R} \right]^{1/3} \tag{6-144}$$

where E_s, σ_s are Young's modulus and Poisson's ratio, respectively, of the solid sphere material (e.g., quartz). For a cubic array of spheres the average pressure $\bar{p}$ and the force P are related to each other by

$$P = (2R)^2 \bar{p} \tag{6-145}$$

The deformation is

$$\varepsilon = \frac{\Delta R}{R} \tag{6-146}$$

Substitution of these equations leads to the stress-strain relation and the expression for the corresponding elastic constant

$$\frac{d\bar{p}}{d\varepsilon} = C_{33} = \left[\frac{3 \cdot E_s^2 \cdot \bar{p}}{8 \cdot (1-\sigma_s^2)^2}\right]^{1/3} \qquad (6\text{-}147)$$

The average density of the model is the mass of a sphere divided by the volume of the surrounding cube (dry case)

$$d = \frac{\pi}{6} \cdot d_s \qquad (6\text{-}148)$$

The *velocity of the longitudinal wave* along an axis of a cubic array with acting pressure $\bar{p}$ is

$$v_p = \left[\frac{C_{33}}{d}\right]^{1/2} = \left[\frac{3 \cdot E_s^2 \cdot \bar{p}}{8 \cdot (1-\sigma_s^2)^2}\right]^{1/6} \cdot \left(\frac{6}{\pi \cdot d_s}\right)^{1/2} \qquad (6\text{-}149)$$

In a similar manner the velocity of the *transverse wave* along the axis of the cubic array can be calculated as (see White, 1983)

$$v_s = [3 \cdot (1-\sigma_s^2) \cdot E_s^2 \cdot \bar{p}]^{1/6} \cdot \left[\frac{3}{(2-\sigma_s)(1+\sigma_s) \cdot \pi \cdot d_s}\right]^{1/2} \qquad (6\text{-}150)$$

Expression of overburden pressure by depth z results in

$$v_p = \left[\frac{27}{2\pi^2} \cdot \frac{E_s^2 g}{d_s^2 (1-\sigma_s^2)^2} \cdot z\right]^{1/6} \qquad (6\text{-}151)$$

where g is the earth's gravity.
In a comparable manner, the velocity for *hexagonal packing* (Gassmann, 1951) or *face-centered cubic* packing (Duffy and Mindlin, 1957) can be derived. The result can be written in a *general form*

$$v_p = K \cdot \left[\frac{E_s^2 \cdot g}{(1-\sigma_s^2) \cdot d_s^2}\right]^{1/6} \qquad (6\text{-}152)$$

where the factor K depends on the sphere array:

simple cubic array $K = (27/2\pi^2)^{1/6} \approx 1.0536$

hexagonal array $K = (16/\pi^2)^{1/6} \approx 1.0838$

This derivation is valid for dry packing, i. e. without effects of pore filling.

These equations show three remarkable results:

1) The velocity-pressure or velocity-depth function is nonlinear; in agreement with experiments (section 6.3.4.4) the relations are

$$v \sim p^{1/6} \qquad\qquad v \sim z^{1/6} \qquad (6\text{-}153)$$

The reason for this agreement is originated by deformation theory at the contacts of spherical particles.

2) There is no agreement between the experimental and calculated porosity dependence. This definite dependence results, after the model calculation, in a velocity increase by a factor of 1.029 only as a result of a porosity decrease from 0.48 (cubic) to 0.26 (hexagonal). This disagreement is obviously caused by the assumption of a crystal lattice array or a particular internal structure.

3) Velocity is primarily independent of grain size (sphere radius); the influence of grain size for natural sediments is a secondary grain size - porosity effect (see Schön, 1969).

Points 1) and 2) lead to

- ► the use of Hertz, Cattaneo and Mindlin theory for the deformation of spheres,
- ► replace the regular lattice array by another model for the internal structure.

Schön (1969, 1983) used a *statistically determined internal structure* for a sphere model. The structure is described by the coordination number η_K (number of contacts of one sphere with its neighbours), and the "structure angle α"; this is the mean angle between the vertical axis x_3 and the line which connects the centres of two neighbouring spheres (Figure 6.62).

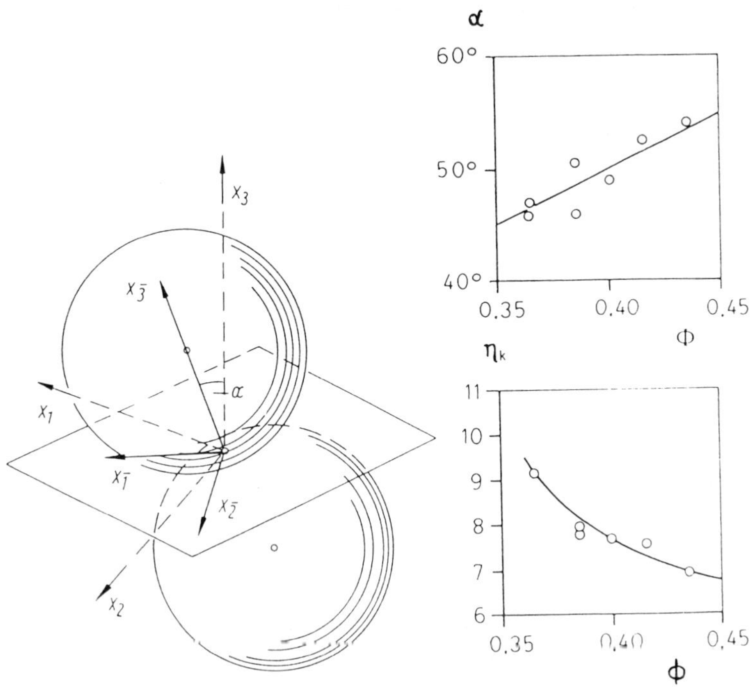

Figure 6.62. Geometric properties of the sphere model
a) Definition of the structure angle α;
b) Mean structure angle α and coordination number η_k as function of porosity; determined for a random glass sphere sediment (Schön, 1969)

This derivation results in the stress-strain relations for the sphere aggregate and velocity equations with a *common form* for the main directions and longitudinal and transverse waves (Schön, 1969):

$$v = \left[6 \cdot \frac{1}{d_s^3} \cdot \left(\frac{E_s}{1-\sigma_s^2}\right)^2 p\right]^{1/6} \cdot \left[\frac{\eta_K}{G(\Phi)}\right]^{1/3} \cdot \left[s_{ik}^{lm}(\alpha, f)\right]^{1/2} \tag{6-154}$$

where E_s, σ_s, are the Young's modulus, Poisson's ratio, and density of the solid sphere material, p is the acting pressure, η_K is the mean coordination number, $G(\Phi)$ is a function depending only on porosity, and $s_{ik}^{lm}(\alpha, f)$ is the component of a tensor with the symmetry of a transversely isotropic medium.

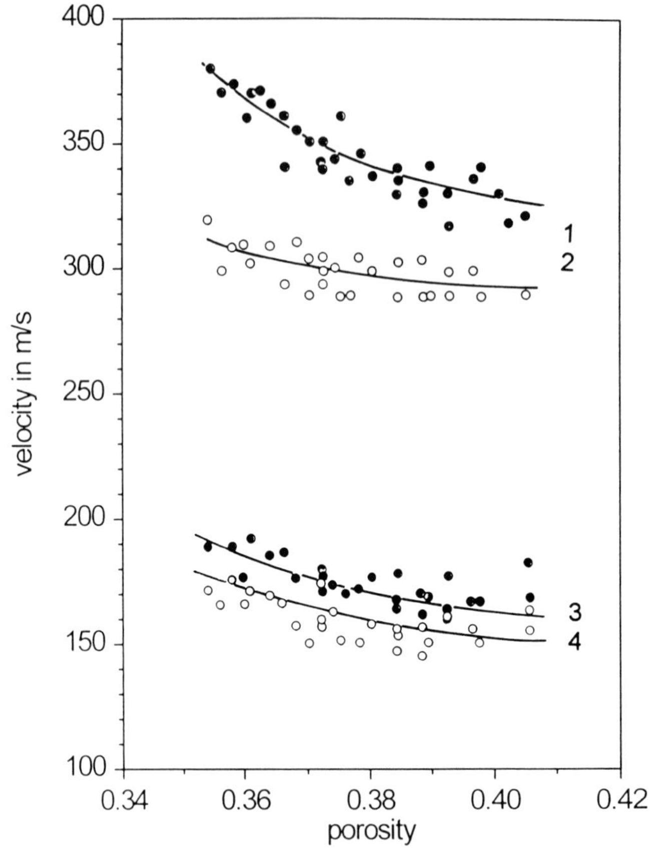

Figure 6.63. Porosity dependence of the elastic wave velocities;
curves: calculated with $E_s = 30$ GPa $\sigma_s = 0.2$ $d_s = 2.55 \times 10^3$ kg m^{-3} $f = 0.3$
points: measured for a random dry glass sphere sediment
1 - P-wave, vertical propagation 2 - P-wave, horizontal propagation
3 - S-wave, vertical propagation 4 - S-wave, horizontal propagation and polarization.

$s_{ik}^{lm}(\alpha, f)$ depends on the structure angle α and the friction coefficient f at the surface of the spheres. This tensor is called "*structure tensor*" (Schön, 1969, 1981; see also section 6.4.8).

Velocities of longitudinal and transverse waves in different directions vary only in the last term. For example, the longitudinal wave in the vertical direction must use the component s_{33}^{33} (see also Table 6.27).

For a velocity calculation, knowledge of the parameters α and η_K is necessary. Both parameters have been determined for a synthetic glass sphere sediment and show a distinct dependence on porosity (Figure 6.62).
Using this dependence, a numerical calculation of the velocity-porosity relationship at a given pressure is possible. There is a satisfactory agreement with experimental data measured at a random glass sphere sediment (Figure 6.63).
The morphology of the grain surface or the contact region can be described also as small spherical undulations on the (large) grain sphere; with this modification the extremely low velocities of the lunar regolith (about 100 ... 130 m/s) can be explained (Schön, 1987).

6.4.5. Wave propagation in saturated porous rocks - the theory of Gassmann, Biot and Geertsma and Smit

In 1951, Gassmann published his theory of the elasticity of porous media. He started with the elastic properties of the dry porous rock (skeleton). He derived the expressions for the corresponding properties of the rock, when it is saturated with any fluid of known properties under the assumption that any *relative motion between fluid and solid is negligible.*

The following parameters are used:

d_s, k_s	- density and bulk modulus of the solid matrix material,
d_f, k_f	- density and bulk modulus of the pore fluid,
$\bar{k}, \bar{\mu}, \overline{M}$	- bulk modulus, shear modulus, and plane wave modulus of the rock skeleton (evacuated porous rock),
k, μ, M	- bulk modulus, shear modulus, and plane wave modulus of the saturated rock,
Φ	- rock porosity.

The assumption that the fluid and solid move in phase results in the inertial density for the saturated rock of

$$d = (1 - \Phi) \cdot d_s + \Phi \cdot d_f \qquad (6\text{-}155)$$

It is assumed that the fluid does not change the properties of the solid skeleton (for example by changing the cementation or grain contact properties), in particular there is no change of rigidity, i.e.,

$$\bar{\mu} = \mu \qquad (6\text{-}156)$$

and that the *skeleton bulk modulus* is the same for the dry and the saturated state.
The *bulk modulus of the saturated porous rock* is defined as the ratio of total pressure p and the resulting relative change of the total volume $\Delta V/V$:

$$k = \frac{p}{\Delta V/V} \qquad (6\text{-}157)$$

The total pressure is the sum of skeleton (or effective) pressure p_{eff} and fluid (or pore) pressure p_{pore}

$$p = p_{\text{eff}} + p_{\text{pore}} \qquad (6\text{-}158)$$

and the total volume change is the sum of the change in solid and in pore fluid volume

$$\Delta V = \Delta V_s + \Delta V_f \qquad (6\text{-}159)$$

The change of the fluid volume is simply

$$\Delta V_p = -V \cdot \Phi \cdot \frac{p_{\text{pore}}}{k_f} \qquad (6\text{-}160)$$

The skeleton pressure and pore pressure act on the solid matrix material of a porous rock. Gassmann (1951) has given the relation for the change of solid volume due to the skeleton pressure p_{eff} and due to the part of the fluid pressure $(1 - \Phi) \cdot p_{\text{pore}}$ acting on the solid material

$$\Delta V_s = -V \cdot \frac{p_{\text{eff}} + (1 - \Phi) \cdot p_{\text{pore}}}{k_s} \qquad (6\text{-}161)$$

Considering an elementary cube of porous rock, its volume change can be described as the effect of

- the skeleton properties and the effective pressure, and
- the additional deformation of the solid material due to the pore pressure:

$$\frac{\Delta V}{V} = -\left(\frac{p_{\text{eff}}}{\bar{k}} + \frac{p_{\text{pore}}}{k_s}\right) \tag{6-162}$$

Solving for the bulk modulus of the saturated porous rock defined by equation (6-157) yields for the bulk modulus of the saturated porous rock

$$k = \frac{p}{\Delta V/V} = \frac{\Phi \cdot (1/k_s - 1/k_f) + \left(1/k_s - 1/\bar{k}\right)}{\left(\Phi/\bar{k}\right) \cdot (1/k_s - 1/k_f) + (1/k_s) \cdot \left(1/k_s - 1/\bar{k}\right)} \tag{6-163}$$

or

$$k = \bar{k} + \frac{\left(1 - \bar{k}/k_s\right)^2}{(1 - \Phi)/k_s + \Phi/k_f - \bar{k}/k_s^2} \tag{6-164}$$

The second term contains the modulus increase by the fluid
With (6-156), we get the following equation for the plane wave modulus:

$$M = \overline{M} + \frac{\left(1 - \bar{k}/k_s\right)^2}{(1 - \Phi)/k_s + \Phi/k_f - \bar{k}/k_s^2} \tag{6-165}$$

Then the velocities for *evacuated (dry)* porous rock are

$$v_p = \left(\frac{\overline{M}}{d}\right)^{1/2} \tag{6-166}$$

$$v_s = \left(\frac{\bar{\mu}}{d}\right)^{1/2} \tag{6-167}$$

and for *fluid saturated* porous rock

$$v_p = \left(\frac{M}{d}\right)^{1/2} \tag{6-168}$$

$$v_s = \left(\frac{\mu}{d}\right)^{1/2} \tag{6-169}$$

Since the shear modulus is not affected by the fluid in the pore space, the velocity of the shear wave is only influenced by the density (equation 6-155). Hence, the speed of shear waves in a fluid saturated rock is moderately less than in an empty skeleton. Equations (6-163) and (6-164) show that the fluid bulk modulus affects the plane-wave modulus and the compressional wave velocity through the factor Φ/k_f. The influence of fluid should decrease with decreasing porosity.

In 1956, Biot published his fundamental papers about the theory of propagation of *elastic waves in a fluid saturated porous solid*. This theory is an extension of Gassmann's theory. Biot included dynamic effects through connected pores in his theory and, additionally, introduced fluid viscosity and hydraulic skeleton permeability into his model. He assumed that the pore fluid may flow relative to the solid, causing viscous frictional losses (thus Biot's theory also describes an attenuation mechanism, see chapter 7.4).
Biot derived a pair of differential equations describing the coupled *motion of the solid-fluid system*. He showed the existence of two dilatational waves, the (normal) seismic wave ("wave of the first kind") and a "wave of second kind" which is, at low frequencies, a highly attenuated diffusion "wave". The existence of the wave of second kind was experimentally confirmed by Plona (1980). Theoretical investigations of de la Cruz and Spanos (1985) also predicts two rotational waves for a viscosity (or high shear modulus) of the pore fluid.
The viscous properties of the model result in a *frequency dependence* of wave velocities (and attenuation). A detailed description is given by Bourbie et al. (1987). In the Biot theory, a *characteristic frequency* f_c is defined

$$f_c = \frac{\eta_f \cdot \Phi}{2\pi \cdot d_f \cdot k_{hydr}} \qquad (6\text{-}170)$$

where η_f is the viscosity of the pore fluid, Φ is the porosity, d_f is the density of the pore fluid, and k_{hydr} is the hydraulic permeability as defined in chapter 2.
The characteristic frequency divides the elastic behaviour into two frequency ranges relating to the dominating effect of fluid motion:

- in the *"low frequency range"* (i.e. at frequencies far below the critical frequency), the motion is controlled by viscous (laminar) drag against the solid,
- in the *"high frequency range"* (i. e. at frequencies far above the critical frequency), the inertia of the fluid dominates.

Table 6.26 shows some values for the characteristic frequency.

Table 6.26 Values for the characteristic frequency in Biot's theory for various pore fluids (after Bourbie et al., 1987)

Material	Φ	k_{hydr} in mD	f_c Water	f_c Normal oil	f_c Heavy oil
Fontainebleau	0.05	10^{-1}	80 MHz	0.8 - 4 GHz	8 - 40 GHz
sandstone	0.20	10^{3}	30 kHz	0.3 - 1.5 MHz	3 -15 MHz
Tight sand	0.08	2 x 10^{-2}	1 GHz	10 - 50 GHz	100 - 500 GHz
Cordova Cream Limestone	0.245	9	4.5 MHz	45 - 230 MHz	0.45 - 2.3 GHz
Sintered glass	0.283	10^{3}	42 kHz	0.42 - 2.1 MHz	4.2 - 21 MHz

Based on Biot's theory, Geertsma and Smit (1961) developed equations for the velocities in terms of compressibility and found an approximate solution based on high values of the characteristic frequency (or low frequency range) for commonly encountered rocks. For the compressional wave of the first kind results in:

$$v_p^2 = \frac{v_{p\infty}^4 + v_{p0}^4\left(\frac{f_c}{f}\right)^2}{v_{p\infty}^2 + v_{p0}^2\left(\frac{f_c}{f}\right)^2} \tag{6-171}$$

where $v_{p0}, v_{p\infty}$ are the zero frequency and infinite frequency velocities respectively. The *compressional wave* velocity in a fluid saturated porous solid at zero frequency follows Gassmann's equation

$$v_{p0} = \left(\frac{k + \frac{4}{3}\cdot\mu}{d}\right)^{1/2} \tag{6-172}$$

The compressional wave velocity in a fluid saturated porous solid at infinite frequency is given by the relation

$$v_{p\infty} = \left(d - d_f \cdot \frac{\Phi}{a}\right)^{-1}$$

$$\cdot \left(\left(\bar{k} + \frac{4}{3}\mu\right) + \left[\frac{\Phi}{a}\frac{d}{d_f} + \left(1 - \frac{\bar{k}}{k_s}\right)\left(1 - \frac{\bar{k}}{k_s} - 2 \cdot \frac{\Phi}{a}\right)\right] \cdot \left[\frac{1 - \bar{k}/k_s - \Phi}{k_s} + \frac{\Phi}{k_f}\right]^{-1} \right)^{1/2} \quad (6\text{-}173)$$

where a is a factor (between one = no coupling and infinity = perfect coupling) describing the degree of coupling between pore fluid and solid skeleton.

At *low frequencies*, the coupling between pore fluid and matrix is perfect; then $a \to \infty$ and equation (6-173) becomes the Gassmann's formula. This shows that Gassmann's theory is a special case (static case or zero frequency, perfect coupling between pore fluid and skeleton motion) of the more general theory of Biot and Geertsma and Smit.

Shear waves result in (see also King, 1966)

$$v_s^2 = \mu \cdot \left(d - d_f \cdot \frac{\Phi}{a}\right)^{-1} \quad (6\text{-}174)$$

which for zero frequency is

$$v_s^2 = \mu \cdot d^{-1} \quad (6\text{-}175)$$

Extensive literature exists regarding the theory of Biot, Geertsma and Smit, directed at velocity as well as attenuation (see chapter 7.). The following list is only a small excerpt: Bourbie et al. (1987), Domenico (1977), Plona (1980), White (1983), Thomson (1985), Murphy (1985), Mörig and Burkhardt (1989), White and Wang (1989), Schröder (1991).

There are four significant aspects of these theories:

1) The equations (e.g. 6-165, 6-173) contain elastic properties, densities and porosity. But one central parameter of the Gassmann-Biot theory is the *skeleton bulk modulus* $\bar{k}$. This modulus itself is a function of the porosity and, also, of other influences (e.g., pressure, internal structure etc.). The velocity calculation needs the determination of this skeleton bulk modulus

- either experimentally (measurement of the evacuated porous rock),
- or theoretically, for example for the sphere pack model.

The unknown $\bar{k}$ is a fundamental problem of the applicability of these theories.

Geertsma and Smit (1961) empirically derived a relation between skeleton modulus, porosity and the solid matrix modulus for clean sandstones (see also White, 1983):

$$\bar{k} = \frac{k_s}{1 + 50 \cdot \Phi} \tag{6-176}$$

Murphy et al. (1993) derived from ultrasonic measurements of sandstone and sand samples under a pressure of 50 MPa the following empirical relationships for the skeleton bulk moduli:

- for $\Phi \leq 0.35$

$$\bar{k} = 38.18 \cdot (1 - 3.39 \cdot \Phi + 1.95 \cdot \Phi^2) \tag{6-177}$$

$$\bar{\mu} = 42.65 \cdot (1 - 3.48 \cdot \Phi + 2.19 \cdot \Phi^2) \tag{6-178}$$

- for $\Phi > 0.35$

$$\bar{k} = \exp(-62.60 \cdot \Phi + 22.58) \tag{6-179}$$

$$\bar{\mu} = \exp(-62.69 \cdot \Phi + 22.73) \tag{6-180}$$

where the moduli results in GPa.

Figure 6.64 compares the decrease of the skeleton bulk modulus after the two equations 6-176 and 6-177. The difference is caused by the different skeleton properties of the investigated sample collections (rock types)

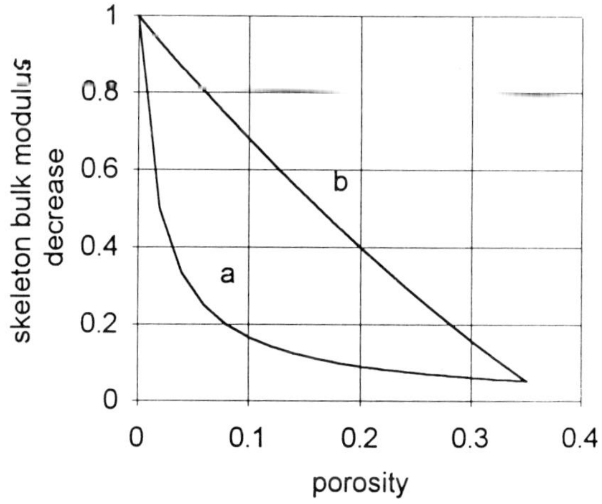

Figure 6.64 Decrease of the skeleton bulk modulus with increasing porosity after equation 6-176 (curve a) and 6-177 (curve b).

2) The equations (e.g., 6-165) can be used for a transformation of the velocity value in the *dry state into the fluid saturated state*; i.e. the "dry velocity" is used to calculate the skeleton modulus

$$\overline{M} = v_{\mathrm{p,dry}}^2 \cdot d \qquad (6\text{-}181)$$

Then with the fluid properties, the calculation of the velocity in a saturated state $v_{\mathrm{p,sat}}$ is possible following equation (6-165).
Figure 6.65 shows two examples:
a.) The traveltime or velocity values of the dry sandstone (curve I) were transformed into a curve for the saturated rock (curve II) using the following properties:
$k_s = 37$ GPa $d_s = 2.65 \times 10^3$ kg m^{-3} $k_f = 0.22$ GPa $d_s = 1.00 \times 10^3$ kg m^{-3}
A reasonable agreement resulted.
b.) The same procedure was applied to data from a sandstone with a clay content of about 4...10 %. The transformation of the "mean dry curve" (curve I) into the "saturated curve" (curve II) resulted in velocities higher than measured. The reason is that the water had softened the skeleton and decreased the skeleton modulus compared with the dry state; in this case the premise of an uninfluenced skeleton modulus is not given. This leads to the assumption that an application is possible particularly in the case of clean porous rocks.

3) From the equations of Gassmann, Biot, Geertsma and Smit some other relations can be derived as *special case.* For a suspension, the skeleton bulk and shear modulus can be neglected. This changes immediately (6-158) into Wood's equation (Wood, 1930)

$$v_{\mathrm{p}}^2 = \left\{ d \cdot \left[\frac{1-\Phi}{k_{\mathrm{s}}} + \frac{\Phi}{k_{\mathrm{f}}} \right] \right\}^{-1} \qquad (6\text{-}182)$$

4) A systematic study published by Domenico (1977) shows the influence of differential- (or effective) pressure and of the *brine-gas saturation* and distribution particularly on the value of the coupling factor.

Besides Biot's consideration of a "*global flow mechanism*" within the pore space a second model exists. Murphy (1984, 1985), Mavko and Jizba (1991) and Mörig (1993) discussed a "*local flow model*" with a mechanism of fluid flow at the grain scale. The model was applied on the problem of seimic wave attenuation (see chapter 7).

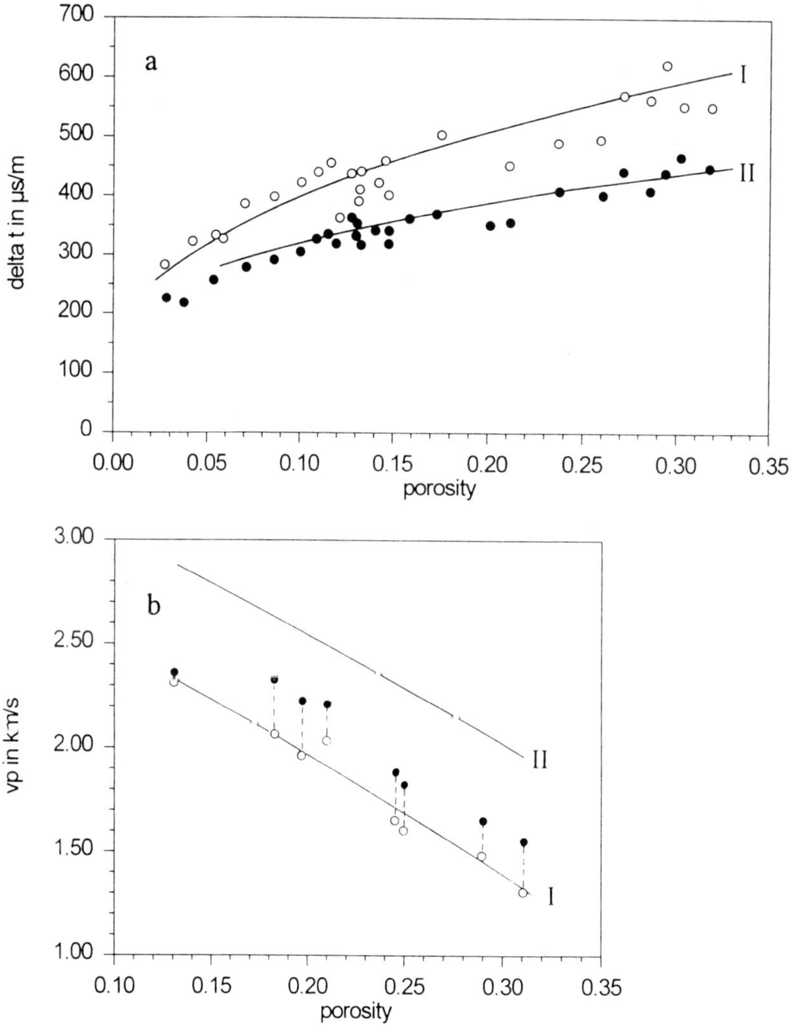

Figure 6.65 Comparison of experimentally determined and transformed velocity values
a - Sandstone, experimental data after Wyllie et al. (1956); b - Shaly sandstone;
curve I: mean curve for dry sandstone; curve II: transformed curve for saturated sandstone

6.4.6. Inclusion models

A homogeneous solid matrix (e.g., porous carbonate rock) containing more or less isolated pores, vugs or cracks that can be modeled as an elastic solid containing *ellipsoidal inclusions*. A variation of the shape of inclusions (from spheres to elliptic cracks or needles) and the properties of the inclusion (empty, gaseous, fluid, solid) opens a broad spectrum of cases. White (1983) distinguished two approaches with respect to the voluminous literature treating these related cases:

- static derivation
- dynamic derivation from the scattering of plane waves.

In this chapter, only a brief description is given including the relationship of these two groups.

6.4.6.1. Static derivation

The fundamental derivation is given by Eshelby (1957). He calculated the change of the elastic energy due to a cavity of ellipsoidal shape. The ellipsoidal cavity has two semi-axes equal to a and a smaller semi-axis b. The *aspect ratio* is then defined as (see Figure 6.66).

$$\textit{Aspect ratio} = \alpha = \frac{b}{a} \tag{6-183}$$

It is unity for spheres and very small for cracks.

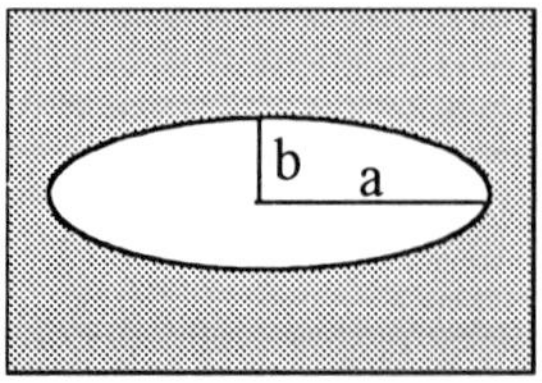

Figure 6.66 Ellipsoidal inclusion and parameters of the aspect ratio $\alpha = b/a$

For the derivation, the assumption is made that interactions between different inclusions can be neglected (so called "first-order" approximation, Hudson and Knopoff, 1989).

The special case of a fluid-filled spherical inclusion results for the moduli in

$$\bar{k} = \frac{k_s}{1 + A \cdot \Phi} \approx k_s \cdot (1 - A \cdot \Phi) \qquad (6\text{-}184)$$

$$\bar{\mu} = \frac{\mu_s}{1 + B \cdot \Phi} \approx \mu_s \cdot (1 - B \cdot \Phi) \qquad (6\text{-}185)$$

with

$$A = \frac{k_s - k_f}{k_s} \cdot \frac{4\mu_s + 3k_s}{4\mu_s + 3k_f} \qquad (6\text{-}186)$$

$$B = 15 \cdot \frac{1 - \sigma_s}{7 - 5\sigma_s} \qquad (6\text{-}187)$$

where k_s, μ_s, σ_s are the bulk modulus, shear modulus, and Poisson's ratio of the solid, k_f is the compressional (bulk) modulus of the pore fluid, and Φ is the porosity.

Eshelby (1957) noted that porosity must be small, since cavities must be far enough apart so that mutual *interaction is negligible*. Hence, the approximations in (6-184) and (6-185) are valid.

For an empty spherical cavity ($k_f = 0$) of radius a in a solid cube of the dimension L, equation (6-184) has the simple form

$$\bar{k} = k_s \cdot \left[1 + 2\pi \frac{1 - \sigma_s^2}{1 - 2\sigma_s} \cdot \left(\frac{a}{L}\right)^3\right]^{-1} \qquad (6\text{-}188)$$

Starting from these fundamental equations two directions can be taken:

(1) In a somewhat different way Walsh (1965) derived a relationship for a low concentration of empty spherical cavities, which is identical to equation (6-188). If there are spheres of different radius, a is replaced by

$$\bar{a} = \left(\frac{1}{N} \sum_{n=1}^{N} a_n^3\right)^{1/3} \qquad (6\text{-}189)$$

or more general with parameters A and B (see Bourbie et al., 1987):

$$\bar{k} = k_s \left(1 + A \frac{\Phi}{a}\right)^{-1} \qquad (6\text{-}190)$$

$$\bar{\mu} = \mu_s \left(1 + B\frac{\Phi}{a}\right)^{-1} \tag{6-191}$$

For "*penny-shaped*" cracks (circular cracks with radius a and thickness $b < a/50$), Walsh derived (see also White, 1983)

$$\bar{k} = k_s \cdot \left[1 + \frac{16}{9} \cdot \frac{1-\sigma_s^2}{1-2\sigma_s} \cdot \left(\frac{a}{L}\right)^3\right]^{-1} \tag{6-192}$$

A study of the effect of non elliptical cracks on the compressibility of rocks using dislocation theory was published by Mavko and Nur (1978). The shape of the empty crack is given by the half crack length l and the half crack width w. If in a volume V there are N cracks, we get an equation for the modulus (6-192) with the term $(a/L)^3$ replaced by $(N \cdot l^2 \cdot w \cdot V^{-1})$.

(2) Budiansky and O'Connell (1976) derived the "self consistent method" for calculation of elastic properties of a material with inclusions. The principle is demonstrated, e.g., by White (1983) as follows: The fundamental equations for the case of a set of empty flat circular cracks are

$$\bar{k} = k_s \cdot \left[1 - \frac{16}{9} \cdot \frac{1-\bar{\sigma}^2}{1-2\bar{\sigma}} \cdot \left(\frac{\bar{a}}{L}\right)^3\right] \tag{6-193}$$

$$\bar{\mu} = \mu_s \cdot \left[1 - \frac{32}{45} \cdot \frac{(1-\bar{\sigma})(5-\bar{\sigma})}{2-\bar{\sigma}} \cdot \left(\frac{\bar{a}}{L}\right)^3\right] \tag{6-194}$$

with

$$\left(\frac{\bar{a}}{L}\right)^3 = \frac{45}{16} \cdot \frac{(\sigma_s - \bar{\sigma})(2-\bar{\sigma})}{(1-\bar{\sigma}^2)(10\sigma_s - \bar{\sigma} - 3\bar{\sigma}\sigma_s)} \tag{6-195}$$

where $\bar{\sigma}$ is the average Poisson's ratio.

For a given distribution of radii and number of cracks, the left term in eq. (6-195) is known and allows the determination of $\bar{\sigma}$. This calculated Poisson's ratio will be inserted into equations (6-193) and (6-194) to obtain the moduli $\bar{k}, \bar{\mu}$ of the cracked medium.

Hudson (1981) also developed a theoretical expression for the effects of aligned cracks on the elastic constants of cracked solids. The equations result for velocities in a general form

$$v = v_{\mathrm{m}}[1 - f(E_{\mathrm{s}}, E_{\mathrm{p}}, \mu_{\mathrm{s}}, \mu_{\mathrm{p}}, \alpha, \varepsilon)]^{1/2} \tag{6-196}$$

where v is the velocity of the cracked material, v_{m} is the velocity of the uncracked solid matrix, E_{s} and E_{p} are the Young modulus for the solid material and pore fluid, μ_{s} and μ_{p} are the shear modulus for the solid material and pore fluid, α is the aspect ratio, and ε the crack density.

The equations 6-184, 6-185, and 6-193, 6-194, and 6-196 show the same general form as the simple model in section 6.4.3.4 ; they give a physical description of the parameter D.

6.4.6.2. Dynamically determined properties

The theory of scattering gives another possibility. Kuster and Toksöz (1974) and Toksöz et al. (1976) developed a method to calculate the average elastic and anelastic properties of a model by studying *plane wave scattering* by inclusions. Their model describes the porous rock as a solid matrix containing isolated spherical or oblate spheroidal inclusions of different aspect ratios α in various concentrations. Interactions between pores are ignored; therefore the theory is valid only for a "dilute pore or crack concentration" (porosity much less than pore aspect ratio).

A solution for higher porosities is given with the *effective medium theory*: The total number of pores and cracks can be divided into N sets; the pore concentration of each set is so small that the assumption of a dilute pore concentration is valid. The calculation of the effective elastic properties is possible now for the medium with the dilute pore concentration as a first step. The next set of pores is introduced into this effective medium in a similar way, and the effective elastic properties are then updated. This procedure is repeated until the total pore concentration (porosity) is reached. In general, the larger N is, the more accurate are the results (Xu and White, 1995).

For the general case of n different pore filling components with incompressibilities $k_{\mathrm{f},n}$ and m aspect ratios $\alpha_{m,n}$, the following equations result for the bulk modulus $\bar{k}$ and shear modulus $\bar{\mu}$ of the effective medium (Toksöz et al., 1976, Endres and Knight, 1989):

$$\frac{\bar{k} - k_{\mathrm{s}}}{3\bar{k} + 4\mu_{\mathrm{s}}} = \frac{1}{3} \sum_{n=1}^{N} \frac{k_{\mathrm{f},n} - k_{\mathrm{s}}}{3k_{\mathrm{s}} + 4\mu_{\mathrm{s}}} \cdot \left\{ \sum_{m=1}^{M} C(\alpha_{m,n}) T_{iijj}(\alpha_{m,n}) \right\} \tag{6-197}$$

$$\frac{\bar{\mu}-\mu_s}{6\bar{\mu}(k_s+2\mu_s)+\mu_s(9k_s+8\mu_s)}=\sum_{n=1}^{N}\frac{\mu_{f,n}-\mu_s}{25\mu_s(3k_s+4\mu_s)}$$

$$\cdot\left\{\sum_{m=1}^{M}C(\alpha_{m,n})\left[T_{ijij}(\alpha_{m,n})-\frac{1}{3}T_{iijj}(\alpha_{m,n})\right]\right\} \tag{6-198}$$

where

$\bar{k}, \bar{\mu}, \bar{d}$	are the elastic moduli and density of the effective medium,
k_s, μ_s,	are the elastic moduli and density of the solid matrix material,
$k_{f,n}, \mu_{f,n}, d_{f,n}$	are the elastic moduli and density of the n pore phases,
$C(\alpha_{m,n})$	is the volume fraction of the inclusions filled with fluid n with the aspect ratio $\alpha_{m,n}$; thus, the porosity is

$$\Phi=\sum_{n=1}^{N}\sum_{m=1}^{M}C(\alpha_{m,n}) \tag{6-199}$$

$\Phi_{f,n}$	are the fractional void volumes occupied by the n- th pore phase,
N	is the number of pore fluids within the rock,
M	is the number of different aspect ratio spheroids which are used to describe the pore structure.

The terms T_{iijj} and T_{ijij} are scalar quantities, functions of aspect ratio, pore and matrix elastic properties (see Kuster and Toksöz, 1974, Toksöz et al., 1976).

For water- and gas-saturated rocks, Figure 6.67 shows normalized compressional- and shear wave velocities as a function of porosity for different aspect ratios. The normalized velocity is the velocity divided by the matrix velocity. From Figure 6.67 it is seen that (Toksöz et al, 1976):

1) the velocities decrease with increasing porosity,
2) for a given porosity, flatter pores have a relatively greater effect on velocities than rounder or spherical pores,
3) the properties of the saturating fluid have greater effects on compressional velocities than on shear velocities,
4) for rounder pores, velocities are higher for the gas-saturated case than for the water-saturated case; for flatter pores, gas-saturated velocities are lower than those of the water-saturated case. (For natural rocks, the compressional velocities for the dry or gas-saturated cases are mostly lower than for the water-saturated case. Shear velocities are about the same; for highly porous rocks, the velocity for the gas-saturated case is somewhat higher).

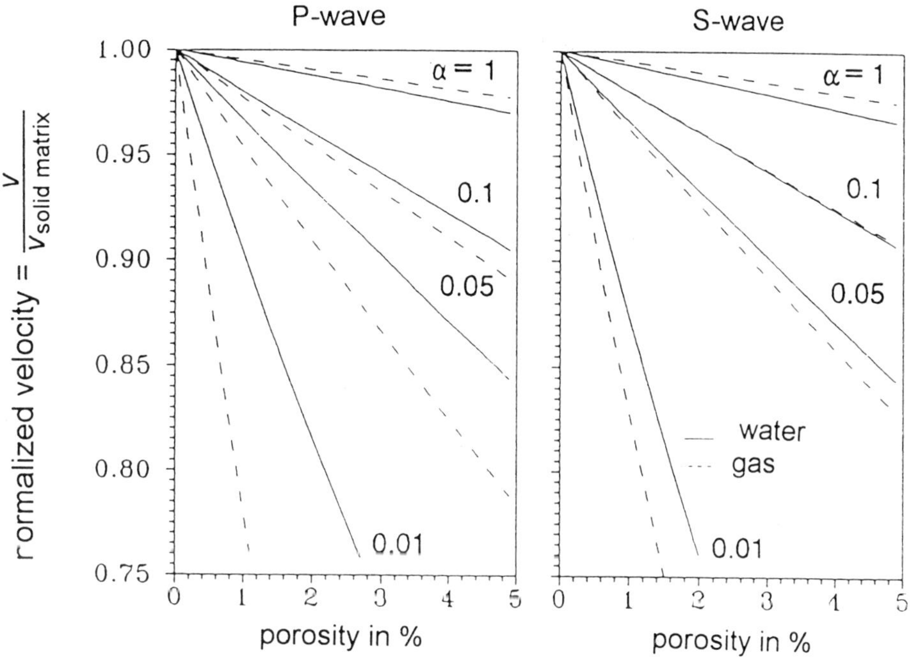

Figure 6.67. Normalized compressional and shear wave velocities versus porosity (volume concentration of inclusions) for different aspect ratios (after Toksöz et al., 1976)

For spherical inclusions ($\alpha = 1$) with fluid pore content ($\mu_{f,n} = 0$) and a fractional volume (porosity) Φ the equations 6- 197 and 6-198 reduce to (Tao and King, 1993):

$$\frac{\bar{k} - k_s}{3\bar{k} + 4\mu_s} = \Phi \cdot \frac{k_f - k_s}{3k_p + 4\mu_s} \tag{6-200}$$

$$\frac{\bar{\mu} - \mu_s}{6(k_s + 2\mu_s) + \mu_s(9k_s + 8\mu_s)} = -\Phi \cdot \frac{1}{9k_s + 8\mu_s} \tag{6-201}$$

One significant result is that the above equations are identical to the Hashin-Shtrikman upper bounds (see next section) for a two-phase medium (Tao and King, 1993).

Zimmerman and King (1986) have applied the theory of Kuster and Toksöz to interpret the influence of freezing in unconsolidated sediments on the elastic wave velocities.

Tao and King (1993) modified the model to approximate a pore aspect ratio spectrum. For the "equivalent spectrum approximation" equidimensional pores are approximated by spheres, while grain boundary pores and flat pores are approximated by low aspect ratio cracks. In a further study, Tao et al., (1995) have derived the pore aspect ratio spectrum from the measured velocities of sandstones; this iterative *inversion* algorithm based on the scheme developed by Cheng and Toksöz (1979). The aspect ratio spectrum of the sandstones are compatible with the impression of SEM photographs.

Xu and White (1995) developed a velocity model for clay-sand mixtures in *combined* terms of the Kuster and Toksöz, effective medium and Gassmann theories:

In a first step the elastic properties of the dry rock (skeleton or frame elastic bulk and shear modulus) were calculated using the Kuster and Toksöz theory. For the calculation the total pore space consists of two parts:

(1) pores associated with sand grains,

(2) pores associated with clay.

The essential feature of the model is the assumption that the geometry of pores associated with sand grains is significantly different from that associated with clays. Clay particles create pores with very small aspect ratios.

The result of the calculation are compressional and shear wave velocities as a function of porosities and aspect ratios of the two parts (sand, clay). From the velocities of the dry rock the bulk and shear modulus of the rock skeleton can be derived.

In the second step the Gassmann (1951) model (see 6.4.5) is used to simulate the fluid effect.

The model was applied on experimental data. A comparison with laboratory measurements by Han et al. (1986) - see Figure 6.23 - results in aspect ratios of 0.15 for sand-related pores and 0.04 for clay-related pores. The application on log measurements and a generalization show that the mean aspect ratios are in the range 0.02 ... 0.05 for clay-related pores, around 0.12 for sand-related pores, and 0.1 for carbonate-related pores.

Toksöz et al (1976) also studied the case of a porous rock saturated with a *mixture of fluids* and found that

► immiscible gas bubbles inside pore water very rapidly lower the compressional velocity,

- if gas and water occupy different pores, the component occupying the thinner pores has greater effect on velocities at low pressures, whereas, at higher pressures the component occupying the rounder pores dominates (as a result of closing the thin, flat pores).

Also, Endres and Knight (1989) have studied the influence of the fluid distribution within the pore space upon the elastic behavior and found that "the nature of v_p as a function of S_w is strongly dependent on the saturating scheme ... In all cases, the relationship ... is determined by microscopic fluid distribution within the pore structure."

6.4.7. Bounds

In section 6.4.1, the relatively simple assumption of uniform stress or strain, yields the Voigt and Reuss estimates, respectively, as *bound values* for a composite medium. Particularly for cases of large contrast, the unsuitability of Voigt and Reuss bounds is evident. A very small but finite amount of one of the components determines the order of the resulting modulus

- in the case of the *Voigt model* if the modulus of this small component is very high in comparison to the other components,
- in the case of the *Reuss model* if the modulus of this small component is very small in comparison to the other components.

Hashin and Shtrikman (1961) have used *variational principles* to find upper and lower bounds for the moduli of composite materials, which are more restrictive than the Voigt and Reuss bounds. A concentrated description and discussion of the derivation of the "Hashin-Shtrikman-bounds" is given by Hudson and Knopoff (1989). The derivation is carried out with the assumption that the spatial distribution of each component is isotropic, but without regard for the geometry of the particles or inclusions ("arbitrary phase geometry", Hashin, 1964).

In the two-phase system, characterized by

V_1, V_2	volume fractions of the components ($V_1 + V_2 = 1$),
k_1, k_2	bulk (compression) moduli of the two components,
μ_1, μ_2	shear moduli of the two components

the following boundaries for the bulk modulus $\bar{k}$ and shear modulus $\bar{\mu}$ of the composite result

$$k_1 + V_2 \cdot \left(\frac{1}{k_2 - k_1} + \frac{3V_1}{3k_1 + 4\mu_1} \right)^{-1} < \bar{k} < k_2 + V_1 \cdot \left(\frac{1}{k_1 - k_2} + \frac{3V_2}{3k_2 + 4\mu_2} \right)^{-1} \qquad (6\text{-}202)$$

$$\mu_1 + V_2 \cdot \left(\frac{1}{\mu_2 - \mu_1} + \frac{6}{5\mu_1} \cdot \frac{V_1 \cdot (k_1 + 2\mu_1)}{3k_1 + 4\mu_1} \right)^{-1} < \bar{\mu}$$

$$< \mu_2 + V_1 \cdot \left(\frac{1}{\mu_1 - \mu_2} + \frac{6}{5\mu_2} \cdot \frac{V_2 \cdot (k_2 + 2\mu_2)}{3k_2 + 4\mu_2} \right)^{-1} \qquad (6\text{-}203)$$

where $k_1 < k_2, \mu_1 < \mu_2$ is assumed.
The lower bound for $\bar{k}$ is equal to the compound bulk modulus of a composite in which spherical inclusions of component 2 are embedded in a matrix of component 1 and the spheres have no contact with each other (Hashin, 1962; Hudson and Knopoff, 1989).

A simple numerical example may demonstrate the range of the Reuss-Voigt and Hashin-Shtrikman bounds:

For	$k_1 = 50$ MPa	$k_2 = 10$ MPa	or $\frac{k_1}{k_2} = \frac{5}{1}$
	$k_1 = \mu_1$	$k_2 = \mu_2$	or $\frac{\mu_1}{\mu_2} = \frac{5}{1}$
	$V_1 = 0.6$	$V_2 = 0.4$	

the following bounds result:

Modulus	Reuss-Voigt		Hashin-Shtrikman	
	lower	upper	lower	upper
bulk modulus	19.2 MPa	34.0 MPa	24.2 MPa	29.9 MPa
shear modulus	19.2 MPa	34.0 MPa	23.2 MPa	28.8 MPa

6.4.8. Models with variable internal structure

6.4.8.1. Principle of derivation

In order to derive a simple model which allows description of the different influences and dependencies, we look at a real porous rock, e.g., a sandstone (Figure 6.68a). It may represent the behaviour of a given homogeneous part of the rock. For the derivation, we distinguish three different parts of the rock:

- the solid matrix material (index s),
- the pore filling substance (index p), and
- the contact region (index c) with properties that may be different from those of the solid matrix material as result of a different substance and/or the particular shape of the contact.

These three parts are shown in Figure 6.68b as a rock element and connected with two Cartesian coordinate systems:

In the *"macrosystem"* $x_1 - x_2 - x_3$ the x_3- axis is vertical (up).

In the *"microsystem"* $x_{\bar{1}} - x_{\bar{2}} - x_{\bar{3}}$ is related to the microscopic structure of the two rock particles and their connected pore space. The $x_{\bar{1}} - x_{\bar{2}}$ - plane is the plane of the particle contact and the $x_{\bar{3}}$- axis the normal direction upwards on this plane. The microsystem related to the macrosystem is described by an angle α (α is one of the three Euler angles; the other two are constant 45° and 225°, see Figure 6.68b). This angle α is a *measure* only for the internal structure of the rock and, therefore, called *"structure angle"*. It has a statistical characteristic.

For further consideration, the rock element is idealized as a model with rectangular cross section (Figure 6.68c). Thus, the model consists of a *rectangular block with a surrounding pore channel*, a special *contact region* and a *spatial orientation*.

The following influences can be described:

- by varying the relative volume of the block and pore channel, the porosity changes,
- by variation of the cross section of the block, the "grain shape" and of the pore channel, the "pore shape" varies; the dimensions simulate the "grain size";
- by variation of the angle α, the internal structure (texture) is modified; the model has the geometry of a transverse isotropic material;
- by variation of the material properties for the solid block and the pore channels, the influence of matrix and pore fluid properties is introduced;

► by the particular properties of the contact region, a consideration of different kinds of "bonding" (cementation, etc.) and, also, of the pressure influence is possible.

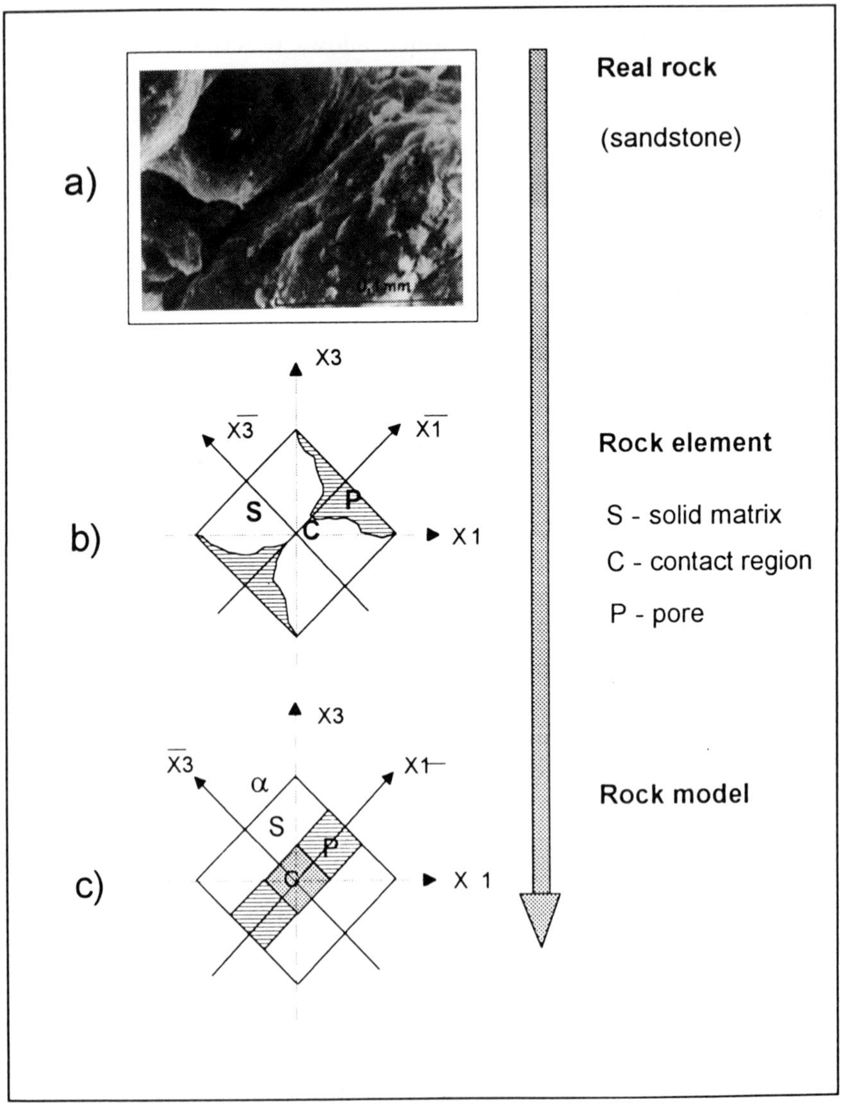

Figure 6.68 A model for porous rocks including a variable internal structure - principle of derivation

Figure 6.69 shows the model with the pore channel and the signature for the geometrical description. The following parameters are used to describe the "pore and grain geometry":

$2a$ - grain size

$2b$ - pore size

α_{grain}- grain aspect ratio (flattness of the grain)

$$\alpha_{grain} = \frac{\text{grain dimension in } x_{\bar{1}} - x_{\bar{2}} - \text{ direction}}{\text{grain dimension in } x_{\bar{3}} - \text{ direction}} \tag{6-204}$$

α_{pore}- pore aspect ratio (flattness of the pore)

$$\alpha_{pore} = \frac{\text{pore dimension in } x_{\bar{1}} - x_{\bar{2}} - \text{ direction}}{\text{pore dimension in } x_{\bar{3}} - \text{ direction}} \tag{6-205}$$

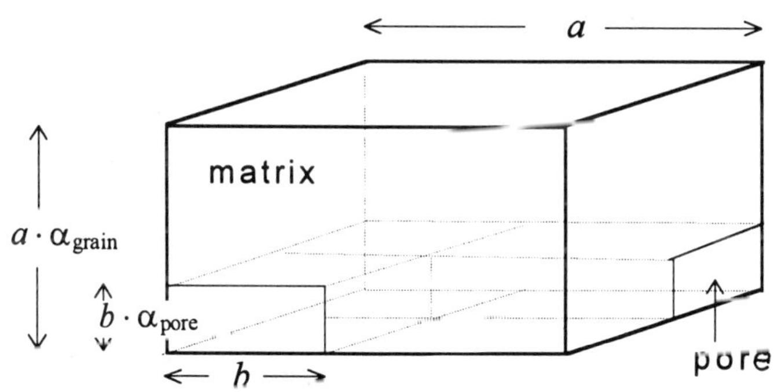

Figure 6.69 Model element with geometrical dimensions and signature; a and $a \cdot \alpha_{grain}$ are the dimensions of the matrix block, b and $b \cdot \alpha_{pore}$ are the dimensions of the pore channel.

The porosity of the model is

$$\Phi = \left(\frac{b}{a}\right)^3 \cdot \left(2 \cdot \frac{a}{b} - 1\right)\left(\frac{\alpha_{grain}}{\alpha_{pore}}\right) \tag{6-206}$$

Any calculation of physical properties follows two steps

(1) Calculation of properties with respect to the *microsystem*,

(2) Transformation of this properties into the *macrosystem*.

For the calculation of the elastic properties in the *microsystem*, we consider the (dry) porous rock as a parallel and series arrangement of springs with the properties of the matrix and the contact region. For a normal pressure $p_{\bar{3}}$ acting in $x_{\bar{3}}$-direction, a deformation $\varepsilon_{\bar{3}}$ results

$$\varepsilon_{\bar{3}} = M_{\mathrm{N}}^{-1} \cdot p_{\bar{3}} = \left\{ \frac{1}{2}\left(\frac{1}{M_{\mathrm{s}}} + \frac{1}{M_{\mathrm{c}}} \right) \right\} \cdot G \cdot p_{\bar{3}} = M_{\mathrm{s}}^{-1} \cdot \left\{ \frac{1 + M_{\mathrm{s}}/M_{\mathrm{c}}}{2} \right\} \cdot G \cdot p_{\bar{3}} \tag{6-207}$$

where

M_{N} is the plane wave modulus of the rock model for a pressure normal to the contact plane,
M_{s} is the plane wave modulus of the solid matrix material,
M_{c} is the plane wave modulus of the contact region,
G is a geometric parameter, it contains only geometric properties (ratio of "pore diameter and grain dimension" (b/a) and aspect ratios) of the model:

$$G = \left[1 - \frac{\alpha_{\mathrm{grain}}}{\alpha_{\mathrm{pore}}} \cdot \frac{b}{a} \cdot \left(1 - \left(1 - \frac{b}{a} \right)^{-2} \right) \right]^{-1} \tag{6-208}$$

For simplification, it is assumed that the solid part of the model is divided into one half of the contact region (modulus M_{c}) and one half of pure solid matrix material (modulus M_{s}). For this consideration, the dry case is assumed (no contribution of the pore fluid).

The derivation for the other components is possible in the same manner and results in the strain-stress relations for the microsystem. The transformation into the macrosystem (see Schön, 1974, 1983) gives the *tensor of elasticity* for the model, consisting of a scalar and a tensorial term

$$C_{mn} = \left(M_{\mathrm{s}} \cdot \frac{2}{1 + M_{\mathrm{s}}/M_{\mathrm{c}}} \cdot \frac{1}{G} \right) \cdot S_{mn}(\alpha, f) \tag{6-209}$$

The components of the tensor S_{mn} depend only on the angle α and on a parameter f defined as the ratio of the compressional and shear deformation properties of matrix and contact region:

$$f = \frac{M_{\mathrm{s}}}{\mu_{\mathrm{s}}} \cdot \frac{1 + (\mu_{\mathrm{s}}/\mu_{\mathrm{c}})}{1 + (M_{\mathrm{s}}/M_{\mathrm{c}})} \tag{6-210}$$

where μ_s, μ_c are the shear moduli of the solid matrix material and of the contact region, respectively.

The tensor C_{mn} is the tensor of elasticity for the rock skeleton and, thus, it represents the skeleton bulk properties in Biot's theory (see section 6.4.5). Figure 6.70 shows the dependence of four components of the tensor S_{mn} on the angle α and the parameter f. These four elements are necessary for a velocity calculation related to the vertical and horizontal direction of wave propagation; a detailed description is given by Schön (1974, 1983).

For the elastic wave velocities in the *dry porous rock*, one has

$$v_{...} = \left\{ \left[\frac{M_s}{d_s} \right] \cdot \left[\frac{2}{1 + M_s/M_c} \right] \cdot \left[\frac{G}{1 - \Phi} \right] \cdot [s_{mn}(\alpha, f)] \right\}^{1/2} \tag{6-211}$$

The velocity is given as a product of *four factors*; each of them contains a special "field of influence":

► the first contains only the *solid matrix properties* and is identical to the compressional wave velocity for the pore-free matrix,

► the second contains the *influence of a contact region* with its different properties related to the matrix material, this influence is expressed by the ratio of the two moduli; for $M_c \to M_s$ this factor becomes unity.

This term also contains the *pressure dependence* of the velocity controlled by a strong pressure dependence of the contact elasticity (modulus M_c). For this modulus, a power dependence on the pressure can be assumed as a generalization of the Hertz theory (see equation 6-147 and by comparison with experimental results by Schön, 1983) in the form

$$M_c(p) = M_{c,0} \cdot \left(\frac{p}{p_0} \right)^m \tag{6-212}$$

where $M_{c,0}$ is the contact modulus for a (initial) pressure p_0.

► the third contains the influence of *porosity* Φ and model- (or grain-) *shape* expressed by the parameter G;

► the fourth is an element of the structure tensor (Figure 6.70) and depends on the internal *structure* (expressed by the angle α) and the parameter f, which is controlled by the "*bonding* properties" of the contact (see later). Compressional and shear waves with propagation in vertical and horizontal direction differ only in this last term. Which component we use depends on the wave-type of interest (Table 6.27).

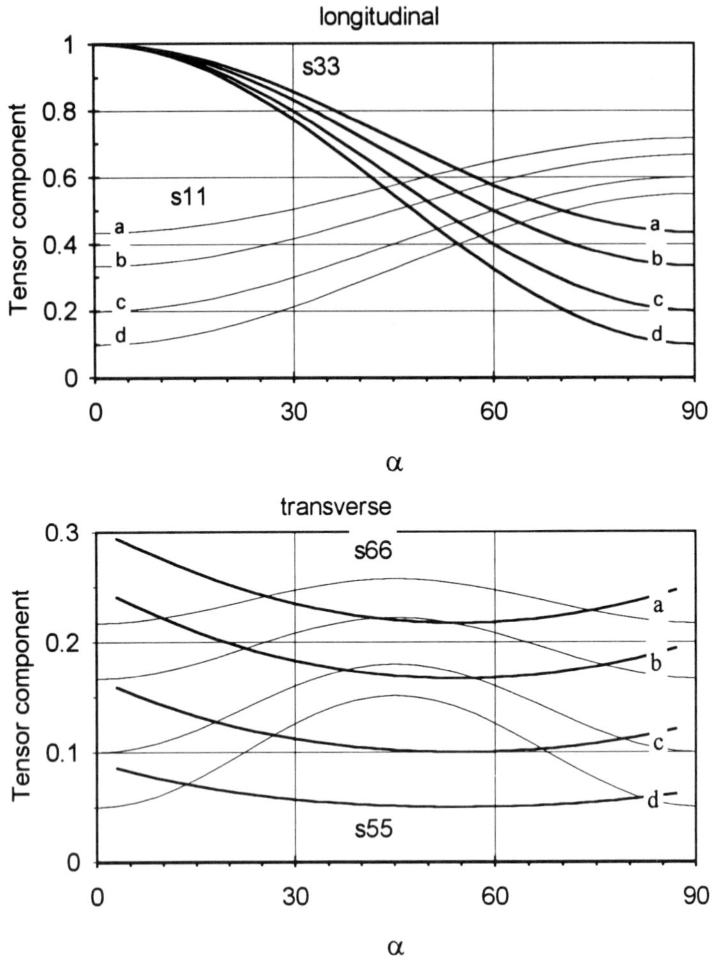

Figure 6.70. Components of the structure tensor as a function of the structure angle α and the contact bonding parameter *f*; a - *f* = 2.3 b - *f* = 3.0 c - *f* = 5.0 d - *f* = 10

It follows immediately that velocity *ratios* (e.g., v_p/v_s, anisotropy parameters) depend only on *structural and bonding* properties. A more general version of equation (6-211) gives the velocities as a function of

- properties and volume fractions of the constituents,
- internal geometry (grain shape, structure),
- bonding or contact properties.

The velocity for a considered volume as a macroproperty results, therefore, as

macroproperty = function of microproperties, structure, and bonding

Table 6.27. Components of the structure tensor for different wave-types and directions (the index transformation - see section 6.1 - is applied)

Signature	wave type	direction of propagation	direction of polarization	tensor component
v_{11}, (v_{22})	longitudinal	x_1, (x_2)		s_{11}
v_{33}	longitudinal	x_3		s_{33}
v_{12}, (v_{21})	transverse	x_1, (x_2)	x_2, (x_1)	s_{66}
v_{13}, (v_{23})	transverse	x_1, (x_2)	x_3	s_{55}
v_{31}, (v_{32})	transverse	x_3	x_1, (x_2)	s_{55}

The above equations are valid for dry porous rock without influence of the pore filling. A transformation for the additional effect of a pore fluid is possible by an extension of the model including the elastic contribution of the pores or by application of the Biot-Gassmann equations.

6.4.8.2. Examples for application

In the following, we discuss some applications and possibilities of the structural model. Such a discussion is made easier by the *decomposition* of the equation (6-211) as a product, which allows a separate discussion of the four factors. The discussion concentrates on

- the influence of porosity and "grain shape" $\left[\frac{G}{1-\Phi}\right]^{1/2}$
- the influence of structure and bonding properties $[s_{ik}(\alpha, f)]^{1/2}$

Figure 6.71 shows the third factor of equation (6-211) as a function of the *porosity* for different model cross sections. The model cross section type (or grain shape) is characterized by the ratio $(\alpha_{grain}/\alpha_{pore})$

The curves show the relative decrease of the velocity $v(\Phi)/v_0$ with increasing porosity. The velocity is normalized with the value v_0 (it includes the terms 1, 2, and 4). But the figure also shows the strong controlling *influence of the "grain shape"* on the character of the velocity-porosity-relation.

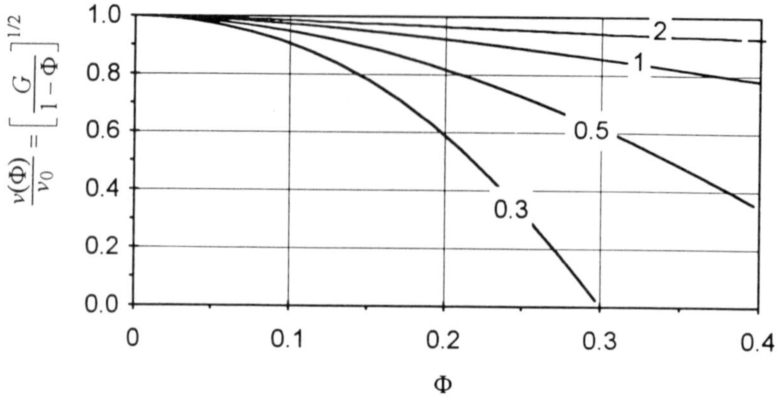

Figure 6.71 Calculated decrease of the velocity (expressed by $\frac{v(\Phi)}{v_0} = \left[\frac{G}{1-\Phi}\right]^{1/2}$) on the porosity Φ for different model cross sections, expressed by the curve parameter ($\alpha_{grain}/\alpha_{pore}$)

A *comparison with experimental results* confirms its applicability: Figure 6.72 shows the comparison of calculated dependences (Figure 6.71) and experimental results from Figure 6.23 for four classes of shale content. Iteratively the optimal approximation was determined by variation of v_0. Obviously v_0 decreases with increasing shale content.

As an example for the shale content Vsh = 0...0.1 the best approximation was obtained using a model cross section with $\alpha_{grain}/\alpha_{pore} \approx 0.5$. The curves correspond to the following relation for the velocity in km/s:

$$v_p = 5.3 \cdot \left[\frac{G}{1-\Phi}\right]^{1/2} \tag{6-213}$$

For this particular cross section is (see 6-206, 6-208)

$$G = \left[1 - 0.5 \cdot \frac{b}{a} \cdot \left(1 - \left(1 - \frac{b}{a}\right)^{-2}\right)\right]^{-1} \tag{6-214}$$

$$\Phi = \left(\frac{b}{a}\right)^3 \cdot \left(\frac{a}{b} - \frac{1}{2}\right) \tag{6-215}$$

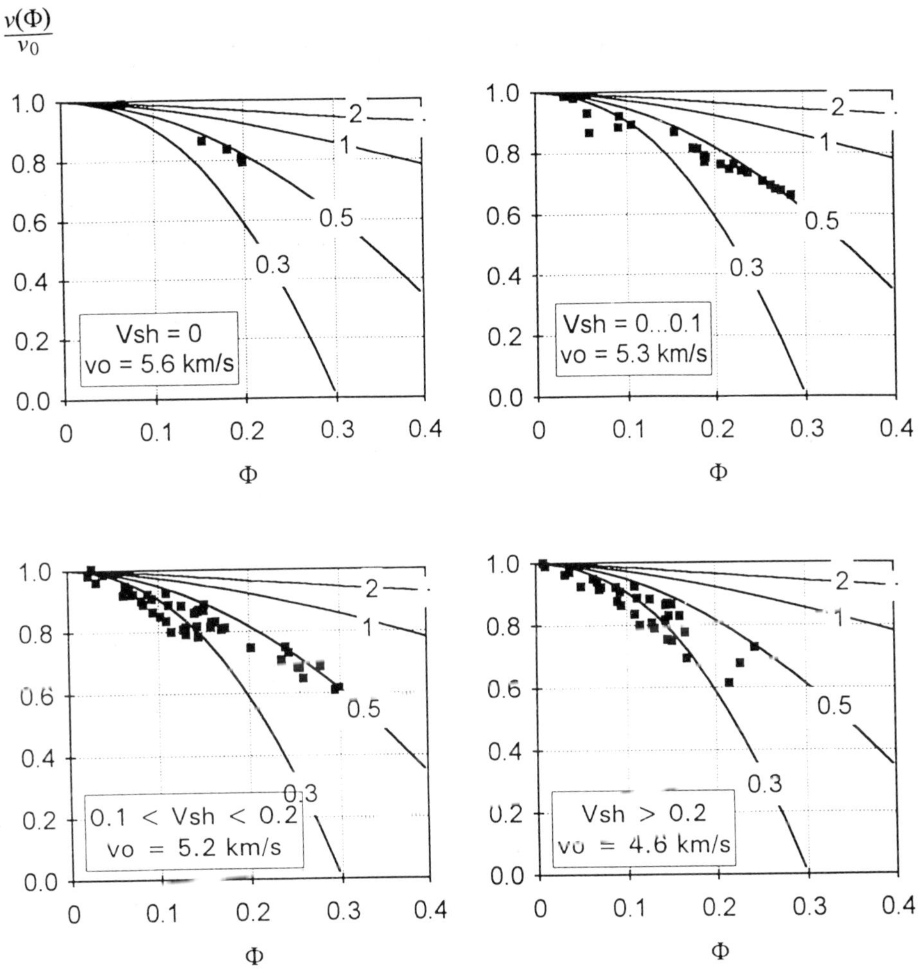

Figure 6.72. Comparison between measured and calculated porosity dependence; sandstone with various shale content (data from Figure 6.23 after Han et al., 1986)

The fourth term in equation 6-211 contains the *structure and bonding* properties, which particularly control the *velocity-ratios*. Therefore, for the following considerations such ratios are studied.

Figure 6.73 shows the v_p/v_s- ratio for vertical and horizontal propagation and the anisotropy ratio for longitudinal and transverse waves as a function of the structure angle and the parameter f.

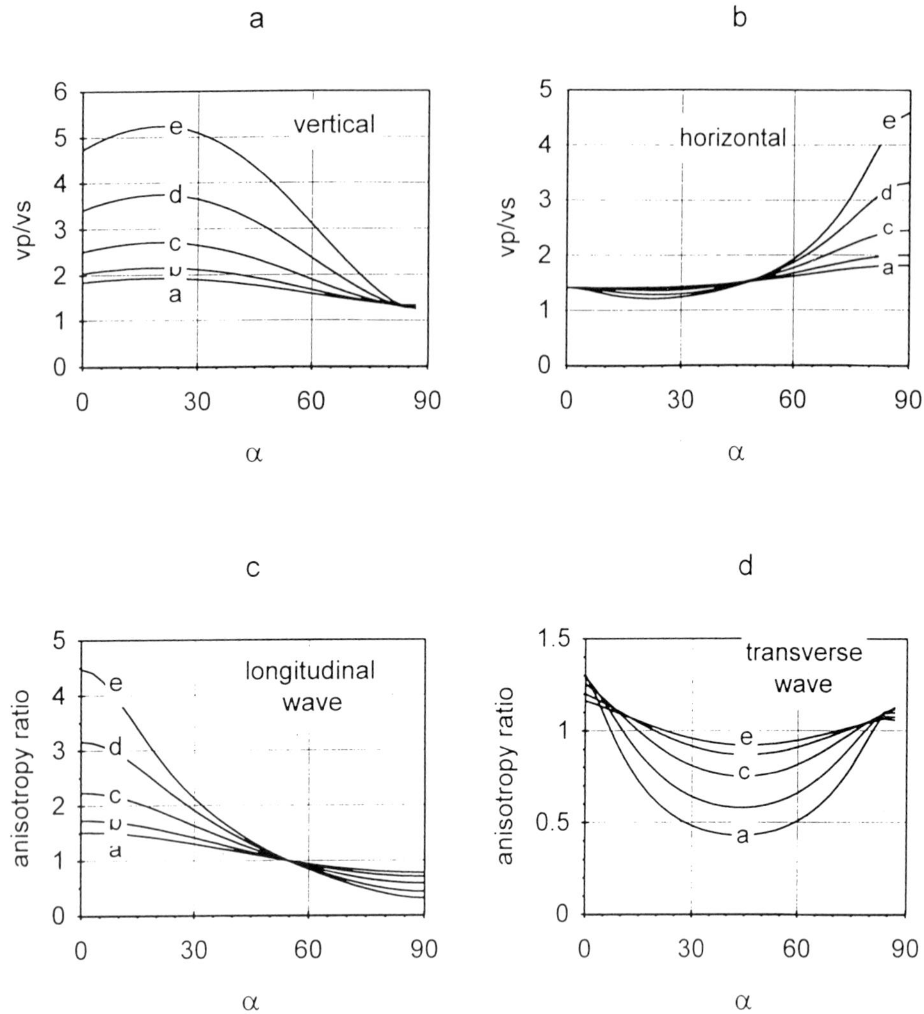

Figure 6.73 Velocity ratio as a function of the structure angle and the parameter *f*:
a) v_p/v_s for vertical wave propagation b) v_p/v_s for horizontal wave propagation (v_s is the velocity of the horizontally polarized wave)
c) anisotropy ratio for the longitudinal wave d) anisotropy ratio for the transverse wave

For the following considerations, a grid v_p/v_s versus v_s is useful. This grid related to vertical propagation and following equation 6-211 becomes

$$\frac{v_p}{v_s} = \frac{v_{33}}{v_{31}} = \left(\frac{s_{33}}{s_{55}}\right)^{1/2} \quad \text{and} \quad v_s = v_{31} \sim (s_{55})^{1/2} \quad \text{or} \quad v_s = A \cdot (s_{55})^{1/2}$$

where A contains the factors 1 to 3 of equation 6-211.

Figure 6.74 shows such a $\frac{v_p}{v_s} = \left(\frac{s_{33}}{s_{55}}\right)^{1/2}$ versus $v_s/A = (s_{55})^{1/2}$ plot. Curve parameters of the resulting "spider web" controlling the velocity ratio are the structure angle α and the bonding property f.

For the bonding parameter f there are two extreme situations:

a) For a *completely cemented material* the contact properties and solid matrix properties are equal:

$$M_c = M_s \qquad \mu_c = \mu_s$$

or $$\frac{M_s}{M_c} = \frac{\mu_s}{\mu_c}$$

In this case is $f = \frac{M_s}{\mu_s} \approx 2.3...2.5$. This represents the low limit value for the parameter f.

b) for a *"soft" cemented or unconsolidated porous rock* the contact modulus for a tangential stress μ_c decreases much more than the modulus for the normal stress M_c at the contact; the extreme case is a normal stiffness and a tangential sliding at the contact. Therefore in such formations it is

$$\frac{M_s}{M_c} < \frac{\mu_s}{\mu_c}$$

This results in higher values for the parameter f, i.e. $f > 2.3$.

Thus, we obtain the *rule*: Increasing cementation and bonding quality is related to decreasing f.

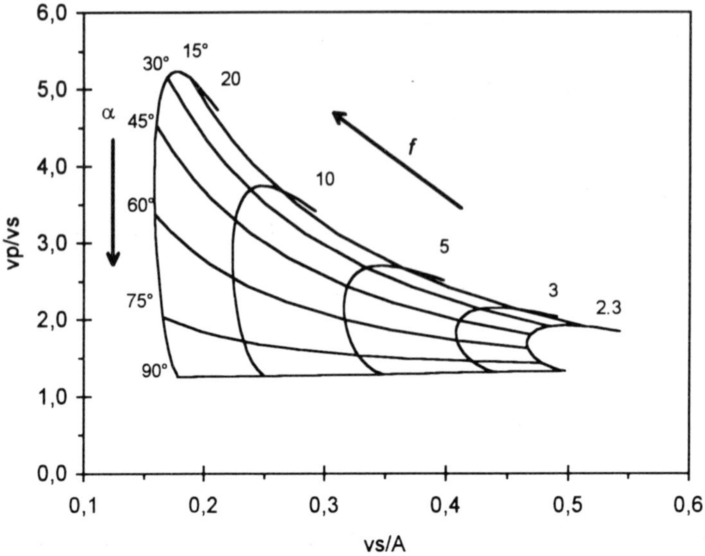

Figure 6.74. Calculated plot $\frac{v_p}{v_s} = \frac{v_{33}}{v_{31}} = \left(\frac{s_{33}}{s_{55}}\right)^{1/2}$ versus $(s_{55})^{1/2} = v_{31}/A = v_s/A$ for the interpretation of experimental data; curve parameters are the structure angle α and the bonding property f

Based on this, the following examples are discussed:

1) A comparison between the experimental investigations in Figure 6.40 with the grid in Figure 6.74 confirms the *general tendency* of the correlation and makes it possible to interpret the distribution of individual points as a result of different structure and bonding properties of the rocks.

2) In section 6.3.3.2, the investigations by Han et al., (1986), were discussed. The results of four selected samples are plotted in a v_p/v_s versus v_s plot (Figure 6.75). From high to low velocities the hydrostatic pressure during the measurement is 5 MPa, 10 MPa, 20 MPa, 30 MPa, 40 MPa and marked by symbols.

This plot allows the following *interpretation*:

- the individual samples are characterized by specific structure and bonding properties and their change under pressure,

- the hydrostatic pressure creates a pure variation of the contact bonding; with increasing pressure the contact bonding becomes better and the parameter *f* decreases,
- there is a significant difference in the velocity and the velocity ratio between clean (sample 1 and 5) and shaly sandstones (sample 50 and 58) caused by a significant difference in the contact or bonding properties.

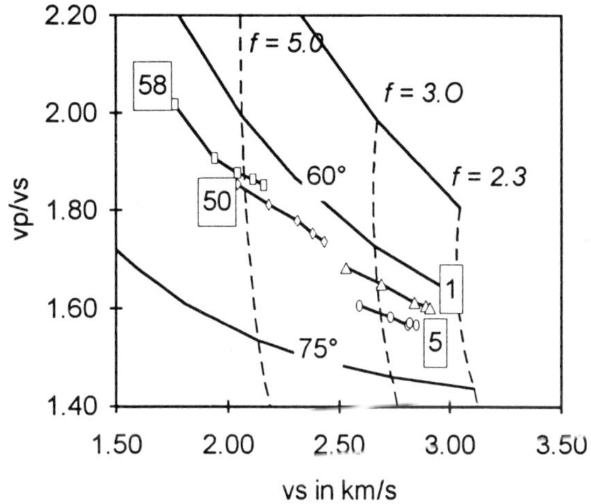

Figure 6.75. Ratio v_p/v_s versus v_s for the interpretation of sample measurements at sand and shaly sandstones;
curve grid was calculated for different α and f (it was assumed empirically $v_s = 6.5 \cdot (s_{66})^{1/2}$ or $A = 6.5$ km/s);
points are measured results after Han et al. (1986)

sample 1 : porosity at 5 MPa 18.5%; clay content 0%;
sample 5 : porosity at 5 MPa 19.9%; clay content 0%;
sample 50 : porosity at 5 MPa 17.7%; clay content 11%;
sample 58 : porosity at 5 MPa 15.4%; clay content 27%;

hydrostatic pressure increases (left to right) 5 MPa, 10 MPa, 20 MPa, 30 MPa, 40 MPa

3) Figure 6.76 shows in a comparable manner an analysis of samples from the Joffre Field, Alberta, Canada. The samples were investigated by Arestad and Mattocks (1993) a discussion of the seismic characterization is given by Arestad et al. (1995). The samples 1 and 9 show an increase of the "bonding quality" with increasing pressure without a change of the internal structure. Sample 10 shows a tremendous structure change between 6.3 and 14.7 MPa; the petrographic description includes the note that there are microfractures

present. The change of internal structure is a change of the dominant direction of active fractures and other defects.

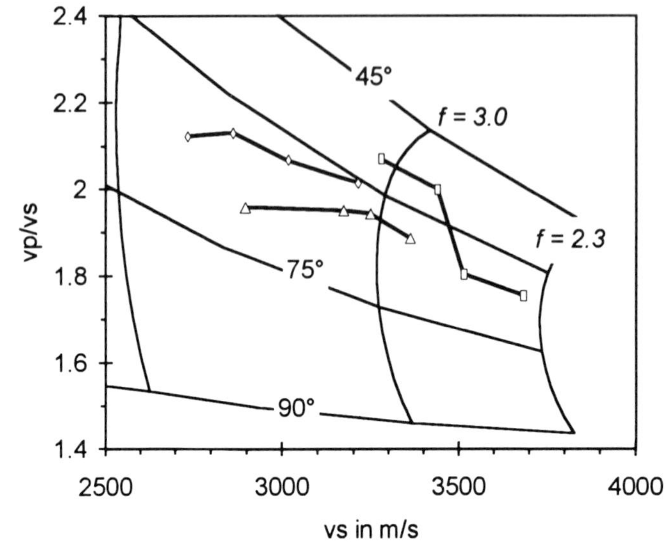

Figure 6.76. Ratio v_P/v_S versus v_S for the interpretation of measurements at samples from the Joffre Field, Alberta;
curve grid was calculated for different α and f (it was assumed empirically $A = 7.1$ km/s)
points are measured results after Arestad and Mattocks, 1993

sample 1 : porosity 4.2 %; dolorudstone;
sample 9 : porosity 1.95 %; dolomud-wackestone;
sample 10 : porosity 5.6 %; dolorud-floatstone;

pressure increases (from left to right) 2.75 MPa, 6.2 MPa, 14.5 MPa, 34.5 MPa.

4) In section 6.3.4.3 (Figure 6.52), the results of field measurements at Pierre shale (White et al., 1983) were discussed. If we transform the two anisotropy parameters for the compression and the shear wave into the terminology of the grid (Figure 6.77), we get a remarkable arrangement of the individual data points. They are all very close to the curve with a structure angle of $\alpha \approx 60$. The following interpretation is possible:

- the internal structure of the Pierre shale can be described statistically by an angle of $\alpha \approx 60$
- the variation in the anisotropy for longitudinal and transverse waves is controlled by the different bonding properties alone,

◆ the constant structure is the cause for the strong linear correlation between the anisotropy for the compressional and the shear wave velocity.

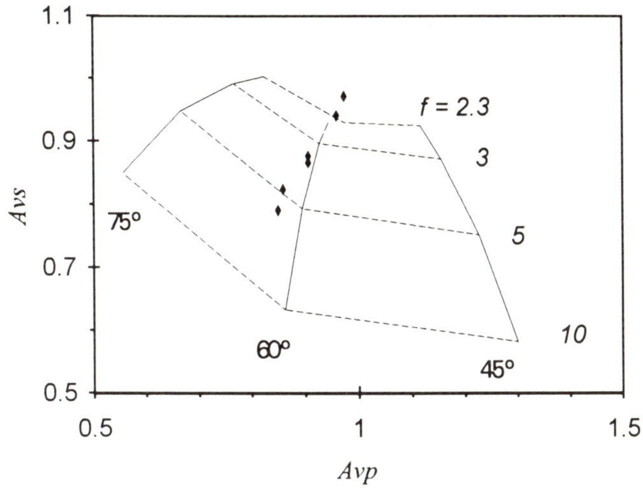

Figure 6.74. $A^*_{vs} = \left(\frac{s_{55}}{s_{66}}\right)^{1/2}$ versus $A^*_{vp} = \left(\frac{s_{33}}{s_{11}}\right)^{1/2}$ - plot for the interpretation of anisotropy investigations and its application to field measurement results on Pierre shale; the curve grid was calculated for different α and f, the points are measured results after White et al., 1983 (transformed values from Figure 6.52).

This simple model with an internal structure is an attempt to include the internal geometry into physical theory of rock properties. The theory as well as the practical application needs further development. The examples include some simplifications and omissions, for example the pore fluid influence or the porosity change with increasing pressure and different matrix properties for quartz and shaly matrix.

But on the other hand it is applicable to other properties such as the thermal conductivity (chapter 8) and the electrical formation factor (chapter 9). A further application is the contribution to the understanding of changes of the physical properties as they relate to deformational and rock failure processes (chapter 10.).

At least the idea of such a structural model is also applicable to unconsolidated rocks (see section 6.4.4) and to fractured rocks; for example, the model described in section 6.4.3.4 must be expanded by the orientation of the internal defects.

$$A(x) = A(x_0) \cdot \exp[-\alpha(x - x_0)] \tag{7-2}$$

where α is the attenuation coefficient.
Frequently equation (7-2) is written in the form

$$A(x) = A_0 \cdot \exp(-\alpha \cdot x) \tag{7-3}$$

where A_0 is the initial amplitude at the distance $x = 0$ from the source or at a reference position; x is the distance between source and measuring point.

Mathematically, these equations can be derived from the general equation for (plane) waves in a homogeneous medium introducing a complex wave number. For the case of a plane wave, the amplitude is

$$A(x,t) = A_0 \cdot \exp[i(kx - \omega t)] \tag{7-4}$$

where t is the time, ω the angular frequency, and k the wave number. Attenuation can be introduced by a complex wave number (or complex frequency)

$$k = k_{\text{real}} + k_{\text{imag}} = k_{real} + i\alpha \tag{7-5}$$

so that

$$A(x,t) = A_0 \cdot \exp(-\alpha \cdot x) \cdot \exp[i(k_{\text{real}} \cdot x - \omega t)] \tag{7-6}$$

where the first term gives the initial amplitude, the second term the exponential decay, and the third term the harmonic oscillation.
With (7-3) the attenuation coefficient is

$$\alpha = -\frac{1}{A(x)} \cdot \frac{\mathrm{d}A(x)}{\mathrm{d}x} = -\frac{\mathrm{d}}{\mathrm{d}x} \ln A(x) \tag{7-7}$$

or for two different positions x_1 and x_2 with respective amplitudes $A(x_1)$ and $A(x_2)$

$$\alpha = \frac{1}{x_2 - x_1} \cdot \ln\left[\frac{A(x_1)}{A(x_2)}\right] \tag{7-8}$$

From (7-7) and (7-8), α is given as an inverse length; it is given in Nepers/m. The unit dB/m (decibel/meter) results from the following form of the equation

$$\alpha[\text{in dB/m}] = \frac{1}{x_2 - x_1}\,{}^{20}\log\left[\frac{A(x_1)}{A(x_2)}\right] \tag{7-9}$$

The conversions are

$$\alpha\ (\text{dB/m}) = 8.686\ \alpha\ (\text{Nepers/m})$$
$$\alpha\ (\text{Nepers/m}) = 0.115\ \alpha\ (\text{dB/m})$$

If the amplitude ratio is related to the distance of a wavelength λ, from (7-8) results the definition of the logarithmic decrement:

$$\delta = \ln\left(\frac{A\,(x)}{A\,(x+\lambda)}\right) = \alpha\,\lambda = \alpha \cdot \left(\frac{v}{f}\right) \tag{7-10}$$

where $A(x)$ and $A(x+\lambda)$ are the amplitudes of the two consecutive cycles, v the velocity of wave propagation, and f the frequency.

In many cases, the measure of attenuation properties are the dimensionless quality factor Q (Knopoff, 1964, 1965) and its inverse Q^{-1} (dissipation factor):

$$Q^{-1} = \alpha \frac{v}{\pi f} \tag{7-11}$$

As an intrinsic property of rock, Q is the ratio of stored energy to dissipated energy. O'Connell and Budiansky (1978) and Johnston and Toksöz (1981) discussed the various physical definitions of Q and their relationship to the viscoelastic constitutive equations for materials. The various definitions of Q are equivalent to intrinsic Q if losses are small ($Q > 10$). Fortunately, under most conditions of interest in geophysics, the small-dissipation assumption is valid (Johnston and Toksöz, 1981). The intrinsic Q is defined as (see Johnston and Toksöz, 1981)

$$Q = \frac{\omega E}{-\mathrm{d}E/\mathrm{d}t} = \frac{2\,\pi\,W}{\Delta W} \tag{7-12}$$

where E is the instantaneous energy in the system, $-\mathrm{d}E/\mathrm{d}t$ is the rate of energy loss, W is the elastic energy stored at maximum stress and strain, and ΔW is the energy loss (per cycle) of a harmonic excitation.

An alternative definition may be found from the stress-strain relations for nearly elastic or low-loss linear solids. Given a sinusoidally varying stress, the strain response will also be sinusoidal. The two are related by the appropriate elastic modulus M and the phase lag φ of strain behind stress. Allowing M to be complex where

$$M = M_{real} + i\, M_{im} \tag{7-13}$$

it can be shown (White, 1965) that

$$Q^{-1} = \frac{M_{im}}{M_{real}} = \tan\varphi \tag{7-14}$$

Equation (7-13) results in an elliptical stress-strain curve under sinusoidal loading. The equivalence of the definitions of Q given in equations (7-12) and (7-14) principally determines the area of the ellipse which is proportional to the energy loss per cycle; the energy stored is given by the total area under the loading curve.

For lossy material, the equation (7-11) must include a second-order term. In this case, the stored energy depends upon the derivative of the complex modulus with respect to frequency, as well as, the value of the modulus (O'Connell and Budiansky, 1978; Hamilton, 1972):

$$Q^{-1} = \alpha \frac{v}{\pi f - \frac{\alpha^2 v^2}{4\pi f}} \tag{7-15}$$

Under the low-loss assumption ($M_{im} >> M_{real}$), the $\alpha^2 v^2 / 4\pi f$ term is negligible and equation (7-11) results immediately from (7-15).

Equation (7-11) shows that if α is a linear function of frequency (see section 7.3.1), then Q is independent of frequency, and it can be used as a dimensionless parameter to describe the absorption attenuation properties of rocks without reference to the particular frequency under consideration.

In summary, for low-loss materials (as are most rocks) the relationship connecting the different measures α, δ, and Q^{-1} for attenuation properties is

$$Q^{-1} = \frac{\delta}{\pi} = \alpha \frac{v}{\pi f} \tag{7-16}$$

7.2. Anelastic properties of the fluid and gaseous rock constituents

Gases and fluids show attenuation properties which are influenced by the composition and thermodynamical conditions (temperature and pressure). *Viscous effects* result in a frequency dependency including a quadratic frequency term for the attenuation coefficient α (or a frequency dependent Q-factor).
Bergmann (1954) gives the mean relation for water at 20 °C

$$\alpha = 8.5 \cdot 10^{-15} f^2 \tag{7-17}$$

where α is in Nepers/m and the frequency f in Hz.

Table 7.1 contains some attenuation properties for air and various fluids. A detailed discussion on the influence of fluid properties (viscosity, dipole moment, and molecule area) was published by Burkhardt et al. (1990, 1991).

Table 7.1. Attenuation properties for air and fluids; References: J - Johnston (1981); T - Tagiev and Mustafaev (1975)

Material	Attenuation	Frequency	Additionally	Ref.
Air, dry	Q = 582	100 Hz	v_p = 343 m/s	J
	Q = 3485	10 kHz		
100% humidity	Q = 4139	100 Hz	= 345 m/s	
	Q = 1434	10 kHz		
Water, fresh 17 °C	Q = 210 000	100 kHz	v_p = 1480 m/s	J
salt 36 ppm	Q = 63 000	100 kHz	= 1520 m/s	
Water	$\alpha = 1.0 \cdot 10^{-5}$	20 kHz		T
	$\alpha = 2.5 \cdot 10^{-4}$	100 kHz		
Oil	$\alpha = 10^{-3}$	100 kHz	d = 0.902 g/cm^3	T
	$\alpha = 10^{-1}$	1000 kHz	d = 0.911 g/cm^3	

Podio and Gregory (1990) have investigated the attenuation (and velocity, see chapter 6.) of various mud types in the frequency range between 100 and 900 kHz and found:

- Attenuation in muds increases as the mud density and the frequency increase.
- There appears to be a linear relationship between attenuation and the volume percent solids in the mud for the range between 200 and 400 kHz.
- Increasing pressure from 690 kPa to 41 MPa caused only a minor decrease of attenuation coefficient in the investigated frequency range.

With a correlation coefficient of 0.999, a least square analysis results in the following empirical relations for the attenuation coefficient in (Nepers/m) as a function of the frequency in MHz:

$\alpha = 8.338 + 2.15 \cdot f + 6.49 \cdot f^2$ for a mud density of 1.20×10^3 kg/m^3
$\alpha = 9.93 + 13.844 \cdot f + 23.01 \cdot f^2$ for a mud density of 1.44×10^3 kg/m^3
$\alpha = 20.364 + 23.37 \cdot f + 107.71 \cdot f^2$ for a mud density of 1.68×10^3 kg/m^3.

7.3. The attenuation of seismic waves in rocks - results of experimental investigations

7.3.1. Overview

Seismic wave attenuation has been measured on many rock types over wide ranges of physical conditions and frequencies, and by various techniques. A collection of reprints, with introductory and summary comments including a state-of-the-art summary by Johnston is given in the SEG Reprint Series No. 2 (1981).

Generally, the attenuation properties of natural rocks are significantly higher than the properties of the minerals. The following is an example:

Calcite (mineral) $Q \approx 1900$ or $Q^{-1} \approx 5 \cdot 10^{-4}$
Limestone $Q \approx 200$ or $Q^{-1} \approx 5 \cdot 10^{-3}$
(after Peselnik and Zietz, 1959)

This shows that attenuation is strongly controlled by defects, inhomogenities and the structure and bonding properties of the rock.

The Figures 7.1 and 7.2 show two characteristics of attenuation:

a) the attenuation coefficient is a *frequency dependent* parameter. It grows with increasing frequency (low-pass filter effect). As a first approximation, Figure 7.1 shows the proportionality

$$\alpha \sim f \tag{7-18}$$

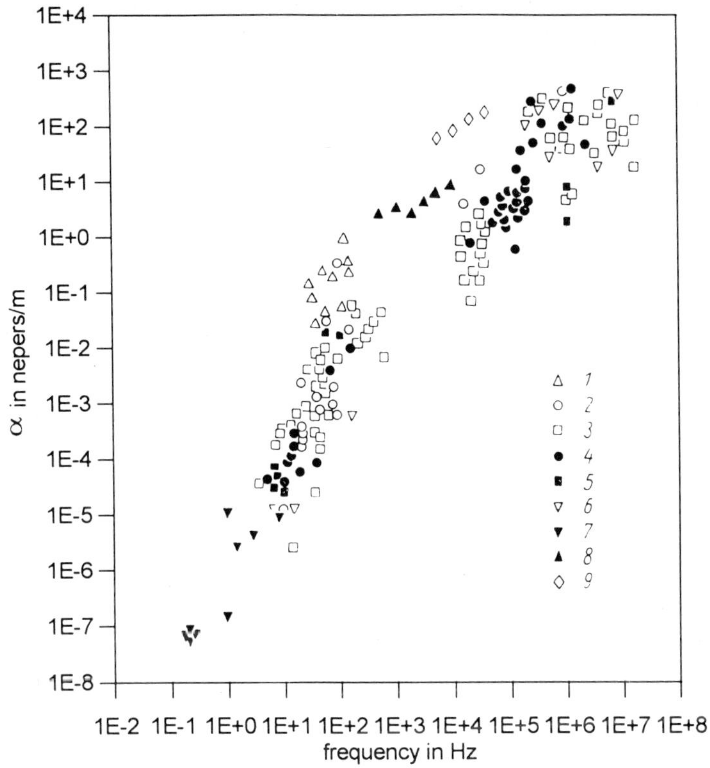

Figure 7.1 Attenuation coefficient of various rock types as a function of frequency (drawn on the basis of a figure after Attewell and Ramana, 1966 and Militzer et al., 1978)

1 - sediments, unconsolidated 2 - sediments, low consolidated
3 - sediments, consolidated 4, 5 - eruptive rocks
6 - metamorphic rocks 7 - from deep seismic reflections
7 - limestone (Germany) 8 - sand, dry

b) The attenuation decreases with increasing rock cementation and depth. This can be expressed as a general rule of thumb: *velocity and attenuation show opposite behaviours* with respect to their dependence on many factors of influence (e.g. porosity, pressure, consolidation).

Berzon (1977) gives the following mean values:

$\alpha = (10^{-6}...10^{-7}) \cdot f$ for mantle and core,

$\alpha \approx 10^{-5} \cdot f$ for unweathered igneous rocks,

$\alpha \approx 10^{-3} \cdot f$ for dry, unconsolidated rocks at the earth's surface.

Figure 7.1 shows the general frequency dependence of the attenuation coefficient - determined at the various frequencies of seismic measurements - as well as the influence of the rock type. The summary figure confirms the general linear frequency dependence, but also indicates the great scatter of the values.

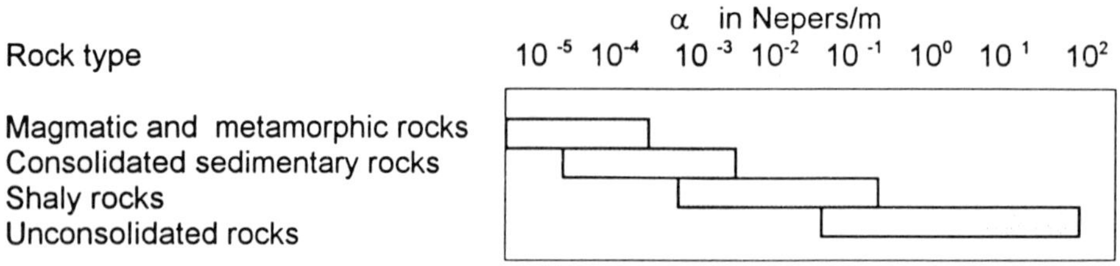

Figure 7.2. Mean value range of attenuation coefficient for some rock types at a frequency of about 50 ... 100 Hz

The general trend of the rock type and porosity dependency is given in Figures 7.2 and 7.3. The data summary for rocks under surface pressure conditions shows the wide variability of attenuation values in rocks and the general trend of Q being inversely proportional to porosity. This decrease of the quality factor Q with increasing porosity denotes an increase of the attenuation properties with increasing porosity and confirms the opposite behaviour related to the elastic properties.The various influences are discussed more in detail in the following.

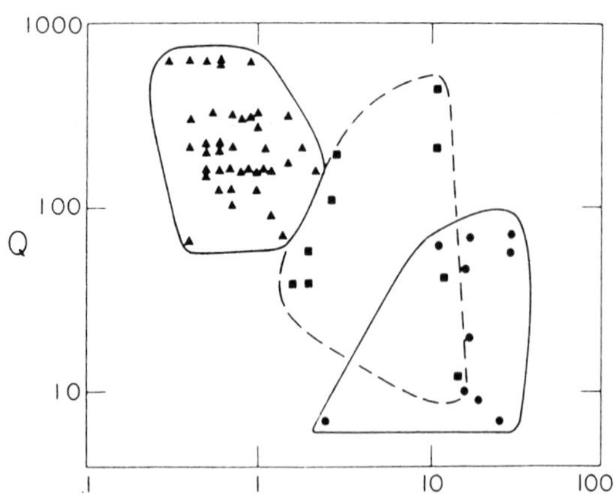

Figure 7.3. Q_p as a function of porosity. Data for igneous and metamorphic rocks (triangles), limestones (squares), and sandstones (circles) are taken from Bradley and Fort and cover wide-frequency and saturation range (after Johnston et al., 1979)

7.3.2. Frequency dependence

Experimental data - particularly laboratory experiments - show that Q^{-1} is nearly independent of frequency (or α proportional to frequency) over a broad frequency range ($10^{-2}..10^{7}$ Hz), especially for dry rocks (Born, 1941; McDonal et al., 1958, Attewell and Ramana, 1966; Karus and Pasecnik, 1954; Bulatova et al., 1970; Silaeva and Schamina, 1960; Berzon, 1977). For example, Berzon (1977) found that the attenuation is nearly linear increasing with frequency within the frequency range between 10^{-1} and 10^{7} Hz (a dependence on a quadratic frequency term was found for some silty-porous rocks).
The following figures give examples of this frequency dependence. Figure 7.4. shows the attenuation coefficient for Pierre shale measured by field experiments from the publication of McDonal et al. (1957) with a *nearly linear dependence on frequency* between about 80 and 550 Hz.

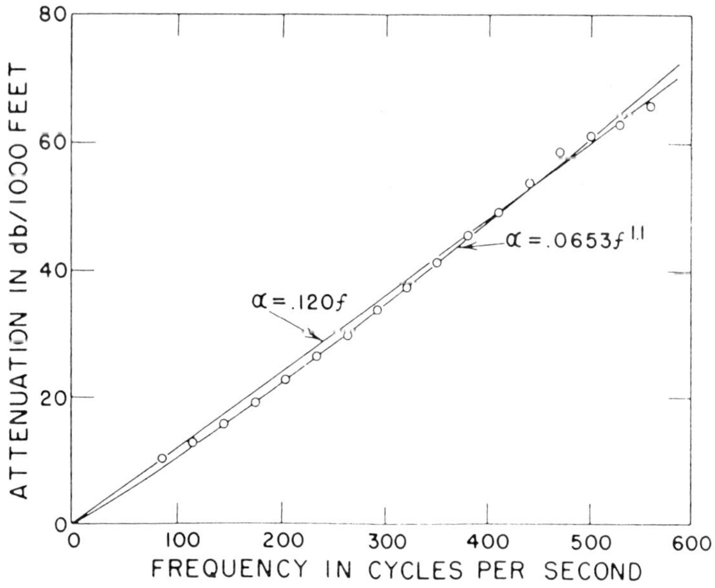

Figure 7.4. Attenuation of vertically traveling compressional waves in Pierre shale (Limon, Colorado); after McDonal et al. (1957)

The two curves show two different approximations:

$$\alpha = 0.120 \cdot f \text{ and } \alpha = 0.0653 \cdot f^{1.1}$$

For horizontally-traveling shear waves with vertical particle motion (SV - wave) in a frequency range from 20 to 125 Hz, the results are expressed by $\alpha = 1.0 \cdot f$. Corresponding to the units used in the original publication the attenuation is given in decibels per 1000 ft of travel and the frequency in Hz.

The linear frequency dependence of the attenuation coefficient corresponds to a frequency independent Q^{-1}. Figure 7.5. shows some examples from a data collection published by Merkulova (1968).

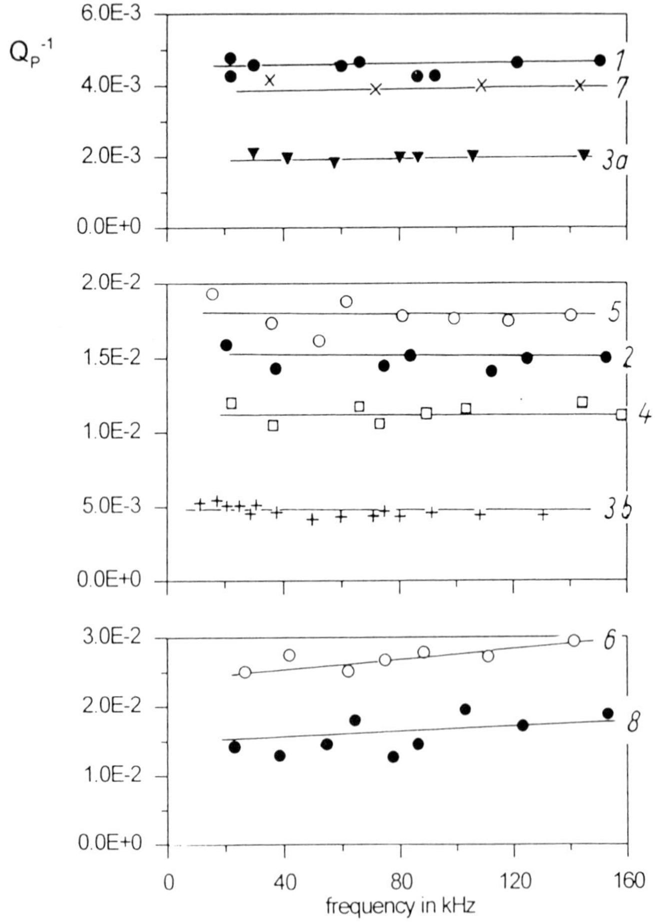

Figure 7.5. Q^{-1} versus frequency for various rocks in the frequency range f = 20...160 kHz, after Merkulova (1968)

1 - Gabbro-diabase 2 - Gabbro-porphyrite 3a, 3b - Quartzite
4 - Gneiss 5 - Granite 6 - Sandstone, water saturated
7- Schist (metamorphic) 8 - Schist (with clay and coal).

Over a frequency range from about 20 to 160 kHz, most rock types confirm frequency independence of Q^{-1} . On the other hand, *water saturated* sandstone and schist (with clay and coal) have the trend of increasing Q^{-1} with frequency. This behaviour of water bearing rocks has been referred by Born (1941).

Martin et al. (1992) studied the attenuation properties and modulus dispersion of Sierra White granite over a frequency range from 10^{-2} to 10^{6} Hz and also found frequency independence for dry rock, but for water saturated rock, a strongly frequency dependent Q^{-1}-value (increase from 0.012 to 0.060).

Figure 7.6 also confirms the nearly linear dependence of the attenuation coefficient on frequency for various granite types. A comparison of curve 1 and curve 2 shows the effect of *anisotropy* for attenuation properties.

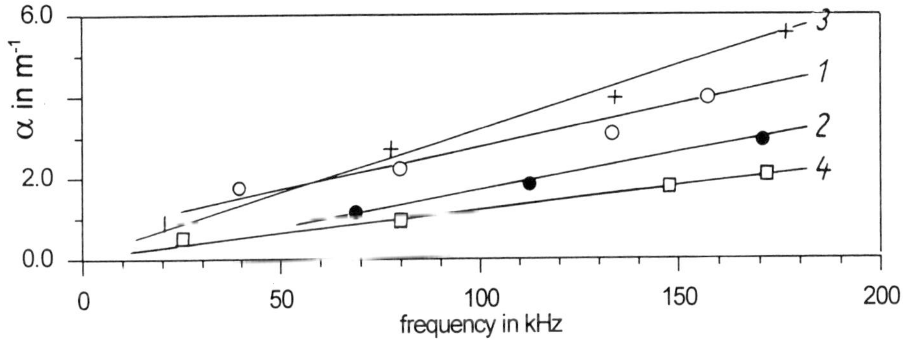

Figure 7.6. Attenuation versus frequency for granites (after Silaeva and Schamina, 1960)
1 - pink colour, wave propagation parallel texture 3 - red colour
2 - pink colour, wave propagation perpendicular texture 4 - gray colour

Hamilton (1972, 1976) and McCann and McCann (1969) have systematically determined attenuation properties for *marine sediments*. The publication of Hamilton, 1972 contains an extended overview of experimental data from the literature.
Also, the nearly linear frequency dependence of α is typical for marine sediments. For the analyses of the frequency dependence, Hamilton (1972) used an empirical relationship

$$\alpha = \alpha_1 \left(\frac{f}{f_1}\right)^n \qquad (7\text{-}19)$$

where α_1 is the attenuation coefficient extrapolated for a reference frequency of $f_1 = 1$ kHz (Hamilton has used some different symbols); here f is the frequency in kHz, and n is the exponent of frequency. Table 7.2. shows these attenuation properties and some petrographic data for different fractions and types of marine sediments.

Table 7.2. Seismic wave velocity, attenuation, and petrographic properties of water saturated sediments; References: Ha - Hamilton (1972): Data determined for marine sediments off San Diego; f = 3.5...14 kHz; MC - McCann and McCann (1969). Data determined for sediments from land and beach; f = 5...50 kHz; ML - McLeroy and DeLoach (1968). Data determined for beach sand and St. Andrew Bay, Florida; f =15...1500 kHz (all data taken from Hamilton, 1972)

Sediment type	D	Φ	v_p	α_1 in dB/m	n	Ref.
	mm	percent	m/s			
Sand, coarse	0.5704	38.0	1817	0.53	0.96	Ha
medium	0.4931	39.2	1798	0.41+0.12	1.00 ± 0.14	Ha
fine	0.1708	45.6	1686	0.45+0.07	1.04 ± 0.07	Ha
	0.1770	41.0	1742	0.13	1.26 ± 0.13	MC
very fine	0.1015	43.3	1716	0.38+0.05	1.11 ± 0.06	Ha
	0.1210	47.0	1708	0.072	1.17 ± 0.13	MC
	0.0720	42.3	1613	0.56	1.00 ± 0.01	MC
Silty sand		82.2	1.540	0.072	1	ML
Clayey silt	0.0058	77.8	1459	0.19	0.94	Ha
		89.2	1530	0.118	1	Ha
Silty clay		93.3	1.525	0.072	1	ML

In some highly porous rocks, Q^{-1} may have a frequency dependent component (e.g., Born, 1941; Wyllie et al., 1962). This effect was also found for clay bearing sediments (Berzon, 1977). The frequency dependent component of Q^{-1} is caused by the frequency dependence of the attenuation of the fluid; viscoelastic behaviour results in a Q^{-1} proportional to frequency (or α proportional to the square of frequency). The additional *fluid influence* on the frequency dependence was formulated by Wyllie et al. (1962) for the logarithmic decrement as follows:

$$\delta = A + B \cdot f \tag{7-20}$$

Also, a form related to Q or α is possible:

$$\alpha = \alpha_1 \cdot f + \alpha_2 \cdot f^2 \tag{7-21}$$

At high frequencies, if the wavelength is of the same order as the pore, grain, or fracture dimensions *Rayleigh scattering* appears at these inhomogenities. This results in an attenuation mechanism (see chapter 7.4.) which follows the proportionality

$$\alpha_{\text{Rayleigh}} \sim D^3 f^4 \quad (7\text{-}22)$$

where D is the mean dimension of the inhomogenities.
Figures 7.7. and 7.8. give examples of this phenomenon. Figure 7.7. shows the attenuation coefficient for Westerly granite in the frequency range between 100 kHz and 800 kHz.

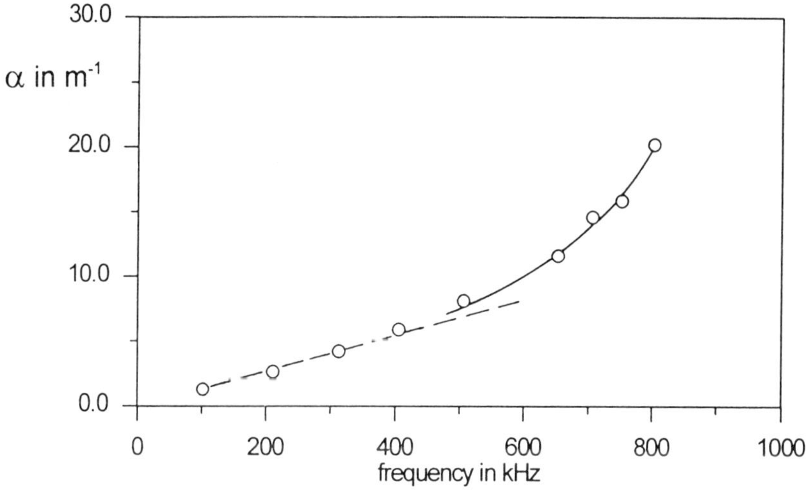

Figure 7.7. Attenuation coefficient as a function of frequency for Westerly-granite (Knopoff and Porter, 1963)

Whereas, in the lower range (100...500 kHz) a linear dependence exists

$$\alpha = 1.34 \cdot 10^{-5} \cdot f \quad (7\text{-}23)$$

a nonlinear relation with a f^4-term is found for the higher range (500...800 kHz)

$$\alpha = 1.34 \cdot 10^{-4} \cdot f + 1.36 \cdot 10^{-23} \cdot f^4 \quad (7\text{-}24)$$

where f in Hz and α in m^{-1}. For a mean velocity of 5 km/s and a frequency of 500 kHz, the wavelength is 1 cm (at 800 kHz, it is 0.6 cm).

A comparable sandstone example is given in Figure 7.8. for the frequency range between 10 kHz and 10 MHz. The plot shows Q_p^{-1} versus frequency f. There are two significant parts:

- below 0.7..1 MHz there is a small increase of Q_p^{-1} with frequency; this is probably a result of the viscous attenuation mechanism of the pore filling material,
- above 0.7...1 MHz there is a stronger dependence on frequency as a probably result of an additional scattering effect.

The compressional wave velocity of the sandstone is 3600 m/s. For 1 MHz, a wavelength of 3.6 mm results; this is of the same order as the mean grain diameter (1 mm).

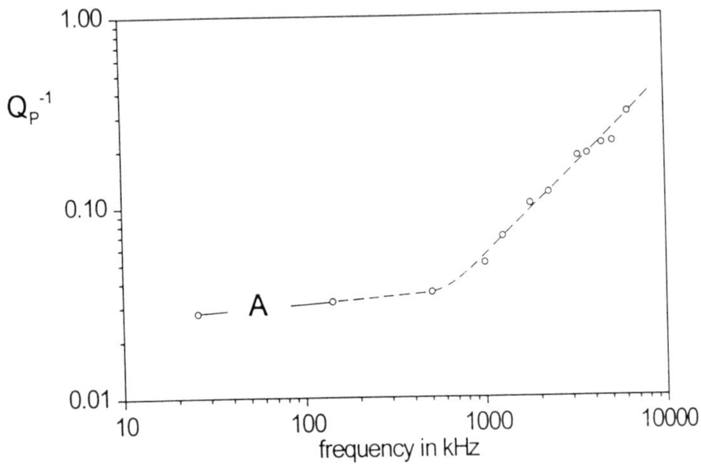

Figure 7.8. Q_p^{-1} versus frequency f for a sandstone sample; plotted after data from Merkulova (1966 and 1968); A is the curve in Figure 7.5. (No. 6).

For unconsolidated sand of various grain size fractions between 0.18 and 2.18 mm Plona and Tsang (1979) experimentally confirmed the Rayleigh scattering in the MHz frequency range.

7.3.3. Attenuation of seismic waves in igneous rocks

Dense, pore- and fracture-free igneous rocks show a correlation of their attenuation or anelastic properties with their elastic properties. Figure 7.9a shows attenuation coefficient versus density for the P-wave and Figure 7.9b shows the attenuation coefficient versus velocity for dense basic rocks. Obviously, there is a tendency for opposite behaviour of attenuation and velocity; the velocity vs. density relation was discussed earlier in this book (chapter 6).

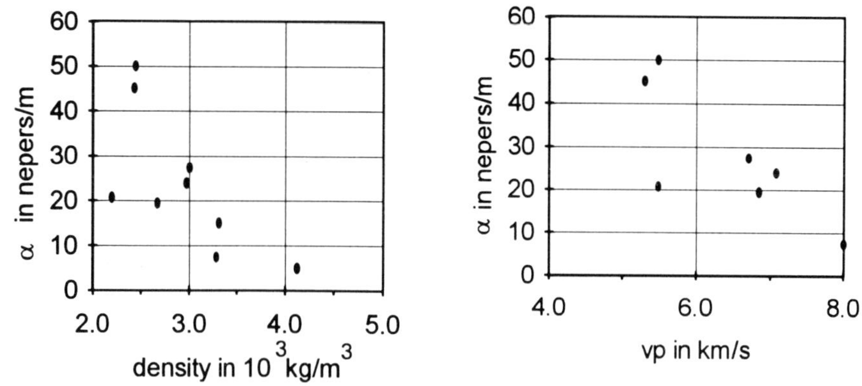

Figure 7.9. Attenuation coefficient of dense basic rocks (peridotite, olivine, andesite, dunite); experimental data from Volarovic et al. (1978);
frequency 2...5 MHz
a) Attenuation coefficient versus density
b) Attenuation coefficient versus velocity (the attenuation was measured at T = 773 K and the velocity at T = 293 K)

On the other hand, fractures, microcracks, texture and other influences act with different effects on both properties. This may result in significant divergence of these trends within one rock type as demonstrated in Table 7.3 for granite.

Table 7.3. Density, velocity, and quality factor of granite samples; after Merkulova (1965)

d in 10^3 kg m^{-3}	v_p in m s^{-1}	Q_p
2.42	4690	160
2.46	4700	190
2.48	4710	170
2.48	4726	180
2.54	4860	50
2.62	4750	140

The presence of *microcracks, fractures and other defects* in the solid rock material results

► in increasing attenuation (and decrease of velocity),

► in significant pressure dependence of attenuation (decrease with increasing pressure) and velocity (increase with increasing pressure) as a result of fracture closing and improved contacts.

Figure 7.10 shows these effects.

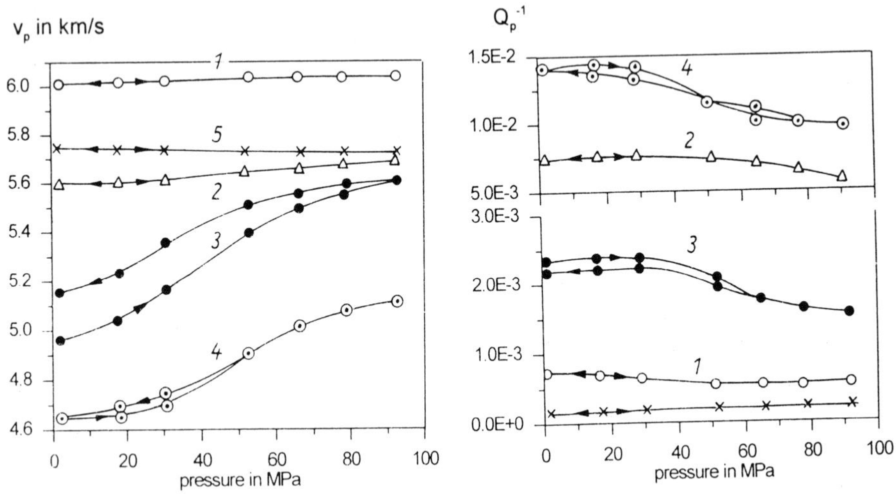

Figure 7.10. Dependence of attenuation coefficient and velocity of compressional wave on hydrostatic pressure; frequency f=32 kHz; after Merkulova et al. (1972).
1 -gabbro 2 - quartzite 3 - quartzite (thermal cracked) 4 - granite 5 - quartz (amorphous)

Four features may be emphasized:

1) Rocks with a small pressure dependence of velocity show a small pressure dependence of attenuation (samples 1, 2, 5); rocks with a strong dependence of velocity on pressure also show a strong dependence of attenuation (samples 3 and 4).
2) The fractional changes in attenuation are greater than those of velocity for the same sample and pressure difference.
3) The pressure dependence of attenuation (and velocity) is nonlinear.
4) In some cases, attenuation and velocity show the phenomenon of "hysteresis".

These features and the correspondence of velocity and attenuation behaviour confirm that both physical properties are controlled by the "mechanical rock constitution" and mainly its change due to defects and pressure.

Figure 7.11 shows the change of attenuation coefficients for compressional and shear waves as a function of pressure for gneiss samples. Additionally, a difference between the attenuation coefficients for compressional and shear wave is seen. In these experiments, the compressional wave attenuation shows a stronger decrease with increasing pressure than shear wave attenuation.

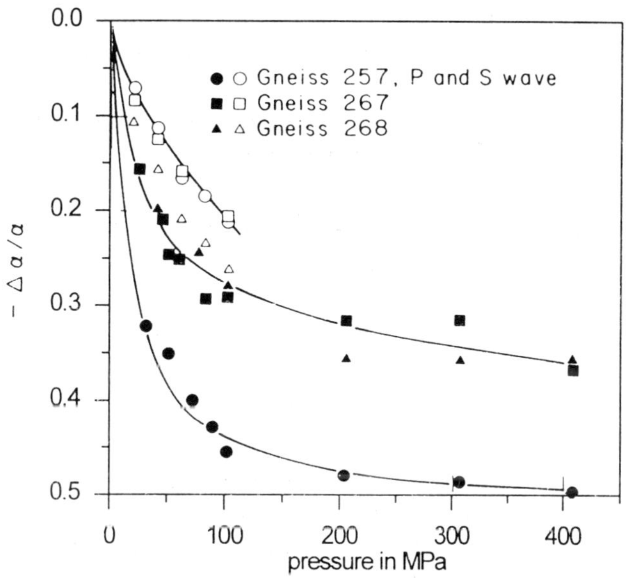

Figure 7.11. Change in the attenuation coefficients of compressional (P, closed symbols) and shear (S, open symbols) waves as functions of pressure for several gneisses (air-dry; frequency 1 MHz). After Johnston et al. (1979); data from Levykin (1965).

The figures confirm that, in general, the fractional changes in attenuation are greater than those of the velocity measured in the same type rocks under the same conditions (Pros et al., 1962; Johnston et al., 1979).

Similar experimental results were published by Klima et al., 1964 (P-wave attenuation for diabase and graywacke up to 4 kbar = 400 MPa) and Gordon and Davis, 1968 (fluid-saturated granite up to 4 kbar = 400 MPa).

Under nonhydrostatic pressure, attenuation appears to be *anisotropic* (Merkulova et al., 1972). For shear waves polarized normal to the axis of maximum compression, attenuation is lowest due to the closure of cracks and improvement of contacts with faces normal to this axis (Johnston et al., 1979).

7.3.4. Attenuation of seismic waves in porous sedimentary rocks

The attenuation properties of porous sediments are mainly influenced by
- properties of the rock skeleton, particularly particle bonding, cementation and by porosity,
- properties of the pore fluid and by interactions between the different rock constituents (particularly interactions between solid and fluid components),
- pressure and temperature.

The strong influence of particle bonding requires separate discussions of consolidated and unconsolidated sedimentary rocks.

7.3.4.1. Attenuation of seismic waves in consolidated sedimentary rocks

Some significant characteristics of the attenuation behaviour of sedimentary rocks are shown in the example of a Berea sandstone (Figure 7.12). The quality factor Q is plotted versus the differential pressure Δp.
These experimental results show that:
a) attenuation decreases (Q increases) with increasing pressure both for P- and S-waves in all cases of saturation. This dependence is nonlinear. Distinct changes are present in the lower pressure range. In the higher pressure range, the change of grain contacts and porosity is smaller,
b) the attenuation in brine- or water saturated rocks is higher than for dry or gas saturated rocks,
c) in dry or gas-saturated rocks, Q_P is slightly lower than Q_S. In water-saturated rocks, Q_P is distinctly higher than Q_S.

The results of Figure 7.12d are comparable those of Figure 7.12c for the low pressure range. At a pressure of about 100 MPa, there is a definite increase in Q due to a collapse of some of the thicker pore spaces and complete locking of some grain boundaries (Toksöz et al., 1979).

Experimental results with comparable dependencies were also published by Winkler and Nur (1982). Resonating bar measurements on Massilon sandstone samples show the contrasting behaviour of shear and bulk compressional attenuation:

- starting with the dry state, the attenuation for both wave types increases, but the compressional energy loss is about twice as large as the shear energy loss,
- for shear attenuation, the loss continues to increase to full saturation (S_W=1), whereas, for bulk attenuation the loss reaches a maximum at approximately 95% saturation (S_W = 0.95) and then decreases with further saturation.

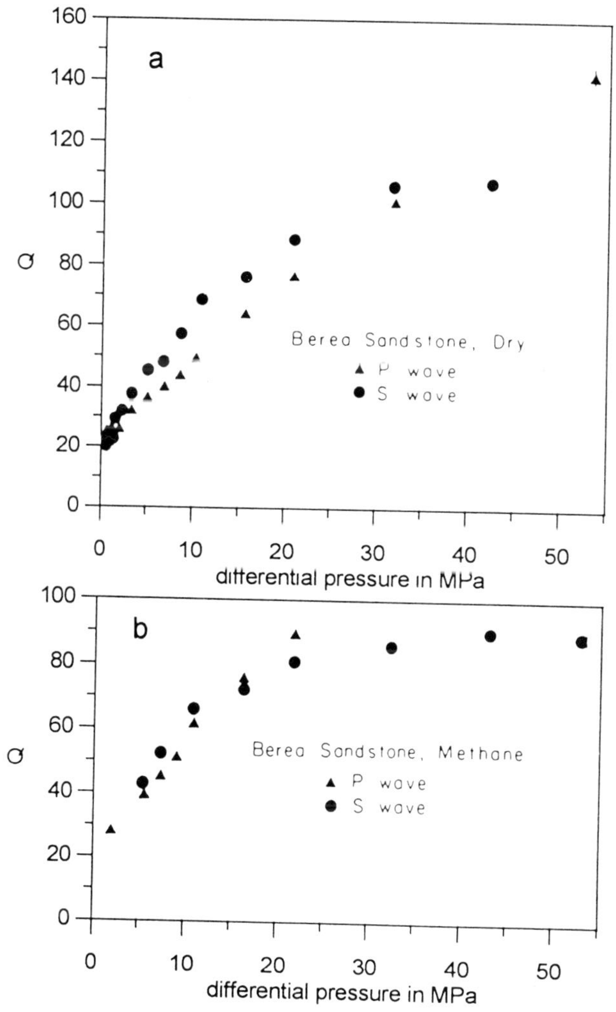

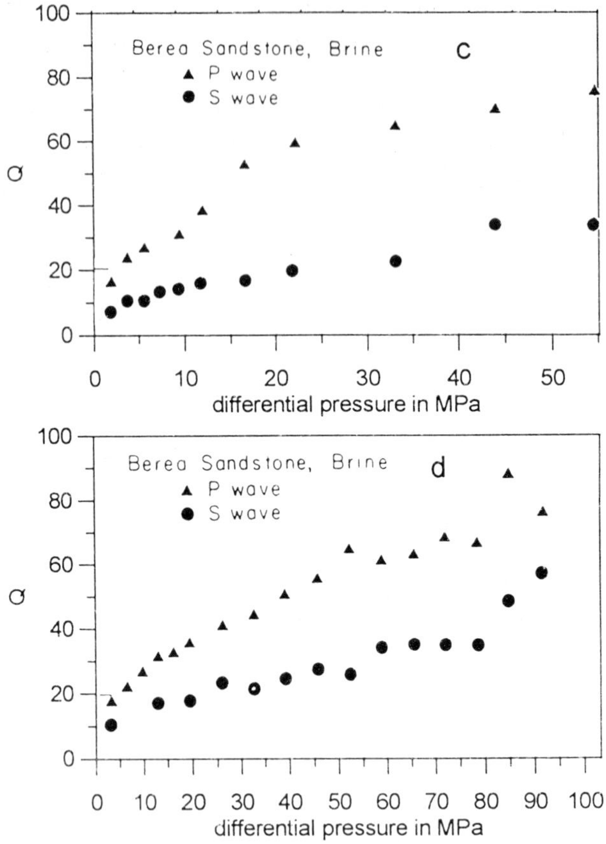

Figure 7.12. Q-values of P- and S-waves of Berea sandstone as a function of differential pressure under various pressure and saturation conditions
p - confining pressure p_{pore} - pore pressure $\Delta p = p - p_{pore}$ differential pressure
a) dry sandstone; pore pressure $p_{pore} = 0$ (confining pressure = differential pressure),
b) methane saturated sandstone; $p_{pore} = 0.465\,p$,
c) brine saturated sandstone (NaCl concentration = 67.191 ppm); $p_{pore} = 0.465\,p$,
d) brine saturated sandstone (NaCl concentration = 161.334 ppm); p was fixed (1035 bars = 1.035 MPa) and p_{pore} was decreased from 1000 bars = 1 MPa.
Measurements were taken with a pulse technique at frequencies between 0.1 and 1.0 MHz; after Toksöz et al., (1979).

In *summary*, the shear attenuation is minimum in dry rock, is greater in partially saturated rock, and is maximum in fully saturated rock. The compressional attenuation is also minimum in dry rock and greater in partially saturated rock, but is significantly reduced by total saturation (Winkler and

Nur, 1979). A detailed description of the influence of partial water saturation and various effects of saturation is given by Bourbie et al. (1987). The interrelationships between water saturation and the character of frequency dependence was investigated by Murphy (1982) and by Winkler and Nur (1982).

The general tendency of the porosity effect on attenuation properties was shown in Figure 7.3 and is also demonstrated in Table 7.4. The two *sandstone* samples show decrease of the velocities v_p and v_s. The velocity ratio v_p/v_s as a "lithology indicator" (see chapter 6) is nearly constant. The quality factor - particularly Q_p - is higher for the low porosity Navajo sandstone than for the porous Boise sandstone. This confirms the increase of attenuation with increasing porosity.

Table 7.4. Porosities, velocities, and quality factors of three rock samples at room temperature; data after Toksöz et al (1979)

Rock type	Φ	v_p in km/s	v_s in km/s	v_p/v_s	Q_p	Q_s	Q_p/Q_s
Navajo sandstone	0.125	4.25	2.38	1.79	7.3	6.2	1.18
Boise sandstone	0.25	3.42	1.90	1.80	6.9	6.1	1.13
Spergen limestone	0.148	4.70	2.49	1.89	14.9	12.1	1.23

The *limestone* generally shows higher velocities and a different v_p/v_s ratio as compared with sandstone. Limestone has a more compact rock skeleton than the granular skeleton of sandstone. This may be the cause for the higher Q-value or lower attenuation of limestone as compared with sandstone. Again it confirms the trend of an opposite dependence of velocity and attenuation on the kind of bonding of the solid rock skeleton. It is also of interest that the ratio Q_p/Q_s seems to be more influenced by rock type than by porosity variation (see also section 7.3.4).

In addition to porosity, the *clay content* is a strong influence on the attenuation properties like it is on the velocity. The combined influence of porosity and clay content have been investigated by Klimentos and McCann (1990). The 42 sandstone samples from wells in the UK are differentiated into 3 sets with respect to location and petrographic features, grain size sorting and clay mineral type. Figure 7.13 shows the correlation between attenuation coefficient, porosity, and clay content for all samples at a pressure of 40 MPa (equivalent to a depth of burial of about 1.5 km) and at a frequency of 1 MHz.

A multivariate statistical analysis results in the relation

$$\alpha = 0.0315\Phi + 0.241C - 0.132 \qquad (7\text{-}25)$$

where α is the attenuation coefficient in dB/cm at 1 MHz, Φ is the porosity in percent, and C is the clay content by volume in percent; the correlation coefficient is 0.88.

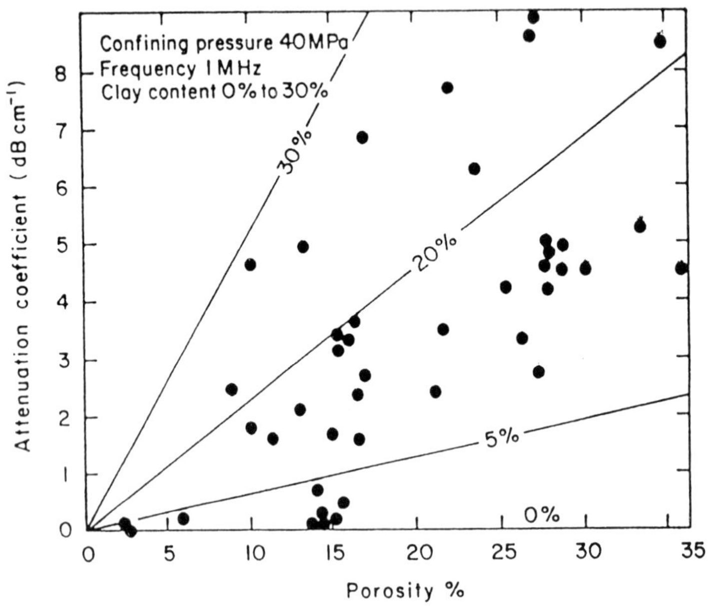

Figure 7.13 Attenuation coefficient versus porosity for 42 sandstone samples. The solid lines separate samples of given percent clay content; after Klimentos and McCann (1990)

The clay effect is an order of magnitude greater than the porosity effect. This dominant influence of clay content on attenuation properties leads for the 39 samples with porosities between 6 and 36 percent (and non zero clay content) to a strong correlation between clay content and quality factor

$$Q = 179 \cdot C^{-0.843} \qquad (7\text{-}26)$$

with a correlation coefficient of 0.91.

A comparable result is shown in Figure 7.14 for sandstones from the North Sea and the data of Klimentos and McCann (1990) with a nearly linear relation between clay content and quality factor.

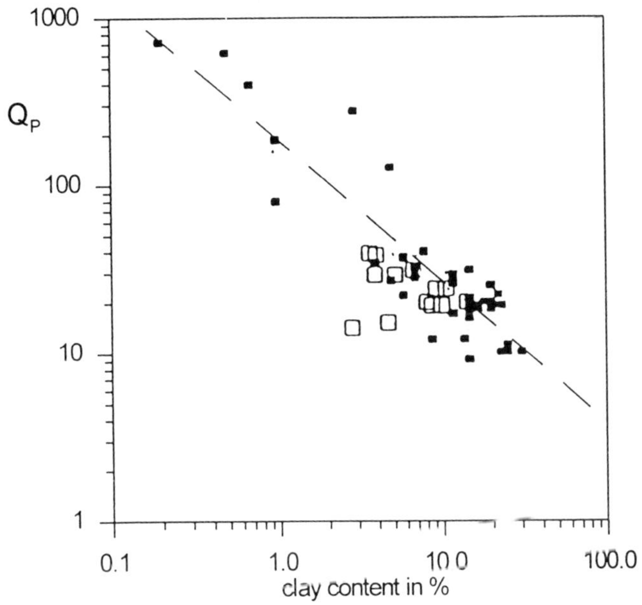

Figure 7.14. Quality factor Q_p versus clay content for sandstone samples
open squares - data from Marks et al. (1992),
solid squares - data from Klimentos and McCann (1990), curve - equation 7.26

The correlation between attenuation and *permeability* was also studied, but this relation is less well defined. Some observations are from a limited dataset (Klimentos and McCann, 1990):

- rocks with low permeability caused by small porosity/small pores have a low attenuation,
- rocks with low permeability caused by authigenic clay filling have high attenuation,
- rocks with medium permeability caused by medium porosity and clean pores have a very low attenuation coefficients (generally > 1 dB/cm)

(under the conditions of measurements: $f = 1$ MHz, pressure 40 MPa.).

7.3.4.2. Attenuation of seismic waves in unconsolidated sediments

Unconsolidated sediments generally have the *highest attenuation properties* among rocks. The physical cause is the non compact nature of the grain skeleton and the *high energy loss at the grain contacts* with poor energy transmission properties.
This is the cause for the grain size and grain shape effects on the attenuation properties. *Scattering* and loss as a result of *fluid - matrix interaction and fluid motion* in the pore space are additional effects.

Table 7.5. shows some mean values for quality factors and velocities.

Table 7.5. Velocity (in m/s) and quality factor of unconsolidated sediments
References: K- Kudo and Shima, 1977 (in situ measurements; frequency range 20...80 Hz); P - Prasad and Meissner, 1992 (laboratory measurements, frequency about 100 kHz, values are taken for a pressure p_{eff} = 2 MPa); M - McCann et al, 1989 (in situ measurements in a Oxford clay profile between two boreholes; frequency range 0.5...2.5 kHz);
the following references are taken from Hamilton, 1976, data converted into Q - values: V - Vassilev and Gurevich (1962), M - Molotova (1966), W - Warrick (1974)

Sediment type	Q_p	Q_s	v_p	v_s	Remarks	Ref.
Sand, diluvial/Japan		8		260		K
Silt, alluvial/Japan		20		102		K
Loam, Kwanto/Japan		6.5		420		K
Mudstone, tertiary/Japan		5		150		K
Weathered clay	37 ± 8		1707		depth 10 m	M
Sand layer	21 ± 4		2187		30 m	
	64 ± 30		1975		40 m	
Clay,silt	44 ± 7		2076		50 m	
	21 ± 3		2132		60 m	
Clay		8...13			saturated	V
		5				M
Mud, San Franc. Bay		17				W
Sand, rounded	13...20	15...60	750...850	320...480	dry	P
	20...60	18...50	1900...2100	400...460	saturated	
Sand, angular	16...30	50...65	700...780	360...400	dry	
	30...100	30...60	1800...2000	400...500	saturated	

- porosity and clay content,
- grain properties (grain size, grain shape, grain size distribution),
- pore fluid properties, saturation, interactions between pore fluid and grains,
- pressure.

From studies of marine sediments, Figure 7.15a and Figure 7.15b show the attenuation coefficient α_1 normalized to a frequency of 1 kHz (compare equation 7-19) versus porosity and mean grain diameter, respectively.
The increase of attenuation with increasing *porosity* up to about 50 % can be explained by an increase in the number of inhomogenities (pores) related to the surrounding matrix. On the other hand, the increase of attenuation from a porosity of 90 % to a porosity of 50 % may be the result of increasing number of inhomogenities (grains) related to the surrounding water.
The attenuation versus *grain size* includes two effects:

- the primary grain size effect and
- the secondary porosity effect resulting from the porosity - grain size correlation (see Figure 2.2).

Mean Q - values for sediments of various grain sizes are given in Table 7.6, and also show the mean range and the trends Comparable results as the attenuation vs. porosity and attenuation vs. grain size were published by Shumway (1956, 1960) .

Table 7.6. Q_p and Q_p^{-1} of water saturated marine sediments off San Diego; data after Hamilton (1972)

Sediment type	Q_p			Q_p^{-1}		
	14 kHz	24 kHz	100 kHz	14 kHz	24 kHz	100 kHz
Sand,						
coarse	32		29	0.031		0.034
medium	31		35	0.032		0.028
fine	31		31	0.032		0.032
very fine	32		29	0.031		0.034
Sandy silt		23	54		0.044	0.018
Clayey silt	111...437	118	104	0.002...0.009	0.009	0.010
Sand-silt-clay	368	31		0.003	0.033	

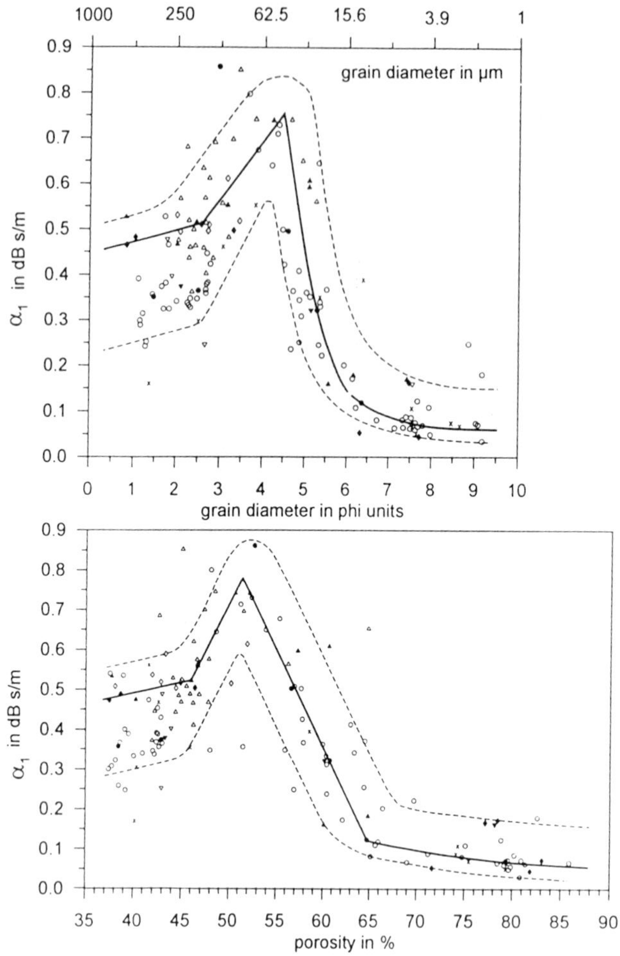

Figure 7.15 Attenuation parameter α_l (definition see equation 7-19) of marine sediments after Hamilton (1972); a) α_l versus mean grain size b) α_l versus porosity (the various symbols denote values from different areas and authors)

The grain size effect in Hamilton's investigations also includes the *clay influence*. The data in Table 7.5 and 7.6 already show the increasing effect of clay on the attenuation properties. McCann and McCann (1969) measured the attenuation coefficients of deep sea sediments at a frequency of 368 kHz: The Q_p- factor decreases from about 200 (at 20% sand/silt content) to about 30 (at 100 % sand/silt content).

Increasing *pressure* results in decreasing attenuation properties as a result of the sediment compaction, improvement of grain-grain contacts and energy transfer, decreasing porosity. Figure 7.16 shows the decrease of attenuation and the increase of velocity with increasing pressure.

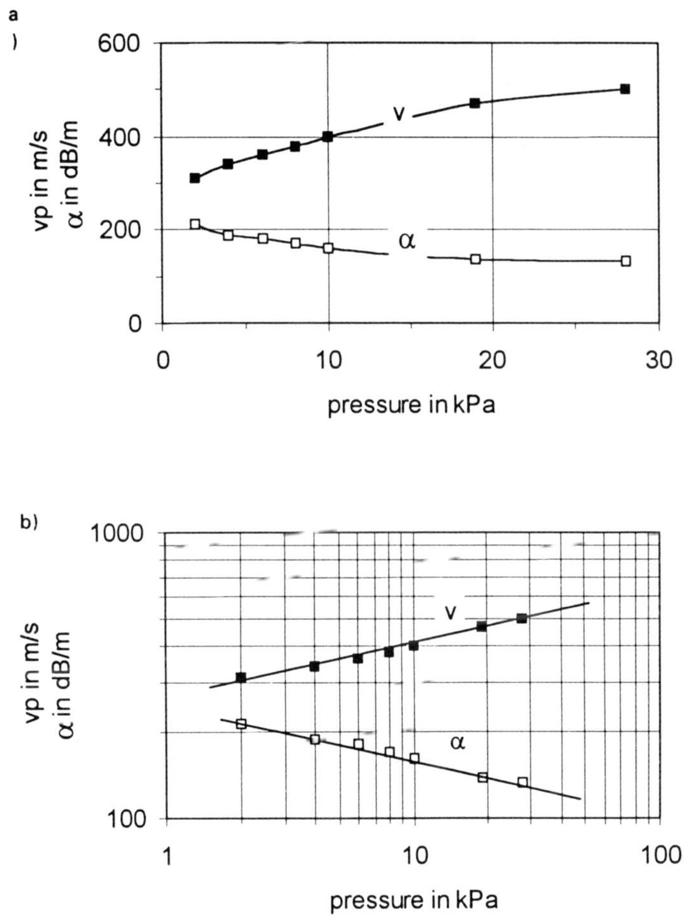

Figure 7.16 Dependence of compressional wave velocity and attenuation on pressure for dry sand, frequency 25 kHz, data after Hunter et al. (1961)
a) linear scaled axes b) logarithmic scaled axes

Both dependencies follow a *non linear relation*. The attenuation coefficient vs. pressure dependence can be described by a power law, comparable to the velocity vs. pressure correlation (equation 6.80):

$$\alpha = \alpha_0 \left(\frac{p}{p_0} \right)^{-n} \qquad (6\text{-}27)$$

where p is the pressure, p_0 the reference pressure (e.g., 1 kPa), α_0 the attenuation coefficient at this pressure, and n an exponent. The data from Hunter et al. (1961) lead to a value of $n = 1/6$. Hamilton (1976) gives values between 1/6 and 1/2 for water saturated sediments.

Prasad and Meissner (1992) have studied the *influence of grain size, grain shape, and differential pressure* on compressional and shear wave velocity and attenuation at frequencies of 100 kHz. Figure 7.17 shows an example of the pressure dependence of the quality factor. A non linear dependence is significant for this coarse material except for the compressional quality factor in the saturated case. Similarly to the velocity behaviour, attenuation is also dominantly determined by the *grain skeleton* properties for shear waves in the dry and water saturated case, and for the compressional wave in the dry case. The compressional properties in water saturated case are dominantly controlled by the *pore fluid*. This dependence is not characteristic of finer fractions.

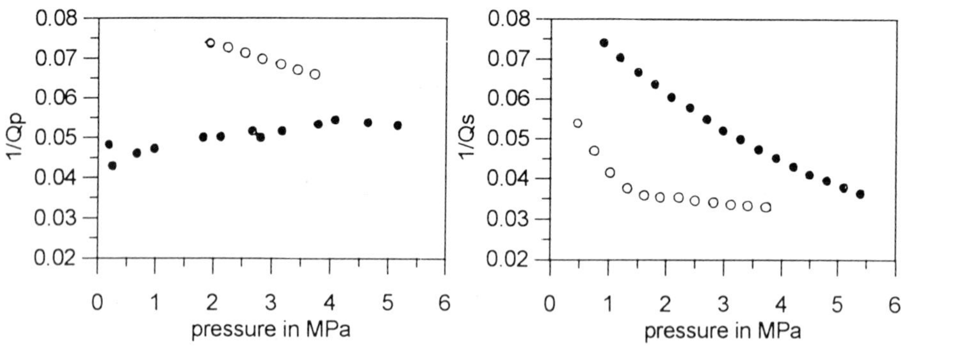

Figure 7.17. Q_p^{-1} and Q_s^{-1} versus pressure for a dry and water saturated sand samples; data selected from Prasad and Meissner (1992); sample No P1 (rounded sand, effective grain size 530 μm); open symbols - dry solid symbols - water saturated
a) Q_p^{-1} versus pressure b) Q_s^{-1} versus pressure

To summarize, the measurements show that

- coarser grains increase the velocity and attenuation of compressional waves and the attenuation of shear waves; the shear wave velocity seems to be unaffected by grain size of the experiments (from about 5 to 600 μm mean grain size),
- angularity of grains causes a decrease in velocity and attenuation.

The attenuation pressure dependence for in situ measurements results in an *attenuation versus depth* relation. For unconsolidated sediments of the Lower Rhine Basin, Ahorner and Budny (1985) plotted the Q vs. depth distribution (Figure 7.18) and derived an empirical relation

$$Q = 2.66 \cdot z^{0.36} \tag{6-28}$$

where the depth z is in m.

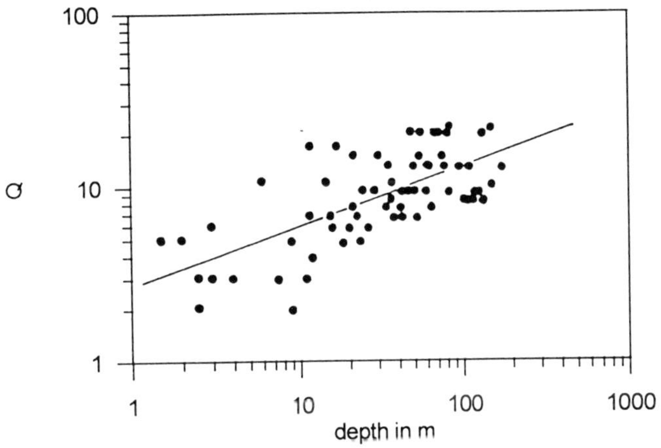

Figure 7.18. Variation of Q-values with depth for Quaternary and Tertiary soft soil deposits in the Lower Rhine Basin, Germany; after Ahorner and Budny (1985)

7.3.5. Relations between the ratios Q_p/Q_s and v_p/v_s

Various experimental data show a tendency for some correlations between velocity and attenuation properties. This may be produced by common influential parameters on both properties (for example pressure, porosity). But on the other hand, velocity and attenuation depend on different rock properties (mainly properties of the rock microstructure) or on these rock properties in different ways.

Figure 7.19 shows separation between dry, partly, and fully saturated Massilon sandstone in a Q_p/Q_s versus v_p/v_s plot. Winkler and Nur (1982) note that "Q_p/Q_s is much more sensitive to degree of saturation than is v_p/v_s ".

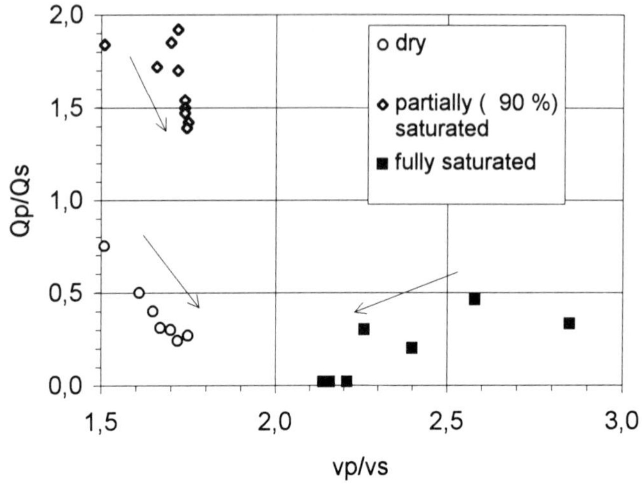

Figure 7.19 Attenuation and velocity data for Massilon sandstone plotted as Q_p/Q_s versus v_p/v_s . Arrows indicate directions of increasing effective pressure (from 0 to 34.5 MPa), after Winkler and Nur (1982)

Burkhardt et al. (1991) have investigated the Obernkirchner sandstone and did not find three such distinct fields on a comparable plot, but also found that the Q_p/Q_s-ratio is more strongly influenced by varying saturation than the velocity ratio. The frequency range was 0.03 ... 200 Hz. Burkhardt et al. (1991) note that their result is "valid only for the rock investigated, but probably not for other rocks with significantly differently microstructure or for other frequency ranges."

Prasad and Meissner (1992) found a separation for unconsolidated sediments with different saturations (dry and fully saturated) and grain shapes in a "loss diagram" of Q_p/Q_s versus v_p/v_s.

The common dependencies on the one hand and the different behaviour on the other hand give the potential to obtain information about rock structure and saturation based on a combined interpretation of velocity and attenuation data.

7.4. Attenuation mechanisms and theories

"The phenomenon of attenuation is complex. While elastic wave propagation is generally well understood, anelasticity is not. As observed from attenuation data, anelastic variations with changes in physical state are complicated and probably cannot be explained by a single model or mechanism."
M.N. Toksöz and D.H. Johnston (1981) Attenuation Mechanisms, Preface Chapter 3 SEG Geophysics reprint series No. 2 Seismic Wave Attenuation, pp.136-139. SEG Tulsa/Oklahoma.

7.4.1. Overview

The nature and source of attenuation is very complex. There are various theories and models to explain the attenuation effects and the various dependencies. Obviously, anelastic behaviour cannot be expressed by one simple model or mechanism. We must assume *complex action and interaction of various mechanisms*, known and may be unknown physical features. Nevertheless, in the literature - and also in this book - the various sources of energy loss are described individually.

The discussion of attenuation mechanisms is started by a (partially complete) overview in Figure 7.20. Toksöz and Johnston (1981) distinguish *two lines* of the mathematical and physical descriptions for seismic energy loss:

The first method is to explain the nature of attenuation in terms of *generalized equation of linear elasticity* (Hookes`Law) or by modified equations allowing certain nonlinearities. "These phenomenological models ... have been well studied but yield little information concerning the microscopic properties of rock." In may cases, the stress-strain-behaviour of models is described by rheological models - in the simplest case represented by a combination of a spring (elastic) and a damping (viscous) element, see Figure 7.21.

The second type of theories and models uses physical and mathematical descriptions of possible *attenuation mechanisms*. These mechanisms are related to the microscopic properties of the rock and their behaviour during elastic wave propagation. Therefore, they generally give us more insight and information about the rock properties.

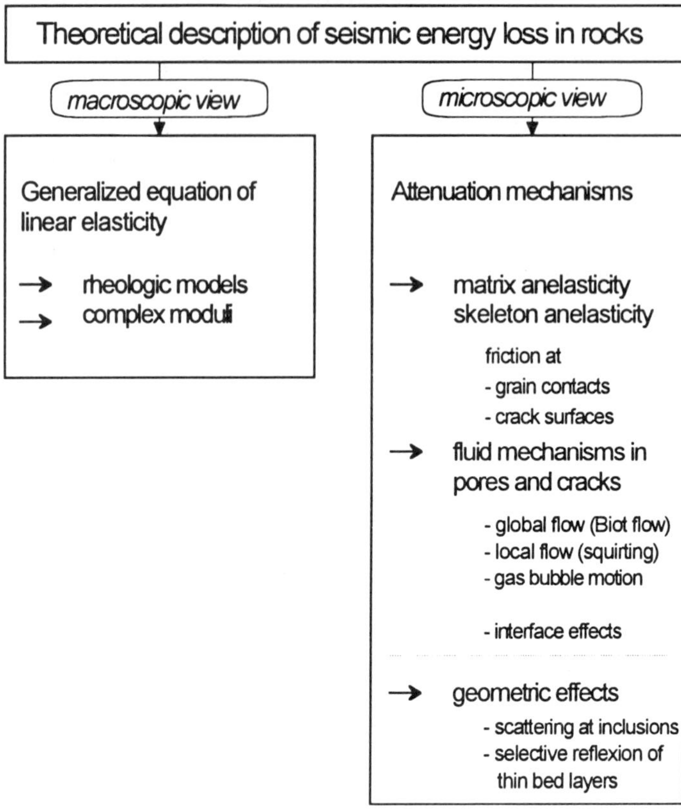

Figure 7.20 Theoretical description of seismic energy loss - models and mechanisms

The main types and groups are related to their "source location" within the rock as for example:

- *matrix anelasticity* including frictional dissipation owing to relative motions at grain boundaries (White, 1966, 1983) and across crack surfaces (Walsh, 1966, Walsh and Grosenbaugh, 1979),
- *fluid mechanisms* in the pores and cracks, these include relaxation owing to shear motions at pore-fluid boundaries (Walsh, 1968), dissipation in a fully saturated rock because of relative motion of the frame with respect to the fluid inclusions (Biot, 1956), squirting phenomena (Mavko and Nur, 1975; O'Connell, R.J., B. Budiansky, 1977, 1978; Mavko and Kjartansson, 1979), and the effect of "gas bubbles" or spherical gas pockets (White, 1975),

- *interface effects* at the solid - pore fluid interface (Schopper, 1983; Schopper and Riepe, 1986) and also
- *geometric effects* as a source of energy loss such as scattering off small pores (Kuster, Toksöz, 1974) or selective reflection of thin beds (O'Doherty and Anstey, 1971; Schoenberger and Levin, 1974, Shapiro et al., 1994).

The following sections show some examples and representative results only of the groups of theories.

7.4.2. Generalized equation of linear elasticity and rheologic models

The macroscopic deformation behaviour is described by a combination of elastic and viscous parts or elements. In the model description, these two parts are symbolized by a spring and a damping element and called *"rheologic models"*.

Figure 7.21 shows two simple combinations: the Maxwell and the Kelvin-Voigt model. Combinations of more than one spring and viscous elements (linear standard solid or Zener model, Burgers body, Nakamura body) are described and discussed by Karato and Spetzler (1990), Qaisar (1989), Kogan (1966), Stoll (1971), and Rösler (1975).

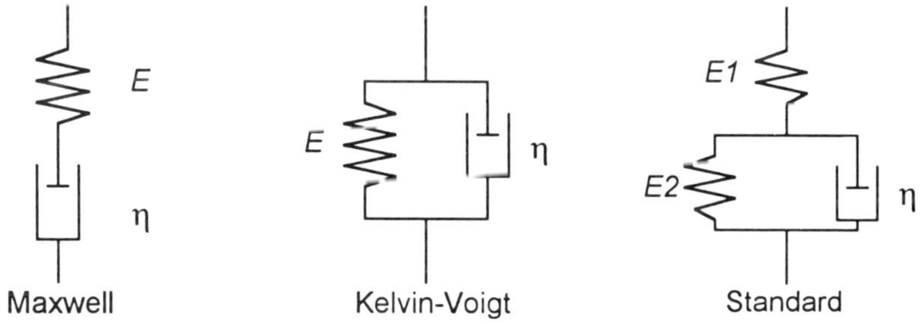

Figure 7.21 The Maxwell, Kelvin-Voigt, and Standard or Zener model

For example, the *Kelvin-Voigt model* is discussed for the homogeneous and isotropic case. The parallel array of the spring (elastic) and viscous element results in a *homogeneous strain* ε. The stress consists of two parts:

- elastic: $$\sigma_{\text{spring}} = E \cdot \varepsilon \tag{7-29}$$
- viscous: $$\sigma_{\text{viscous}} = \eta \cdot \frac{d\varepsilon}{dt} \tag{7-30}$$

Which for the model results in

$$\sigma = \sigma_{\text{spring}} + \sigma_{\text{viscous}} = E \cdot \varepsilon + \eta \cdot \frac{d\varepsilon}{dt} \tag{7-31}$$

For a solid material with elastic and viscous properties, the stress-strain relation 6-4 has a modified form:

$$\sigma_{ik} = \lambda \delta_{iklm} \varepsilon_{lm} + 2\mu \varepsilon_{ik} + \lambda' \theta_{iklm} \frac{d\varepsilon_{lm}}{dt} + 2\mu' \frac{d\varepsilon_{ik}}{dt} \tag{7-32}$$

Introducing the wave equation for a damped wave into the equation of motion yields complex moduli, frequency dependent velocity (dispersion) and attenuation. For the longitudinal wave, we get (see White, 1983):

$$v_p = \left(\frac{M}{d}\right)^{1/2} \left[2 \cdot \frac{1 + \left(\frac{\omega}{\omega_0}\right)^2}{\sqrt{1 + \left(\frac{\omega}{\omega_0}\right)^2} + 1} \right]^{1/2} \tag{7-33}$$

$$\alpha_p = \frac{\omega_0 (\omega/\omega_0)^2}{\left[2 \cdot \left(\frac{M}{d}\right) \left(1 + (\omega/\omega_0)^2\right) \left(\sqrt{1 + (\omega/\omega_0)^2} + 1\right) \right]^{1/2}} \tag{7-34}$$

where

$$M = \lambda + 2\mu \qquad \omega = 2\pi f \tag{7-35}$$

ω_0 is the characteristic frequency of the model and defined as

$$\omega_0 = \frac{\lambda + 2\mu}{\lambda' + 2\mu'} \tag{7-36}$$

This is the ratio of the real and imaginary component of the plane wave modulus. For the low frequency range, ω^2 is much less than ω_0^2 and an approximation for a frequency independent velocity follows:

$$v_p = \left(\frac{M}{d}\right)^{1/2} \tag{7-37}$$

and an attenuation coefficient which increases with the *square of frequency*:

$$\alpha_p = \left[\frac{\omega_0}{2 \cdot (M/d)^{1/2}}\right] \cdot \left(\frac{\omega}{\omega_0}\right)^2 = \left[\frac{1}{v_p} \cdot \frac{1}{2\omega_0}\right] \cdot \omega^2 \tag{7-38}$$

The proportionality

$$\alpha \sim f^2 \tag{7-39}$$

is not in agreement with most experimental results $(\alpha \sim f)$.

A detailed discussion of the elastic-anelastic behaviour of the standard model is given e.g., by Bourbie et al. (1987). The study includes a description of series networks of Kelvin-Voigt models. Arts et al. (1992, 1992) developed for a viscoelastic anisotropic medium the complete viscoelastic tensor.

7.4.3. Matrix inelasticity - frictional dissipation

Matrix attenuation can be caused by

- intrinsic anelasticity of the solid matrix material (minerals),
- frictional dissipation due to relative motions at the grain boundaries and across crack surfaces.

The intrinsic anelasticity of minerals is generally small (see section 7.3.1); in crystals Q - values are higher than a few thousand, while in the rock, they are normally lower than a few hundred - thus, the intrinsic attenuation in minerals can be mostly neglected compared with the attenuation across grain surfaces and cracks (Johnston et al., 1979) or at defects in the solid rock matrix.

Walsh (1965, 1966) has approximated the *cracks* within the rock by *ellipsoides* in plane strain.

The calculation for a random orientation of cracks results in equations for Q_p^{-1} and Q_s^{-1} with the form (see also Johnston et al., 1979):

$$Q_p^{-1} = \frac{E_{eff}}{E_s} \cdot \frac{(1-\sigma_{eff})}{\left(1-2\sigma_{eff}^2\right)} \cdot \frac{l^3 N}{V} \cdot F(\kappa, \sigma_{eff}) \tag{7-40}$$

$$Q_s^{-1} = \frac{E_{eff}}{E_s} \cdot \frac{1}{1+\sigma_{eff}} \cdot \frac{l^3 N}{V} \cdot F(\kappa) \tag{7-41}$$

where E_s and E_{eff} are the solid matrix and the effective (rock) Young's moduli, respectively; σ_{eff} is the effective Poisson ratio, κ is the friction coefficient at the crack surfaces. N is the number of cracks with a half-length l in a volume V. These properties and parameters were partially introduced in section 6.4.6.1. The function $F(\kappa, \sigma_{eff})$ is implicitly dependent on the angle between the normal to the crack plane and the direction of wave propagation.

One important result of this theory and model is *frequency independent Q*. However, the calculation of absolute values requires knowledge or specifics about many mostly unknown parameters, particularly the crack geometry and the friction coefficients, which also depend on saturation. Walsh (1966) notes in his paper, that, although the attenuation cannot be calculated on an absolute basis, the validity can be checked by comparing the ratio Q_p/Q_s after the model with experimental data. Figure 7.22 shows this ratio as a function of the friction coefficient and the effective Poisson ratio as curve parameter. Based on data from Horn and Deere (1962) and Byerlee (see Walsh, 1966), the friction coefficient can be assumed to have a range between 0.1... 0.2 (polished surface) and maximum 0.5...0.6 (rough surface).

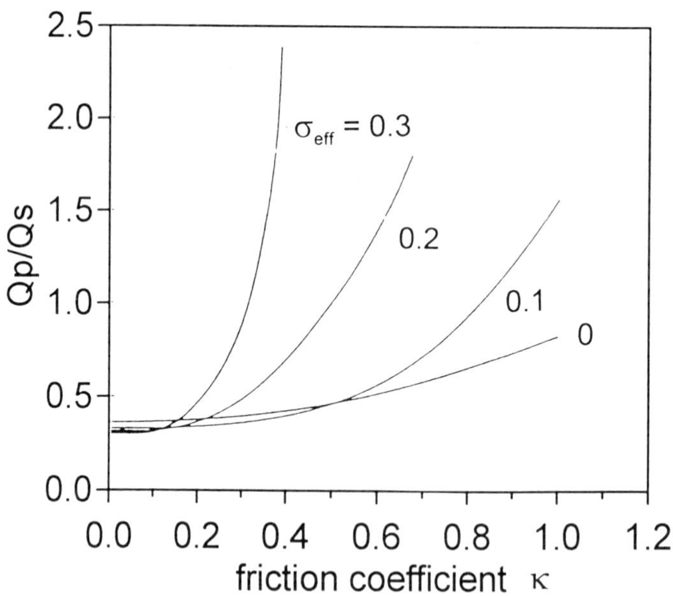

Figure 7. 22. The ratio Q_p/Q_s as a function of the friction coefficient κ ; curve parameter is the effective Poisson ratio σ_{eff} ; after Walsh (1966)

For mean values of Poissons ratio between 0.15 and 0.25 this results in a Q_P/Q_S ratio between about 0.4 and 1.6. This is in a good agreement with the experimental data particularly for dry rocks (compare e.g., section 7.3.4. and Figure 7.19).

The Walsh mechanism and model can be used to describe the *effect of pressure* on the Q value (see Johnston et al., 1979). Utilizing the assumption of spheroids with a small aspect ratio and a pressure independent friction coefficient, a relation results of the form

$$Q_P^{-1} = Q_{P,0}^{-1} \cdot \frac{E_{eff}}{E_{eff,0}} \cdot \exp(-A \cdot p) \qquad (7\text{-}42)$$

where the parameter A depends on the aspect ratio, the ratio $N \cdot l^3/V$, the Poisson ratio, and the effective static frame bulk modulus. The equation 7- 42 describes an exponential decrease of Q^{-1} or attenuation with increasing pressure.

For porous rocks - particularly unconsolidated porous rocks - *sphere models* are applied not only for the calculation of elastic, but also for the anelastic behaviour. Duffy and Mindlin (1959), Deresiewics (1958), and White (1975) have studied the energy loss in a sphere packing as a result of Coulomb *friction* and frictional sliding at the sphere contacts. These studies resulted in relations for Q_S^{-1} depending on pressure, sphere radius, frictional coefficients (for the case "slip" and "no slip"), and the elastic material properties of the spheres.

A modification of this model type was published by Digby (1981), assuming initially deformed contacts. A comparison with experimental results, particularly of Meissner and Prasad (1992), shows that this model type can express some tendencies, such as the decrease of attenuation with increasing pressure (non linear relation), and the general effect of grain size.

In general, for sphere models with frictional mechanisms, a frequency independent Q results.

7.4.4. Fluid mechanisms in the pores and cracks

The fluid flow within the pore space induced by the stress of the passing elastic wave is one cause of attenuation. The fluid motion results in shearing stresses in the fluid and, consequently, in a *viscous dissipation* of wave energy. The viscous mechanism includes the fluid viscosity η and is proportionally related to attenuation by the *square of the frequency*:

$$\alpha \sim f^2 \tag{7-43}$$

The flow mechanisms fall into two categories:

- inertial flow (Biot, 1956),
- squirting flow (Mavko and Nur, 1975; O'Connell and Budiansky, 1977; Murphy et al., 1986).

Biot's theory of the elastic and anelastic behaviour for porous and permeable rock results in equations for the wave velocities (see chapter 6.4.). The oscillating motion of the viscous pore fluid results in frequency dependent energy loss for the seismic waves. Below the characteristic frequency, this motion can be described as a Poiseulle (laminar) flow. The attenuation coefficient for the P-wave (1-st kind) and the S-wave varies as the square of the frequency:

$$\alpha_p, \alpha_s \sim f^2 \tag{7-44}$$

or

$$Q_p, Q_s \sim f \tag{7-45}$$

For the low frequency range the equations for the attenuation coefficient are:

$$\alpha_p = \frac{1}{v_p} \cdot 2\pi^2 \cdot \frac{d_f^2}{d} \cdot \frac{K}{\eta_f} \cdot \left(1 - \frac{d}{d_f} \cdot \frac{ks}{\overline{M}} \cdot \frac{1 - \bar{k}/k_s}{1 - \Phi + \Phi \cdot k_s/k_f - \bar{k}/k_s}\right)^2 f^2 \tag{7-46}$$

$$\alpha_s = \frac{1}{v_s} \cdot 2\pi^2 \cdot \frac{{d_f}^2}{d} \cdot \frac{K}{\eta_f} \cdot f^2 \tag{7-47}$$

where d, d_f are the rock and fluid density, Φ is the porosity, η_f is the fluid viscosity, K is the permeability, and $\overline{M}, \bar{k}, k_s, k_f$ are the plane wave modulus and compression modulus of the rock skeleton, solid matrix material, and fluid, respectively.

The dependence on the square of the frequency does not agree with the general practical experience (see section 7.3.1.).

At higher frequencies, flow is turbulent, and the effects of viscosity are felt only in a thin boundary layer (Johnston et al., 1979) where a frequency dependence of the type

$$\alpha_p, \alpha_s \sim f^{1/2} \tag{7-48}$$

results.

White (1965) has numerically calculated attenuation values for a sandstone and found

$$\alpha_p = 9 \cdot 10^{-10} \cdot f^2 \qquad \alpha_s = 65 \cdot 10^{-10} \cdot f^2 \tag{7-49}$$

where the attenuation is in Nepers per meter and the frequency in Hz. Particularly for low frequencies (< 100 Hz), the attenuation based on this is lower than experimental results. This and the poor frequency dependence leads to the conclusion, that Biot mechanism only gives a (small) contribution to the attenuation, particularly in highly porous and permeable water saturated rocks at relatively high frequencies. This may be one explanation for departures from frequency independent Q value for such rock types (see Figure 7.6).

The Biot mechanism treats bulk flow in the pores (global flow), but ignores the intercrack and -pore flow (*local flow*). This intercrack fluid flow, sometimes called *"squirt flow"*, was first investigated as an attenuation mechanism by Mavko and Nur (1975, 1979). Their model consists of a distribution of partially saturated cracks or pores. It is assumed that the liquid is segregated into "drops" within each pore and flow as the pore is deformed, the separate pores do not interact. For the mathematical derivation, an individual pore is considered to be a two-dimensional inclusion utilizing a plane strain and low frequencies.

The model is described by the following parameters:

- a - pore half width
- d - pore dimension in the third dimension
- D - drop length
- α - aspect ratio of the pore
- N - number of pores in a volume V

The Q^{-1}- values are:

$$Q_p^{-1} = \frac{32}{15} \cdot \frac{d \cdot D^3}{a \cdot \alpha^2} \cdot \frac{N}{V} \cdot \eta_f \cdot \frac{E_{eff}}{E_s^2} \cdot \frac{\left(1-\sigma_{eff}^2\right)^2}{1+\sigma_{eff}} \cdot F(\sigma_{eff}) \cdot 2\pi f \tag{7-50}$$

$$Q_s^{-1} = \frac{32}{15} \cdot \frac{d \cdot D^3}{a \cdot \alpha^2} \cdot \frac{N}{V} \cdot \eta_f \cdot \frac{E_{eff}}{E_s^2} \cdot \frac{\left(1-\sigma_{eff}^2\right)^2}{1+\sigma_{eff}} \cdot 2\pi f \tag{7-51}$$

with

$$F(\sigma_{eff}) = \frac{1-\sigma_{eff}}{2 \cdot (1-2\sigma_{eff})} \cdot \left[3 + 4 \cdot \frac{\sigma_{eff}}{1-\sigma_{eff}} + 8 \cdot \left(\frac{\sigma_{eff}}{1-\sigma_{eff}}\right)^2\right] \tag{7-52}$$

and for the ratio we also find

$$\frac{Q_P}{Q_s} = \{F(\sigma_{eff})\}^{-1} \tag{7-53}$$

This represents a relation between the ratio of the attenuation properties Q_p/Q_s and the effective Poisson ratio or the velocity ratio v_p/v_s (see Figure 7.23).

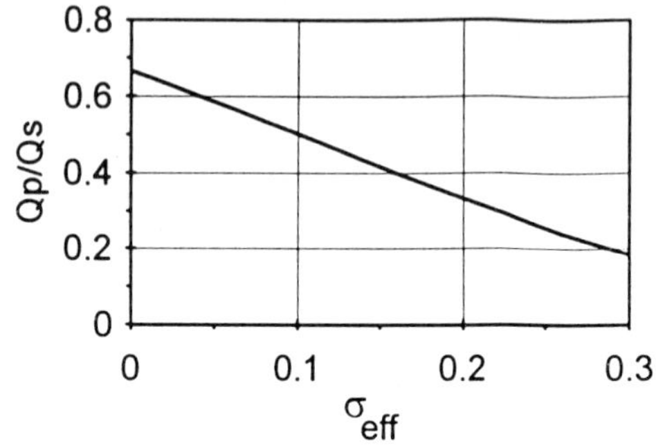

Figure 7.23 The ratio Q_p/Q_s versus effective Poissons ratio for low frequencies (after Mavko and Nur, 1978).

The Mavko and Nur model describes the high sensitivity of attenuation to

- the aspect ratio,
- the fluid drop shape

or more generally of the sensitivity to the pore geometry and fluid distribution.

For porous sedimentary rocks - particularly sandstones - Murphy et al. (1986) developed the local flow model (contact relaxation model) with respect to the specific sandstone microstructure: "The grain contacts are the key element." The sandstone grains are not perfectly cemented across the entire area of the apparent contact; a fraction of the grain contacts is permeated by sheet-like gaps. This incomplete solid-solid contact allows an interconnected fluid film to exist between the grain surfaces. In the model, a narrow gap is connected to a finite annular pore. The stress of the elastic wave passing the model drives the fluid film out of the narrow contact region and into the adjacent pore. This viscous flow results in a dissipation of energy. Numerical calculations, compared with experimental results (Berea sandstone, Fort Union sandstone) show the strong influence of water saturation upon absorption and dispersion.

CHAPTER 8

Thermal Properties of Rocks

"To understand the thermal structure of the earth, knowledge of the thermal properties of the material that constitutes the interior of the earth is indispensable." (KI-ITI HORAI, 1971)

8.1. Physical basis and units

Geothermal investigations are related to many geoscience questions, ranging from studies of the physical state of the earth, tectonics, seismicity and volcanism to practical problems in mining, drilling, geothermal resources and geothermal methods used in exploration, mining and engineering and environmental geophysics. The following petrophysical properties are of main interest for these investigations:

- λ thermal conductivity
- a temperature conductivity (thermal diffusivity)
- c specific heat

Specific heat

The specific heat (or heat capacity per mass unit) c is defined as the ratio of the heat input Q to the product of the mass m and the resulting temperature increase ΔT.

$$c = \frac{Q}{m \cdot \Delta T} = \frac{Q}{d \cdot V \cdot \Delta T} = \frac{\bar{c}}{d} \qquad (8\text{-}1)$$

The SI unit of the specific heat is the J kg^{-1} K^{-1};
$\bar{c} = c \cdot d$ is the *"volumetric heat capacity"*, with density d.

The specific heat depends on the type of thermal process; a distinction must be made between the specific heat at constant volume c_V and at constant pressure c_p. Their difference for isotropic materials is generally given by the relation

$$c_p - c_V = (3\alpha_T)^2 \cdot k \cdot \frac{T}{d} \tag{8-2}$$

where α_T linear coefficient of thermal expansion
k modulus of compression (bulk modulus)
d density
T temperature.

The low values of the coefficient of thermal expansion for solid materials and most fluids results in very small differences between c_p and c_V. Skinner (see Clark, 1969) and Baldridge and Simmons (1971) give a review to this matter. The values are mostly in the range $\alpha_T = (1...3) \cdot 10^{-5} \mathrm{K}^{-1}$ for rock forming minerals.

Thermal conductivity and diffusivity

The heat transfer (transfer of heat energy) is realized by physical processes mainly

- conduction: heat transfer due to the crystal lattice interactions, i.e. heat passes through the substance of the body itself,
- convection, in which heat is transferred by relative motion of portions of the heated body, and
- radiation, in which heat is transferred directly between distant portions of the body by electromagnetic radiation (Carslaw and Jaeger, 1982).

Radiation usually is negligible for lithospheric conditions. Also, convection is absent or negligible in most processes. Therefore, for geophysical studies of the earth's lithosphere, only conduction is important and the most substantial thermal property of rocks is their thermal conductivity (Cermak and Rybach, 1982).

The *thermal conductivity* λ of a material determines its properties as a conductor of heat energy. The temperature gradient ∇T and the heat flow j are connected in the form of the Fourier-equation, which gives the definition of thermal conductivity λ:

$$j = -\lambda \nabla T \tag{8-3}$$

The SI unit of the thermal conductivity is $\mathrm{W\,m^{-1}K^{-1}}$.

The *thermal diffusivity* a is related to the thermal conductivity by the specific heat c and density d

$$a = \lambda \cdot (d \cdot c_p)^{-1} \tag{8-4}$$

Thermal diffusivity has the SI unit $m^2\ s^{-1}$. Thermal conductivity and diffusivity are generally tensors of rank two.

Table 8.1 summarizes units, symbols, and conversions for these thermal properties.

Table 8.1 Thermal properties: units and conversion

Symbol	SI-unit	cgs-unit	conversions
λ	$W\ m^{-1} K^{-1}$	$cal\ cm^{-1}\ s^{-1}\ C^{-1}$	$1\ W\ m^{-1} K^{-1} = 2.388\ 10^{-3}\ cal\ cm^{-1}\ s^{-1}\ C^{-1}$
			$1\ mcal\ cm^{-1}\ s^{-1}\ C^{-1} = 0.4187\ W\ m^{-1}\ K^{-1}$
α	$m^2\ s^{-1}$	$cm^2 s^{-1}$	$1\ m^2\ s^{-1} = 10^4\ cm^2\ s^{-1}$
			$1\ cm^2\ s^{-1} = 10^{-4}\ m^2\ s^{-1}$
c	$J\ kg^{-1} K^{-1}$	$cal\ g^{-1} C^{-1}$	$1\ J\ kg^{-1}\ K^{-1} = 2.388\ 10^{-3}\ cal\ g^{-1}\ C^{-1}$
			$1\ cal\ g^{-1}\ C^{-1} = 4.187\ kJ\ kg^{-1}\ K^{-1}$

There are essentially two fundamental processes of thermal conductivity in solid materials:

- electron-conductivity
- lattice- or phonon-conductivity.

In pure metals electron-conductivity is dominant; in silicates lattice- or phonon-conductivity dominates. Horai (1971) wrote in his classic paper: "At temperatures below several hundred centigrade, lattice vibration is the principal mechanism of thermal conduction in dielectric solids like rocks. The mode of lattice vibration varies according to the lattice dynamical character of solid. Accordingly, the thermal conductivity of rock is determined by the mineral composition and the lattice dynamical character of the individual rock forming minerals."

The *theory of thermal conductivity* in crystalline solid materials was established by Debye in 1912. Heat transport in a dielectric or semiconductor is explained by motions of the crystal lattice or phonons. Phonon scattering is possible between phonons, at crystal boundaries, defects etc.. This fact points to structural influences on the process of conduction. On the basis of the Debye-model, thermal conductivity itself seems to be a temperature dependent property. The characteristic parameter for this dependence is the Debye-temperature. For temperatures that exist in the crust and upper mantle (0...1500 °C), the thermal conductivity decreases with increasing temperature. For the *lattice conductivity* the following proportionality is valid

$$\lambda \sim T^{-1} \tag{8-5}$$

Amorphous materials (such as glasses), as well as, feldspars are characterized by an increase in the thermal conductivity with increasing temperature in contrast to this behaviour of crystalline materials (note, that also the elastic wave velocity shows an opposite tendency for crystalline and amorphous materials in its temperature dependence). Figure 8.1. illustrates this behaviour.

The different temperature dependence may also be demonstrated by comparing empirical relations from Ratcliffe (1959) derived for quartz (see also Cermak and Rybach, 1982):

$\lambda = 1.323 + 0.00193\,T - 0.67.10^{-5}\,T^2$ (fused quartz, -150...60 °C)

$\lambda = (0.1450 + 0.578.10^{-3}\,T)^{-1}$ (crystalline quartz, 0...120 °C)

8.2. Thermal properties of minerals and pore contents

8.2.1. Thermal properties of minerals

8.2.1.1. Mean values and ranges

Table 8.2 gives values from various references, mostly from a data collection of Cermak and Rybach (1982) from which the mineral system used is adapted. Among the rock forming minerals, quartz and minerals which are contained in metamorphic rocks (kyanite, andalusite) have relatively high thermal conductivities. Ore-minerals and some accessories (rutile, spinel) have extremely high values of thermal conductivity. Low values are found among the mineral group of mica (biotite), nepheline and polyhalite.

Table 8.2. Thermal conductivity and specific heat of minerals, organic and construction materials; References: Data collections published by Melnikov et al., 1975 (M); Dortman, 1976 (D); Cermak and Rybach, 1982 (C); Quiel, 1975 (Q); Brigaud et al., 1989, 1992 (B); Huenges et. al., 1989; Kobranova, 1989 (K); Western-Atlas, 1992 (W); The data for λ are related to room temperatures and for a to 0 °C in Ref. C

Mineral	λ in W $m^{-1}K^{-1}$	a in kJ $kg^{-1}K^{-1}$	Ref.
Halide			
Halite	5.3...10	0.79...0.84	M
Sylvite	6.7...10	0.55...0.63	M
Fluorite	9...10.2	0.9	M
	9.5	0.85	C
Phosphates			
Apatite	1.37, 1.38		C
	1.4	0.7	M
Monazite	1.10		C
Carbonates			
Calcite	3.57	0.793	C
	3.25...3.9	0.8...0.83	M
	3.30		B
Aragonite	2.23	0.78	C
		0.78...0.79	M
Dolomite	5.5	0.93 (at 60 °C)	C
	3.9...5.5	0.86...0.88	M
	5.30		B
Siderite	3.00	0.683	C
	3.0	0.72...0.76	M
Sulfates			
Anhydrite	4.76	0.52	C
	4.6...5.7	0.55...0.62	M
	4.91...5.75		D
Barite	1.33	0.45	C
	1.5...1.8	0.48...0.6	M
Gypsum	1.05...1.3	1.07	M
Sulfides			
Pyrite	19.2	0.50	C
	19.2		H
Galena	2.28	0.207	C
Oxides			
Hematite	12.1...14.7	0.61	C
	11.2...13.9	0.62	M
	12.1...14.7		D
Magnetite	5.10	0.6	C
	4.7...5.28	0.6	M
	5.3	0.586	D

Mineral		λ in W m^{-1}K^{-1}	*a* in kJ kg^{-1}K^{-1}	Ref..
Silica minerals				
	Quartz (α)	7.69	0.698	C
		6.6		H
		8.0...13.0	0.74	M
		7.70		B
		6.5...7.2		D
	Vitreous silica	1.36	0.70	C
Ortho- and ring-silicates				
	Forsterite (Fo)	6.0	0.98	M
	Fayalite (Fa)	3.0	0.84	M
	Olivines ($Fo_{98}Fa_2$)	5.06		C
	(Fo_4Fa_{96})	3.16	0.55	C
	Garnets			
	Almandine	3.31		C
		3.3		M
	Sillimanite	9.09	0.743	C
		9.1		H
	Epidote	2.82		C
			0.78	M
	Kyanite	14.2	0.70	C
		14.2	0.7	M
Chain silicates				
	Enstatite	4.34	0.80 (at 60 °C)	C
	Diopside	5.02	0.69	C
		4.1...5.1	0.7...0.75	M
	Augite	4.03	0.67	C
		3.8		M
	Hornblende	2.85		C
		2.3...3	0.67...0.75	M
Sheet silicates				
	Muscovite	2.32		C
		2.32		H
		1.7	0.76	M
	Biotite	1.17	0.77	C
		1.73		H
	Chlorite	5.14		C
		4.2	0.6	M
		3.30		B
	Smectite	1.90		B
	Illite	1.90		B
	Kaolinite	2.60		B
	Mixed-layers	1.90		B
	Clay min. (mean)	2.9		Q

Mineral	λ in W m^{-1}K^{-1}	a in kJ kg^{-1}K^{-1}	Ref.
Framework silicates			
Feldspar	2.30		B
Microcline	2.49	0.680	C
	2.9	0.67...0.7	M
Orthoclase	2.31	0.61	C
	2.49		H
	3.2	0.6...0.75	M
Plagioclase, mean	2.31		H
Albite ($Ab_{99}An_1$)	2.31	0.709	C
	2.05...2.4	0.7...0.75	M
Anorthite(Ab_4An_{96})	1.68	0.70	C
	1.88...1.95	0.7...0.73	M
Labradorite	1.75...1.85	0.8...0.82	M
Nepheline	1.73	0.8	M
Leucite	1.15	0.745 (at 60 °C)	C
	1.5	0.74...0.77	M
Organic materials (mean)	0.25		Q
	1.00		B
Construction materials			
Steel	46		W
Cement	0.29		W

Thermal properties of minerals (and rocks) are summarized and tabulated in several handbooks and papers (e.g. Clarke, 1969; Horai, 1971; Kappelmeyer and Haenel, 1974; Melnikov et al., 1975; Dortman, 1976; Cermak and Rybach, 1982).

After Kobranova (1989), the principal mineral classes can be arranged in order of decreasing mean thermal conductivity values in the following series despite their large fluctuations (mean values in brackets):

- native metals and such elements as graphite and diamond (120 W m^{-1} K^{-1})
- sulfides (19)
- oxides (11.8)
- fluorides and chlorides (6)
- carbonates (4.0)
- silicates (3.8) - sulfates (3.3)
- nitrates (2.1)
- native elements/nonmetals (selenium, sulphur ≈ 0.85).

8.2.2. Thermal properties of pore contents

8.2.2.1. Mean range and overview

A look at the mean values of the thermal properties for air (dry), water and oil given by Table 8.4 shows two facts:
1) a distinct difference between thermal conductivity of matrix materials (minerals) and pore filling materials:

$$\lambda_{matrix} > \lambda_{pore}$$

2) differences between the properties of the individual pore filling materials:

$$\lambda_{water} > \lambda_{oil} > \lambda_{gas,air}$$

Table 8.4. Thermal properties of pore contents; after data from Clark (1966), Kappelmeyer and Haenel (1974), Ebert (1975), Woodside and Messmer (1961), Gearhart (1978), Kobranova (1989), Western-Atlas (1992)

Medium	T °C	λ W m^{-1} K^{-1}	c kJ kg^{-1} K^{-1}	a m^2 s^{-1}
Air (mean)		0.025		
Gas (mean)		0.027		
Air (dry)	0	0.0243	1.005	$1.87\ 10^{-5}$
	20	0.0257	1.005	$2.12\ 10^{-5}$
	40	0.0271	1.009	$2.49\ 10^{-5}$
	100	0.0314	1.013	$3.38\ 10^{-5}$
Water (mean)		0.50...0.59		
	0	0.5602	4.220	$1.33\ 10^{-7}$
	20	0.5992	4.166	$1.51\ 10^{-7}$
	40	0.6281	4.182	$1.62\ 10^{-7}$
	70	0.6619	4.192	$1.68\ 10^{-7}$
	100	0.6787	4.219	
Oil (mean)	20	0.14...0.15		
Crude oil	20	0.13...0.14	1.88...2.76	
Kerosene	30	0.149		
n-Pentane	20	0.113		
	100	0.081		
Ice	0	2.20	2.040	$1.18\ 10^{-6}$
	- 10	2.32	1.997	
	- 20	2.43	1.946	
	- 40	2.66	1.817	
Snow, dense		0.46		

These differences are reflected in the thermal conductivities of porous rocks (see Figure 8.9.).

Table 8.4. contains values for ice in addition to fluid media values. Further detailed data on the thermal properties of minerals, water and air are summarized in the form of tables by Clark (1966), Dortman (1976), Kappelmeyer and Haenel (1974), Horai and Winkler (1975).

8.2.2.2. Influence of pressure and temperature

The influence of pressure on thermal properties is relatively small in relation to the influence of temperature. For example a pressure increase of up to 10 MPa only results in an increase of the thermal conductivity from 0.61 to 0.64 W m K^{-1} for water (at about 30 °C). For air (at about 18 °C), the increase is 0.025 to 0.031 W m^{-1} K^{-1}.

Kobranova (1989) described the behaviour of thermal conductivity of distilled water as follows: "...λ_w at zero temperature and atmospheric pressure equals 0.582 W m-1 K^{-1}. With increasing the temperature its values first increase (λ_w= 0.589 at 20 °C and 0.67 W m^{-1} K^{-1} at 75 °C), then starting at approximately 120 °C, diminish since the attraction between the molecules of the fluid decreases. The λ_w values increase with pressure: at p = 1200 MPa the coefficient has a value half as great as the atmospheric pressure (0.1 MPa). At $p < 3...4$ MPa the effect of pressure on λ_w is very small."

Kaye and Laby, 1968 (see Griffiths et al., 1992) derived for temperature dependence of thermal conductivity of water the relationship

$$\lambda_{water} = 0.56 + 0.002 \cdot T - 1.01 \cdot 10^{-5} \cdot T^2 + 6.71 \cdot 10^{-9} \cdot T^3 \qquad (8\text{-}8)$$

where thermal conductivity λ is in W/mK and temperature T in °C.

8.3. Thermal properties of rocks

8.3.1. Review

Figure 8.2 gives an overview of the mean value range in thermal conductivity for abundant rock types; some more detailed data is contained in Tables 8.5 and 8.7.

Similarly to other rock properties, thermal conductivity is characterized by a broad range of values within one rock type. This range is primarily controlled by variations in mineral content, pores and cracks, their thermal properties, and their volume fractions and spatial distribution within the rock. The value ranges for individual rock types overlap. Kappelmeyer and Haenel (1974) note that for "...more detailed and quantitative geothermal studies it is necessary to measure the thermal conductivities of the individual rocks involved in the problem".

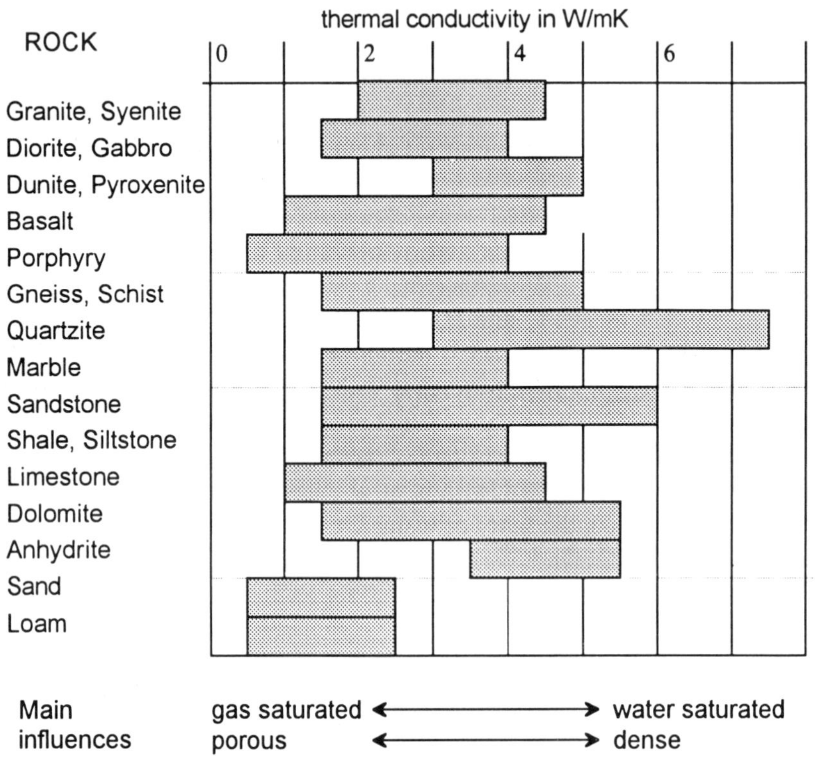

Figure 8.2. Mean range of thermal conductivity for abundant rock types

Five qualitative features should be noted:

(1) the highest values are observed on quartzite and rock salt;

(2) crystalline rocks have medium values with a relatively small range of values compared with the distinct broader ranges of sedimentary rocks (as result of changing porosity);

(3) within igneous rocks, we find a tendency of increasing thermal conductivity from acid/intermediate to basic and ultrabasic rocks;
(4) within the sedimentary rocks, an increasing thermal conductivity tendency can be seen for the series clay - sandstone - limestone, dolomites - rock salt;
(5) loose materials (dry sand, soil etc.) show the lowest values.

The correlation between thermal conductivity and density has been frequently discussed. For example Dachnov and Djakonov (1952) derived an empirical relation within the density range (1.4...3.0) x 10^3 kg m^{-3}

$$\lambda = 0.51 \cdot d^{3.1} \tag{8-9}$$

Cermak (1967) and Cermak and Rybach (1982) recommend a more general simple form:

$$\lambda = a + b \cdot d \tag{8-10}$$

where a and b are empirically determined (compare with equation 8-6 for minerals).

In the following two sections the thermal properties - particularly thermal conductivity - are discussed for the igneous and metamorphic rocks and for sedimentary rocks.

8.3.2. Thermal properties of igneous and metamorphic rocks

8.3.2.1. Mean values and ranges

Table 8.5 contains (in addition to Figure 8.2.) ranges and mean values of thermal conductivity and specific heat for different igneous and metamorphic rocks.

8.3.2.2. Dependence on mineral composition

In the dense, unweathered and nonfractured state, thermal conductivity is controlled mainly by its

- mineral content and the thermal properties of the minerals (see Figure 8.3),
- internal structure.

Table 8.5. Thermal properties of igneous and metamorphic rocks; References: C - Cermak and Rybach, 1982 (value ranges and mean values from Figure 1 of their paper); J - Jessop, 1990 (data after Roy et al., 1981); D - Dortman, 1976; K - Kobranova, 1989.

Rock type	λ in W m^{-1} K^{-1}			*c* in kJ kg^{-1} K^{-1}			*a* in m^2 s^{-1}		Ref.
	n	range	mean	*n*	range	mean	range	mean	
Granite	356	1.25...4.45	3.05	102	0.67...1.55	0.958			C
	153	2.3 ...3.6	3.43						J
	174	1.34...3.69	2.40	84	0.74...1.55	0.946	3.33...15.0	9.27	D
		1.12...3.85			0.25...1.55		3.33...16.5		K
Granodiorite	89	1.35...3.40	2.65	11	0.84...1.26	1.093			C
	194	2.00...3.50	2.63						J
	23	1.64...2.48	2.11	10	0.74...1.26	1.057	3.05...7.5	5.15	D
Syenite	50	1.35...5.20	2.31						C
	31	1.74...2.97	2.26				5.4...7.9	6.30	D
		1.80...2.97					5.4...7.9		K
Diorite	185	1.72...4.14	2.91	3	1.13...1.17	1.14			C
	137	2.02...3.33	2.50						J
	43	1.38...2.89	2.20	3	1.12...1.17	1.14	3.32...8.64	6.38	D
		1.38...2.89			1.12...1.17		3.34...6.64		K
Gabbro	71	1.62...4.05	2.63	9	0.88...1.13	1.005			C
	64	1.98...3.58	2.57						J
	7	1.80...2.83	2.28		0.88...1.13	1.01	9.32...12.2	9.72	D
		1.59...2.98			0.90...1.13		9.32...12.2		K
Peridodite	23	2.30...4.80	3.83	4	0.92...1.09	1.01	11.9...14.1	13.26	D
	14	3.78...4.85	4.37		0.96...1.09		11.97..14.1		K
		3.78...4.85			0.92...1.09				C
Pyroxenite	71	3.30...5.05	4.27	64	0.88...1.21	1.00			C
	64	3.48...5.02	4.33		0.88...1.21	1.01	9.44...14.86	12.86	D
		3.48...5.02			0.88...1.21		9.44...14.86		K
Dunite	700	3.53...5.14	4.41						C
					0.59...0.80		7.16...8.48		K
Diabase,	115	1.55...4.30	2.64	22	0.75...1.00	0.905			C
Dolerite	43	1.38...2.89	2.29	5	0.79...0.93	0.860	9.0...10.16	9.51	D
		1.71...3.25			0.79...0.93		5.04...12.0		K
Andesite	644	1.55...4.65	3.20						C
	108	1.35...4.86	2.26						J
		1.42...2.79			0.81...0.82		6.17...6.44		K
Obsidian		1.38...1.56			0.96...1.21		5.47...5.87		K

Rock type	λ in W m⁻¹ K⁻¹			*n*	*c* in kJ kg⁻¹ K⁻¹		*a* in m² s⁻¹		Ref.
	n	range	mean		range	mean	range	mean	
Basalt	64	1.40...5.33	1.95	3	0.88...0.89				C
	72	1.12...2.38	1.69						J
	22	0.51...2.03	1.45	12	0.76...2.14	1.23	4.33...6.77	5.34	D
		0.44...3.49			0.54...0.82		6.17...6.44		K
Lava	85	0.20...4.60	2.50	5	0.67...1.38	1.08			C
	5	0.25...0.73	0.49	5	0.67...1.38	1.12	2.35...4.13	2.89	DK
Pumice		0.25(dry)...							K
		0.5(moist)							K
Quartzite	186	3.10...7.60	5.26	8	0.71...1.34	1.013			C
	140	2.33...7.45	5.03						J
	9	2.68...7.60	5.26	8	0.72...1.33	0.991	13.6...20.9	17.9	D
Eclogite	4	2.40...3.40	3.10						C
	5		3.35						D
		2.31...3.50							K
Amphi-	78	1.35...3.90	2.46						C
bolite	10	1.57...2.89	2.22	4	1.06...1.20	1.134	5.25...8.14	6.69	DK
Gneiss									
all data	388	1.16...4.75	2.44						C
Switzerland	55	1.2...2.6 a	2.74	55	0.46...0.92	0.75	6.2..12.8 a	8.7	C
		1.2...3.1 b	2.12	55			6.0...15.7 b	10.6	C
Russia	40	0.94...4.86	2.02	7	0.75...1.18	0.979	6.30...8.26	7.32	D
Schist									
all data	122	1.0...5.2	3.14	23					C
Switzerland	18	1.0...3.1 a	2.05	18	0.67...1.05	0.80	4.5...14.1 a	9.3	C
	18	1.7...4.1 b	2.88	18			7.8...18.3 b	13.1	C
	144	1.03...4.93	2.46	95	0.75...1.73	1.096	4.9...16.2	9.6	D
		0.65...4.76			0.70...1.64		2.87...22.5		K
Marble	26.00	1.59...4.00	2.56	8	0.75...0.88	0.86	7.80...12.0	11.00	D
Hornfels		2.12...6.10			1.47...1.48		13.4...15.6		K

a - perpendicular schistosity, b - parallel schistosity.

Roy et al. (1981) have analysed the correlation between thermal conductivity and quartz content of 100 granite and quartz monzonite samples and derived a relation for large scattering of the individual points (Figure 8.4.)

$$\lambda = 0.0245 \cdot V_{\text{quartz}} + 2.59 \qquad (8\text{-}11)$$

where λ is in W m^{-1} K^{-1} and V_{quartz} is the fraction volume content of quartz.

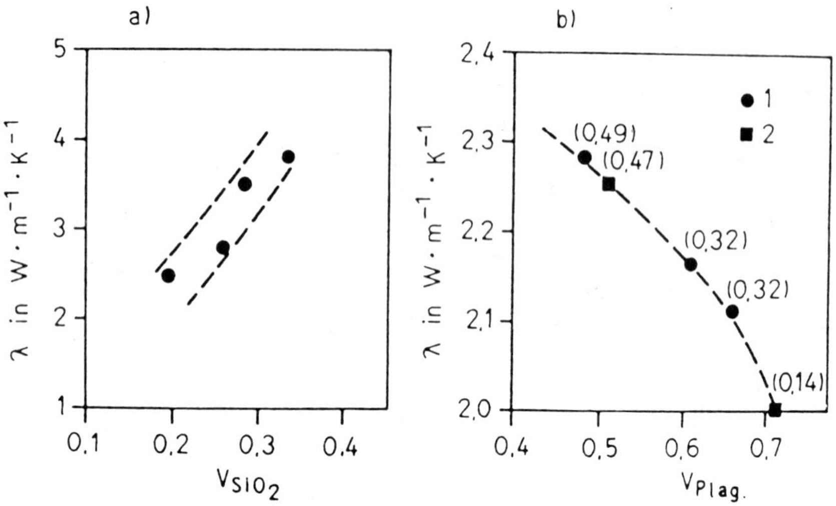

Figure 8.3. Influence of the mineral content on the thermal conductivity of magmatic rocks; after data from Birch and Clarke (1940)
a) Granite - influence of quartz content,
b) Diabase (1) and gabbro (2) - influence of plagioclase content; in brackets is the sum of olivine and pyroxene content.

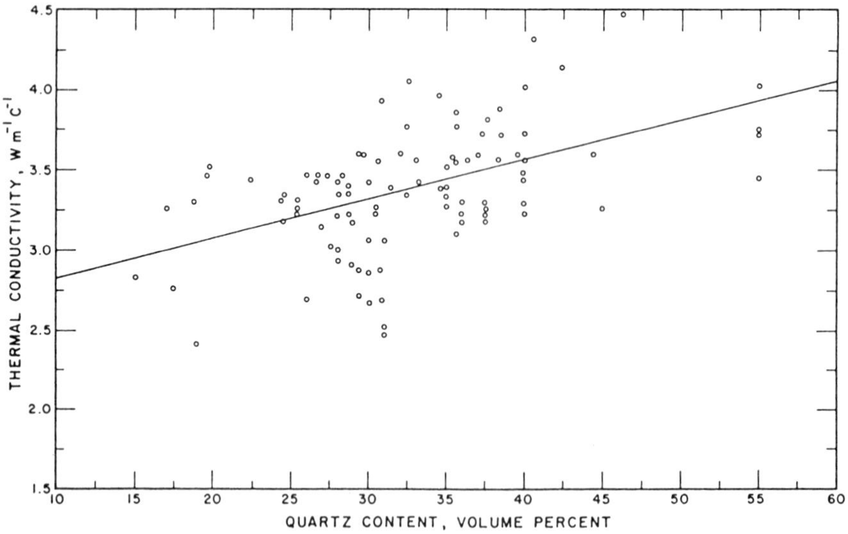

Figure 8.4. Thermal conductivity of granite and quartz monzonite as a function of quartz content (from Roy et al., 1981); 100 samples

8.3.2.3. Influence of fractures and pressure dependence

In fractured or cracked rocks, thermal conductivity is additionally influenced by
- ► the content and properties of crack filling materials and
- ► the crack geometry and distribution.

This results in a pronounced pressure dependence originated by the stress strain behaviour of the crack systems and their effect on the thermal conductivity.
At low pressures (up to about 100 MPa), the thermal conductivity increases due to closure of cracks, fractures, pores etc. and the improvement of the contact conditions (at grain and crack boundaries). It creates a non-linear dependence of thermal conductivity on pressure and the phenomenon of a "hysteresis" as a result of irreversible deformations.

Figure 8.5. illustrates the non-linearity of a gneiss sample G from the KTB borehole. The temperature dependence is also plotted. In comparison with gneiss, a more compact amphibolite sample A obviously does not show the typical crack conditioned behaviour at lower pressures.

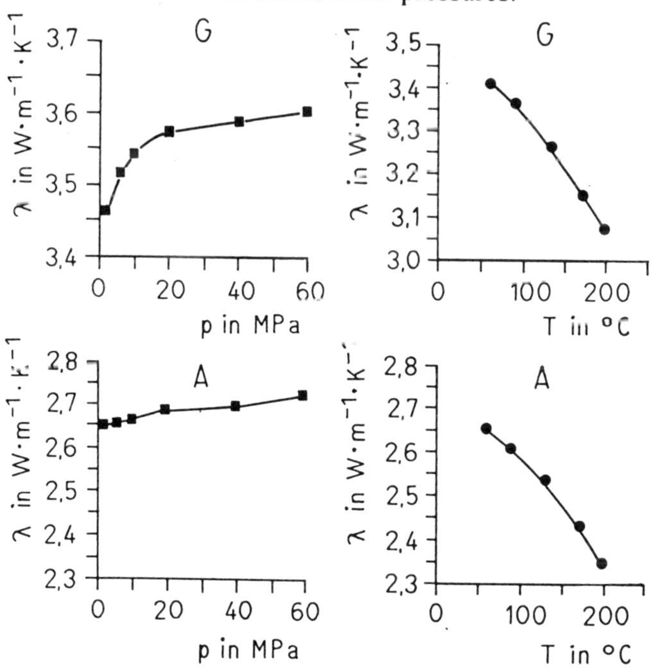

Figure 8.5. Thermal conductivity of samples from the KTB borehole as a function of uniaxial pressure (left) and temperature (right); pressure dependence measured at T = 54 °C, temperature dependence measured at p = 10 MPa, after Huenges et al. (1990). G - gneiss from a depth of 1793 m; A - amphibolite from 147 m;

At higher pressures (for example gneiss) or for only slightly fractured rocks (e.g. amphibolite sample), a linear approximation is possible (Cermak and Rybach, 1982):

$$\lambda = \lambda_0 \cdot [1 + \alpha_p \cdot (\frac{p}{p_0})] \qquad (8\text{-}12)$$

where λ_0 thermal conductivity at normal pressure p_0
λ thermal conductivity at actual pressure p
p actual pressure
α_0 pressure coefficient.

This type of equation was already derived in 1924 by Bridgeman (see Haalck, 1959) e.g. for basalt under hydrostatic pressure up to 1200 MPa (after conversion in SI units):

$$\lambda = 1.69 \cdot (1 + 4.7 \cdot 10^{-2} \cdot p) \quad \text{at } T = 348 \text{ K} \qquad (8\text{-}13)$$
$$\lambda = 1.73 \cdot (1 + 2.2 \cdot 10^{-2} \cdot p) \quad \text{at } T = 303 \text{ K} \qquad (8\text{-}14)$$

where the pressure is given in GPa.

Seipold (1990, 1995) investigated the pressure and temperature dependence of thermal properties of collections of granite (Saxonian Ore Mountains, Germany) and amphibolites (Saxonian Granulite Mountains, Germany). The temperature dependence was studied up to 1000°C and the pressure dependence up to 500 MPa. For the analyse of the experimental data different relationships were tested. He investigated granite samples from two intrusion complexes in the Saxonian ore mountains (Germany) up to 500 MPa and analyzed the experimental data using equation (8-12) with the following results:

- range for λ_0 = 3.46...4.31 W m^{-1}K^{-1},
- range for α_p = 0.007...0.249 GPa^{-1}.

A further analysis shows a difference of mean values between the older intrusion (Kirchberg granite) with λ_0 = 3.65 W m^{-1} K^{-1} and α_p = 0.09 GPa^{-1} and the younger intrusion (Eibenstock granite) with λ_0 = 4.01 W m^{-1} K^{-1} and α_p = 0.144 GPa^{-1}. This may give a hint on the different composition and the different magmatic-tectonic history of the two intrusions.

For an amphibolite sample from the KTB borehole (sample number 323 B 1d) the thermal conductivity versus pressure dependence was described also by equation 8-12 (Seipold, 1993)

$$\lambda = (2.51 \pm 0.03) \cdot [1 + (0.\ 13 \pm 0.04) \cdot p \qquad (8\text{-}15)$$

where λ is in W $m^{-1}K^{-1}$ and p in GPa; thermal diffusivity shows a some different dependence:

$$a = (7.16 \pm 0.05) \cdot [1 + (0.03 \pm 0.02) \cdot p] \qquad (8\text{-}16)$$

The different pressure dependence of thermal conductivity and diffusivity results in a pressure dependent specific heat.

8.3.2.4. Dependence on temperature

The temperature dependence of thermal properties is controlled by the mineral composition, internal structure and the related processes described in chapter 8.1.. With respect to the dominant process, Dortman (1976) distinguished three types of rocks:

Type I : crystalline formed (e.g. granite, diorite)
Type II : amorphous formed (e.g. obsidian)
Type III: crystalloamorphous formed (e.g. diabase, porphyrite).

The temperature dependence of type III is determined by the combined characteristics of crystalline and amorphous substances as described in part 8.1.. Figure 8.5. shows the temperature dependence of two samples.

For most rock types, a decreasing *thermal conductivity* with increasing temperature results from the dominating crystalline behaviour. This also determines the character of some empirical relations. Haenel, 1973 (cited by Cermak and Rybach, 1982) has derived the temperature dependence for the broad temperature range of 0...1200 °C

$$\lambda(T) = 3.6 - 0.49 \cdot 10^{-2} \cdot T + 0.61 \cdot 10^{-5} \cdot T^2 - 2.58 \cdot 10^{-9} \cdot T^3 \qquad (8\text{-}17)$$

where the temperature is given in K.

For many practical questions, a more simplistic relation in the temperature range of 0...600 °C is recommended by Cermak and Rybach (1982)

$$\lambda(T) = \lambda_0 \cdot (1 + a_T \cdot T)^{-1} \qquad (8\text{-}18)$$

where λ_0 is the thermal conductivity under surface conditions,
λ_T is the thermal conductivity at temperature T, and
a_T is an empirical factor.

the dip angle curve. High dip angles result in higher conductivity values because the heat flow direction is closer to the foliation direction: at a dip angle of 90 degree the value is $\lambda_{//}$.

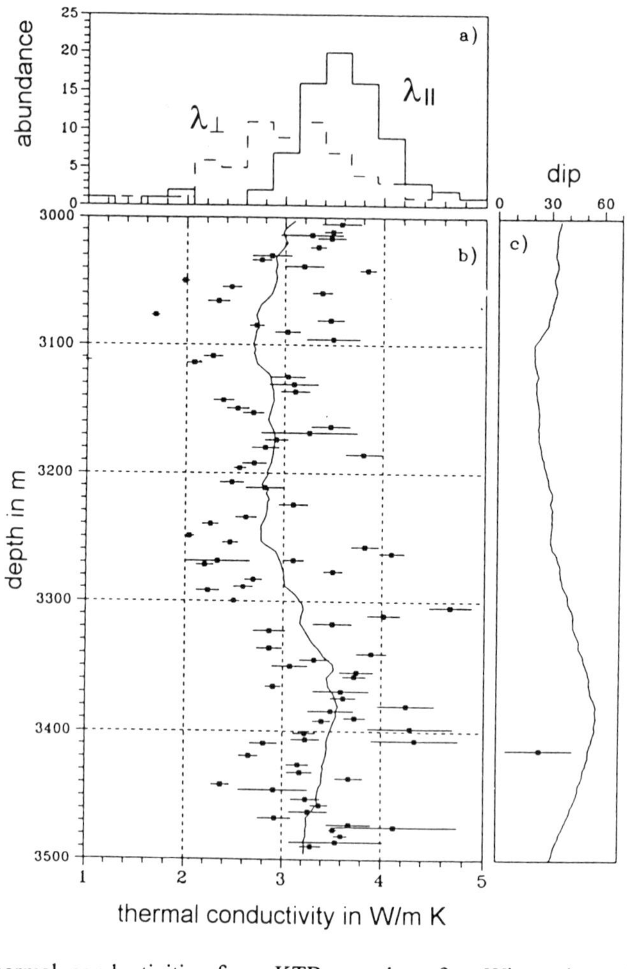

Figure 8.6. Thermal conductivities from KTB samples after Wienand et al., 1989; The geological profile intersection consists of biotite gneiss and alternating layers of metabasite and gneiss;
a) Distribution of thermal conductivities measured parallel (full line) and perpendicular (dotted line) to foliation,
b) Thermal conductivity, measured at cores parallel to the core axis (points with error bar); the curve represents the moving average over 100 m,
c) Dip angle of foliation (moving average over 100 m).

8.3.3. Thermal Properties of Sedimentary Rocks

8.3.3.1. Mean values and ranges

Table 8.7 gives an overview of range and mean values for thermal conductivity of sedimentary rocks.
It is useful to distinguish for further discussions between

- dense, i. e. pore and crack free sedimentary rocks and
- porous sedimentary rocks

resulting from the difference of thermal properties between minerals and pore filling materials.

Table 8.7. Thermal properties of sedimentary rocks;
References: B - Blackwell and Steele, 1989 ; C - Cermak and Rybach, 1982; D - Dortman, 1976; J - Jessop, 1990 (data after Roy et al., 1981); K - Kobranova, 1989; S - Schuch, 1980;

Rock type	λ in W m^{-1} K^{-1}			c_p in kJ kg^{-1} K^{-1}			a in m^2 s^{-1}		Ref.
	n	range	mean	n	range	mean	range	mean	
Anhydrite	77	1.00...6.05	4.00		0.81...0.94	0.98			C
	3	4.89...5.73	5.43						J
		4.90...5.75							D
		4.90...5.80							B
		2.50...5.80			0.58...0.61				E
Gypsum		0.60...1.67			0.90...1.10				K
Salt[1]	70	1.40...7.15	4.00			0.84/			C
	15	1.67...5.50	3.64			0.92			D
		4.80...6.05							B
		1.67...5.50					11.2...17.7		K
Dolomite	29	1.60...5.50	3.62	21	0.84...1.55	1.00			C
	8	3.43...5.73	4.68						J
	72	1.63...6.50	3.24	35	0.65...1.47	1.088			DK
		3.75...6.30							B
Limestone	487	0.62...4.40	2.29	38	0.82...1.72	0.933			C
	445	1.30...6.26	3.44						J
	216	0.92...4.40	2.40	92	0.75...1.71	0.887	3.91...16.9	11.3	D

Rock type	λ in W m^{-1} K^{-1}			c_p in kJ kg^{-1} K^{-1}			a in m^2 s^{-1}		Ref.
	n	range	mean	n	range	mean	range	mean	
Sandstone	1262	0.90...6.50	2.47	41	0.75...3.35	1.64			C
	11	1.88...4.98	3.72						J
	447	0.38...5.17	1.66	113	0.67...3.35	0.97	2.54...20.4	9.86	DK
		2.50...4.20							B
- dry		0.67...6.49							K
- moist		1.10...7.41							K
- oil sat.		0.84...4.24							K
Shale	377	0.55...4.25	2.07	17	0.88...1.44	1.18			C
		0.88...1.25							B
Clay-Siltst.	19	1.70...3.40	2.46	9	0.82...0.93				C
Claystone	242	0.60...4.00	2.04						C
Siltstone	3	2.47...2.84	2.68						J
Argillite	346	0.80...3.70	2.09						C
	203	0.25...3.01	1.21	17	0.74...0.99	0.87	5.94...15.3	9.46	DK
Marl,	191	0.50...4.00	2.21	48	0.78...3.10	1.15			C
marlstone	38	0.50...3.91	1.92		0.84...3.09	1.63	3.14...13.9	7.13	DK
Sand	24	0.10...2.75	1.44	8	1.97...3.18	2.30			C
		1.70...2.50							B
- dry		0.18...3.49							K
- moist		1.63...4.75							K
Clay		0.60...2.60	1.53	24	0.84...1.00	0.85			C
	292	0.38...3.02	1.49	79	0.75...3.55	1.24	2.54...11.6	6.60	DK
- dry	193	0.14...0.24							K
Soil)2		0.60...3.33							S
Peat		0.60...0.80							S
-dry			0.11						K
-moist			0.46						K
Coal		0.04...1.50	0.29						C
	97	0.13...0.30	0.21						D
		0.13...2.24			0.86...1.53		0.70...7.02		K
	6								
Ocean sed.	648	0.40...1.70	0.95						C

)1 see Table 8.8;)2 see fig 8.14

8.3.3.2. Non porous sedimentary rocks

Dense sedimentary rocks show the same main influences and properties which control the thermal properties of magmatic rocks:

- ► mineral content and thermal properties of minerals,
- ► internal structure of the rock.

Typical members of this group are non porous anhydrite, carbonates and the different types of salt. Table 8.8. contains some examples of the thermal conductivity of salt.

Table 8.8. Some examples of thermal conductivity of salt; References: C - Clarke (1966); Cr - Creutzburg (1965); M - Meinke (1967); P - Plewa (1966); H - Herrin and Clarke (1965); L - Loney and Wilson (1966) - all cited by Cermak and Rybach, 1982; D - Dortman (1976); K - Kobranova (1989).

Salt type, location		λ in W $m^{-1}K^{-1}$		Ref.
	n	range	mean	
Rock salt		1.67 ... 5.71		K
pure rock salt		6.1 ... 7.1		K
Germany	14	4.5 ...5.7	5.52	Cr
Thuringia/Germany	45	1.9 ...4.0	3.15	M
Poland			3.63	P
New Mexico, Oklahoma	4		5.34	H
Colorado, Michigan	4		5.56	L
Russia	15	1.7 ...5.5	3.64	D
Polyhalite				
Carlsbad/N. Mexico	2		1.55	C
Russia			1.55	D
Sylvinite	10	4.6 ...5.8	5.29	Cr
		2.6... 5.8		K

Thermal conductivity of salt usually decreases with increasing temperature and increases with increasing pressure (see for example Cermak and Rybach, 1982). Figure 8.7 shows this temperature dependent decrease for thermal conductivity and thermal diffusivity.

It must be noted that, impurities or admixtures e.g. of anhydrite or clay change the properties of the "pure" salt. Figure 8.8 illustrates this in a plot of thermal conductivity vs. density for rock salt containing anhydrite and clay; both parameters are influenced by impurities in the same direction.

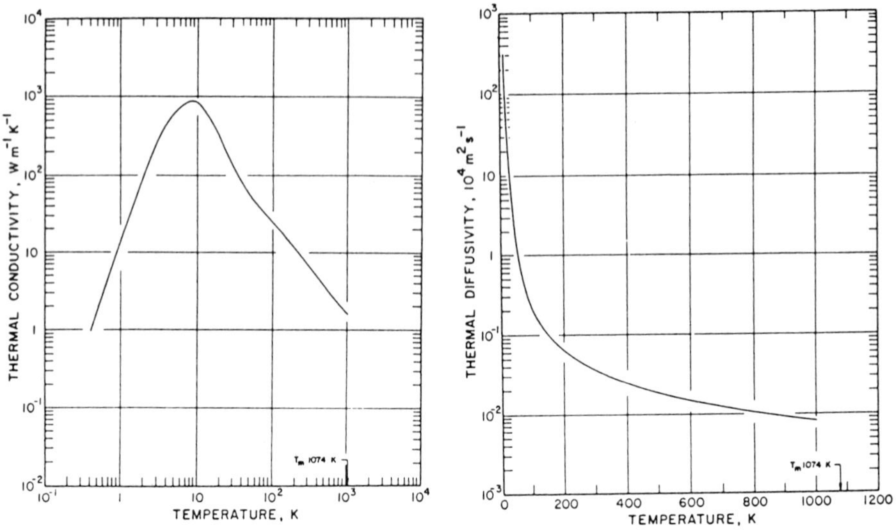

Figure 8.7. Thermal conductivity and thermal diffusivity of crystalline rock salt as a function of temperature (after Yang, 1979 cited by Roy, 1981)

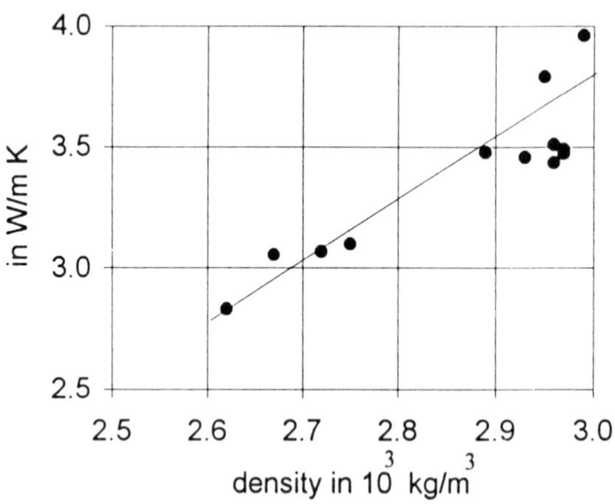

Figure 8.8. Thermal conductivity vs. density for anhydritic and clayey rock salt (halite) from Thuringia/Germany; after measured data from Meinke et al. (1967).

8.3.3.3 Porous sedimentary rocks

In porous sedimentary rocks, "porosity and moisture content effect the thermal conductivity of rocks to a degree, which can be dominant" (Kappelmeyer and Haenel, 1974). The thermal behaviour is strongly influenced by the distinct difference of thermal properties of the solid matrix material (minerals) and the various pore filling materials.

The principle dependencies can be understood by examining the schematic picture in Figure 8.9.: The thermal conductivities of rock constituents are shown on the top and in the lower part, the resulting dependence on porosity (left) and water content (right).

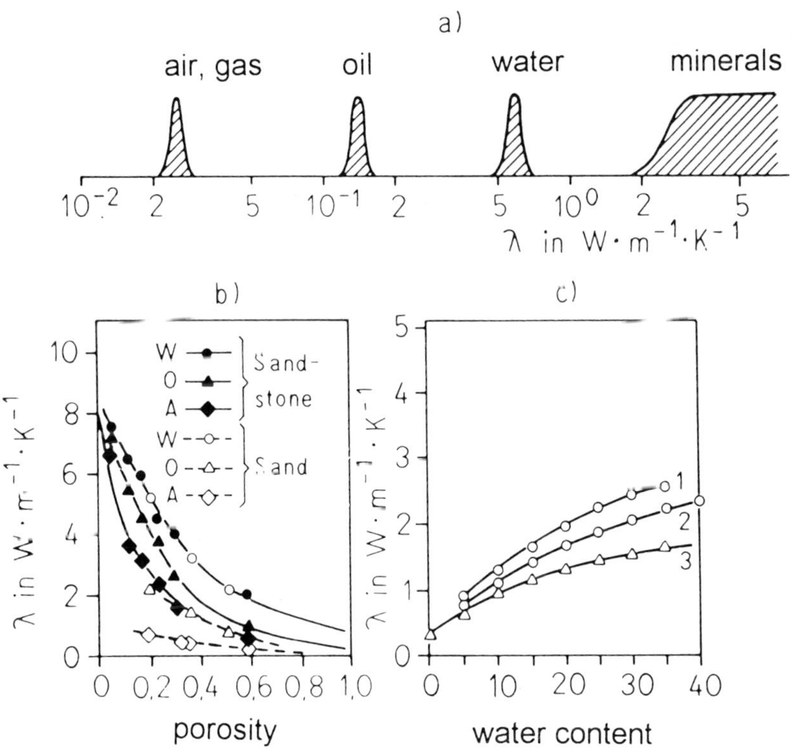

Figure 8.9. Thermal conductivity of porous sedimentary rocks - principal dependencies
a) mean ranges of the constituents (A - air, gas; O - oil; W - water; M - matrix, minerals),
b) thermal conductivity λ vs. porosity Φ for sand and sandstone at different pore filling materials (after Woodside and Messmer, 1961)
c) thermal conductivity vs. water content for different solid components of solid-fluid mixtures: 1 - quartzite (Φ = 0.366), 2 - quartzite (Φ = 0.423), 3 - limestone. (Φ = 0.389)

Generally, the thermal conductivity increases with

- decreasing porosity (Figure 8.9 b.),
- increasing thermal conductivity of pore content (compare different curves in Figure 8.9 b., 8.9 c., and Table 8.9),
- increasing water content (Figure 8.9 c.),
- increasing thermal conductivity of the solid mineral substance (compare quartzite and limestone in Figure 8.9 c.),
- improvement of grain bonds or cementation. This influence is especially remarkable in case of low thermal conductivity pore contents with high contrast of thermal conductivity between matrix and pore material (compare sand and sandstone in Figure 8.9 b.). Rhzewski and Novik (1971) and Griffith et al. (1992) discussed the effect of grain size upon thermal conductivity. Decreasing grain size increases the number of grain contacts per unit volume and thus, decreases thermal conductivity (for example a reduction of 27 % of the monocrystalline thermal conductivity value at a grain size of 0.1 mm, and a 50 % reduction at 0.05 mm).

Table 8.9 Thermal conductivity of sandstones with various pore fluids (after data from Woodside and Messmer, 1961)

Material	Porosity	Thermal conductivity in W/m K with pore fluid			
		vacuo	air	n-heptane	water
Pore fluid		0	0.026	0.128	0.628
Berkeley s.	0.3	2.90	6.49	7.11	7.41
St. Peters s.	0.11	2.49	3.56	5.34	6.36
Tensleep s.	0.155	2.62	3.04	4.37	5.56
Berea s.	0.22	1.68	2.39	3.74	4.48
Teapot s.	0.29	1.09	1.54	2.65	4.05

These dependencies will be discussed in succession.

The correlation between porosity and thermal conductivity has been experimentally investigated by several authors. The general tendency, as well as, the influence of pore filling material and cementation is seen in the results of Woodside and Messmer (1961) in Figure 8.9 b..

The separation between the values for pore filling materials (air, oil and water) is clearly visible for unconsolidated sand, whereas for the cemented sandstone the values are much closer. This is a result of the dominant part of heat transfer by the cemented matrix skeleton.

Plewa (1976) has used a linear empirical relation between thermal conductivity and porosity to fit data, measured on Jurassic carbonates (λ in W m^{-1} K^{-1}, Φ in fractions)

$$\lambda = 4.12 - 9.82 \cdot \Phi \quad (8\text{-}24)$$

or more generalised

$$\lambda = a_1 - b_1 \cdot \Phi \quad (8\text{-}25)$$

Frequently, an empirical linear relation between thermal conductivity λ and density *d* is used (for example Cermak, 1967, Cermak and Rybach, 1982)

$$\lambda = -a_2 + b_2 \cdot d \quad (8\text{-}26)$$

where a_2, b_2 are empirically.
As an example of this application Figure 8.10. shows the results obtained by Poulsen et al. (1981) for compact Silurian limestones.

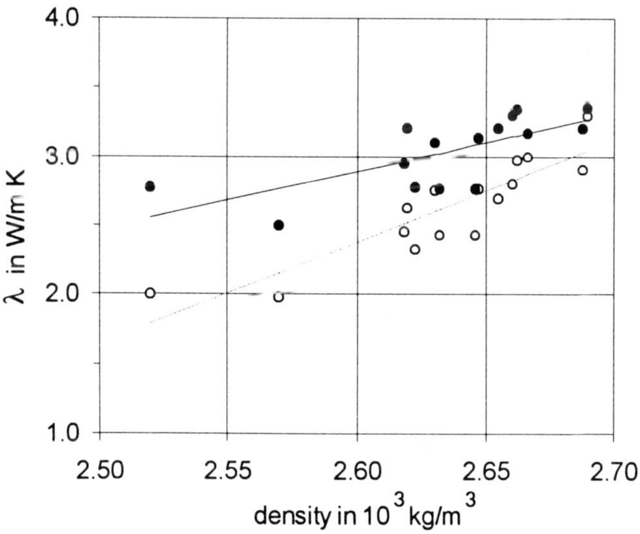

Figure 8.10. Thermal conductivity vs. density for limestones (○ - dry; • - water saturated) data after Poulson et al. (1981); equations for the regression curves see text.

An analysis of the data following equation (8-26) results in the relationship for dry rock:

$$\lambda = 0.00735 \cdot d - 16.73 \quad (8\text{-}27)$$

and for water saturated rock:

$$\lambda = 0.0042 \cdot d - 8.030 \tag{8-28}$$

where the density d is in kg m^{-3} and the thermal conductivity λ in W m^{-1} K^{-1}.

Table 8.10 contains some parameters for the relation (8-26) acquired from experimental investigations of different authors. Kappelmeyer and Haenel (1974) referred to the fact of a high scattering of such relations because the individual samples differ in its quartz content. Cermak (1967) notes the tendency of growing value a_2 with increasing geological age of the sediment.

Table 8.10. Empirical coefficients of the equation 8-26 for sedimentary rocks using data from the following references: C67 -Cermak (1967), C82 - Cermak and Rybach (1982), H76 -Hurtig and Schlosser (1976), P81 - Poulsen et al. (1981).

Rock type	a_2 in W m^{-1} K^{-1}	b_2 in W m^2 kg^{-1} K^{-1}	Ref.
Sandstone	1.65	1.52	C82
Permian, Carbon	2.06	1.71	C67
Cretaceous	2.03	1.70	C67
Neogen	1.84	1.51	C67
Mesozoic	3.66	3.33	H76
Siltstone	2.49	1.69	C67
	2.35	1.63	C82
Claystone	2.69	1.83	C82
Limestone	2.49	1.60	C67
dry	16.7	7.35	P81
water saturated	8.03	4.20	P81
Marlstone	4.02	2.32	C67

The equations 8-25 and 8-26 are equivalent; by using the density - porosity relation (equation 3-3) we get

$$\lambda = -a_2 + b_2[(1 - \Phi) \cdot d_m + \Phi \cdot d_p] \tag{8-29}$$

it follows the conversion

$$\begin{aligned} a_1 &= -a_2 + b_2 \cdot d_m \\ b_1 &= b_2 \cdot \Phi \cdot (d_m - d_p) \end{aligned} \tag{8-30}$$

Based on a geometric mean model (see section 8.4.2.2) Balling et al. (1981) derived by least squares fits relationships between thermal conductivity and porosity of the form (see equation 8-46) for Cenocoic and Mesozoic sediments (Denmark)

$$\lambda = a^{\Phi} \cdot b^{1-\Phi} \qquad (8\text{-}31)$$

Table 8. 11 shows mean values for the parameters a and b.

Table 8.11 Empirically determined parameters a and b (in W/mK) for equation 8-31, References: B - Balling et al., 1981; L - Lovell, 1985 (see also Griffith et al., 1992)

Sediment type	a	b	Ref.
Clay, claystone, shale	0.46	3.43	B
Carbonates	0.54	3.24	B
Sandstone	0.69	4.88	B
Quartz sand	0.64	8.58	L

The parameter b is controlled by the matrix properties (see section 8.4.2.2). Therefore, it is relatively high for sandstone and especially for quartz sandstone. An increasing clay content (e.g. in case of shaly sands) results in a decreasing thermal conductivity. A study of this problem - with a theoretical consideration - was published by Brigaud et al. (1989).

Figures 8.9 and 8.10 show the influence of the thermal *properties of pore filling material* on the thermal conductivity of porous rock; it is even more distinct for unconsolidated rocks than for consolidated (see Figure 8.9b.).
Somerton (1958) has published an empirical relationship for the ratio of the thermal conductivity of dry (λ_{dry}) and fluid saturated (λ_{sat}) porous rock related to the thermal conductivity of the pore fluid (λ_{fluid})

$$\frac{\lambda_{sat}}{\lambda_{dry}} = \left(\frac{\lambda_{fluid}}{\lambda_{dry}}\right)^{c \cdot \Phi} \qquad (8\text{-}32)$$

where Φ is the porosity and c an empirical exponent in the range of 0.9...1.3.

The increasing influence of pore material conductivity with growing porosity may also be illustrated by Figure 8.11. Using data from Kunii and Smith (1960; see Roy, 1981) and Woodside and Messmer (1961), the ratio of rock conductivity for pores filled with water and air $\lambda_{water}/\lambda_{air}$ and, also, the ratio for pores filled with water and oil $\lambda_{water}/\lambda_{oil}$ is plotted versus porosity Φ

A linear regression leads to the following relations for the rock conductivity ratio:

$\lambda_{water}/\lambda_{air} = 4.00 \cdot \Phi + 0.97$ (data after Kunii and Smith, 1960)

$\lambda_{water}/\lambda_{air} = 5.50 \cdot \Phi + 1.01$ (data after Woodside and Messmer, 1961)

$\lambda_{water}/\lambda_{oil} = 0.94 \cdot \Phi + 1.0$ (data after Kunii and Smith, 1960)

$\lambda_{water}/\lambda_{oil} = 1.55 \cdot \Phi + 1.01$ (data after Woodside and Messmer, 1960).

We note a strong influence of the "conductivity contrast" between the analysed pore filling materials (see also Figure 8.9.).

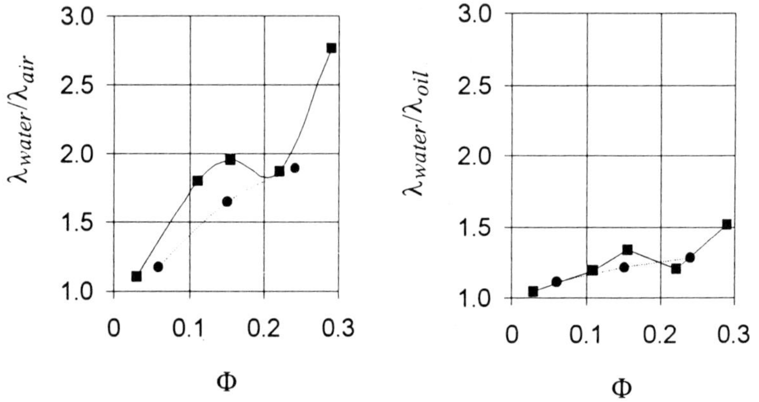

Figure 8.11. Ratio of thermal conductivity of sandstones with pore filling water and air $\lambda_{water}/\lambda_{oil}$ (left) and ratio for pore filling water and oil $\lambda_{water}/\lambda_{air}$ (right) versus porosity Φ • - data after Kunii and Smith (1960, see Roy, 1981); ▪ - data after Woodside and Messmer (1961).

Cermak and Rybach note that there is an increase of conductivity with increasing water content, which reaches a maximum at a porosity of about 20...30 %. They derived an empirical relation between this "wet" conductivity and "dry" conductivity

$$\lambda_{wet} = \lambda_{dry} \exp(2.4 \cdot C_w) \qquad (8\text{-}33)$$

where C_w is the water content of the rock in volume fractions. This relation contains the combined effect of changing porosity and changing degree of water saturation ($C_w = \Phi \cdot S_w$).

Dachnov and Djakonov (1952) and Djakonov and Jakovlev (1969) have found a comparable relation with an exponent of 0.8 C_w.

8.3.3.4. Unconsolidated sediments

Some properties and dependencies are discussed in section 8.3.3.3, the following supplements this discussion.

Marine unconsolidated sediments are a particular type of sediments, which are characterised by high porosity and very "loose" bonding between the individual grains. Experimental results have been published by Somerton and Boozer (1960), Ratcliffe (1960), Hutt and Berg (1968), v. Herzen and Maxwell (1969), and in the cited data collections.

Figure 8.12 demonstrates the correlation between thermal conductivity and water content (weight %), plotted after data from Ratcliffe (1960), and the derived correlation between thermal conductivity and porosity.

Bullard and Day (1961) suggested for seafloor sediments the empirical relationship between sediment thermal conductivity and porosity

$$\lambda = [(0.161 \pm 0.014) + (0.651 \pm 0.030) \cdot \Phi]^{-1} \quad (8\text{-}34)$$

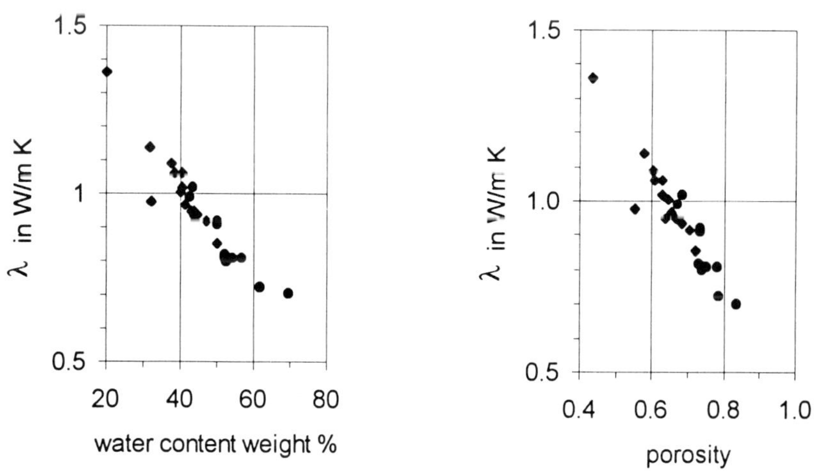

Figure 8.12 Thermal conductivity of deep-sea sediments (red clay, globigerina ooze, and glacial clay); Clark (1966) after data from Ratcliffe (1960).
a - thermal conductivity vs. water content (weight %); b -thermal conductivity vs. porosity. the derived thermal conductivity vs. porosity relationship is $\lambda = 2.24 - 1.85\ \Phi$

Another special type of unconsolidated rocks are *soils*. Their thermal conductivity is essentially influenced by the soil moisture (Figure 8.13) and also show a non-linear correlation. This strong correlation was discussed by Winterkorn in 1962.
Schuch (1982) may be cited for the general behaviour of soils: "The thermal conductivity of dry soils is small (0.2...0.8 W $m^{-1}K^{-1}$), reaches a maximum by 20 to 30 wt% of water content (2 or 3 W $m^{-1}K^{-1}$), decreases for higher contents of water, e.g. wet bog, and draws near the value of thermal conductivity of water (0.6 W $m^{-1}K^{-1}$). This decrease is originated by an increasing porosity in that range and a connected decrease of the heat transfer by the skeleton of the solid parts of the soil".

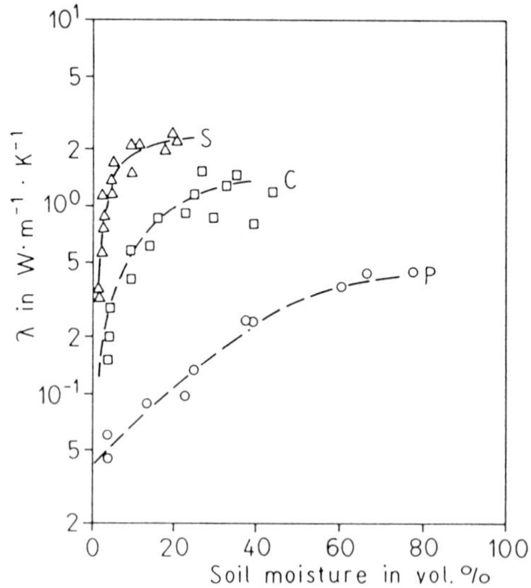

Figure 8.13. Thermal conductivity of unconsolidated sediments vs. soil moisture (volume %) after Quiel (1975); converted in SI units. S - sand; C - clay; P - peat

8.3.3.5 Dependence of thermal conductivity on pressure and depth

Increasing *pressure* yields an increasing thermal conductivity in sedimentary rocks (Figure 8.14) due primarily to

- improved heat transport at grain-grain contacts and also at cracks, microcracks or other defects within the compact solid matrix,
- decrease of porosity.

Therefore, pressure conditioned variations of the thermal conductivity are more evident in compressible rocks (unconsolidated sediments, consolidated sediments with high porosity) rather than in rocks with zero or small compressibility (dense carbonates, anhydrite). The effective pressure is the decisive variable (equation 2-12). The direct dependence of thermal conductivity variations on the deformation behaviour explains the phenomena of nonlinearity and partial irreversibility (*"hysteresis"*) of the thermal conductivity vs. pressure curve.

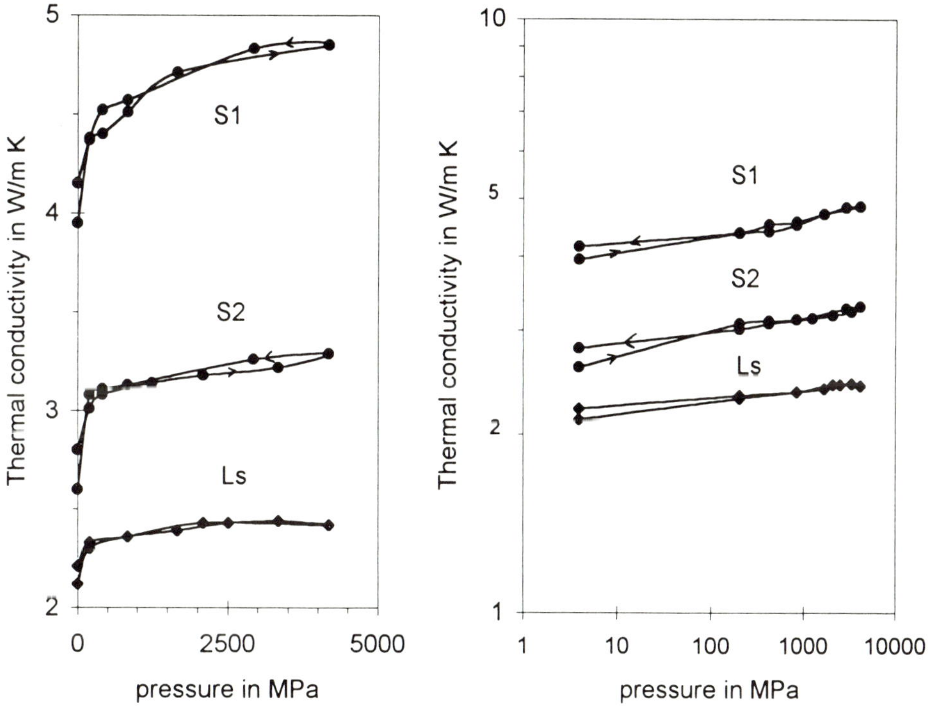

Figure 8.14. Thermal conductivity vs. uniaxial pressure (Hurtig and Brugger, 1970; Cermak and Rybach, 1982) in linear and log-log presentation
S2 - sandstone (sample 234); S1 - sandstone (sample 313); Ls - limestone (sample. 19)

A log-log presentation of thermal conductivity versus pressure results in a linear correlation as a first approximation. This corresponds to a power law of the form

$$\lambda = \lambda_0 \cdot (p/p_0)^m \qquad (8\text{-}35)$$

where in this particular case the exponent m is between 0.02 and 0.04.

8.3.3.6. Anisotropy of thermal conductivity (sedimentary rocks)

The anisotropy of thermal conductivity correlates with the structural-textural properties of sedimentary rocks. Carbonates and sandstones usually show only a small anisotropy (mostly in the range up to 1.3). Higher anisotropy in carbonates (up to about 1.3...1.4) can be originated by a distinct bedding or crack orientation. Typical anisotropic rocks are shales (up to 2.5) and claystones (up to 4).

8.4. Theories and models

"Several mathematical models have been proposed to predict the conductivity of a rock from a knowledge of its constituents. All depend on a knowledge of conductivity of the minerals, and so all begin with the same disadvantage. Each one employs a different mathematical formulation to account for the distribution of the conductivities within the mineral matrix." (Jessop, 1990)

8.4.1 Review

The theories for the calculation of the thermal properties (especially thermal conductivity) can be classified following the schematic picture of Figure 1.6. Sheet or laminated models are also dominant for thermal properties.

The *specific heat* as a scalar property can be described by a simple relation of type equation (3-2) as follows:

$$c_p = \frac{1}{d} \sum_{i=1}^{n} V_i \cdot d_i \cdot c_{p,i} \qquad (8\text{-}37)$$

where V_i - volume fraction of component i
$c_{p,i}$ - specific heat of component i
d_i - density of component i.

This relationship is valid for rocks consisting of n components (minerals, pore filling materials).

The *thermal conductivity* as a tensor depends on the volume fractions and thermal conductivity of rock components, but also on their distribution, on geometry and internal structure and on the heat transfer conditions at the contacts between them. This complex feature makes the problem of a theoretical treatment difficult.
The following presents an overview of three groups of models related to thermal conductivity:

- sheet or laminated models, their modifications and comparable mixing rules,
- sphere or inclusion models,
- a model with internal structure.

A discussion on the pressure dependence of thermal conductivity is presented with the first group.

8.4.2. Theories based on sheet models, their modifications and comparable mixing rules

8.4.2.1. The series and parallel model

A rock consisting of n components (minerals, pore filling materials etc.) can be idealised in the simplest case as a sheet model in analogy to Voigt's and Reuss's concept (see section 6.4.2). The sheets represents the individual rock components. The relative thickness of each sheet is given by the volume fraction of each rock component (see Figure 8.18.). This results a *"series model"* and a *"parallel model"*.
For the case of the series model, the heat flow passes perpendicular to the boundary planes between the components and for the case of the parallel model, it passes parallel. Figure 8.17 shows these models for of a two component rock (such as a porous sandstone).
The corresponding relations for n components are:

- parallel model $$\lambda_{//} = \sum_{i=1}^{n} V_i \cdot \lambda_i \qquad (8\text{-}38)$$

- series model $$\lambda_{\perp}^{-1} = \sum_{i=1}^{n} V_i \cdot \lambda_i^{-1} \qquad (8\text{-}39)$$

where V_i is the volume fraction and λ_i the thermal conductivity of the component i. These equations represents the upper ($\lambda_{//}$) and lower ($\lambda_{\perp}$) bound of thermal conductivity for a rock of a given composition.

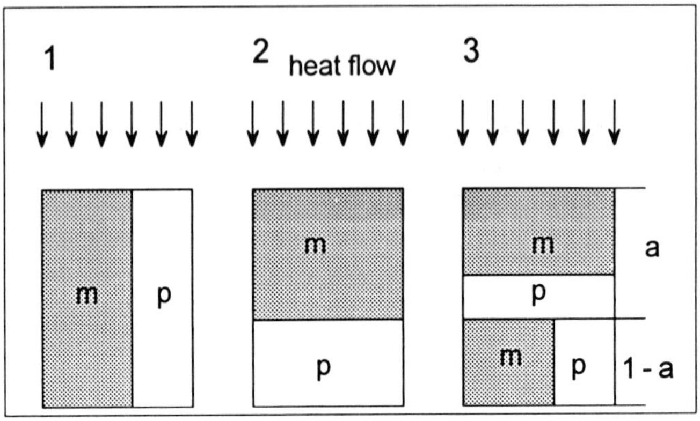

Figure 8.17. The family of sheet models for calculation of the thermal conductivity
1 - parallel model 2 - series model 3 - model after Krischer and Esdorn (1956)
m - matrix p - pore fluid

For a *porous rock* - consisting of matrix (subscript m) and pore content (subscript p) - the equations become

$$\lambda_{//} = (1 - \Phi) \cdot \lambda_m + \Phi \cdot \lambda_p \tag{8-40}$$

$$\lambda_{\perp} = [(1 - \Phi)/\lambda_m + \Phi/\lambda_p]^{-1} \tag{8-41}$$

These porosity dependencies for the series ($\lambda_{\perp}$) and the parallel ($\lambda_{//}$) model are plotted in the Figures 8.18 and 8.19.

Equation (8-40) was extended to a two component matrix by Hurtig and Schlosser (1976) to analyse the thermal properties of sedimentary rocks (Mesozoic and Cenozoic) from the Northgerman-Polish-basin. By assuming mean values, the following relations result

sandstone-silitstone-beds $\lambda = 7.1 \cdot V_{\text{quartz}} + 2.3 \cdot V_{\text{silt}} + 0.63 \cdot \Phi$ (8-42)

limestone-siltstone-beds $\lambda = 3.3 \cdot V_{\text{calcite}} + 2.3 \cdot V_{\text{silt}} + 0.63 \cdot \Phi$ (8-43)

8.4.2.2. Modifications and comparable mixing rules

The real, experimentally determined data are mostly situated between the two boundaries given by the series and the parallel model. There are different theoretical developments to obtain a better approximation between calculated and measured values:

a) A simple combination of the two fundamental models is their *arithmetic mean* (in analogy to "Hills value" after equation 6-111, 6-112)

$$\lambda_H = \tfrac{1}{2}(\lambda_{//} + \lambda_{\perp}) \tag{8-44}$$

This relation was applied e.g. by Horai and Simmons (1969).

b) A model without this simple geometric significance, but with a simple mathematical expression is the *geometric mean*

$$\lambda_g = \prod_{i=1}^{n} \lambda_i^{V_i} \tag{8-45}$$

In particular for the case of a porous rock, eq. 8-45 becomes

$$\lambda_g = \lambda_m^{1-\Phi} \cdot \lambda_p^{\Phi} \tag{8-46}$$

or

$$\log\lambda_g = (1-\Phi)\cdot\log\lambda_m + \Phi\cdot\log\lambda_p \tag{8-47}$$

More complex models have been developed by Hashin and Shtrikman, 1962 (compare chapter 6.4.7, see also Jessop, 1990).
Figure 8.18 gives a comparison of experimental and calculated data using equations 8-40, 8-41, 8-44 and 8-46. There is a fairly good fit of the geometric mean equation (8-46) with the sandstone data, whereas for highly porous unconsolidated marine sediments, the series model (8-41) yields better results.
Applications utilising the geometric mean model with good results have been published by Woodside and Messmer (1961) and Griffith et al. (1992).
Merkle et al., 1976 (cited by Blackwell and Steele, 1989) have determined the in situ thermal conductivity on the basis of the mineral composition (derived from well logging results) of the Edwards Limestone (San Antonio/Texas) and mean thermal conductivities of the minerals using equation (8-45).

Jessop (1990) notes that "this model has achieved a certain popularity, and to the level of accuracy that may expected, it is probably the best simple model available".

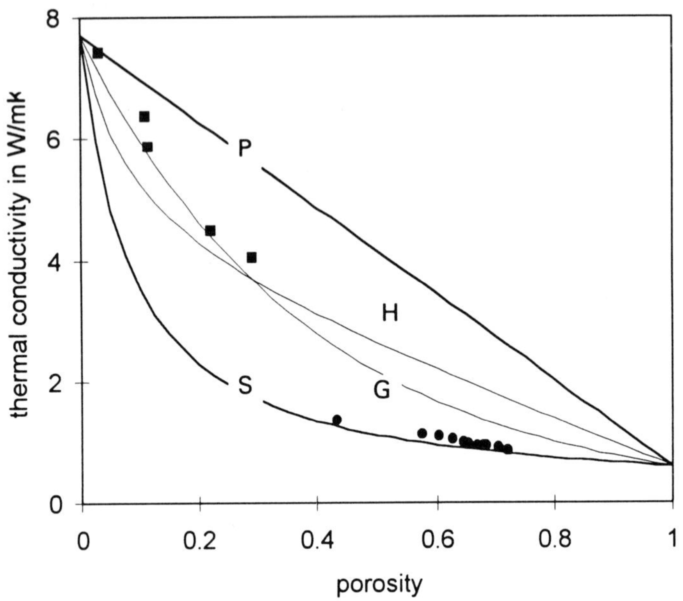

Figure 8.18. Comparison of calculated and experimental determined thermal conductivity vs. porosity dependencies for water saturated sediments curves are calculated after equations 8-40: P - parallel; 8-41: S - series; 8-44: H - Hill mean; 8-46: G - geometric mean λ_{matrix} = 7.7 W/m K; λ_{fluid} = 0.6 W/mK
experimental data: ■ - sandstone, after Woodside and Messmer (1961)
• - marine red clay, after Ratcliffe (1960), see Figure 8.12.

Brigaud et al. (1989), Griffith et al. (1992), and Brigaud et al. (1992) have applied a detailed model based on the serial and the geometric model to calculate thermal conductivity of sedimentary rocks using well log analysis results. The model consists of

- porosity
- sandstone (quartz)
- carbonate
- shale (smectite, illite, kaolinite and mixed-layers).

The fundamental relations can be combined in a some different way so that the range between the curves of the upper and lower boundaries can be "filled" by varying one (or more) additional parameter. Three examples are given; they are related, particularly for porous rocks:
(1) Krischer and Esdorn (1956) have combined the two fundamental models

$$\lambda_{KE} = \left(\frac{a}{\lambda_{\perp}} + \frac{1-a}{\lambda_{//}} \right)^{-1} \tag{8-48}$$

An additional parameter a is introduced (Figure 8.17, 8.19).

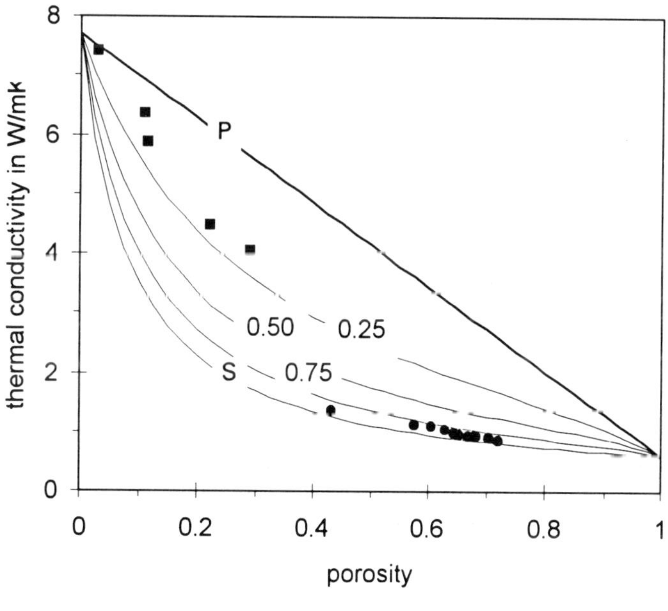

Figure 8.19. Comparison of calculated and experimentally determined thermal conductivity vs. porosity for water saturated sediments;
curves are calculated after equation 8-48 parameter a; P - parallel S - series model
experimental data: ■ - sandstone, after Woodside and Messmer, 1961
• - marine red clay, after Ratcliffe, 1960; see Figure 8.12.

This parameter a describes the volume fraction of the series model of the whole combined model. By varying this parameter from its minimum value (identically with the parallel model) to the maximum value (identically with the series model), the range between the extremes is described. The authors have detected a tendency for this parameter to decrease with increasing degree of compactness or cementation for construction materials. Figure 8.19 gives a comparison with experimental data in analogy to Figure 8.18, which confirms this tendency.

2) Sugawara and Yoshizawa (1961) started with the parallel model for porous rocks. But porosity was substituted for with a porosity-function Ψ and equation (8-40) thus becomes

$$\lambda_S = (1-\Psi)\cdot\lambda_m + \Psi\cdot\lambda_p \tag{8-49}$$

where

$$\Psi = \frac{2^n}{2^n-1}\{1-(1+\Phi)^{-n}\} \tag{8-50}$$

n is an exponent, which must be determined experimentally and is dependent on the structure, pore size and pore shape, emission properties and also from the properties of the pore contents. A comparison with the sandstone data yields values of about $n = 1.3$.

3) Xu and Desbrandes (1991) started with the geometric mean equation (8-45) in a some modified version (Somerton et al., 1974)

$$\lambda = \lambda_{\mathrm{m}}^{1-k\cdot\Phi}\cdot\lambda_{\mathrm{f}}^{k\cdot\Phi} \tag{8-51}$$

where k is an empirical coefficient. Somerton (1958) found that the coefficient k is usually > 1; Xu and Desbrandes (1991) give values between 0.9 and 2.3. Based on this, an equation for multiphase saturating fluids (w - water, o - oil, g - gas) is proposed for thermal conductivity

$$\lambda_X = \lambda_{\mathrm{m}}^{1-k\cdot\Phi}\cdot(\lambda_{\mathrm{w}}\cdot S_{\mathrm{w}} + \lambda_{\mathrm{o}}\cdot S_{\mathrm{o}} + \lambda_{\mathrm{g}}\cdot S_{\mathrm{g}})^{k\cdot\Phi} \tag{8-52}$$

Note that the "mean conductivity" of the saturating fluid is calculated after the arithmetic mean.

The corresponding equation for volumetric heat capacity as a scalar property is

$$d\cdot c = (1-\Phi)\cdot d_{\mathrm{m}}\cdot c_{\mathrm{m}} + \Phi\cdot(S_{\mathrm{w}}\cdot d_{\mathrm{w}}\cdot c_{\mathrm{w}} + S_{\mathrm{o}}\cdot d_{\mathrm{o}}\cdot c_{\mathrm{o}} + S_{\mathrm{g}}\cdot d_{\mathrm{g}}\cdot c_{\mathrm{g}}) \tag{8-53}$$

where is density, c heat capacity, S saturation and λ thermal conductivity. Xu and Desbrandes (1991) applied equations 8-52 and 8-53 for calculation of a crossplot of the (logarithm of) *thermal conductivity versus volumetric heat capacity* to the interpretation of porosity and saturation. A comparison with experimental data from 45 sandstone samples (Somerton et al., 1974) show a good match between the calculated and measured porosities (the percentage of error for most samples is < 6%). The coefficient in this case was $k = 1.5$.

The philosophy of the models may also be applied to other materials, e.g. on polymineralic igneous rocks with an adapted relation.
For example analogous to the Krischer and Esdorn model, the relation becomes:

$$\lambda = \left[(1-a) \cdot \left(\sum_{i=1}^{n} V_i \cdot \lambda_i \right)^{-1} + a \cdot \left(\sum_{i=1}^{n} V_i \cdot \lambda_i \right)^{-1} \right]^{-1} \tag{8-54}$$

Drury and Jessop (1983) investigated different models and their applicability for *igneous rocks*. A comparison between measured and calculated conductivity for granites and granodiorites shows that the geometric model (equation 8-45) generally gives an estimate of conductivity to better than 10 %. The highest correlation results from an additional consideration of crack porosity using a model of Waff (1974), derived for electrical conductivity: It assumes that the porosity in a crystalline rock is mainly along grain boundaries. The model consists of a system of solid cubic grains in a continuous fluid matrix (as matrix conductivity the geometric mean value is used):

$$\lambda_W = \left[\lambda_m \lambda_p (1-\Phi)^{2/3} \right] \cdot \left[\lambda_m (1-\Phi)^{1/3} + \lambda_p \Phi^{1/3} \right]^{-1} + \lambda_p \cdot \left[1 - (1-\Phi)^{2/3} \right] \tag{8-55}$$

The authors conclude that "it appears, therefore, that at least for massive crystalline rocks the cubic array model is most appropriate for modelling the effects of porosity."

8.4.2.3. A simple model for fractured rocks

In section 6.4.3.4 a simple model for rocks with internal defects (cracks, fractures etc.) was introduced. The same model with identical assumptions now is applied to thermal conductivity:

$$\lambda = \lambda_m(1 - D) \tag{8-56}$$

where λ_m is the thermal conductivity of the solid and compact matrix material and D is the "defect parameter" related to the effect on thermal properties. Using the relation between the defect parameter and pressure (see equation 6-138), the pressure dependence of thermal conductivity is:

$$\lambda = \lambda_m\left[1 - D_0 \cdot \exp\left(-A \cdot \frac{p}{p^*}\right)\right] \tag{8-57}$$

where D_0 is the initial value of the defect parameter D at the pressure $p = 0$ and A a deformation property parameter. This model is applied to analyse the experimental results of the gneiss sample presented in Figure 8.5.

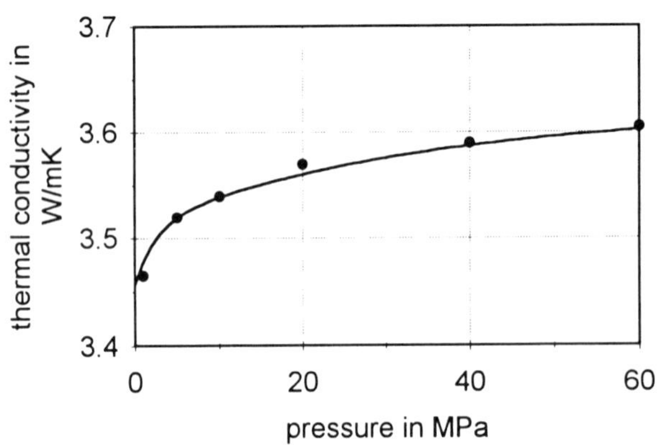

Figure 8.20. Application of the model on experimental results; points - experimental data from Figure 8.5 (gneiss sample), curve - calculated with equation 8-58

A satisfactory approximation between calculated and measured values (Figure 8.20) needs two exponential terms

$$\lambda = 3.62 \cdot [1 - 0.0015 \cdot \exp(-0.4 \cdot p) - 0.03 \cdot \exp(-0.03 \cdot p)] \tag{8-58}$$

The two exponential terms express that the rock behaviour is characterised by at least two fracture systems comparable to the case of the elastic properties (see section 6.4.3.4).

8.4.3. Sphere or inclusion models

The classic Maxwell-model consists of dispersed spherical particles in a homogeneous substance. The assumption of *spherical pores* in a solid matrix yields to the equation (see also Parrott and Stuckes, 1975):

$$\lambda = \lambda_m \cdot \frac{(2\eta + 1) - 2\Phi(\eta - 1)}{(2\eta + 1) + \Phi(\eta - 1)} \qquad (8\text{-}59)$$

where $\eta = \lambda_m / \lambda_p$ is the ratio of the thermal conductivities of the two components (matrix and pore fluid).
If we utilise *spherical solid particles* in an fluid medium (pore content) the following equation results:

$$\lambda = \lambda_p \cdot \frac{3\eta - 2 \cdot \Phi(\eta - 1)}{3 + \Phi \cdot (\eta - 1)} \qquad (8\text{-}60)$$

With regard to the supposition that there is no influence or distortion of the thermal field of one inclusion by any neighbouring inclusion, we can expect that

- equation (8-59) is useful mainly for rocks with small porosity,
- equation (8-60) is useful mainly for high porous sediments (marine sediments).

A comparison with experimental data (Figure 8.21) confirms this. The limitation of the applicability for both equations is determined by the ratio η. Equation 8-59 is applicable up to a porosity of about $\Phi = 0.15$, if $\eta \geq 10$(after Beck, 1976). In the case of Figure 8.21 the ratio is about 15.

Brailsford and Major (1964) developed an equation for a model in which *two solid components* (1, 2) of different conductivity are immersed in a continuous fluid third phase (3):

$$\lambda = \left[C_1 + \frac{3C_2\lambda_2}{2\lambda_1 + \lambda_2} + \frac{3C_3\lambda_3}{2\lambda_1 + \lambda_3} \right] \cdot \left[\frac{C_1}{\lambda_1} + \frac{3C_2}{2\lambda_1 + \lambda_2} + \frac{3C_3}{2\lambda_1 + \lambda_3} \right]^{-1} \qquad (8\text{-}61)$$

where C_i and λ_i denote the volume content and thermal conductivity of each component. For two components, equation (8-61) equals equation (8-60).

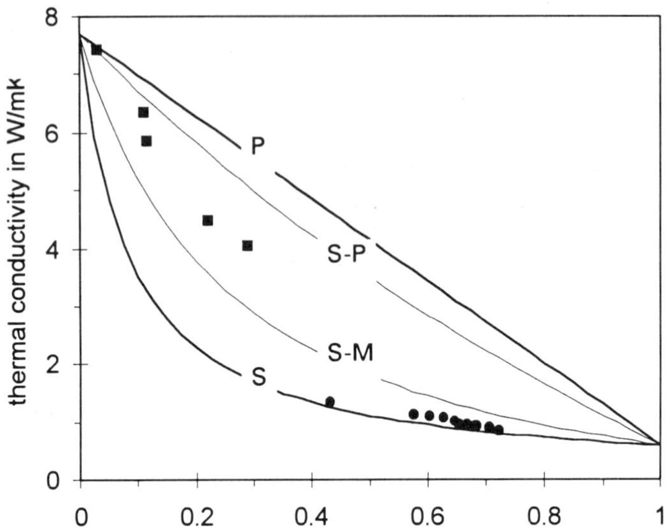

Figure 8.21. Comparison of calculated and experimental determined thermal conductivity vs. porosity dependency for water saturated sediments; λ_{matrix} = 7.7 W/m K λ_{fluid} = 0.6 W/m K

P - parallel S - series S-P - spherical pores S-M - spherical matrix; experimental data: ■ - sandstone, after Woodside and Messmer (1961), • - marine red clay, after Ratcliffe (1960).

The applicability of the Maxwell model and equation is not limited to mixtures of solid matrix material and liquid pore content. It is also useful for a mixture of two different solid components, such as for solid spheres in another solid material or cement. In this case, the porosity must be substituted with the volume fraction of the other material or cement.

Kobranova (1989) has applied the Maxwell equation sequentially to a polymineralic rock. The solid matrix consists of 70% quartz, 20% feldspar, and 10% kaolinite. In the first step, a solid conductivity is calculated for the quartz-feldspar mixture. In the next, this solid material is combined with the kaolinite, and in the final step, this three-component matrix is combined with the pore material.

The Maxwell theory was developed for *ellipsoidal particles* by de Vries (1952) and applied to the properties of soils by Woodside and Messmer (1961). Further relations - also based on the Maxwell theory - were developed by Russell and Eucken (see Francl and Kingery, 1954).

Artemieva and Chesnokov, 1991 have derived theoretical relationships for thermal conductivity of an anisotropic medium based on the inclusion model. The model consists of a matrix with ellipsoidal inclusions; the main assumptions are

- The intrinsic anisotropy of inclusions and matrix material is negligible; the heat field inside the inclusion is homogeneous.
- There is no interaction between different inclusions and, thus, for calculations it is possible to use relations obtained for a model of a medium comprising only one isolated inclusion in a matrix.

The thermal characterization of the model is given by the ratio between thermal conductivity of inclusions and matrix, and the geometric characterisation is given by the volume concentration of inclusions (or porosity) and the aspect ratio of inclusions.

The results show, that the effective thermal conductivity of the composite material greatly depends on

- the porosity and the material saturating the pores,
- the aspect ratio; for a case with oblate inclusions (cracks) with $\alpha < 1$, anisotropy is much more pronounced than for a case of prolate inclusions $\alpha > 1$

When $\alpha \geq 10^2$ or $\alpha \leq 10^{-3}$, the calculated conductivity tends to approach the asymptotic values obtained for the limiting cases described by equations 8-38 and 8-39 ("fibred" and "layered" media respectively). Thus, a medium with $\alpha \leq 10^{-3}$ behaves macroscopically like a layered medium.

A comparison with experimental data for sandstone (measurements of Woodside and Messmer, 1961) confirms a good approximation can be achieved with calculated thermal conductivity vs. porosity curves under assumption of an aspect ratio between 0.05 and 0.1.

8.4.4. Effective medium theory

The effective medium theory - first published by Bruggeman (1935) - in many cases can give a very useful method for calculation the properties of a composite material if the volume fractions V_i of the i individual components and their properties (λ_i) are known. The effective medium theory predicts the thermal conductivity of the composite material (rock) λ by assuming that the distribution of the grains is random and that on a scale much larger than the components (grain) size, the composite is homogeneous and isotropic. The basic result of this theory is (see also Palciauskas, 1986):

$$\lambda^{-1} = \sum_{i=1}^{n} 3V_i \cdot (2 \cdot \lambda + \lambda_i)^{-1} \tag{8-62}$$

A porous rock consisting of matrix and pore filling material results in

$$\lambda^{-1} = 3\Phi \cdot (2\lambda + \lambda_p)^{-1} + 3 \cdot (1 - \Phi) \cdot (2\lambda + \lambda_m)^{-1} \tag{8-63}$$

where $\lambda, \lambda_m,$ and λ_p are the thermal conductivities of the rock, the matrix, and the pore filling material. Figure 8.22 shows a comparison of the experimental data for sandstone and the result of calculations using the equations for the series model, the parallel model, and the effective medium theory.

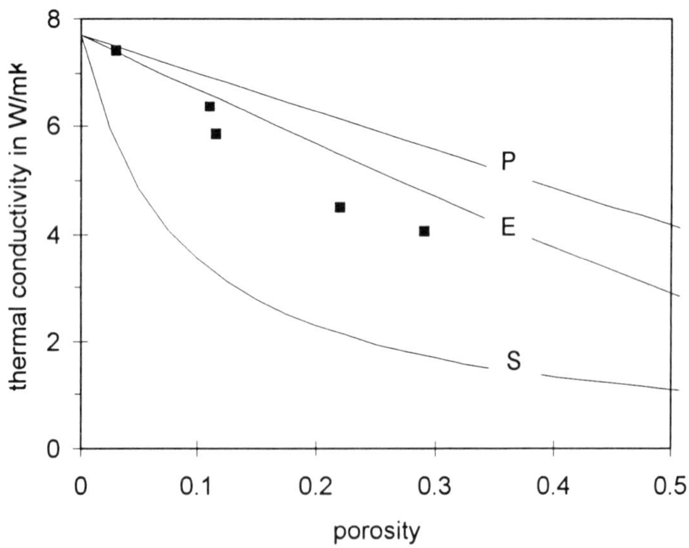

Figure 8.22. Comparison between experimental and theoretical thermal conductivities vs. porosity; P - parallel S - series E - effective medium theory;
λ_m = 7.7 W/mK λ_p = 0.6 W/mK; ■ - water saturated sandstone; Woodside and Messmer (1961)

8.4.5. Models with variable internal structure

The model with internal structure - expressed by a "structure angle" - was introduced in section 6.4.8. This model can be used also for a calculation of thermal conductivity. For the derivation, the model parts are arranged as thermal conductors in a series and parallel array.

The thermal conductivity of the individual parts solid matrix, contact region and pore filling are λ_s, λ_c, λ. For thermal conductivity related to the microsystem results:

$$\lambda_{\bar{1}} = \lambda_{\bar{2}} = \left(1 - \frac{\alpha_{grain}}{\alpha_{pore}} \cdot \frac{b}{a}\right)\lambda_s + \frac{\alpha_{grain}}{\alpha_{pore}} \cdot \frac{b}{a} \cdot \lambda_p\left[\left(\frac{\lambda_c}{\lambda_p} + \frac{1}{a/b - 1}\right)^{-1} + \frac{b}{a}\right]^{-1} \quad (8\text{-}64)$$

$$\lambda_{\bar{3}} = \left\{\left(1 - \frac{\alpha_{grain}}{\alpha_{pore}} \cdot \frac{b}{a}\right) \cdot \lambda_s^{-1} + \frac{\alpha_{grain}}{\alpha_{pore}} \cdot \lambda_p^{-1}\left[\left(2 - \frac{b}{a}\right)^2 + \left(\frac{a}{b} + \frac{b}{a} - 2\right)\frac{\lambda_c}{\lambda_p}\right]^{-1}\right\}^{-1}$$

(8-65)

The transformation into the macrosystem leads to the following relations for the thermal conductivity

$$\lambda_1 = \lambda_2 = \lambda_{//} = \lambda_{\bar{1}} \cdot (\cos\alpha)^2 + \lambda_{\bar{3}} \cdot (\sin\alpha)^2 \quad (8\text{-}66)$$

$$\lambda_3 = \lambda_{\perp} = \lambda_{\bar{1}}(\sin\alpha)^2 + \lambda_{\bar{3}} \cdot (\cos\alpha)^2 \quad (8\text{-}67)$$

This transformation is not limited to this particular model. It is valid for the general case of transversely isotropy of thermal properties originated for example by foliation or a sheet structure and also by cracks and fractures with a preferred orientation (see Figure 8.23.).

The *anisotropy* ratio for the structured model results in a combined effect of "microanisotropy" ($\lambda_{\bar{1}}/\lambda_{\bar{3}}$) and internal structure ($\tan\alpha$):

$$A_\lambda = \frac{\lambda_{//}}{\lambda\perp} = \frac{\lambda_1}{\lambda_3} = \frac{(\lambda_{\bar{1}}/\lambda_{\bar{3}}) + (\tan\alpha)^2}{(\lambda_{\bar{1}}/\lambda_{\bar{3}}) \cdot (\tan\alpha)^2 + 1} \quad (8\text{-}68)$$

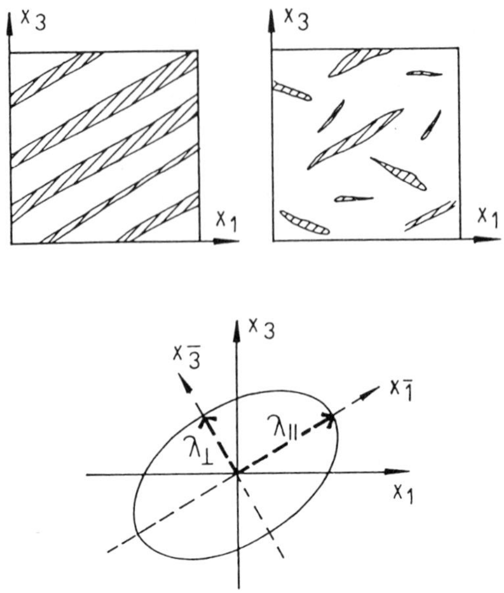

Figure 8.23. Transformation of thermal conductivity values

A *pressure dependence* of thermal conductivity can be incorporated into this model by using a pressure dependent thermal conductivity for the contact region (e.g. in the form of a power dependence).

The equations 8-64 to 8-67 show, that thermal conductivity depends on
(a) thermal conductivity of the constituents,
(b) model dimensions. We note that the dimensions are present only as ratios; this means that the thermal conductivity depends on porosity and grain shape, but not primarily on grain size,
(c) internal structure, expressed by the structure angle.

Figures 8.24 to 8.27 demonstrate some of these factors. The results are presented as plots of normalized vertical thermal conductivity λ_3/λ_s versus porosity Φ (calculated after equation 6-206).

The following values for the thermal conductivity of the constituents were assumed:

$\lambda_s = \lambda_{quartz} = 7.7$ W/mK, $\lambda_p = \lambda_{water} = 0.6$ W/mK, $\lambda_p = \lambda_{air} = 0.025$ W/mK

The ratio of thermal conductivities of contact region and matrix is assumed for two cases :

contact 1 $\lambda_c/\lambda_s = 1$ or $\lambda_c = \lambda_s$ contact 0.5 $\lambda_c/\lambda_s = 0.5$ or $\lambda_c = \lambda_s/2$

The first may represent a well cemented and the second case, a medium cemented porous rock.
To study the influence of the grain geometry, two types are applied:
type I: $\alpha_{grain}/\alpha_{pore} = 1$ type II $\alpha_{grain}/\alpha_{pore} = 0.5$.
Figure 8.24 shows the dependence of the normalised thermal conductivity on *porosity* for type I and type II .

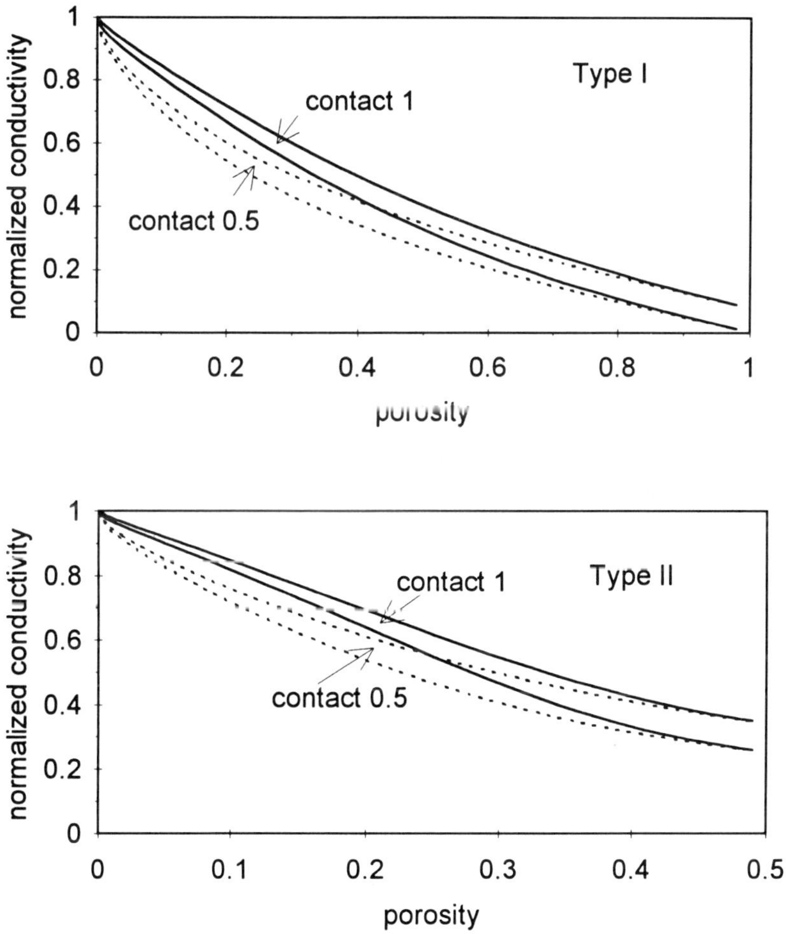

Figure 8.24. Normalized vertical thermal conductivity λ_3/λ_s versus porosity, calculated for a model with internal structure; structure angle $\alpha = 45°$
solid curves: well cemented rock dashed curves: medium cemented rock
upper curves : pore fluid water lower curves: pore fluid air

In both cases, particularly in the porosity range below 0.5, a strong *cementation influence* (contact properties) is seen. In this range, the heat conduction of the rock skeleton is dominant. The thermal conductivity of the *pore filling material* is also of influence.
Figure 8.24 demonstrates the decrease of conductivity with increasing porosity. The dependence on the porosity is more distinct for the dry rock than for the water saturated rock, as was shown by experimental results (Figure 8.9.).

A comparison between the two *grain geometry* types I and II shows the stronger decrease in case of type II as a result of its relatively small "solid contact bridge" between neighbouring particles (note the different porosity scale in Figure 8.24). A decreasing contact conductivity results in a decreasing thermal coupling between the solid constituents and a decrease of rock conductivity. At a porosity of 0.5, the curves for the two different cases of cementation coincide for model type II, because at this value there is no solid contact; thermal conductivity is equal the series model at this point for vertical direction.

Figure 8.25 demonstrates the *structure influence* on thermal conductivity for type II: With decreasing structure angle, the vertical conductivity increases (the normalised horizontal conductivity λ_1/λ_s decreases). The values for α = 90° are identical to $\lambda_{//}$ and the values for α = 0° are identical to $\lambda_\perp$. The decrease of conductivity with increasing porosity is more significant for pore filling air than for water, as a result of the conductivity contrasts with the solid matrix.

These results are compared with experimental data below. For comparison, either numerical approximations between experimental data and calculated curves are applied or a simple graph is used in which a logarithmic scaled axis for the normalised thermal conductivity allows a comparison with experimental results if plotted using the same grid. It can be determined, which individual curve gives the best approximation to the experimental data (calculated curves were applied as "master curves").

Figure 8.26 shows an example for comparison of calculated dependencies with the experimental data of Woodside and Messmer, 1961 (see Figure 8.9) for the water saturated case. There is a good agreement between measured and calculated values with the tendency that for higher porosities, the approximation for the "medium cementation" is better. This expresses a correlation between porosity and the degree of cementation or compaction.

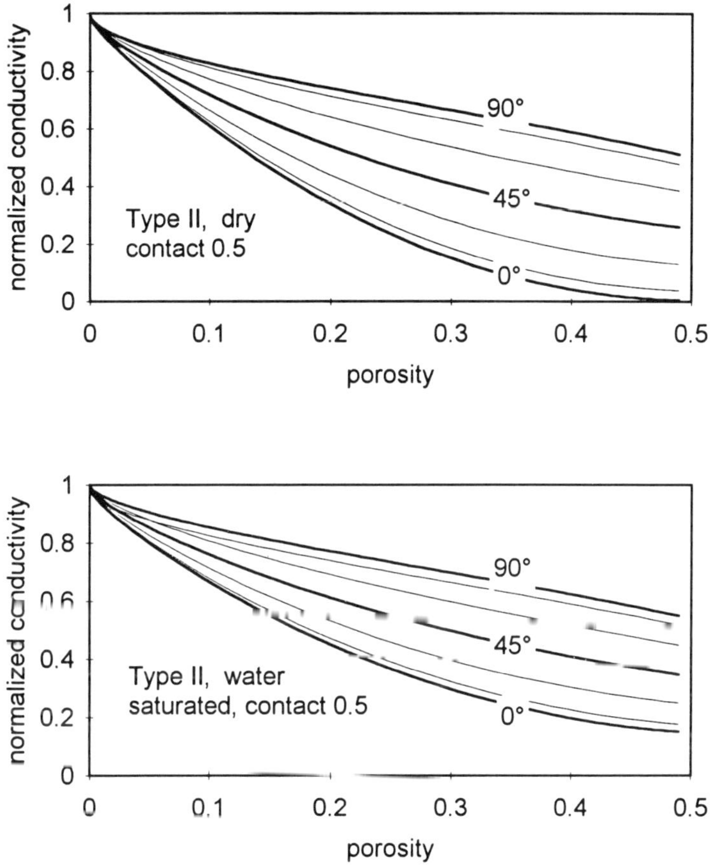

Figure 8.25. Dependence of the normalized vertical thermal conductivity on the porosity and the structure angle for model type II, medium cemented (contact 0.5); curve parameter is the structure angle α (in steps of 15°).

Results of investigations with water saturated marine sediments (Figure 8.27) show a reasonable agreement with these obtained theoretically by the use of grain shape type I (for type II porosities > 0.5 are impossible). For this calculation, a matrix (and contact) conductivity of 4.2 W/mK results in the optimal approximation. This may correspond to the clay and carbonate content of the sediments. The range for the structure angle lies between 45^0 and $75\ ^0$ and includes the measured values.

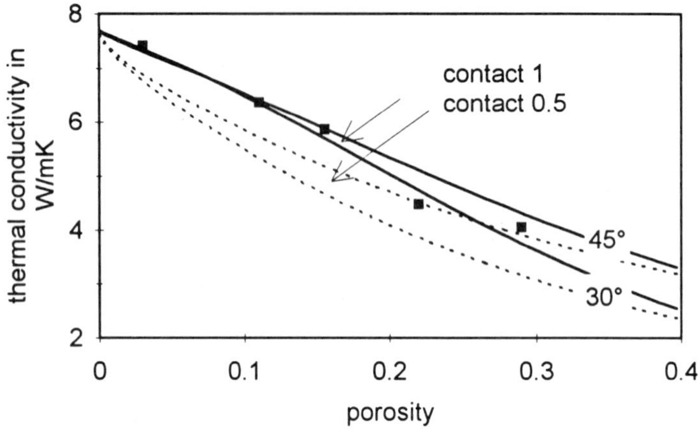

Figure 8.26. Comparison between calculated and experimental dependencies of thermal conductivity on porosity; Points: experimental data for water saturated sandstone (after Woodside and Messmer, 1961); Curves: calculated after the model with internal structure for type II, two different contact properties and two structure angles $\alpha = 30°$ and $\alpha = 45°$

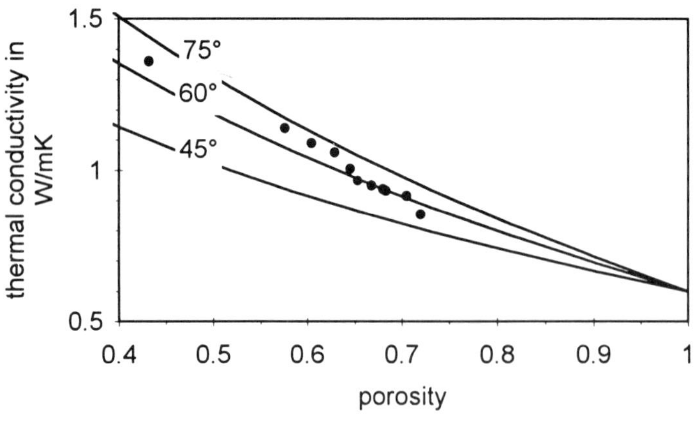

Figure 8.27 Comparison between calculated and experimental dependencies of thermal conductivity on porosity; Points: Red clay, data after Ratcliffe, 1960 (see Clark, 1966). Curves: calculated after the model type I with a matrix conductivity $\lambda = 4.5$ W/mK; curve parameter is the structure angle α.

CHAPTER 9

Electrical Properties of Rocks

"The most influential study of rock resistivity was Archie's (1942) examination of sandstone cores from the gulf coast region." *(Hearst and Nelson, 1985)*

9.1. Physical basis and units

9.1.1. Fundamentals

The electrical properties of a material include the primary properties of electric *conduction* (charge transport) and dielectric *polarization* (charge separation). They are defined by two fundamental relationships

$$j = \sigma E \tag{9-1}$$

$$D = \varepsilon E = \varepsilon_0 \varepsilon_r E \tag{9-2}$$

where

- E is the electric field strength (V/m)
- j is the current density (A/m^2)
- D is the electric displacement (C/m^2)
- σ is the electric conductivity (S/m or mho/m)
- ε is the dielectric permittivity (F/m)
- ε_0 is the free space or vacuum permittivity ($= 8.854 \times 10^{-12}$ F/m)
- ε_r is the relative permittivity or dielectric constant of the material (dimensionless).

In general, E, j and D are vectors and the properties σ and ε are tensors.
With respect to the electromagnetic phenomena described by the Maxwell-equations, the material relationships may be described by the formula using the magnetic permeability μ

$$B = \mu_0 \mu_r H \qquad (9\text{-}3)$$

where are H is the magnetic field
B is the induction
μ_0 is the vacuum permeability ($= 4\,\pi\,10^{-7}$ V s A^{-1} m^{-1})
μ_r is the relative permeability of the material and is dimensionless (see chapter 4).

In geophysics instead of using conductivity σ, its reciprocal value, the specific electric resistivity ρ is often used where

$$\rho = \sigma^{-1} \qquad (9\text{-}4)$$

The SI-unit for the specific electric resistivity is the Ohm meter (Ohm m or Ω m) and for the electric conductivity the Siemens/meter (S/m) or the mho/m (1 S/m = 1 mho/m). The relation to basic SI-Units is given by 1 Ohm m = 1 m^3 kg s^{-3} A^{-2}.

In addition to conductivity and permittivity, further parameters and properties are often used such as chargeability, phase angle, loss tangent etc., which are related to the primary properties and their frequency dependence.

With respect to their conductivity, materials are subdivided mostly into conductors ($\sigma > 10^5$ S m^{-1}), semiconductors ($10^5 > \sigma > 10^{-8}$ S m^{-1}) and insulators ($\sigma < 10^{-8}$ S m^{-1}).

Olhoeft (1985) gives the following classification:

1) *Metallic conductors*: A material throughout which there is a uniform distribution of valence electrons that are not tightly bound or associated with any particular atom. Motions of electrons are constrained by electron-phonon scattering, electron-electron scattering, and scattering from stationary imperfections and impurities. Electrical conductivity increases with decreasing temperature (except near absolute zero).
2) *Nonconductors*: Materials in which electrons are tightly trapped near atoms due to large energy barriers between atoms. Electrical conductivity increases with increasing temperature as thermal activity overcomes the energy barriers. Most rocks and minerals fall within this broad category, with further divisions of:
   a) *Insulators*: Materials with very large energy barriers between atoms such that electrons rarely become charge transport carriers.

b) *Semiconductors*: Materials with energy barriers that are slightly higher than the available energy from thermal activation at room temperature. At higher temperatures, electrons may become sufficiently activated to overcome these barriers: At lower temperatures, charge transport is by electrons or holes moving across barriers that are lowered by impurities within the material.

c) *Electrolytes*: materials with split (dissociated) into oppositely charged particles when treated appropriately (such as dissolving in a liquid solution or melting). The motion of the charged particles is hindered by interparticle interactions (scattering, viscous drag, etc.). Electrical conductivity increases with increasing temperature until near the critical point, above where it decreases with further temperature rise.

A very few natural materials are metallic conductors near room temperature (such as native copper, silver, gold, etc. near 10^{-8} Ohm m). Most geologic minerals are either semiconductors (most sulfides and oxides, ranging from 10^{-6} to 10^{4} Ohm m) or insulators (most silicates, ranging from 10^{9} to 10^{17} Ohm m). Water ranges in conductivity from 5 x 10^{-6} (pure) to roughly 10^{2} mho/m (saturated with salt in solution) or specific resistivities between roughly 10^{-2} Ohm m and 2 x 10^{5} Ohm m.

Mixtures of these materials span the entire range of resistivity from 10^{-8} to 10^{17} Ohm m (Olhoeft, 1985).

9.1.2. Electrical conduction

For electrical conduction in *rocks and minerals*, the following types are important:

- Metallic conduction (native metals e.g. copper and gold, graphite)
- Electronic semiconduction (minerals like ilmenite, magnetite, pyrite, galena)
- Electrolytic conduction:
  - Solid electrolytes (ionic crystals): most (rock-forming) minerals act as solid electrolytes, with the transfer of electric current being by motion of ions through a crystal lattice (Keller, 1989)
  - Electrolyte water solutions: in water-bearing rocks the electrolyte conductivity of the pore water has the dominant influence upon the rock conductivity.

The different kinds of electrical conduction have different *temperature dependencies*:

For the case of metallic conduction, the rock or mineral conductivity decreases with increasing temperature and can be described by an equation (Kobranova, 1989):

$$\rho(T) = \rho_0 \cdot (1 + \alpha \cdot \Delta T + \beta \cdot \Delta T^2) \qquad (9\text{-}5)$$

or

$$\sigma(T) = \frac{\sigma_0}{1 + \alpha \cdot \Delta T + \beta \cdot \Delta T^2} \qquad (9\text{-}6)$$

where ρ_0 is the specific resistivity and σ_0 the conductivity at the beginning of the temperature increment interval ΔT; α and β are the temperature coefficients.

For semiconduction, the conductivity increases with increasing temperature; this dependence can be described by the relationship

$$\sigma(T) = \sigma_0 \cdot \exp\left(-\frac{E_0}{\mathrm{k} \cdot T}\right) \qquad (9\text{-}7)$$

where E_0 is the activation energy, k Boltzman's constant, and T the absolute temperature.

In a graphic presentation of log σ versus $(1/T)$, a straight line results for a temperature dependence according equation (9-7). Figure 9.1 confirms this principal dependency.

Additionally, this graph shows the influence of mineral composition and a small pressure influence. The relatively small dependence on pressure compared with the dependence of temperature has been described by Kobranova (1989) as follows: "... the pressure in the range 0.05 - 5 GPa causes conductivity and resistivity change to a much lesser extent than temperature variation over the range 100 - 1000 °C".

9.1.3. Dielectric polarisation

The dielectric polarisation of rocks includes the following principle mechanisms (Olhoeft, 1985):

"*Electron polarisation*: the distortion of an atom's electron cloud in response to an external electric field in which one side of the atom becomes more positive and the other more negative than normal.

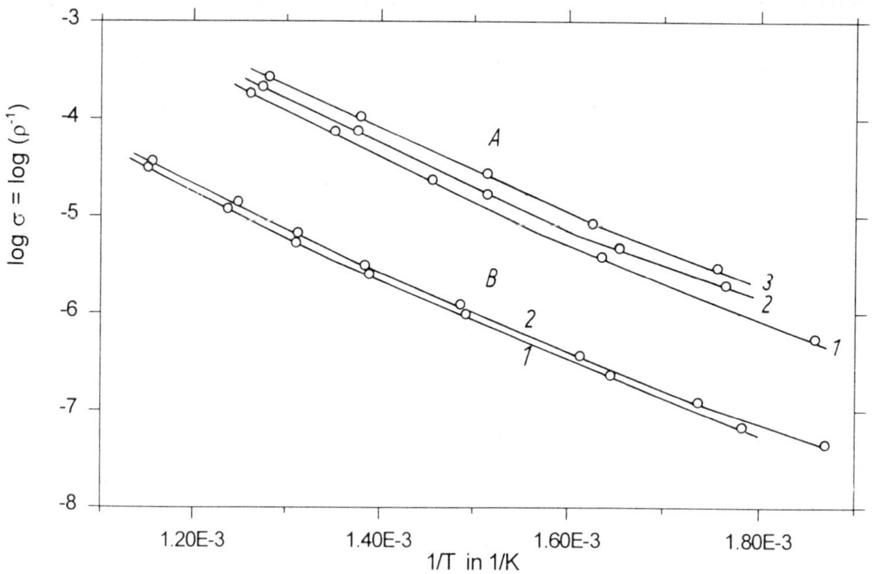

Figure 9.1. Dependence of electrical conductivity on temperature in a $\lg \sigma = \lg(\rho^{-1})$ vs. T^{-1} - plot for olivine (after Hamilton, 1965); σ is in S/m and T in Kelvin

A - fayalite content 17.5 % B - fayalite content 10.4 %

1 - pressure 1.15 x 10^3 MPa 2 - pressure 2.16 x 10^3 MPa 3 - pressure 3.18 x 10^3 MPa

Molecular polarisation: the distortion of an entire molecule in response to an external electric field, in which one part of the molecule becomes more positive than the rest and another part more negative,

Ionic polarisation: a redistribution of ions within a material due to an external electric field, with positive ions migrating toward the negative side of the field, and negative ions the other side,

Orientational polarisation: realignment or (reorientation) of polar molecules (without distortion of shape) in response to an external electric field,

Interfacial polarisation: charge separation and accumulation at local variations in electrical properties during migration of charge in response to an external electric field."

These polarisation mechanisms occur in the order listed from high to low frequency, and are simply additive to each other in going from high to low frequency (see Figure 9.2).

The *Maxwell-Wagner effect* in heterogeneous materials generally results from a non uniform charge distribution at the boundary of regions with different

electrical properties (for example, grain boundaries and other discontinuities also within the solid rock matrix).
Charge separation processes also occur due to coupling between electrical and other forces such as stress (piezoelectric), thermal (pyroelectric), and others.

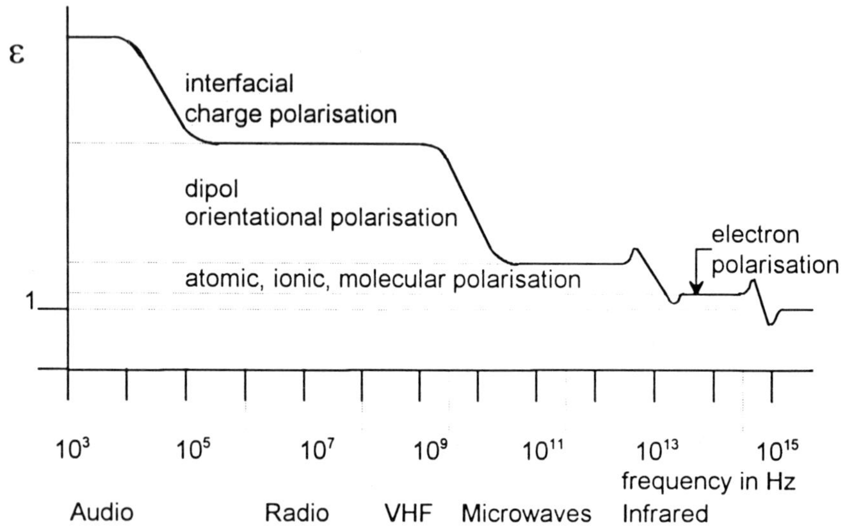

Figure 9.2. Dielectric polarisation - mechanisms (schematic)

9.1.4. Conductivity and permittivity as complex properties

Conductivity and the dielectric permittivity are in general complex (Fuller and Ward, 1970) with a real (sign ') and imaginary (sign ") part:

$$\sigma(\omega) = \sigma'(\omega) - i\,\sigma''(\omega) \tag{9-8}$$

$$\varepsilon(\omega) = \varepsilon'(\omega) + i\,\varepsilon''(\omega) \tag{9-9}$$

where $i = \sqrt{-1}$ and ω the angular frequency ($\omega = 2\,\pi\,f$).

Rocks mostly have *conductivity as well as polarisation properties*. For a harmonic electric field (based on Maxwell's law), an effective conductivity σ^* can be defined:

$$\sigma^* = \sigma^{*\prime} + i\,\sigma^{*\prime\prime} = \sigma + i\,\omega\varepsilon \tag{9-10}$$

with (9-8) and (9-9) results in

$$\sigma^* = \sigma'(\omega) - i\,\sigma''(\omega) + i\,\omega[\varepsilon'(\omega) + i\,\varepsilon''(\omega)]$$

$$= \sigma'(\omega) - \omega\varepsilon''(\omega) + i\,[\omega\varepsilon'(\omega) - \sigma''(\omega)] \tag{9-11}$$

The term [σ' (ω) - i σ" (ω)] is caused by carrier transport due to Ohmic and Faraday diffusion mechanisms and the term [ε' (ω) + i ε "(ω)] is caused by dielectric relaxation mechanisms (Olhoeft, 1979).
The effective specific resistivity is

$$\rho^* = \frac{1}{\sigma^*} \tag{9-12}$$

The resistivity magnitude - one of the most measured properties - results as

$$|\rho^*| = \sqrt{\frac{1}{(\sigma^{*\prime})^2 + (\sigma^{*\prime\prime})^2}} \tag{9-13}$$

The ratio of the imaginary and real part gives the phase angle Ψ

$$\tan\Psi = \frac{\sigma^{*\prime\prime}}{\sigma^{*\prime}} \tag{9-14}$$

The relation between the loss tangent δ and phase angle Ψ is

$$\delta = \Psi + \frac{\pi}{2} \tag{9-15}$$

In a comparable manner, the effective permittivity is

$$\varepsilon^* = \varepsilon^{*\prime} + i\,\varepsilon^{*\prime\prime} = \varepsilon + i\,\frac{\sigma}{\omega} \tag{9-16}$$

with (9-8) and (9-9) results in

$$\varepsilon^*(\omega) = \varepsilon'(\omega) + \frac{\sigma''(\omega)}{\omega} + i\left[\frac{\sigma'(\omega)}{\omega} + \varepsilon''(\omega)\right] \tag{9-17}$$

An examination of equations (9-10) and (9-17) leads to the rule that at zero frequency or low frequencies the conductivity measurement gives the real part of conductivity and at infinite frequency or high frequencies the permittivity gives the real part of the permittivity as approximation:

conductivity measurement: σ^* at frequency $f \to 0$
permittivity measurement: ε^* at frequency $f \to \infty$

Keller (1989) notes that high dielectric permittivity observed at low frequencies is also associated with a substantial amount of conduction, and it is likely that the high dielectric permittivity results from interfacial polarisation (see section 9.5).

9.1.5. Debye relaxation and Cole-Cole model

In equation (9-9) the real and imaginary part of the relative permittivity are frequency dependent. For the case of ideal dielectrics, Debye (1924) has formulated a relaxation model

$$\varepsilon^*(\omega) = \varepsilon_\infty + \frac{\varepsilon_0 - \varepsilon_\infty}{1 + i\,\omega\tau} \tag{9-18}$$

where ε_0 and ε_∞ are, respectively, the dielectric permittivity at the frequencies $f = 0$ (static dielectric permittivity) and $f = \infty$ (optic dielectric permittivity). The single relaxation time τ corresponds to the single relaxation process. At the peak loss frequency $f_c = (2\pi\tau)^{-1}$ is a maximum in $\varepsilon''(\omega)$.

In general, however, such single relaxations are rarely observed. Multiple relaxations or distributions of relaxations are found instead. The most commonly observed simple relaxation distribution is the Cole-Cole distribution (Olhoeft, 1979). For complex permittivity, Cole and Cole (1941) give the following formula to fit data observed on various dielectrics:

$$\varepsilon^*(\omega) = \varepsilon_\infty + \frac{\varepsilon_0 - \varepsilon_\infty}{1 + (i\,\omega\tau)^{1-q}} \tag{9-19}$$

where q is the Cole-Cole distribution parameter ($0 \le q \le 1$), which accounts for the broadening of the distribution:
$q = 0$ describes the single relaxation behaviour of the Debye relaxation model,
$q = 1$ represents an infinitely broad, continuous distribution.

The presentation of permittivity data $\varepsilon^*(\omega)$ in the complex plane $(\varepsilon', \varepsilon'')$ is the so-called Cole-Cole plot. The Debye relaxation results in a semicircle and a Cole-Cole relaxation in a depressed semicircle.

Based on the Cole-Cole model, other relaxation models were developed. Raistrick (see Knight et al., 1985) derived an equivalent circuit model to fit data of measurements on sandstones in the frequency range between 10 kHz and 13 Mhz. This model is composed of a parallel RC circuit (the resistor represents the DC resistance, and the capacitor the high frequency capacitance of the sample); parallel to this circuit is a complex admittance parameter, which models all the superimposed frequency dependence of the system:

Jonscher (1977, 1981) and Dissado and Hill (1979, 1987) developed a further universal relaxation theory for solid materials assuming of a many body model. Other relaxation model types are based on fractal theory (Le Mehaute and Crepy, 1985; see also for example Ruffet et al., 1991, 1995; Börner and Kulenkampf, 1992, 1994).

9.2. Electrical properties of minerals and pore contents

9.2.1. Minerals

Most *rock-forming minerals* - particularly silicates and carbonates - have very high specific resistivities (> 10^9 Ohm m) and are classed as *insulators*. Conductive minerals are comparatively rare in the earth crust, being restricted to the sulfides, some oxides, and native elements.

Kobranova (1989) has classified minerals into three groups:

1) Native metals, their natural paragenetic sequences, graphite. The specific electrical resistivity is of the order between 10^{-8} and 10^{-5} Ohm m.
2) A number of sulfides, some oxides. The specific electrical resistivity is in the range between 10^{-6} and 10^{11} Ohm m.

3) Native nonmetals (sulphur, selenium, diamond) excepting graphite, silicates, carbonates, chlorides, phosphates, and minerals of a number of other classes. The specific electrical resistivity varies between 5×10^7 and 2.7×10^{16} Ohm m; it is mostly greater than 10^{11} Ohm m.

But the individual varieties of minerals show a wide range of scatter; for example, the specific electrical resistivity of pyrite can vary from $> 10^{-6}$ to 10^{2} Ohm m, that from magnetite from 10^{-4} to 10^4 Ohm m. This is accounted for by impurities (Kobranova, 1989).

Table 9.1 shows values for the specific electrical resistivity and the dielectric number of some *minerals*; more detailed compilations are published by Olhoeft (1981), Beblo et al. (1982), Melnikov et al. (1975), Parchomenko (1965), Katsube and Collett (1976), and Keller (1989). In particular, the data published by Melnikov et al. (1975) and Dortman (1976) show some scatter and some different values compared with the data from Olhoeft (1981) and Kobranova (1989); the cause of this is the influence of various impurities, structural effects, and crystalline structures which originate anisotropy effects. Thus, Olhoeft's data may represent the electrical property of the "pure mineral". Figure 9.3 demonstrates the *specific resistivity* ranges resulting from natural influences.

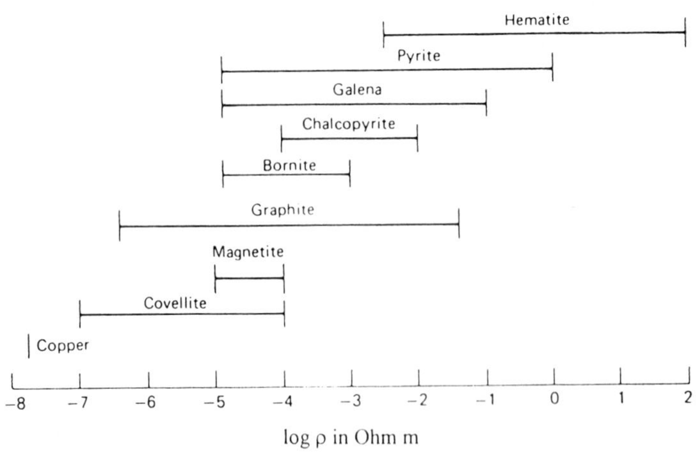

Figure 9.3. Range of resistivity values of some conductive minerals; Hearst and Nelson (1985)

The relative *dielectric permittivities* of the most abundant rock-forming minerals are in the range 4...10. Some minerals have higher values, such as, the sulfide and oxide groups.

Table 9.1. Specific electric resistivity ρ and relative dielectric permittivity ε_r of minerals; References: D - Dortman (1976); K -Keller (1989), ε_r for radio frequencies; O - Olhoeft (1981), calculated from conductivities and rounded, ε_r at 1 MHz; P - Parchomenko (1965).

Mineral	ρ in Ohm.m	ε_r
Elements		
graphite	1.4×10^{-5} (O)	> 81 (D, P)
gold	2.2×10^{-8} (O)	
Halide		
halite	5×10^{13} (O); 10^{10} - 10^{13} (P)	5.9 (O); 5.7 - 6.2 (K, D)
sylvite	10^{11} - 10^{13} (P)	4.4 - 6.2 (K, D)
fluorite	7.7×10^{13} (O); 7.9×10^{13} (P)	6.76 (O); 6.79 (K); 6.26 - 6.79 (D); 6.2 - 8.5 (P)
Phosphates		
apatite	8.3×10^{11} (O)	11.7 (O); 7.4 - 10.5 (K);
monazite		3 - 6.6 (P, M)
Carbonates		
calcite	9×10^{13} (O); 5×10^{12} (P)	6.35 (O); 7.5 - 8.7 (D, P)
dolomite	4.3×10^{13} (O)	7.46 (O); 6.3 - 8.2 (P); 6.8 - 8.0 (K)
aragonite	3.4×10^{12} (O)	8.67 (O); 7.4 (P)
siderite	8.3×10^{9} (O)	9.3 (O); 5.2 - 7.4 (P)
Sulfates		
anhydrite	1×10^{9} (O); 1×10^{9} (P)	6.5 (O, D, P); 5.7 - 6.7 (M)
barite	1.2×10^{7} (O)	10.03 (O); 6.99 - 12.2 (K)
gypsum	1.05×10^{11} (O)	6.39 (O); 5.0 - 11.5 (K)
Sulfides		
pyrite (semicond.)	1.0×10^{-3} (O); 10^{-4} - 10^{-1} (D)	33.7 - 81.0 (O)
galena (semicond.)	1.0×10^{-3} (O); 10^{-5} - 10^{-3} (D)	17 - 81 (D)
sphalerite (semicond.)	2.6×10^{11} (O)	7.5 (O); 6.9 - 12.1 (D)
Oxides		
hematite (semicond.)	1.0×10^{2} (O); 10^{-2} - 10^{6} (D)	25.0 (O, P)
magnetite (metal)	1.0×10^{-4} (O)	33.7 - 81 (P)
rutile (semicond.)	2.1×10^{9} (O)	78.9 (O); 80 - 173 (D, P)
Silica minerals		
quartz	2×10^{14} (O); 10^{12} - 2×10^{14} (P)	4.5 (O); 4.19 - 5.00 (K)
forsterite , fayalite		6.8 (O)
olivines (Fo98Fa2)		6.8 (D)
sillimanite	1.0×10^{11} (O)	11.0 (O)
epidote	7.7×10^{9} (O)	14.4 (O)
kyanite	5.6×10^{11} (O)	7.8 (O); 5.7 (P)
diopside	5.9×10^{9} (O)	8.6 (O); 10 (D, P)
augite	4.8×10^{10} (O)	9.3 (O); 6.9 - 10.3 (K)
amphiboles, hornblende	4.8×10^{10} (O)	8.0 (O); 4.9 6.5 (D, P)
glaucophane	1.03×10^{11} (O)	9.3 (O)

Mineral	ρ in Ohm.m	ε_r
Sheet silicates		
muscovite	2.2×10^{12} (O); 10^{12} - 10^{14} (P)	7.6 (O); 6.2 - 8.0 (K, P)
biotite	8.3×10^{10} (O)	6.3 (O); 6.2 - 9.3 (K);
chlorite	1.6×10^{9} (O)	9.0 (O)
illite		10.0 (O)
kaolinite	3.2×10^{7} (O)	11.8 (O); 9.1 (P)
Framework silicates		
feldspar	10^{11} - 10^{12} (D)	
microcline	1.8×10^{11} (O)	5.48 (O); 5.6 (P)
orthoclase	1.4×10^{12} (O)	5.6 (O); 4.5 - 6.2 (P)
oligoclase	10^{9} (M)	6.06 (K)
albite	4.8×10^{8} (O)	(5.3-5.5) (K), 6.95 (O); 5.63(D)
anorthite	7.7×10^{9} (O)	6.9 (O, P); (7.05-7.24) (K),
labradorite	6.3×10^{8} (O)	5.87 (O); 5.8 (P)
nepheline		6.04 (O); 6.2 (P)
leucite		6.80 (O, P); 6.5 - 7.1 (M)

The variations are attributed also to crystalline structures, which produce electrical *anisotropy* (Hearst and Nelson, 1985). Examples of mineral anisotropy are

a) for graphite, the resistivity ranges from $36 \times 10^{-8} \ldots 10^{-6}$ Ohm m for current flow parallel to cleavage to $28 \times 10^{-6} \ldots 99 \times 10^{-4}$ Ohm m for current flow across cleavage,

b) the dielectric numbers for quartz are 4.96 across optical axis and 5.05 along optical axis (Keller, 1989).

Because the number of charges per unit volume available for displacement is roughly proportional to density, dielectric permittivity correlates to some degree with mass density as shown in Figure 9.4.

Olhoeft (1981) also investigated the dependence of relative permittivity of minerals on other properties and found for silicates (lunar and terrestrial samples), a remarkable correlation with the bulk density d (in g/cm^3):

$$\varepsilon_r = (1.93 \pm 0.17)^{d} \qquad (9\text{-}20)$$

Minerals not fitting this correlation contain water (e.g. montmorillonite) or are semiconducting mineral phases like sulfides.

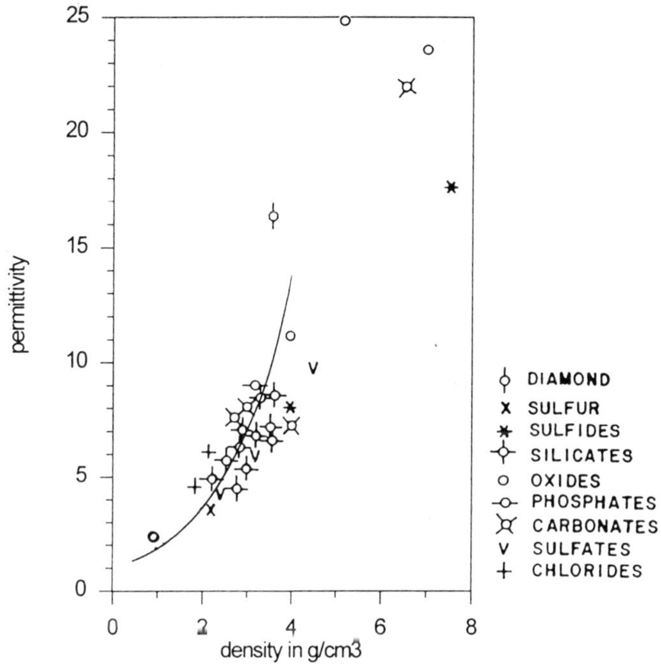

Figure 9.4. Relationship between dielectric permittivity (dielectric constant) and density for mineral groups (Keller, 1989); the curve represents the empirical relationship (9-20)

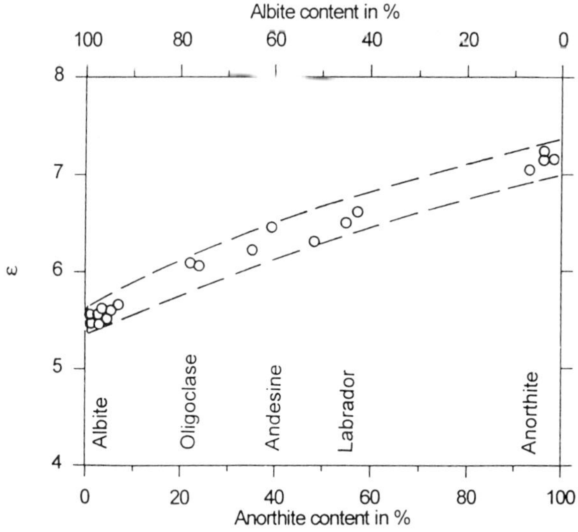

Figure 9.5. Relative permittivity vs. anorthite content for plagioclases (data Clark, 1969)

Figure 9.5 shows the effect of the mineral composition on the dielectric permittivity for plagioclases. With increasing anorthite content, the permittivity value increases (compare Table 9.1).

A special type of matrix material is *clay*; the electrical properties of clay are discussed in section 9.3.3.

9.2.2. Pore fluids

Gases and oil are essentially nonconductors. The average values of the specific electrical resistivity of the atmospheric air near earth's surface is 10^{14} Ohm m. Gaseous hydrocarbons show similar values (Kobranova, 1989). Dortman gives a specific resistivity for oil of $\rho_{oil} = 10^9 ... 10^{16}$ Ohm m .

Aqueous solutions (electrolytes) are ionic conductors. In many cases the ionic conduction of the pore water is the dominant conduction mechanism in a porous or fractured rock and the specific electrical resistivity is controlled mainly by

- the type, concentration, and temperature of the electrolyte in the pores,
- its volume fraction (porosity, saturation) and distribution of the electrolyte in the rock.

There also exists a difference of relative permittivity between water ($\varepsilon_r \approx 81$) and most other fluids ($\varepsilon_r = 1...3$) as shown in Table 9.2. Thus, with respect to the electrical properties, we have *two main types of pore fluids*:

water:	conducting,	$\varepsilon_r \approx 81$
oil, gas:	non conducting,	$\varepsilon_r < 3$

The conductivity of an aqueous *solution* depends on

- concentration of ions (amount of salt/unit solvent, degree of dissociation),
- charge number of ions (kind of substance, salt type) and ion mobility, hydration,
- temperature,
- interactions of ions in the solution (at high concentrations).

Table 9.2b shows the properties of organic chemical contaminants (the table contents the top 10 organic chemical contaminants by frequency of occurrence following the US. EPA Superfund list, Lucius et al., 1990, Olhoeft, 1992).

Table 9.2. Electrical properties of fluids (and ice, snow)

9.2a Relative dielectric permittivity of some pore fluids; References: E - Ebert (1976); Le - Lexikon der Physik (1972); Ly - Lytle (1974); R - Rshewski and Novik (1978); W - Western Atlas (1992); S - Schlumberger (1989); U - Urban et al. (1994).

Fluid	ε_r	Ref.
water	80.8	Le
glycerine	56.2	U
nitrobenzole	36.4	E
ethanol	25.8	U
acetone	21.5	E
diesel	2.0 ... 2.4	U
natural oil	2.0 ... 2.7	R
	2.0 ... 2.4	W, S
petroleum	2.0 ... 2.2	E
air	1.0006	L
ice	3.1	E
ice, from pure distilled water	4.15 (-12°C, 10^6 Hz)	Ly
	3.45 (-12°C, 10^8 Hz)	
	3.20 (-12°C, 10^9 Hz)	
snow, freshly fallen	1.20 (-20°C, 10^6...10^9 Hz)	Ly
hard-packed, followed by rain	1.55 (-6°C, 10^6 Hz)	

9.2b. Properties of organic chemical contaminants, Lucius et al., 1990, Olhoeft, 1992 CAS RN is the Chemical Abstracts Service Register Number of the chemical; *d* - density in 10^3 kg/m^3 = g/cm^3, *sol* - aqueous solubility in weight percent, *vap* - vapor pressure in kPa, ε_r - relative dielectric permittivity; log ρ - logarithm of specific electrical resistivity (in Ohm m); values for 20°C

Chemical	CAS RN	*d* in g/cm^3	*sol* in wt.%	*vap* in kPa	ε_r	log ρ
Trichlorethene (TCE)	79-01-6	1.46	0.11	8	3.42	6
Dichlormethane	75-09-2	1.32	1.5	50	9	8
Tetrachloroethene (PCE)	127-18-4	1.62	0.01	2	2.3	7
Toluene	108-88-3	0.87	0.05	4	2.4	7
1,1 -Dichloroethane	75-34-3	1.17	0.5	10		7
Bis(2-ethylhexyl)phthalate	117-81-7	0.98	4 x10^{-5}	10^{-8}	5	
Benzene	71-43-2	0.879	0.17	10	2.28	9
trans-1,2-Dichloroethene	156-60-5	1.25	0.6	40	2.1	
1,1,1-Trichlooethane	71-55-6	1.33	0.2	14	7.5	6
Chloroform	67-66-3	1.49	0.8	4.8	20	8

The resistivity of pure *water* without ions other than H^+ and OH^- is 2.8×10^5 Ohm m at 17.6 °C (Dorsey, 1940, Hearst and Nelson, 1985). The resistivity decreases with increasing impurity content: The primary charge transport in very pure water is by ionic conduction of H^+ and OH^-, while charge transport in aqueous solutions is due to impuric conduction, such as Na^+ and Cl^- (Falkenhagen, 1971; Olhoeft, 1979).
For *low salt concentrations*, where ion-ion interactions can be neglected, the ionic conductivity can be expressed by the valence z_i, the mobility v_i, the concentration c_i, and the degree of dissociation α_i

$$\sigma_w = \frac{1}{\rho_w} \sim \sum_{i=1}^{n} \alpha_i \cdot c_i \cdot z_i \cdot v_i \tag{9-21}$$

where n components contribute to the conductivity. The parameters α_i and v_i are temperature dependent.

In many cases, a correlation between the conductivity (or specific resistivity), concentration, and temperature for sodium-chloride solution (NaCl) can be applied. Figure 9.6 shows such a correlation in the form of a nomogram. The *temperature dependence* of NaCl solutions is described by an empirical relationship (Arps, 1953; Worthington et al., 1990, Schlumberger, 1989):

$$\rho_{w(T_2)} = \rho_{w(T_1)} \cdot \frac{T_1 + 21.5}{T_2 + 21.5} \qquad \rho_{w(T_2)} = \rho_{w(T_1)} \cdot \frac{T_1 + 6.77}{T_2 + 6.77} \tag{9-22}$$

where T_1 and T_2 the temperatures in Celsius degrees for the first equation, and in degrees Fahrenheit for the second relationship. Dachnov (1975) gives the relation

$$\rho_{w(T)} = \rho_{w(20C)} \cdot \left(1 + 2.16 \cdot 10^{-2} \Delta T + 8 \cdot 10^{-6} (\Delta T)^2\right)^{-1} \tag{9-23}$$

where ΔT is the temperature difference between the measured temperature and the reference temperature 20°C. For many cases, the second term in parenthesis can be neglected. This is in agreement with the formula of Keller and Frischknecht (1966)

$$\rho_{w(T)} = \frac{\rho_{w(T=18C)}}{1 + \beta \cdot (T - 18)} \tag{9-24}$$

where β is the temperature coefficient of resistivity (about 0.025 per degree centigrade).

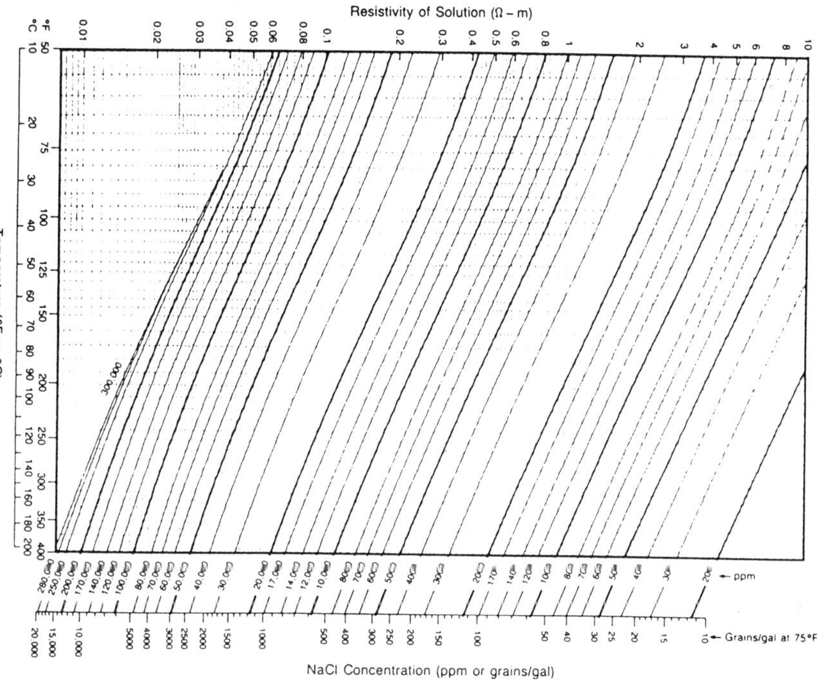

Figure 9.6. Correlation between specific resistivity, concentration, and temperature for (sodium-chloride) solution (NaCl) after Schlumberger (1989);

. Conductivity depends on concentration Llyod and Heathcoke (1985) derived a linear relationship between chloride concentration and electrical conductivity for groundwater from Grand Cayman:

$$\sigma_W = \frac{CCl^- + 129.12}{0.37441} \quad \text{or} \quad CCl^- = 0.37441 \cdot \sigma_W - 129.12 \qquad (9\text{-}25)$$

where σ_w is in μS/cm (= 10^{-4}S/m) and CCl^- in mg/l. More detailed relationships for conversion between resistivity and concentration of NaCl solutions published Worthington et al. (1990); the general equation is

$$\log\left(\frac{1}{\rho_W}\right) = a + b \cdot \log C_{\text{mol}} + c \cdot (\log C_{\text{mol}})^2 + d \cdot (\log C_{\text{mol}})^3 + e \cdot (\log C_{\text{mol}})^4 \qquad (9\text{-}26)$$

where the resistivity ρ_w is in Ohm.m and the concentration C_{mol} in mol/l. Table 9.3 shows the empirical parameters *a, b, c, d, e* for three concentration ranges.

Table 9.3 Empirical parameters a, b, c, d, e in (9-26) for three concentration ranges (Worthington et al., 1990)

	Low concentration range C_{mol} = 0.0001...0.10 mol/l ρ_w = 796...0.94 Ohm.m	Mid concentration range C_{mol} = 0.09...1.4 mol/l ρ_w = 796...0.94 Ohm.m	High concentration range C_{mol} = 1.0...5.53 mol/l ρ_w = 796...0.94 Ohm.m
a	+ 9.42203 x 10^{-1}	+ 9.33134 x 10^{-1}	+ 9.33292 x 10^{-1}
b	+ 8.88900 x 10^{-1}	+ 8.47181 x 10^{-1}	+ 8.45971 x 10^{-1}
c	- 2.72398 x 10^{-2}	- 1.04563 x 10^{-1}	- 9.10632 x 10^{-2}
d	- 2.25682 x 10^{-3}	- 7.05334 x 10^{-2}	+ 2.27456 x 10^{-1}
e	+ 1.46605 x 10^{-5}	- 2.35562 x 10^{-2}	- 3.81035 x 10^{-1}

The conversion between ppm and mol/l is

$$\text{ppm} = \frac{58.443 \cdot 10^3 C}{d} \tag{9-27}$$

where ppm = mass of NaCl per 10^6 mass units of solution, 58.443 = molecular mass of NaCl (g), C = solution concentration (mol/l), and d = solution density (g/cm^3).

If the solutions contain *ions other than* Na^+ *and* Cl^- (for example Ca, K), so-called "*multipliers*" are applied to calculate an equivalent NaCl-concentration and its conductivity (see Schlumberger, 1989). A direct comparison between the resistivity of a NaCl and a KCl solution is shown in Table 9.4. The last row depicts the ratio of the resistivity values for NaCl and KCl solution of the same equivalent concentration.

Table 9.4. Specific electrical resistivity of NaCl and KCl solutions at 20°C as a function of the normal or equivalent concentration C_{val} (gram equivalents or val per litre of solution) and concentration C in mg/l, after Berktold (1982)

	NaCl		KCl		$\frac{\rho_{w,KCl}}{\rho_{w,NaCl}}$
C_{val} in val/l	C in mg/l	ρ_w in Ohm m	C in mg/l	ρ_w in Ohm m	
0.0001	5.85	882	7.456	741	0.840
0.001	58.5	89.4	74.56	75.1	0.840
0.01	585	9.34	745.6	7.81	0.836
0.1	5850	1.03	7456	0.85	0.825
1.0	58440	0.13	74560	0.10	0.769

The ratios are of the same order as data from Dachnov (1975); for the C_{val} range between 0.02 and 0.6 he gives a ratio between 0.87 and 0.90.
The specific resistivity of *natural waters* varies over decades. The highest values are for rainwater (10^2 ... 10^3 Ohm m), the lowest are for saturated brines (about 10^{-1} Ohm m; for example in sedimentary saline facies). Table 9.5 shows average values for rocks of various ages and from different regions.

Table 9.5. Specific resistivities (in Ohm m) of natural waters in rocks at 20 °C; after Keller, 1989; *n* number of samples

Source of water samples	*n*	specific median	resistivity range
Igneous rocks, Europe	314	7.6	3.0 - 40
Igneous rocks, South Africa	175	11.0	0.5 - 80
Metamorphic rocks, South Africa	88	7.6	0.86 - 80
Metamorphic rocks, Precambrian of Australia	31	3.6	1.5 - 8.6
Recent and Pleistocene continental sediments, Europe	610	3.9	1.0 - 27
Recent and Pleistocene sediments, Australia	323	3.2	0.38 - 80
Tertiary sediments, Europe	993	1.4	0.70 - 3.5
Tertiary rocks (Miocene, Oligocene), Australia	240	3.2	1.35 - 10
Mesozoic sedimentary rocks, Europe	105	2.5	0.31 - 47
Palaeozoic sedimentary rocks, Europe	161	0.93	0.29 - 7.1
Chloride waters from oil fields	967	0.16	0.049 - 0.95
Sulfate waters from oil fields	256	1.20	0.43 - 5.0
Bicarbonate waters from oil fields	630	0.98	0.24 - 10

Of particular interest in well logging are the specific resistivities of mud (see also section 9.3.3).

The influence of concentration on the relative *permittivity* of water is relatively small compared with the influence on the conductivity. Olhoeft (1981, see also Hearst and Nelson, 1985) gives an empirical equation for the effect of salt concentration C_{mol}

$$\varepsilon_r = \varepsilon_{r,\text{ pure water}} - 13.00 \cdot C_{mol} + 1.065 \cdot C_{mol}^2 - 0.03006 \cdot C_{mol}^3 \tag{9-28}$$

Utilising the theory of Debye and Falkenhagen for low concentrations, the increase of permittivity is described by a function

$$\varepsilon_{r\,w} = \varepsilon_{r\text{ pure water}} + \xi \cdot \sqrt{C_{mol}} \tag{9-29}$$

where C_{mol} is the molal concentration and $\varepsilon_{r,\text{pure water}} \approx 81$. The property ξ depends on temperature and kind of the electrolyte. For binary electrolytes at 18 °C Cerniak (1964) and Parchomenko (1965) give the value $\xi = 3.79$ for concentrations below 5 g/l (= 5 g/1000 cm^3).

Permittivity decreases with increasing temperature (Figure 9.7) whereas pressure only changes the value slightly (Hearst and Nelson, 1985).

The complex electrical properties of pure water can be described by a Cole-Cole model (Klein, 1978; Rauen, 1991); for a temperature of 20 °C, the following properties (equation 9-19) result:

$\varepsilon_{r,0} = 80.1$ $\quad\quad \varepsilon_{r,\infty} = 4.23 \pm 0.16$ $\quad\quad \tau = 9.3$ ps $\quad\quad 1 - q = 0.987$

A more detailed description of the dielectric pemittivity of water - including temperature and concentration dependence - is given by Olhoeft (1981).

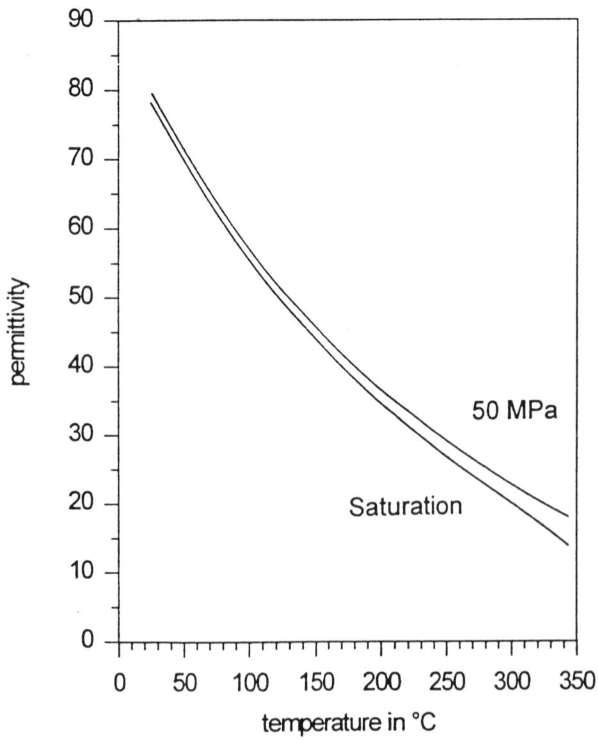

Figure 9.7. Relative dielectric permittivity of water as a function of temperature at two pressures based on polynomial fit by Helgeson and Kirkham (1974); Curve marked "saturation" indicates equilibrium pressure of water and steam (after Hearst and Nelson, 1985).

9.3. Electrical properties of rocks

"... mixing water with other geological materials is the greatest cause of variation in the electrical behaviour of rocks..." (Gary R. Olhoeft, 1985)

9.3.1. Overview

A review of the electrical conductivity of the common minerals shows that most of them are characterised by very high resistivity or low conductivity. Comparable behaviour is expected for an absolutely dry rock.
For of a porous or fractured water-bearing rock, the electrolytic conductivity of the water itself and interactions between solid and fluid components create a enhanced electrical conductivity. Therefore, the consideration of the two cases

- dense, pore and fracture free rocks or absolutely dry rocks
- porous or fractured water bearing rocks

is meaningful.

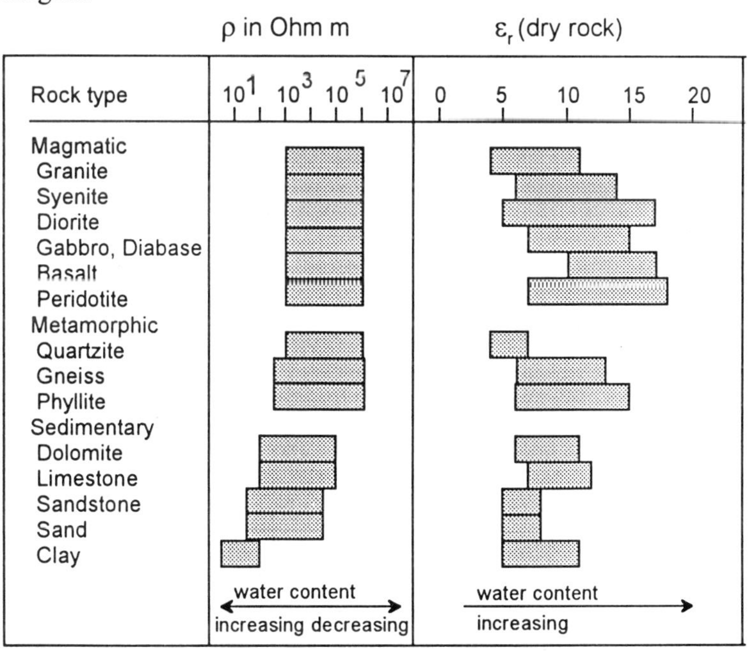

Figure 9.8. Mean value ranges of specific electrical resistivity and relative permittivity for some rock types; the arrows indicate the tendency of changing with increasing and decreasing water content.

Figure 9.8 gives an overview of the mean ranges of the specific resistivity and relative permittivity of some common rock types. The dominant influence of the pore water on the electrical properties produces a broad range of rock properties for each type, which can be scattered over orders of magnitude, and usually overlaps different types. The large resistivity ranges and overlap of these ranges for rock types make rock type identification by resistivity difficult (Ward, 1990).
The tendencies we note are
► the resistivity decreases with increasing porosity and fracturing,
► the permittivity increases with increasing porosity and fracturing.
These are "pore water effects". Additionally, accessory minerals like graphite or ores can influence electrical behaviour.

Ward (1990) presented a diagram, which describes the effect of various geological processes on the resistivity of a rock - most processes lower the resistivity (Figure 9.9).

Clay alteration	↓	Weathering	↓
Dissolution	↓	Induration	↑
Faulting	↓	Carbonate percipitation	↑
Salt water intrusion	↓	Silification	↑
Shearing	↓	Metamorphism	↕

↓ decreases ↑ increases ↕ increases or decreases

Figure 9.9. Effect of various geological processes on resistivity; after Ward (1990)

The *models and theories* describing the electrical properties of the composite "rock" can be classified as follows
- ◆ models and theories describing the individual electrical components (conductivity, permittivity, interface phenomenon),
- ◆ models and theories describing the combined or complex effect of the individual electrical active components (for example, electrolytic and interface conduction including boundary conditions).

These models are discussed in connection with the experimental results in the following sections.

9.3.2. Electrical properties of dense rocks, rocks without water content

Completely dense or dry rocks (i.e. rocks without water) occur infrequently in the near surface crust. These rocks have - except for ores or graphite-containing rocks - very high resistivities (mostly > 10^4 Ohm m). The dielectric permittivity is usually relatively low and in the range of the dominant rock forming minerals (3...5); accessory minerals (for example iron ores, graphite) can produce higher values.

Table 9.6 gives an overview for ores and ore minerals; concentration and distribution of conductive ore minerals in the matrix control the electrical properties in this case. A study of the influence of coal and graphite inclusions upon the electric properties was published by Jödicke, 1991.

Table 9.6. Specific electrical resistivity (in Ohm m) for ore and ore minerals, after Parchomenko, 1965

Main ore mineral	Specific electrical resistivity	
	ore	mineral (pure)
Pyrite	10^{-4}...10^{1}	5×10^{-5}...5×10^{-2}
Chalkopyrite	10^{-4}...10^{-1}	10^{-4}...7×10^{-4}
Pyrrhotite	10^{-5}...10^{-3}	10^{-5}...5×10^{-5}
Arsenopyrite	10^{-3}...10^{-1}	3×10^{-4}
Galena	10^{-2}...3×10^{2}	3×10^{-5}...3×10^{-4}
Magnetite	10^{-1}...10	10^{-4}

The correlation between electrical conductivity, permittivity and various mineral contents were analyzed by Lebedev et al. (1986). For example they found the relationship for the ore content C_{ore} (in %) at a temperature 200 °C

$$\lg \sigma = 0.08 \cdot C_{ore} - 7.15 \quad \text{(with a correlation coefficient of 0.82)} \quad (9\text{-}30)$$

$$\lg \varepsilon_r = 0.017 \cdot C_{ore} + 1.13 \quad \text{(with a correlation coefficient of 0.83)} \quad (9\text{-}31)$$

Conduction in dense and absolutely dry rocks is controlled by *matrix properties* (semiconduction, lattice defects, and conductive accessories). This results in three significant observations:

1) high absolute resistivity values,
2) temperature dependence (exponential decrease of resistivity with increasing temperature; compare section 9.1.),

3) pressure dependence (non linear decrease of resistivity with increasing pressure).

The dependencies related to pressure and temperature are illustrated by the mean values shown in Table 9.7 and in Figures 9.10 and 9.11.

Table 9.7. Variation of the specific electrical resistivity with temperature and pressure (data after Volarovich and Parchomenko, 1976); $\rho_{0.1}$ is the resistivity at 0.1 GPa and ρ_2 at 2 GPa; $\Delta\rho = \rho_{0.1} - \rho_2$.

Rock	*T* in °C	$\rho_{0.1}$ inOhm m (at 0.1 GPa)	ρ_2 in Ohm m (at 2 GPa)	$\Delta\rho/\rho_{0.1}$
Granite	200	1.8×10^7	1.1×10^7	0.39
	300	1.2×10^7	7.2×10^6	0.40
	400	7.7×10^5	5.9×10^5	0.23
	500	1.2×10^5	9.8×10^4	0.14
	550	4.3×10^4	3.0×10^4	0.29
Diorite	200	3.1×10^7	8.3×10^6	0.27
	300	3.5×10^6	4.5×10^5	0.13
	400	5.0×10^5	1.8×10^5	0.36
	500	1.0×10^5	2.7×10^4	0.27
Gabbro	200	1.3×10^8	7.7×10^7	0.59
	300	9.8×10^6	6.3×10^6	0.64
	400	7.1×10^5	5.0×10^5	0.70
	500	8.3×10^4	4.3×10^4	0.52
	600	5.0×10^2	3.4×10^3	0.68
Olivinite	300	5.5×10^7	2.3×10^7	0.42
	400	5.9×10^6	2.4×10^6	0.41
	500	1.8×10^5	5.2×10^4	0.29
	600	1.1×10^4	4.2×10^3	0.38
	650	3.5×10^3	1.0×10^3	0.29

Figure 9.10 shows the *temperature dependence* of the electrical conductivity in form of a graph $\lg(\sigma)$ vs. T^{-1} . Assuming validity of the equation (9-7) such a plot yields a straight line for the temperature dependence. Figure 9.10 confirms this general exponential temperature dependence.

The conductivity versus temperature dependence not always can be described by one exponential term. Olhoeft et al. (1981) investigated the properties of samples from St. Helens dacite dome (eruption of June 1980). The experimental data were fitted by an equation

$$\sigma = \rho^{-1} = 2 \cdot 10^{-10} \exp\left(-\frac{0.07}{k \cdot T}\right) + 0.9 \exp\left(-\frac{0.95}{k \cdot T}\right) + 3 \cdot 10^{14} \exp\left(-\frac{4}{k \cdot T}\right) \tag{9-32}$$

where T is in Kelvin; k is the Boltzmann number. The composite of three terms with various activation energies corresponds to different dominating conduction mechanisms, related to the disequilibrium in the material as a result of rapid cooling and changing of the chemical environment.

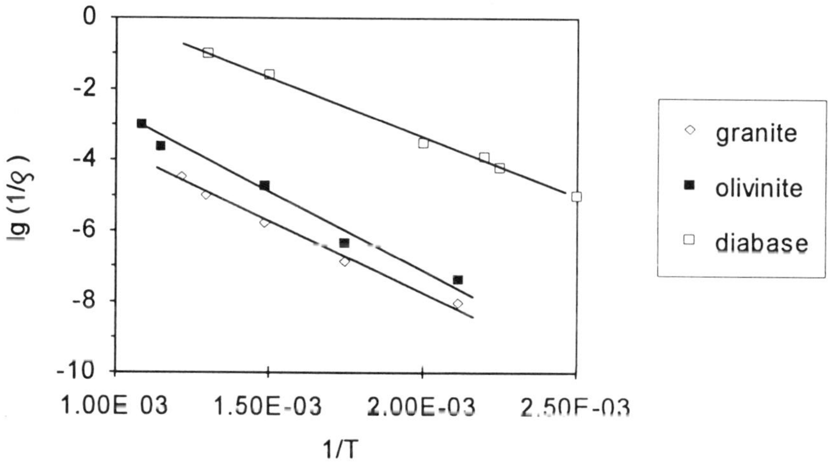

Figure 9.10. Dependence of the logarithm of the electrical conductivity (lg (σ^{-1})) on the reciprocal temperature ((T^{-1})); where T is in Kelvin. Data for granite and olivinite from Table 9.7 for 2 GPa; data for diabase from Parchomenko (1965).

Increasing *pressure* results in decreasing specific resistivity by

- closure of fractures and microcracks (these defects in dry rock are nonconductive parts within the poorly conducting solid mineral matrix),
- decreasing activation energy.

Figure 9.11 illustrates this resistivity decrease with increasing pressure, the "hysteresis" for increasing and decreasing pressure cycle (Figure 9.11a), and again confirms the temperature influence (Figure 9.11b).

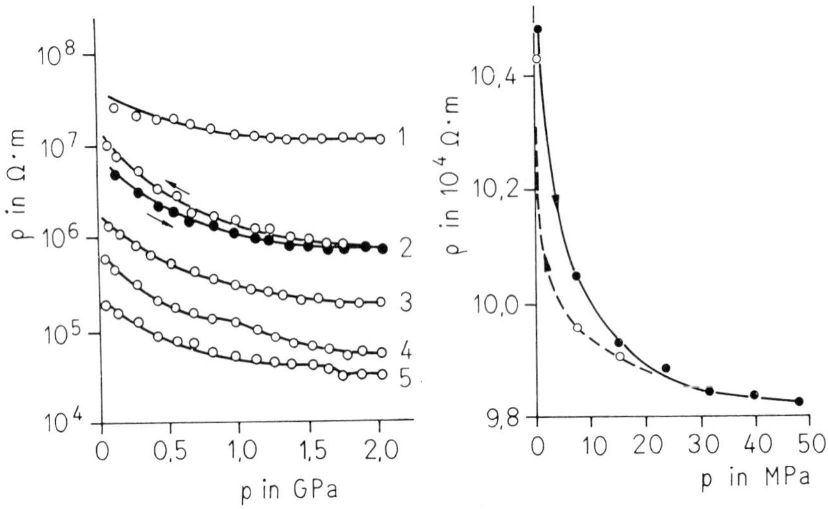

Figure 9.11. Pressure dependence of resistivity for water free rocks:a) Peridotite at various temperatures; Volarovich et al. (1966), Volarovich and Parchomenko (1976);
1 - T = 293 K 2 - T = 623 K 3 - T = 693 K 4 - T = 723 K 5 - T = 793 K
b) Basalt (increasing and decreasing pressure); Parchomenko (1965)

The pressure dependence - originated by closure of defects in the rock matrix - to a first approximation can be described by the simple fracture model (section 6.4). For the conductivity (or inverse specific resistivity), this results in

$$\rho^{-1}(p) = \rho_{\text{matrix}}^{-1} \cdot \left[1 - D_0 \cdot \exp\left(-\frac{p}{p*}\right)\right] \tag{9-33}$$

where $\rho^{-1}(p)$ is the actual conductivity at the pressure p
$\rho_{\text{matrix}}^{-1}$ is the matrix conductivity
$p*$ is a rock deformation property expressed as a pressure
D_0 is the initial defect parameter.

Figure 9.12 shows examples for two dry basalt samples which follow the equations

$$\rho(p)^{-1} = 8.5 \cdot 10^{-8} \cdot \left[1 - 0.4 \cdot \exp\left(-\frac{p}{30}\right)\right] \quad \text{sample I} \tag{9-34}$$

$$\rho(p)^{-1} = 6.0 \cdot 10^{-9} \cdot \left[1 - 0.55 \cdot \exp\left(-\frac{p}{100}\right)\right] \quad \text{sample II} \tag{9-35}$$

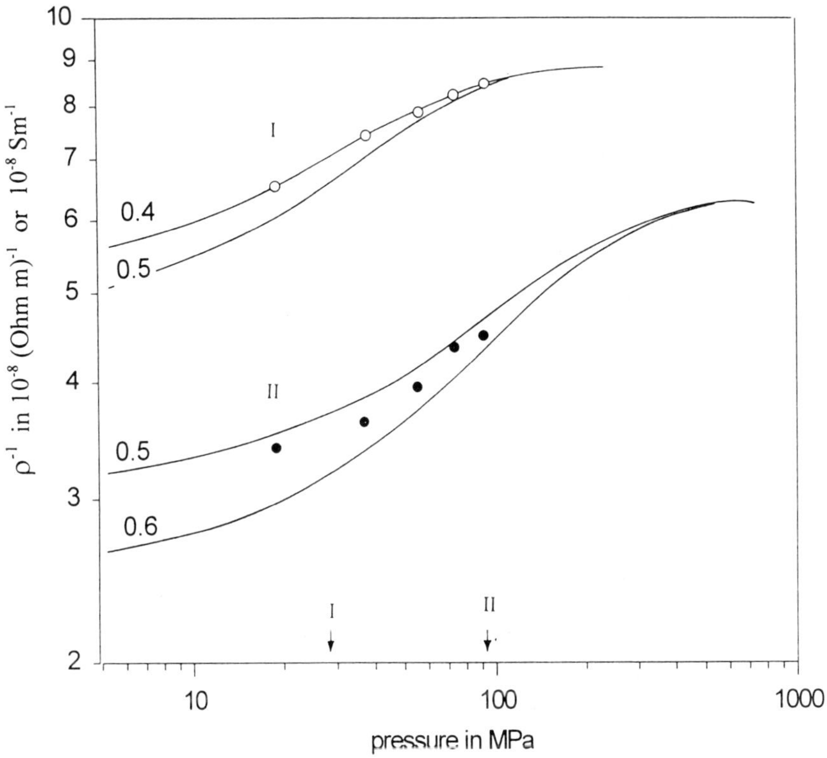

Figure 9.12. Specific electrical resistivity of two dry basalt samples (I and II) as a function of pressure (Schön, 1983); points - experimental data after Parchomenko (1965). curves - calculated on the basis of the simple model for fractured rocks. The numbers at the curves give the parameter D_0.

9.3.3. Electrical properties of porous and fractured rocks containing water

9.3.3.1. Overview

The electrical conductivity of a rock bearing water is controlled mainly by the properties of the *water*, which is an electrolyte. The conductivity of the solid matrix is negligible in most cases, except in the presence of vein type ores, rich disseminated ores or graphite.

The natural pore water contains anions and cations of dissolved salts. They can contribute to the rock conductivity in two ways:

1. by their intrinsic electrolytic conductivity. The geometry of the connected pore or crack space determines the geometry of the conductive path; this includes pore geometrical properties like porosity and tortuosity. The contribution of this rock conductivity component is directly proportional to the *electrolyte conductivity*. The analyses of these relationships are the subject of the Archie-equations.
2. by electrochemical interaction with the solid matrix at the interface fluid-solid. The geometry of the conductive path for this case is described by the properties of the internal surface. The *interface or excess conductivity* can be considered to be practically independent of the electrolyte conductivity. Relationships considering the phenomenon of interface conductivity were derived by Waxman and Smits (1968) and Rink and Schopper (1974).

We often find the case of a dominant electrolytic conductivity component in so called *"clean"* (i.e. clay and shale free) porous rocks. This behaviour is empirically described by the Archie equations.

The existence of a second conducting component was first detected in *shaly sands*. Their description is given by the different shaly-sand models related to the type of clay distribution within the rock.

Also in sands and sandstones, the second component interface conductivity was detected particularly at low pore water conductivities. The investigation of this phenomenon results in a deeper insight into pore geometries.
The study of the interface properties shows that the interface acts both as a conductor and capacitor. This leads to the complex character of the electrical properties related to the interface.

These types are discussed in more detail in the following sections for conductivity and dielectric properties.

9.3.3.2. Rocks with only electrolytic conductivity - the Archie equations

Archie equation for saturated rocks

In his historical 1942 paper about electrical conduction in clean sands, G.E. Archie (see also the historical review in the paper of Thomas, 1992) suggested that the conductivity of the brine saturated rock is proportional to brine conductivity and, also, the specific resistivity of the water saturated rock ρ_0 is proportional to the specific electrical resistivity of the brine ρ_w:

$$\rho_0 \sim \rho_w$$

This proportionality indicates that the brine conduction (electrolytic conduction) is the only conductive mechanism. As a result of this proportionality, Archie introduced the "*formation resistivity factor F* "

$$\rho_0 = F \cdot \rho_w \tag{9-36}$$

The formation resistivity factor expresses the resistivity magnification related to the brine as a result of the presence of a non-conductive matrix (formation). Therefore, it shows a correlation with connected porosity. A graphic presentation of the logarithm of the formation factor F versus the logarithm of porosity Φ results approximately in a straight line with the slope m (Figure 9.13.).

$$\lg F = -m \cdot \lg \Phi \tag{9-37}$$

or "1-st Archie equation"

$$\frac{\rho_0}{\rho_w} = F = \frac{1}{\Phi^m} \tag{9-38}$$

where the exponent m is an empirical quantity.

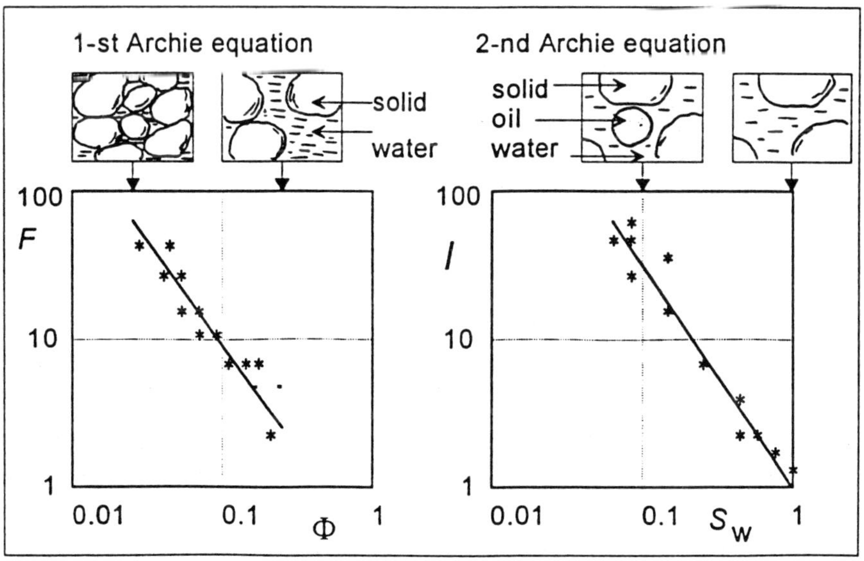

Figure 9.13. Schematic presentation of the Archie equations (9-42) and (9-66)

Archie noted that the exponent has a value of about 1.3 for unconsolidated sands and a range between 1.8 and 2.0 for many consolidated sandstones. Therefore, m was called subsequently *"cementation exponent"*.
For sandstones, the following values presented in many log analysis texts (such as Doveton, 1986):

Unconsolidated sands	$m = 1.3$
Very slightly cemented sandstones	$m = 1.4...1.5$
Slightly cemented sandstones	$m = 1.5...1.7$
Moderately cemented sandstones	$m = 1.8...1.9$
Highly cemented sandstones	$m = 2.0...2.2$

Figure 9.14 shows as example the formation factor versus porosity plot.

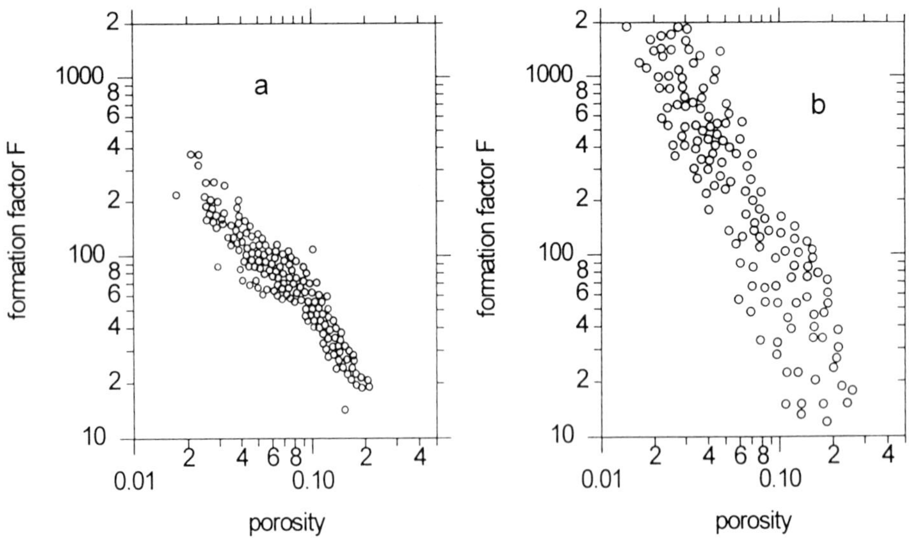

Figure 9.14. Formation factor versus porosity for sandstone samples
a) sandstone samples, after Tunn (1966) b) limestone samples, after Dachnov (1975)

The historical importance of the Archie equation results from the fact that it gave the first practical relationship between measurable properties (resistivities) and an important reservoir property (porosity).

Often used versions of the Archie equation are the Humble or Winsauer (1952) equation for sand with the parameters $a = 0.62$ and $m = 2.15$

$$F = \frac{0.62}{\Phi^{2.15}} \tag{9-39}$$

and the so-called Shell-formula for low porosity carbonates with a porosity dependent exponent m:

$$F = \Phi^{-m} \tag{9-40}$$

with

$$m = 1.87 + \frac{0.019}{\Phi} \tag{9-41}$$

Bigelow (1991) applied this formula for carbonate rocks (Williston Basin, North Dakota, Montana, Saskatchewan) and found a good fit for porosities between 0.04 and 0.25.

Further studies and investigations show that a better fit between experimental data and the analytic expression is given by a relationship:

$$F = \frac{a}{\Phi^{m}} \tag{9-42}$$

with a as a second empirical parameter.
Porter and Carothers (1971) published results of a systematic study of formation-factor porosity correlations of sandstones of different age and region and derived the following expressions:
- for Miocene sands of the Texas-Louisana Gulf Coast (720 data points)

$$F = \frac{1.97}{\Phi^{1.29}} \tag{9-43}$$

- for Pliocene sands from California (1575 data points)

$$F = \frac{2.45}{\Phi^{1.08}} \tag{9-44}$$

Despite the large data sets, the authors emphasize that, "care should be exercised in the transfer of these relations from one geological province to another."

The two parameters a and m are the subject of many studies, which suggest that m and a are controlled by the *pore channel geometry* and, thus, by textural properties. Doveton (1986) compared the bedding structure and the magnitude of *m* determined on core samples from the Skinner Sandstone (Pennsylvanian) in Southeast Kansas. Ripple-bedded zones have a higher value (about 2.2) than cross-bedded zones (about 2.05). In a second study, Doveton (1986) shows the increase of *m* from about 1.6 to 1.9 in a St. Peter Sandstone profile corresponding to a transition from cross-stratified coarse sandstone to ripple-bedded and thin-bedded fine sandstone and siltstone.

The Archie equation also fits formation-factor versus porosity data for *carbonates*. In this case, the porosity type is used as a reference instead of a "cementation" of the granules. Mean values and tendencies are:

- chalky limestones $m = 1.7...1.8$
- crystalline and granular carbonates $m = 1.8...2.0$
- carbonates with vugs $m = 2.1...2.6$

These numbers reflect the different tortuosity of the pore network for each case. "As such, they can be used as indicators of pore morphology, but any genetic implications are a matter for geological interpretation" (Doveton, 1986). Figure 9.15 shows a profile of the exponent *m* for a Pennsylvanian limestone.

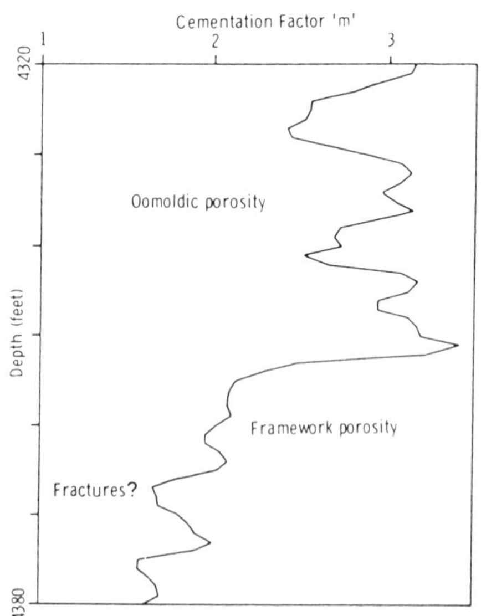

Figure 9.15. Profile of the exponent *m* of a Pennsylvanian limestone from a well in southern Kansas (Doveton, 1986)

The oomoldic porosity of the ooloitic section with its high tortuosity is reflected in the extreme high values. Beneath the oolithic section, a more typical value range suggests a simple matrix porosity in the wackestone section. Individual minima indicate possible fractures (Doveton, 1986). The decrease of m to about 1.4, is caused by fractures (notes Suau and Gartner, 1980).

On the other hand, attempts to correlate the exponent with the porosity (Dachnov, 1975 or Bitterlich and Wöbking, 1972) have not resulted in significant relationships. This confirms that the exponent m is controlled mainly by the *geometry* of the conductive paths and not by the volume fraction of the conducting pore water. Thus, m contains additional information about rock texture.

The textural influence upon the exponent m is the physical basis used in investigations attempting to find a correlation between cementation exponent and permeability (Raiga-Clemenceau, 1977; Hagiwara, 1986).

For the broad variety of rocks Keller (1989) has summarized mean values (Table 9.8):

Table 9.8 Mean values for the Archie parameters a and m (Keller, 1989)

Rock type	a	m
Weakly-cemented detrital rocks, such as sand, sandstone and some limestones, with a porosity range from 25 to 45 %, usually Tertiary in age	0.88	1.37
Moderately well cemented sedimentary rocks, including sandstones and limestones, with a porosity range from 18 to 35 %, usually Mesozoic in age	0.62	1.72
Well-cemented sedimentary rocks with a porosity range from 5 to 25 %, usually Palaeozoic in age	0.62	1.95
Highly porous volcanic rocks, such as tuff, aa and pahoehoe, with porosity in the range 20 to 80 %	3.5	1.44
Rocks with less than 4 % porosity, including dense igneous rocks and metamorphosed sedimentary rocks	1.4	1.58

Pressure dependence

If the conducting pore channels are deformed by pressure, the rock conductivity decreases (resistivity increases) and, thus, the formation factor increases.

The non-linear decrease of the rock conductivity $1/\rho$ with increasing pressure is illustrated by Figure 9.16a. The formation factor also increases (Figure 9.16b) non linear with pressure. This non linear increase of F with increasing pressure results from the non-linear stress-strain behaviour of the rocks related to the reduction of the pore sizes (cross section of the conductor) and change of the tortuosity.

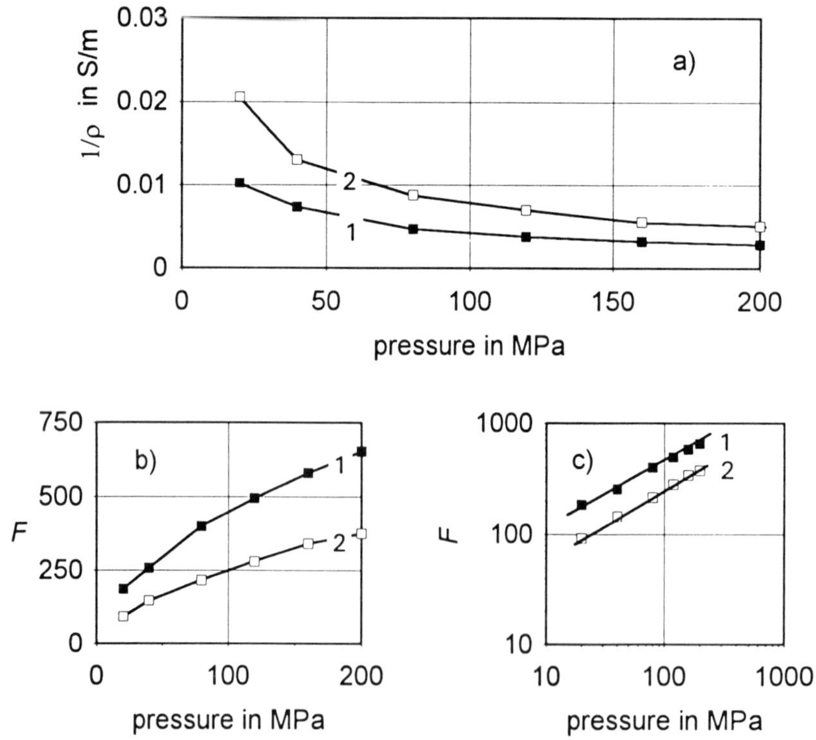

Figure 9.16. Electrical properties as a function of pressure for two water saturated low porous sandstone samples; data after Freund (1987);
a) rock conductivity vs. pressure, linear scales b) formation factor vs. pressure, linear scales c) formation factor vs. pressure, logarithmic scales

In Figure 9.16c, formation factor was plotted versus pressure on a bilogarithmic scale. To a first approximation a straight line results, which expresses the proportionality $\lg F \sim \lg p$ or

$$F \sim p^{g} \tag{9-45}$$

with a pressure exponent g (slope of the straight line).

Palmer and Pallat (1991) analyzed the formation factor versus pressure relationships for rock samples from North Sea and Alaskan reservoirs. The results show a non-linear dependency (Figure 9.17a). They fitted the experimental data with an equation relating the actual formation factor F_p normalized by its value at a pressure of 400 psi F_{400} and the logarithm of pressure

$$\frac{F_p}{F_{400}} = a + b \cdot \log p \tag{9-46}$$

where a and b are determined empirically. For the example in Figurc 9.17a, the resulting relationship is

$$\frac{F_p}{F_{400}} = 0.621 + 0.143 \cdot \lg p \tag{9-47}$$

Figure 9.17. Formation factor vs. overburden pressure dependence
a) linear plot, data after Palmer and Pallatt, 1991 (pressure converted from psi in MPa)
b) fitting by a power law using a bilogarithmic plot.

Figure 9.17b demonstrates an alternative fit using a power law (9-45). A bilogarithmic plot of log F versus log p results in a first approximation to a straight line with the general expression

$$F_p = F_1 \cdot p^g \tag{9-48}$$

where F_1 is the formation factor at a pressure $p = 1$. The exponent g is the slope of the straight line. The data from this example results in an equation for the pressure dependent formation factor of

$$F_p = 16.6 \cdot p^{0.055} \tag{9-49}$$

Archer and Jing (1991) investigated both clean and shaly sedimentary rocks. They found a similar non-linear relationship between formation factor and pressure, as well as, a noticeable hysteresis effect occurring between the loading and unloading cycles. This is a result of non-elastic deformation of the rock skeleton.

A direct comparison of the formation factor and porosity measured at the same pressure provides the opportunity to investigate the pressure dependence of the cementation exponent. Mahmood et al. (1988) and Archer and Jing (1991) reported a small increase of m with increasing pressure. Lewis et al.(1988) indicated that the increase of cementation factor with increasing pressure is not significant.

In the previously described resistivity versus pressure relationship, the porous rock was water saturated. Therefore, a pressure increase results in a decrease of the "conducting component" (pore water). For the case of a *dry rock*, the resistivity is much higher and the "conducting component" is the solid grain skeleton. Thus with increasing pressure, its resistivity (particularly in the contact region) decreases. Figure 9.18a demonstrates for a sandstone sample in the water saturated state (1) and in the dry state (2), this behaviour. The resistivity was normalized with its value at atmospheric pressure (ρ_0). A power law for both cases with positive exponent (water saturated) and negative exponent (dry) fits the data (Figure 9.18b).

Archie's empirical relationship was *primarily derived for reservoir rocks*; but it also has been applied with success on fractured igneous rocks with very low porosity and on highly porous unconsolidated rocks.

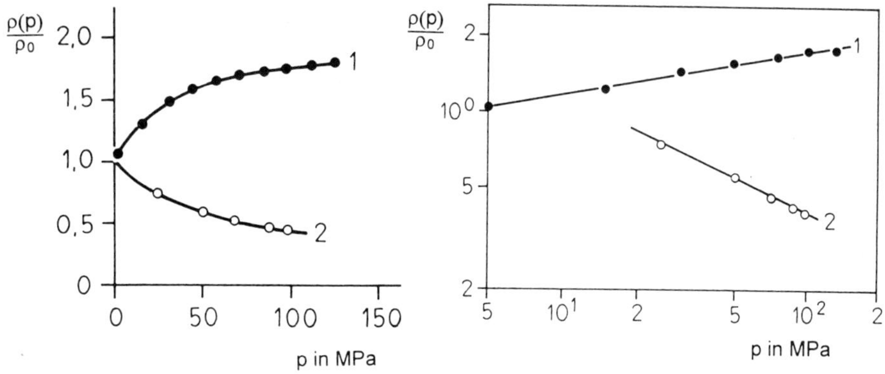

Figure 9.18. The pressure dependence of the specific electrical resistivity of sandstone sample (porosity $\Phi = 0.087$) in the water saturated (1) and dry (2) state, data after Parchomenko (1965)

a) linear plot of the normalized resistivity $\frac{\rho(p)}{\rho_0}$ versus pressure

b) bilogarithmic plot of the normalized resistivity versus pressure

Application on fractured igneous rocks

In 1965, Brace et al. published results of a study about the influence of crack and intergranular porosity with the effect of pressure upon the electrical resistivity of igneous rocks. This yields insight into the complexity of these properties and internal geometries of the rocks. Glover and Vine (1995) studied the electrical conductivity of igneous rocks (granite, granodiorite, pyroxenite, amphibolite, gneisse) at simulated lower crustal conditions of temperature (up to 900°C), pressure (up to 0.4 GPa) , and saturation. The results show the influence of aqueous fluids, carbon dioxide, but also graphite.

For seven rock samples from Brace's paper, Figure 9.19 shows the specific electrical resistivity of water saturated rocks as a function of pressure. Obviously, there is a non-linear increase of the resistivity in the range below 400 MPa. This is caused by the reduction in the porosity, mainly by the decrease of the crack-cross section and the closure of crack parts and, thus, by an increase of tortuosity.
Two additional analyses may be appropriate:
a) *Consideration of a correlation between the relative change of resistivity and the relative change of porosity* in the pressure range below 400 MPa (i.e.

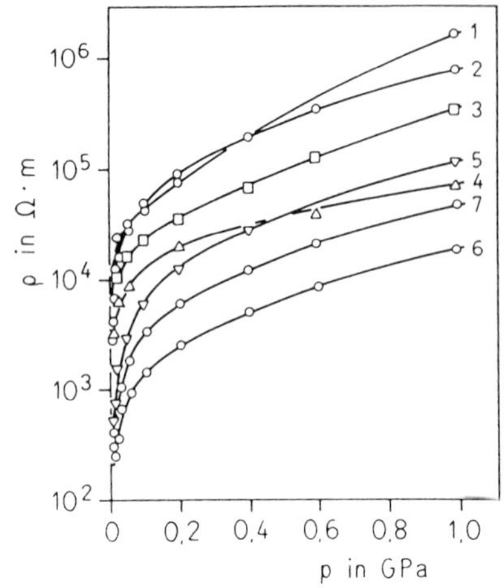

Curve	Rock	Total porosity
1	diabase	0.001
2	dunite	0.003
3	anorthosite	0.002
4	granodiorite	0.004
5	Casco granite	0.007
6	Westerly granite	0.009
7	quartzite	0.005

Figure 9.19. Specific electrical resistivity versus pressure for water saturated igneous rocks (after data and figures from Brace et al., 1965)

in the range of dominant closure process). Two parameters are used for the analyses:

- relative resistivity change $\frac{\Delta\rho}{\rho} = \frac{\rho_{400} - \rho_p}{\rho_{400}}$ where ρ_{400} is the specific electrical resistivity at 400 MPa, and ρ_p its value at the actual pressure p;
- relative change of crack porosity $\left(\frac{\Delta\Phi}{\Phi_0}\right)_c = \frac{\Phi_{c,0} - \Phi_{c,p}}{\Phi_{c,0}}$ where $\Phi_{c,0}$ is the crack porosity at atmospheric pressure, and $\Phi_{c,p}$ its value at the actual pressure.

A bilogarithmic plot of $\frac{\Delta\rho}{\rho}$ versus $\left(\frac{\Delta\Phi}{\Phi_0}\right)_c$ results in a first approximation to straight lines for the individual samples (Figure 9.20) with the proportionality

$$\frac{\Delta\rho}{\rho} \sim \left(\frac{\Delta\Phi}{\Phi_0}\right)_c^{\xi} \tag{9-50}$$

The exponent ξ is controlled by the deformation properties of the material; it is in the range between 0.1 (quartzite) and 0.44 (granodiorite).

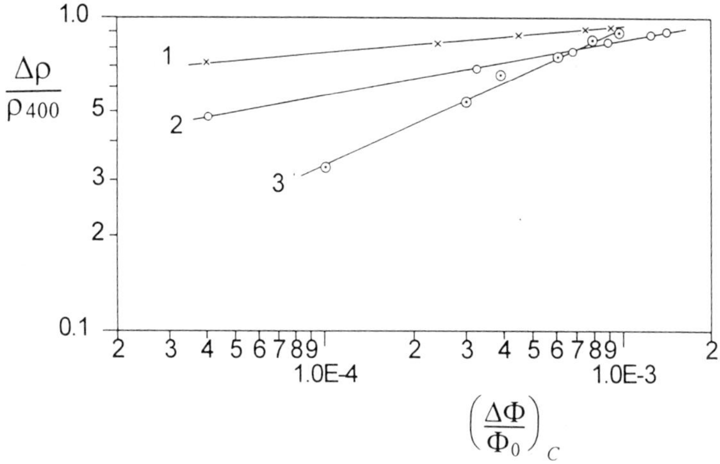

Figure 9.20. Relative change of specific resistivity versus relative change of crack-porosity for the pressure range below 400 MPa; 1 - quartzite 2 - Westerly-granite 3 - granodiorite

b) *Test of the applicability of Archie's formula:* For this, the total porosity Φ_Σ (sum of crack and intergranular porosity) was first plotted against the formation factor calculated as the ratio of specific rock resistivity at 5 MPa and water resistivity ($F_5 = \rho_5/\rho_w$) in a bilogarithmic diagram (Figure 9.21).

There is a scattering of the data with only a general trend of increasing F_5 with decreasing porosity.

A clear correlation exists for a plot of the formation factor calculated as ratio of the rock resistivity at 400 MPa and the water resistivity ($F_{400} = \rho_{400}/\rho_w$) versus intergranular porosity Φ_p.

It can also be assumed that in this case the formation factor is controlled by the intergranular porosity, because most of the cracks are (electrically) closed. Therefore, a better correlation is expected. As a first approximation this results in the relationship

$$F_{400} = \frac{\rho_{400}}{\rho_w} = \frac{1}{\Phi_p^2} \tag{9-51}$$

or $a = 1$ and $m = 2$ in Archie's relationship (9-42.)

Lovell and Pezard, 1990 investigated basalt cores from the Deep Sea Drilling Program (DSDP) hole 504B and found $a = 12.1$ and $m = 1.02$ for fracture-free samples and $a = 6.2$ and $m = 1.05$ for fractured samples. This tendency corresponds to results of Broglia and Moos, 1985 (see Lovell and Pezard, 1990) from another hole (418A9); the samples from the lower, fresher part gave $a = 29.5$ and $m = 1.16$, while samples from the extremely altered upper part gave $a = 11.5$ and $m = 1.85$.

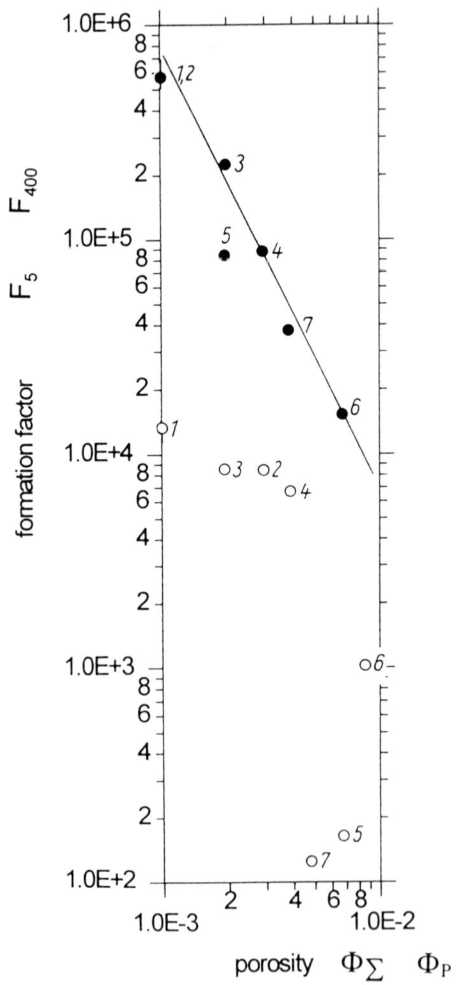

Figure 9.21. Formation factor versus porosity; the number indicates the sample-number in Figure 9.19; closed circles - F_{400} versus Φ_P open circles - F_5 versus Φ_Σ

Application on unconsolidated sediments

Unconsolidated sediments are characterized by relatively low values of m (between 1.1 and 1.3) and a parameter $a \approx 1$.

The effect of *particle shape* on the formation factor versus porosity relationship have investigated Atkins and Smith, 1961. Jackson et al., 1978 published results of measurements on samples of various grain shapes ranging from spheres to shells. Figure 9.22 shows the mean values in a bilogarithmic plot of F versus Φ.

The exponent m correlates with the sphericity S of the sediment grains and follows an equation

$$m = 2.9 - 1.8\,S \qquad (9\text{-}52)$$

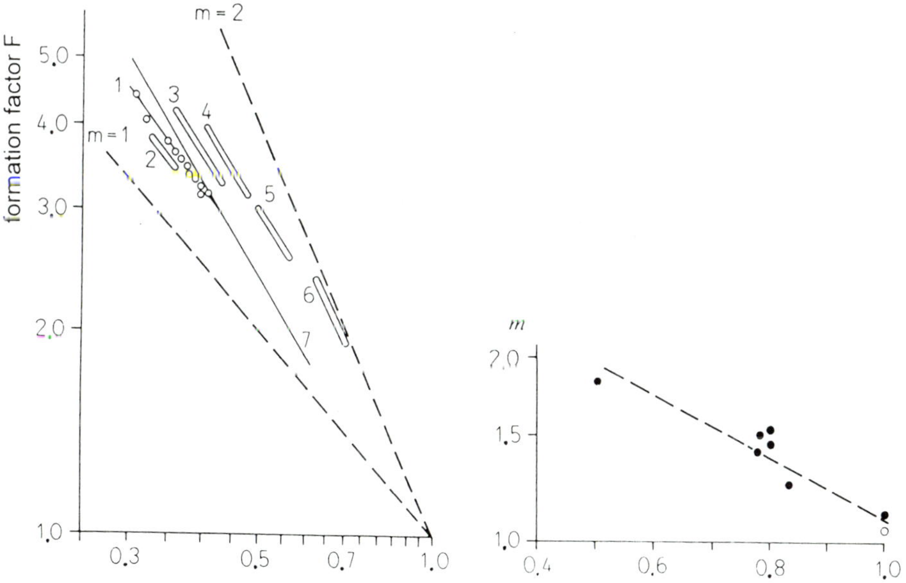

Figure 9.22. Formation factor versus porosity relationship for unconsolidated sediments
a) F versus Φ for sediments of various grain shape; References: 1 - Schön, 1983; 2 to 6 after Jackson et al., 1978; 7 - Dachnov, 1975
b) Correlation between exponent m and sphericity S (Schön, 1983)

Kermabon et al. (1969) analyzed the formation factor versus porosity correlation for 2500 measurements made on samples from the Tyrrhenian abyssal plain. The porosities are between 45 and 85 %. Three types of equations with Φ as a function of F were tested and the following empirical relationships with the corresponding variances were found:

a) Irrational function:

$$\Phi = 144.8934/(F + 0.7193)^{0.6842} \qquad \text{variance } 5.91 \tag{9-53}$$

b) third-degree polynomial:

$$\Phi = -5.9021 \cdot F^3 + 40.0416 \cdot F^2 - 105.3899 \cdot F + 171.2504 \qquad \text{variance } 5.236 \tag{9-54}$$

c) fourth-degree polynomial:

$$\Phi = 2.4611 \cdot F^4 - 25.7247 \cdot F^3 + 97.1580 \cdot F^2 - 174.6992 \cdot F + 200.8948 \qquad \text{variance } 5.178 \tag{9-55}$$

In these equations, the porosity is in percent. The authors conclusion is that the third-degree polynomial has enough freedom to fit the data.

Gerland et al. (1993) applied the Kermabon relationships and other formation factor versus porosity equations to fit data from core measurements from Antarctic marine sediments. Figure 9.23 shows a plot of the specific electrical resistivity versus porosity. The electrical resistivity of the pore water (sea water) was 0.209 Ohm m. Thus, the formation factor results from dividing the plotted specific resistivities of the sediment by $\rho_w = 0.209$ Ohm m. To fit the data the following equations are applied and plotted:

- Archie equation $$\rho_0 = \rho_w \cdot \frac{1}{\Phi^2} \tag{9-56}$$
- Boyce equation (Boyce, 1968) $$\rho_0 = \rho_w \cdot \frac{1.3}{\Phi^{1.45}} \tag{9-57}$$
- Kermabon equation (Kermabon et al., 1969): third-degree polynomial, equation (9-54.)
- Winsauer equation (Winsauer et al., 1952) $$\rho_0 = \rho_w \cdot \frac{0.62}{\Phi^{2.15}} \tag{9-58}$$

As seen in Figure 9.23, the best approximation is given by the Boyce equation; the relatively great difference between measured data and the Winsauer curve is not surprising, because this relationship was derived primarily for consolidated sediments.

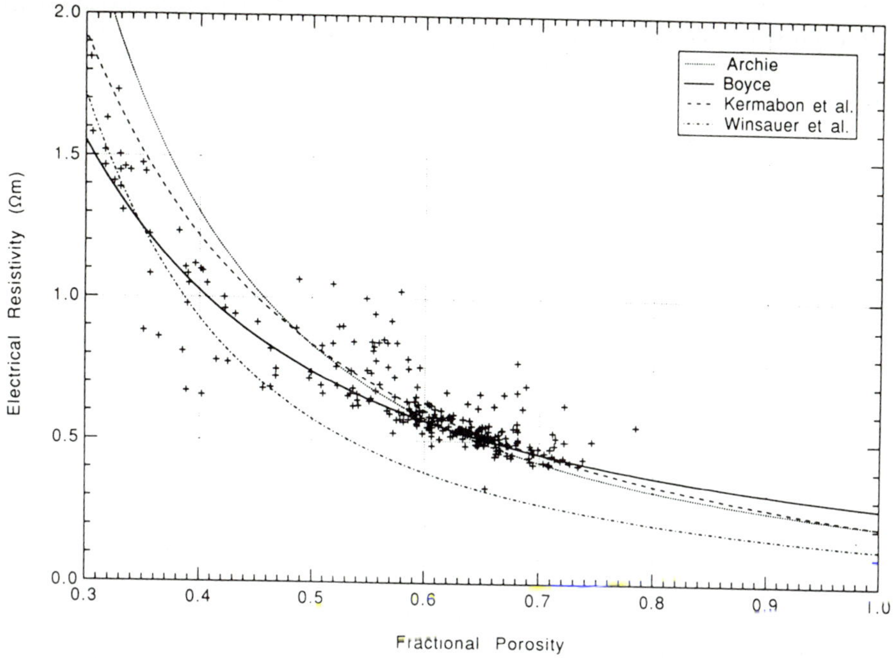

Figure 9.23. Electrical resistivity versus porosity for marine sediments and comparison with various empirical relationships; after Gerland et al. (1993)

Application on drilling mud

The specific resistivity of *drilling mud and mudcake* is of particular interest in well logging. Mud can be considered to be a very "highly porous" suspension. The mud (m) corresponds to the water saturated sediment, and the mud filtrate (mf) to the pore water. Using well logging terms, this means:

$$\rho_0 \rightarrow R_{\mathrm{m}} \qquad \rho_{\mathrm{w}} \rightarrow R_{\mathrm{mf}}$$

The equation of Overton and Lipson (1958)

$$R_{\mathrm{mf}} = K_m \cdot R_{\mathrm{m}}^{1.07} \tag{9-59}$$

therefore corresponds to

$$\rho_0 = \left(\frac{\rho_{\mathrm{w}}}{K_m}\right)^{0.93} \tag{9-60}$$

where K_m decreases with increasing mud density (mud weight) or decreasing "mud porosity" (for example for a mud weight of 1200 kg m^{-3} is $K_m = 0.847$ and for a mud weight of 1560 kg m^{-3} is $K_m = 0.488$; see also Schlumberger, 1989).

Loewe and Dunlap (1986) published a formula for fresh water muds with resistivities in the range between 0.1 and 2.0 Ohm m (at 24 °C or 75 °F)

$$\lg_{10}\left(\frac{R_{\mathrm{mf}}}{R_{\mathrm{m}}}\right) = 0.396 - 0.0475 \cdot d_{\mathrm{m}} \tag{9-61}$$

where d_{m} is the mud density in pounds per gallon (8.345 lb/gal = 1 g/cm^{3}). It is possible to show that this formula corresponds approximately with Archie's law. The "mud porosity " is

$$\Phi_{\mathrm{m}} = \frac{d_{\mathrm{s}} - d_{\mathrm{m}}}{d_{\mathrm{s}} - d_{\mathrm{w}}} \tag{9-62}$$

where d_{s} is the density of the solid components (e.g., silica, barite) and d_{w} is the water density. Insertion of this into equation (9-37) and Taylor expansion of the Archie equation (to 1-st order) results after conversion to decadic logarithms in an equation (Schön, 1987):

$$\lg_{10}\left(\frac{R_{\mathrm{mf}}}{R_{\mathrm{m}}}\right) = m \cdot \left[\lg_{10}\left(\frac{d_{\mathrm{s}}}{d_{\mathrm{s}} - d_{\mathrm{w}}}\right)\right] - 0.4343 \cdot m \cdot \frac{d_{\mathrm{m}}}{d_{\mathrm{s}}} \tag{9-63}$$

With $m = 2$, $d_{\mathrm{s}} = 3.12$ g/cm^3 (barite/silica) and $d_{\mathrm{w}} = 1.00$ g/cm^3 a relationship similar to (9-61) results:

$$\lg_{10}\left(\frac{R_{\mathrm{mf}}}{R_{\mathrm{m}}}\right) = 0.337 - 0.033 \cdot d_{\mathrm{m}} \tag{9-64}$$

Thus, the Archie equation can also be used as a first approximation for mud resistivity estimation.

Partial water saturation

If a part of the pore space is occupied by a nonconducting fluid (oil, gas), the resulting rock resistivity (ρ_t) will be higher than for the brine saturated case (ρ_0). This is expressed by the "*saturation index*" I

$$\rho_t = I \cdot \rho_0 \tag{9-65}$$

where ρ_0 is the specific resistivity of the completely brine-saturated rock,
ρ_t is the specific resistivity of the partially brine-saturated rock.
Also, equation (9-65) assumes that current is conducted by the interstitial electrolyte alone. The saturation index depends on the saturation S_w following an empirical relationship ("2-nd Archie equation") derived also from a logarithmic plot of I vs. S_w (see Figure 9.13):

$$\rho_t = \rho_0 \cdot S_w^{-n} \tag{9-66}$$

For most rocks, the *"saturation exponet"* n is about 2 (see below). The combination with equation 9-36 and 9-42 results in a relationship for the specific electrical resistivity of the partially saturated rock depending on

- brine resistivity ρ_w
- reservoir properties Φ and S_w
- empirical parameters a, m, and n:

$$\rho_t = \rho_0 \cdot S_w^{-n} = \rho_w \cdot a \cdot \Phi^{-m} \cdot S_w^{-n} \tag{9-67}$$

Equation (9-67) offers the possibility of a saturation calculation from resistivity logs (with known porosity).

The saturation exponent n must be determined empirically; it is controlled by the distribution of the conducting brine in the pore space and, thus, depends on the rock texture, wetting properties, and "saturation history" caused by capillary effects.

Therefore, different saturating techniques (imbition, drainage) and techniques of core preparation also have an effect on the laboratory determined saturation exponent (see for example de Waal et al., 1989, Worthington et al., 1988, 1989, 1990).

The application of X-ray and resistivity tomography allows monitoring of the saturation process along the core axis. In the *water-wet* case, water adheres to grain surfaces and builds up a more or less continuous phase in the rock, therefore, conducting the current; the saturation exponent is $n \approx 2$. Departures from the mean $n = 2$ are variously attributed to the influence of wettability, the nature of desaturating fluid and laboratory procedure (core sample-preparation, pressure, and temperature; see also Worthington et al., 1989).

In the *oil-wet* case, the nonconducting oil becomes the continuous fluid phase, and the water occurs mostly as isolated droplets. In this case, the resistivity is much higher and the saturation exponent n is much greater than 2.

A linear relationship between the saturation exponent n and the wettability index was found for sandstones (Berea and Elgin sandstone). The slope of the curve shows that the increase of n with decreasing wettability index is controlled by the rock type (Donaldson et al., 1987).

Table 9.9 contains some values of the saturation exponent for sandstone and limestone.

Table 9.9. Saturation exponent n for sandstones and limestones after various references (data taken from Worthington et al., 1989)

Rock type	n	Reference
Sandstones	1.42...2.55	Wyllie and Spangler, 1952
	1.12...2.52	Pierce and Loewe, 1958
	1.69...2.08	Walther, 1968
	1.65...2.44	Wilson and Hensel, 1982
	1.42...2.24	Hunt et al., 1985
Limestones	2.30 ...2.38	Walther, 1968
	1.10...1.90	Sharma et al., 1980
	1.65...2.22	Swanson, 1980

In some cases, the resistivity index versus saturation curve is nonlinear on a log-log plot. This phenomenon seems to be related to a bi-modal (or multi-modal) pore-size distribution, resulting either from the presence of microporosity or from rough grain surface coatings (de Waal, 1989, Worthington et al., 1989). Worthington and Pallatt (1990) investigated the influence of pore geometry upon the character of the saturation index versus saturation curve and the behaviour of n. For an unimodal pore size distribution (Figure 9.24a), the saturation exponent is effectively constant. In contrast, a trimodal pore size distribution shows a variable exponent n (Figure 9.24b). This results from the heterogeneous brine distribution during the desaturation process. The non-uniform distribution of electrolyte within the pore system results in a variation of the internal geometry of the active current path, described by various tortuosities. In particular for micropores, co-existing effects of surface conduction may also be influential.

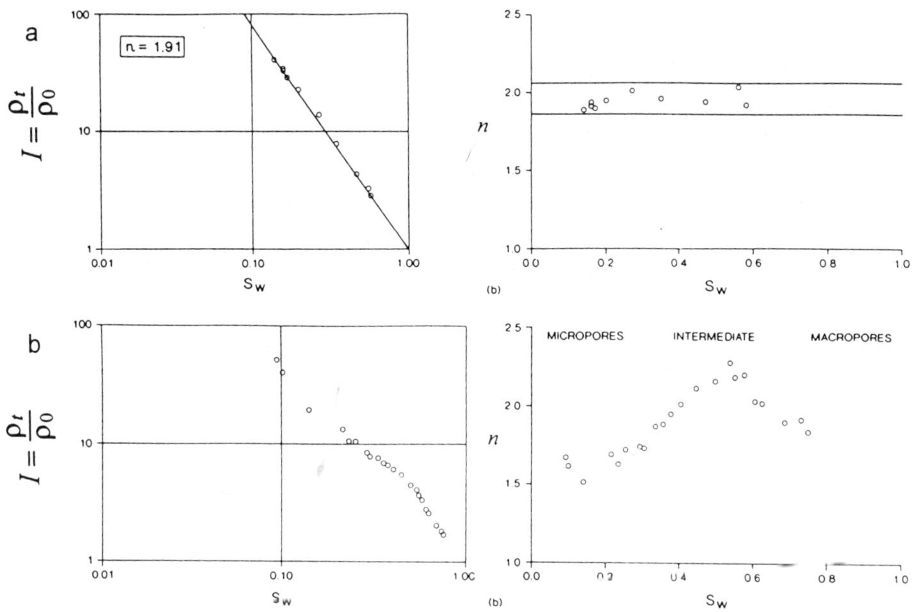

Figure 9.24. Saturation index and saturation exponent as a function of water saturation; after Worthington and Pallatt (1989)

left: Saturation index $I = \rho_t/\rho_0$ versus water saturation S_w

right: Saturation exponent n versus water saturation S_w

a) Sandstone with unimodal pore size distribution

Sample description: porosity $\Phi = 0.22$; Permeability $k = 171$ mD; Formation factor $F = 17.1$

b) Sandstone with trimodal pore size distribution

Sample description: porosity $\Phi = 0.24$ Permeability $k = 127$ mD; Formation factor $F = 14$

fraction of micropores (related to total pore volume) 0.2 ... 0.55

fraction of intermediate pores (related to total pore volume) 0.15 ... 0.3

fraction of macropores (related to total pore volume) 0.3 ... 0.5

The two Archie exponents m and n are the subject of numerous investigations with regard to various parameters (texture, pressure etc.) and also their theoretical derivation (see section 9.6.):

Pressure effects

Lewis et al. (1988) investigated wettability and stress effects on the exponents and found that stress has a relatively minor effect upon the exponents m and n for consolidated sandstones; the trend was an increase with increasing pressure between 0 and 34.5 MPa (0 and 5000 psi). Also, de Waal et al. (1989) found that the stress effect on the exponent n is small.

Temperature effects

For water bearing rocks, temperature changes result in changes of the pore fluid conductivity. In clean water bearing rocks, this is - following Archie's concept - proportional to the rock conductivity.

For the case of a temperature decrease below 0 °C, the water in the pores freezes and the resistivity generally increases. As the temperature decreases through 0° C some of the pore water freezes and resistivity increases (Figure 9.25). However, the adsorbed ions (particularly at the clay surface) and soluted ions in the pore fluid result in a delay of the freezing of the rock. With the freezing process, particularly in unconsolidated sediments or soils, changes of the rock structure can result.

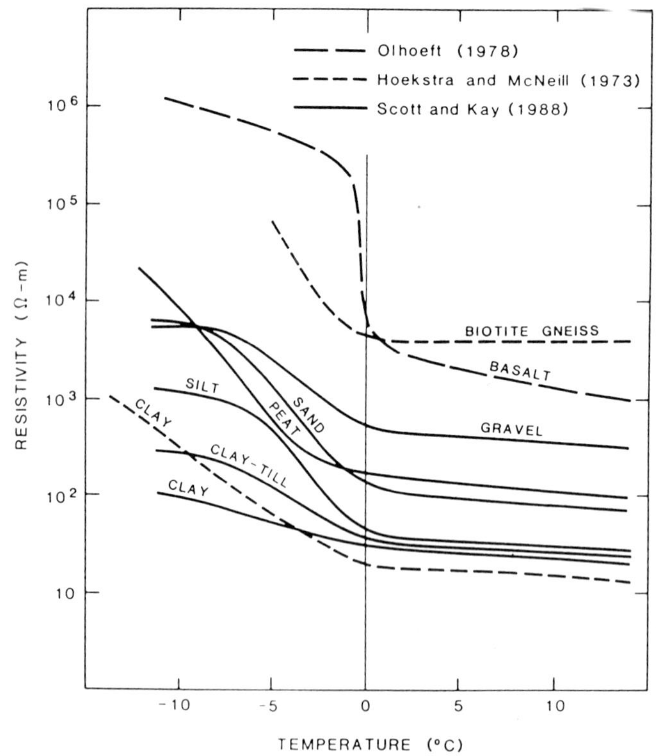

Figure 9.25. Specific electrical resistivity versus temperature for igneous and sedimentary rocks (Scott et al., 1990; data after Hoekstra and McNeill, 1973, Olhoeft, 1978, Scott and Kay, 1988)

9.3.3.3. Models and theoretical considerations

For a rock with *only one conductive* component (pore water), there is a broad spectrum of theories and models to express the dependence of rock conductivity on pore fluid conductivity, saturation, porosity and rock structure. Similarity to the empirical Archie equation is frequently used as a criterion.

For the simple case of one conductive component (i.e. matrix is nonconducting), some models and relationships result from equations used for thermal conductivity by replacing

- thermal conductivity of the matrix with electrical conductivity of the matrix $\lambda_{matrix} \rightarrow \sigma_{matrix}$
- thermal conductivity of the pore fluid with electrical conductivity of the pore fluid $\lambda_{fluid} \rightarrow \sigma_{fluid}$

From this, a fundamental difference between the two kinds of conduction is seen:

► the matrix dominates for thermal conductivity ($\lambda_{matrix} > \lambda_{fluid}$),

► the pore water dominates for electrical conductivity ($\sigma_{matrix} << \sigma_{fluid}$).

Despite the broad spectrum of theories and models, most of them can be classified by the main classes of theories for rock physics (Figure 1.7):

a) Sheet models and their modifications

The simple sheet model (see Figure 8.18) results in the porous rock equations for the specific electrical resistivity

- current perpendicular to bedding plane (series circuit):

$$\rho_{rock,\perp} = (1-\Phi)\cdot\rho_{matrix} + \Phi\cdot\rho_{fluid} \qquad (9\text{-}68)$$

- current parallel to bedding plane (parallel circuit):

$$\rho_{rock,//} = \left\{(1-\Phi)\cdot\rho_{matrix}^{-1} + \Phi\cdot\rho_{fluid}^{-1}\right\}^{-1} \qquad (9\text{-}69)$$

For a water saturated rock without matrix conductivity ($\rho_{matrix} \rightarrow \infty$; Archie type) follows

$$\rho_{rock,\perp} = \rho_{0,\perp} \rightarrow \infty \qquad (9\text{-}70)$$

$$\rho_{rock//} = \rho_{0//} = \frac{\rho_w}{\Phi} \qquad \text{or} \qquad F = \frac{1}{\Phi} \tag{9-71}$$

Both equations have no practical importance - the simplification is too far from any realistic property.

Necai (1964) applied this type of simple model for calculation of the resistivity of *fractured porous* rocks using a modification of this concept. In this case, the porous rock is assumed to consist of a "porous matrix" with a (non infinite) resistivity. The second component is the water filled fracture. The models have one, two, or three sets of fracture plane systems. Figure 9.26 shows the models and resulting equations.

	Equation	Direction	
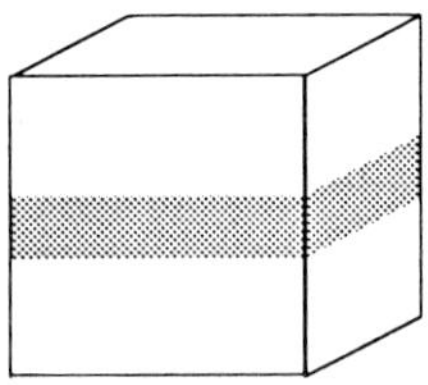	$\rho_0 = (1 - \Phi_c) \cdot \rho_m + \Phi_c \cdot \rho_w$	z	(9-72)
	$\frac{1}{\rho_0} = \frac{1-\Phi_c}{\rho_c} + \frac{\Phi_c}{\rho_c}$	x, y	(9-73)
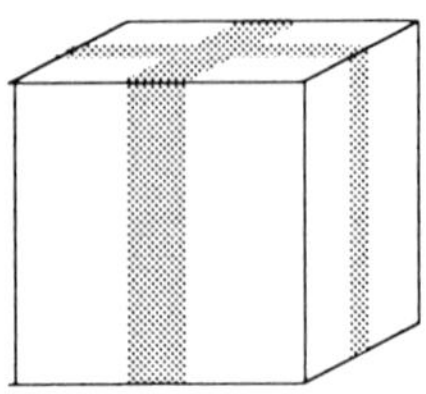	$\frac{1}{\rho_0} = \frac{1-\Phi c}{\rho_m} + \frac{\Phi_c}{2 \cdot \rho_w}$	x, y	(9-74)
	$\frac{1}{\rho_0} = \frac{1-\Phi_c}{\rho_m} + \frac{\Phi_c}{\rho_w}$	z	(9-75)
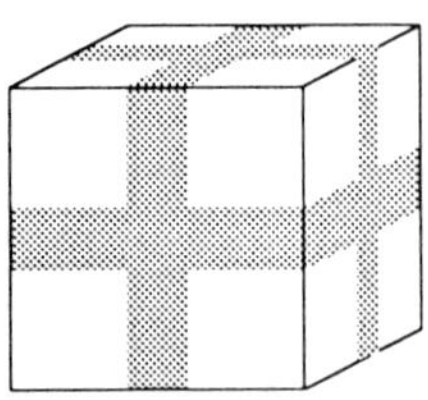	$\frac{1}{\rho_0} = \frac{2 \cdot \Phi_c}{3 \cdot \rho_m} + \frac{3 - 2 \cdot \Phi_c}{(3 - \Phi c) \cdot \rho_w}$	x, y, z	(9-76)

Figure 9.26. Models for fractured rocks with matrix porosity; after Necai (1964)
ρ_m - specific resistivity of the porous matrix (it could be expressed by the Archie equation)
ρ_w - specific resistivity of the water in the fractures (cracks)
Φ_c - fracture (crack) porosity (grey parts of the model)

Sheet models (or equivalent series and parallel circuits) are fundamental constituents of most models for clay containing rocks (shaly sands) as described in section 9.3.4.

b) Pore channel models

As in the case of hydrodynamic permeability - which is also a "pore geometry controlled property" -, the tube model is a preferred idealisation for electrical conductivity of a water saturated clean porous rock. For the derivation Figure 2.15 can be used.

The model with the dimension L has the porosity (compare equation 2-34)

$$\Phi = \pi \cdot \left(\frac{r}{L}\right)^2 \tag{9-77}$$

where r is the radius of the pore channel (tube). The (true) pore length was defined as $l = T\,L$, where T is the tortuosity (see equation 2-36).

For this model, the Ohmic resistance R in direction of the pore channel can be given

- under a "macroscopic aspect" using the dimension L and the specific rock resistivity ρ_0

$$R = \rho_0 \cdot \frac{L}{L^2} = \frac{\rho_0}{L} \tag{9-78}$$

- under a "microscopic aspect" using the pore channel dimensions ($l = TL$ and r) and the specific resistivity of the pore water ρ_w

$$R = \rho_w \frac{l}{\pi \cdot r^2} \tag{9-79}$$

Immediately follows

$$\frac{\rho_0}{L} = \rho_w \frac{l}{\pi \cdot r^2} \tag{9-80}$$

and with 2-37 and 9-77 results for the specific rock resistivity

$$\rho_0 = \rho_w \cdot \frac{T^2}{\Phi} \tag{9-81}$$

A comparison with Archie's equation gives a relationship for the formation factor

$$F = \frac{T^2}{\Phi} \tag{9-82}$$

which expresses the relationship of the Archie - parameter *a* and *m* to the tortuosity.

This simple model was modified and applied in two directions:

(1) Development of the unidirectional capillary tube to tube bundles and networks (see for example Schopper, 1966)

(2) Modification of the capillary tube geometry by variation of the cross section geometry and tortuosity. Schopper (1966) has defined these pore geometrical properties statistically.

Wang and Sharma (1988) developed a three-dimensional pore network utilising methods of percolation theory. Each element of the network is assumed to consist of a converging-diverging pore cross-section. The model was used for the study of resistivity behaviour of partially saturated rocks. With the variation of the pore geometrical properties and model parameters for water wet rocks, saturation exponents (n) between 1 and 2 were calculated.

Herrick and Kennedy (1994) developed the "electrical efficiency model"; the electrical efficiency of a porous rock is the ratio of the average electrical power developed in all water bearing parts of the pore system to the power developed in a straight tube with the same length and water volume as the rock.

c) Sphere or inclusion models (Maxwell equation) and mixing formulas

In 1904, Maxwell derived his equation for the calculation of the properties of a material consisting of *dispersed spheres* in a "host substance". For thermal conduction, equation 8-60 results. Replacing

- thermal rock conductivity λ with electrical rock conductivity σ_0,
- thermal fluid conductivity λ_p with electrical water conductivity σ_w, and
- thermal matrix conductivity λ_m with electrical matrix conductivity $\sigma_m = 0$

results in

$$\sigma_0 = \sigma_w \cdot \frac{2 \cdot \Phi}{3 - \Phi} \tag{9-83}$$

or in terms of specific electrical resistivities

$$\rho_0 = \rho_w \cdot \frac{3 - \Phi}{2 \cdot \Phi} \tag{9-84}$$

The formation factor for the Maxwell model is

$$F = \frac{3 - \Phi}{2 \cdot \Phi} \tag{9-85}$$

For *ellipsoids*, Fricke (1931) derived the relationship

$$F = \frac{(\chi + 1) - \Phi}{\chi \cdot \Phi} \tag{9-86}$$

where χ depends on the shape (for spheres $\chi = 2$ equation 9-85 results) χ decreases with increasing aspect ratio.
Perez-Rosales (1976) has modified this equation for comparison with empirical formulas introducing effective porosity. A compilation of formulas for various particle shapes (spheres, ellipsoids, cubes, octahedrons etc.) is given by Keller (1989).

Bruggeman (1935) and Hanai (1960) have derived general relationships for the calculation of the properties for heterogeneous mixtures by using the properties and fractions of the individual components. For conductivity, the Hanai-Bruggeman equation is

$$\frac{\sigma_0 - \sigma_m}{\sigma_w - \sigma_m} \cdot \left(\frac{\sigma_w}{\sigma_0}\right)^{\psi} = \Phi \tag{9-87}$$

where σ_0 is the rock conductivity, σ_w the water conductivity, and σ_m the matrix conductivity (in the Archie equation assuming $\sigma_m = 0$). ψ is the depolarisation factor of the dispersed particles. The original work of Bruggeman and Hanai considered spheres dispersed in a continuous medium, in which case $\psi = 1/3$. Generally, ψ depends on the shape of the particles and can have values between

$\psi = 0$ for infinitely long cylindrical rods (needles) whose axes are parallel to the electrical field and

$\psi = 1$ for infinitely short cylindrical rods (plates) whose axes are parallel to the electrical field (Sen, 1981; Bussian, 1983).

The first case corresponds to the *"parallel"* sheet model (9-68) with

$$\sigma_0 = (1 - \Phi) \cdot \sigma_m + \Phi \cdot \sigma_w \tag{9-88}$$

and the second to the *"perpendicular"* sheet model (9-69) with

$$\frac{1}{\sigma_0} = \frac{1-\Phi}{\sigma_m} + \frac{\Phi}{\sigma_w} \qquad (9\text{-}89)$$

Using equation 9-87 with the Archie assumption of $\sigma_m \equiv 0$, follows

$$\left(\frac{\sigma_0}{\sigma_w}\right)^{\psi} = \Phi \qquad (9\text{-}90)$$

or

$$\sigma_0 = \sigma_w \cdot \Phi^{1/(1-\Psi)} \qquad \rho_0 = \rho_w \cdot \Phi^{-1/(1-\Psi)} \qquad (9\text{-}91)$$

This gives a direct correlation with Archie's cementation exponent m

$$m = \frac{1}{1-\psi} \quad \text{or} \quad \Psi = 1 - \frac{1}{m} \qquad (9\text{-}92)$$

For spheres with $\psi = 1/3$, it follows that $m = 1.5$.
Figure 9.27 shows the formation factor versus porosity relationships for the Maxwell and Hanai-Bruggeman equations compared with the empirical Archie

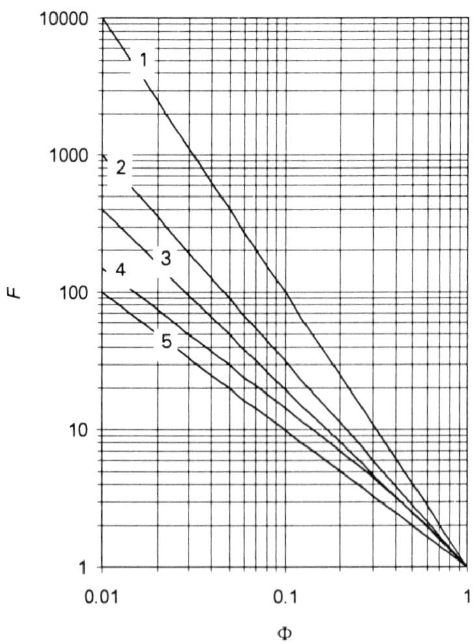

formula. At high porosities both equations tend towards Archie-dependence for low m exponents.

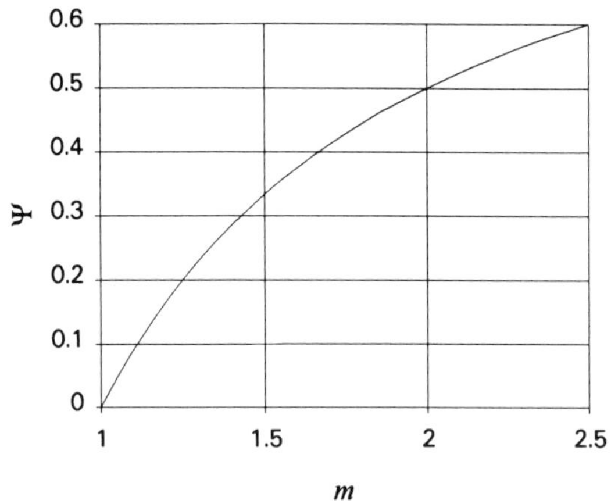

Figure 9.27. Comparison of Archie's formula with Hanai-Bruggeman equation
a (page 432) Formation factor versus porosity calculated after mixing formulas
1 - Archie equation $m = 2.0$ 3 - Archie equation $m = 1.3$ 5 - Archie equation $m = 1.0$
2- Hanai-Bruggeman equation (9-91) with $\psi - 1/3$ 4 - Maxwell equation (9-85)
b (page 433) Relationship between cementation- and depolarization coefficient (equation 9-92)

For clean and shaly hydrocarbon-bearing reservoirs, Spalburg (1988) developed effective medium equations based on these relationships. The derivations also take into consideration the clay conductivity component.

Lima and Sharma (1992) presented models - based on Maxwell equation and Hanai-Bruggeman equation - that incorporates double-layer dielectric enhancement by representing the grains and their associated double layers (see section 9.3.3.4.) by an equivalent sphere that has an effective complex conductivity. The model was successfully used for interpretation of complex impedance spectra in the frequency range 10 Hz to 10 MHz.

d) Models with variable internal structure

A model with internal structure - expressed by a "structure angle α" was introduced in section 6.4.8. In section 8.4.5, the thermal conductivity was derived. For the electrical conductivity in the "microsystem" under Archie' assumptions, the equations result (by the simplifications $\sigma_m = \sigma_c \rightarrow 0$) as follows:

$$\sigma_{\bar{1}} = \sigma_{\bar{2}} = \sigma_w \cdot \left[\frac{\alpha_{pore}}{\alpha_{grain}}\right] \cdot \left[\frac{b}{a}\right] \cdot \left[\frac{b}{a} + \frac{a}{b} - 1\right]^{-1} \quad (9\text{-}93)$$

$$\sigma_{\bar{3}} = 0 \quad (9\text{-}94)$$

The formation factors are

$$F_{\bar{1}} = \left[\frac{\alpha_{grain}}{\alpha_{pore}}\right] \cdot \left[\frac{a}{b}\right] \cdot \left[\frac{b}{a} + \frac{a}{b} - 1\right] \quad (9\text{-}95)$$

$$F_{\bar{3}} = \infty \quad (9\text{-}96)$$

For thc "macrosystem" the formation factor after transformation is

$$F_1 = F_2 = \frac{F_{\bar{1}}}{\cos\alpha} \quad \text{in the two horizontal directions} \quad (9\text{-}97)$$

$$F_3 = \frac{F_{\bar{1}}}{\sin\alpha} \quad \text{in the vertical direction} \quad (9\text{-}98)$$

These results show that the *anisotropy* of the formation factor is controlled by the internal structure alone

$$A_F = \frac{F_{\perp}}{F_{//}} = \frac{F_3}{F_1} = \frac{\cos\alpha}{\sin\alpha} \quad (9\text{-}99)$$

The porosity of this model is (see equation 6-206):

$$\Phi = \left[\frac{b}{a}\right]^3 \cdot \left[2 \cdot \frac{a}{b} - 1\right] \cdot \left[\frac{\alpha_{grain}}{\alpha_{pore}}\right] \quad (9\text{-}100)$$

Using 9-97 (for current direction parallel bedding plane) or 9-98 (for current direction perpendicular bedding plane) and 9-100, a formation factor versus porosity relationship can be calculated.
On a bilogarithmic plot (Figure 9.28) nearly straight lines result for the formation factor vs. porosity relationship.

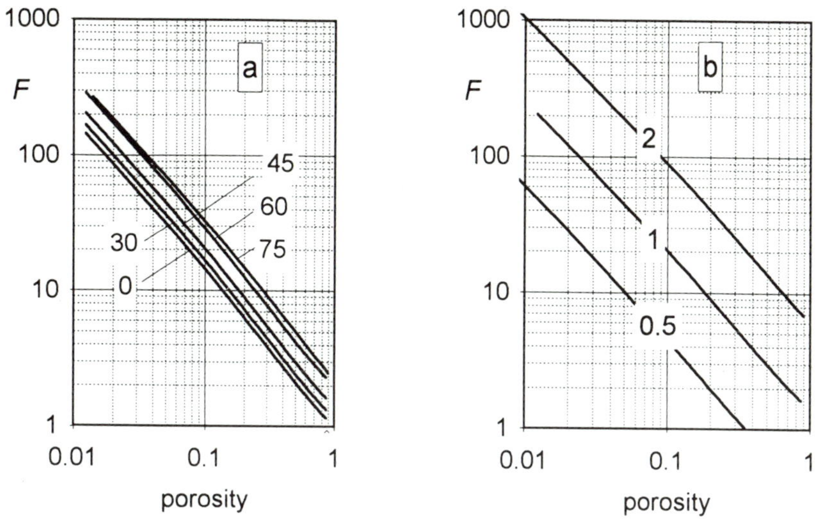

Fig 9.28. Formation factor vs. porosity relationship calculated for the structured rock model
a) F_3 (formation factor for current direction perpendicular bedding plane) vs. porosity for a quotient $\alpha_{grain}/\alpha_{pore} = 1$; curve parameter is the structure angle α
b) F_3 (formation factor for current direction perpendicular bedding plane) vs. porosity for a structure angle $\alpha = 45°$; curve parameter is the quotient $\alpha_{grain}/\alpha_{pore}$

These lines all have the same slope (the same "cementation exponent"), but their position is controlled by

- the internal structure and
- the quotient of their two aspect ratios.

Thus, internal structure and shape of the pore channels are properties which have the strongest secondary influence following porosity.

Figure 9.29 shows a comparison between calculated and experimental results. The curves were calculated for a constant ratio $\alpha_{grain}/\alpha_{pore} = 1$ but with a variable structure angle:
In Figure 9.29 a and b the current direction is *perpendicular* the bedding plane; thus F_3 after equation 9-98 is applied.
In case of the Bunter sandstone (Figure 9.29 c) the measurements were carried out with a current direction *parallel* the bedding, thus F_1 after equation 9-97 was applied. Most experimental data lie in the range between $\alpha = 40°$ and $\alpha = 65°$.

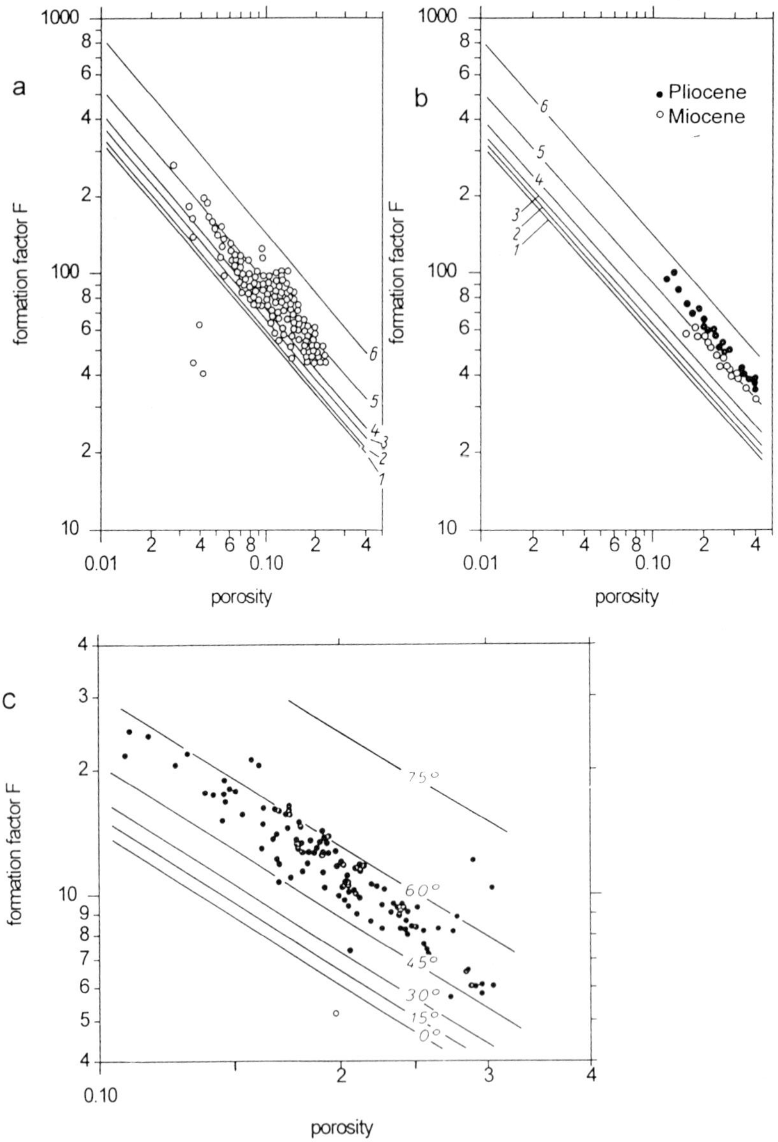

Figure 9.29. Formation factor vs. porosity relationship calculated for the structured rock model and compared with experimental data (sandstone samples).

1 - α = 90 ° 2 - α = 75 ° 3 - α = 60 ° 4 - α = 45 ° 5 - α =30 ° 6 - α =15 °

a) current direction perpendicular bedding. Data after Porter and Carothers (1971)

b) Miocene and Pliocene s. current direction perpendicular bedding. Data after Tunn (1966.)

c) Bunter s. current direction parallel bedding. Data after Barker and Worthington (1973)

9.3.3.4. Rocks with electrolytic and interface conductivity

The previously discussed experimental results and theories for "clean rocks" are based on a two-component model:

component 1 - conducting component: water

component 2 - non-conducting component: solid matrix material, oil, gas.

The presence of conducting minerals and grain- or pore surface conduction produces a *second conducting component* and, therefore, the equations must be modified. This second conductive component was first detected and described for shaly sands by Patnode and Wyllie (1950). They found that Archie's relationships did not apply to shaly sand cores and laboratory made clay slurries, and wrote "... It was apparent, that current was being carried through the core by some medium other than the saturating solution. They called this additional conduction "conductive solids" which include wet clay components in form of disseminated particles or shale streaks" (Oilfield Review, 1)

Clay and grain surface phenomena create an *"double-layer or interface conductivity"*. The electrical double-layer can be described with the Gouy-Chapman model or the Stern model The two components were combined into a *parallel conductor* system

$$\sigma_{\text{shaly rock}} = \sigma_{\text{electrolytic}} + \sigma_{\text{clay}} = \frac{\sigma_{\text{water}}}{F} + \sigma_{\text{clay}} \tag{9-101}$$

The *"electrolytic component"* is given by the Archie equation.

The left side of Figure 9.30 shows the parallel model of the two conductive components. The interface component is nearly independent of the pore water conductivity. However, the electrolytic component is directly proportional to the water (electrolyte) conductivity and, thus, results in the principal dependence of rock conductivity on electrolyte conductivity as shown schematically on the right part of the figure. This offers the possibility of a *separation of the two components* by repeated measurements with different water conductivities or salinities (multiple salinity method).

In the following, various models, equations, and theories are developed to describe the shaly sand properties and the influence of interface effects on the electrical properties. A special "Shaly Sand Reprint Volume" (edited by J.G. Patchett) was published in 1982 by SPWLA. Worthington (1986) published a compilation of relationships for sandstone reservoirs including clay or interface conductivity. Ruffet et al. (1995) give a review of surface conductivity phenomenon and theories.

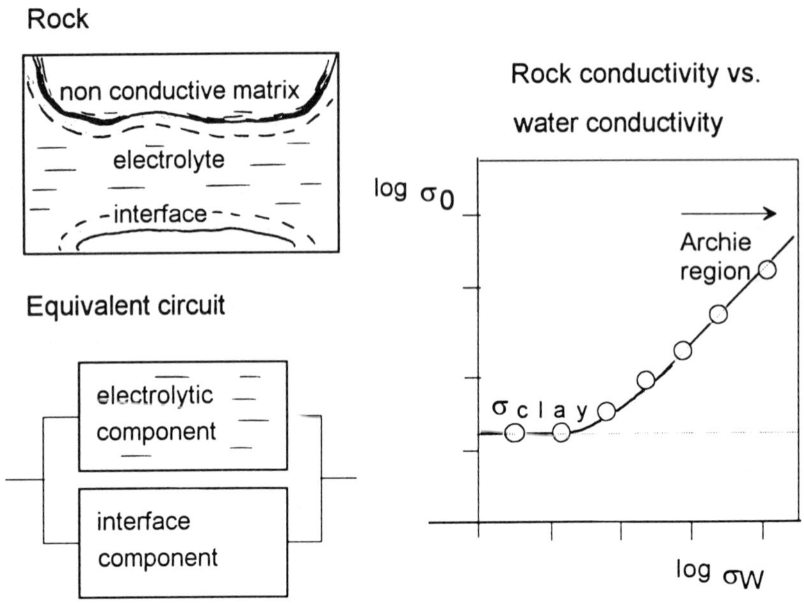

Figure 9.30. Schematic presentation of the two conductive component in porous shaly rocks

Despite the great variety of these models and theories, the fundamental concept of a *parallel conductor system* is used in most cases. The models and theories are mainly directed at 2 subjects:

- consideration of various clay types and clay distributions (for example laminated or dispersed)
- understanding of the physical and chemical phenomena at the grain- or pore surface and interlayer.

Figure 9.31 shows two fundamental types of clay distribution in a porous rock:

a) Laminated shale model

For this model a series of sand-shale laminations is used. The two parts (shale and sand) behave electrically as parallel conductors, then the rock conductivity is

$$\sigma_{lam,t} = (1 - V_{sh}) \cdot \sigma_{sd} + V_{sh} \cdot \sigma_{sh} \qquad (9\text{-}102)$$

where V_{sh} is the shale fractional volume, σ_{sd} and σ_{sh} are the sand and shale conductivity. Archie's equation is applied for the sand conductivity, and it results in

$$\sigma_{lam,t} = \frac{(1 - V_{sh})}{F_{sd}} \cdot \sigma_w + V_{sh} \cdot \sigma_{sh} \qquad (6\text{-}103)$$

where F_{sd} is the formation factor related to the sand layer, its porosity is Φ_{sd}. The sand porosity is related to the total rock porosity Φ and the clay content V_{sh} by

$$\Phi_{sd} = \frac{\Phi}{1 - V_{sd}} \qquad (6\text{-}104)$$

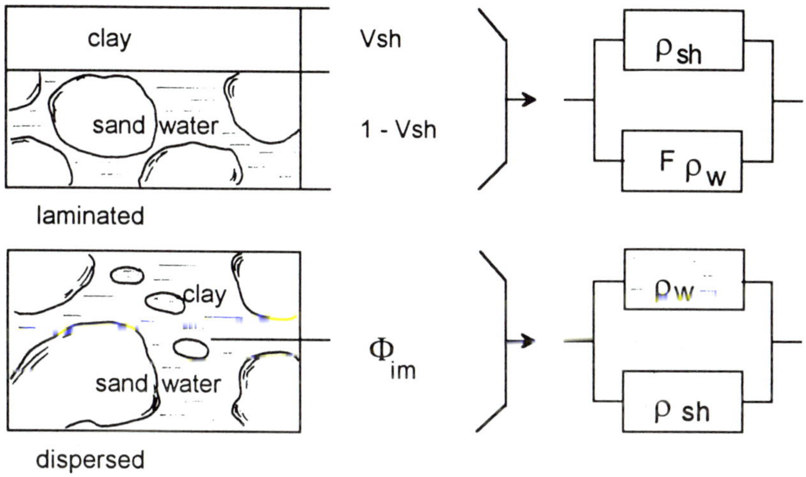

Figure. 9.31. Laminated and dispersed shale model

b) Dispersed shale model

In the dispersed shale model the electric current flows through a mixture ("slurry") of pore water and dispersed clay in the pores. The model is characterised by the following parameters (compare e.g. Schlumberger, 1989):

Φ_{im} - intermatrix porosity, which includes all the space occupied by fluids and dispersed shale,

S_{im} - is the fraction of the intermatrix porosity occupied by the formation-water, dispersed shale mixture,

q - is the fraction of the intermatrix porosity occupied by the dispersed shale,

The intermatrix porosity corresponds to a formation factor of

$$F_{im} = \frac{a}{\Phi_{im}^{m}} \qquad (9\text{-}105)$$

and for the specific resistivity of the water-shale mixture a parallel model (weighted arithmetic mean for conductivity) is applied:

$$\rho_{\text{slurry}} = S_{\text{im}} \cdot \left(\frac{S_{\text{im}} - q}{\rho_{\text{w}}} + \frac{q}{\rho_{\text{sh}}} \right)^{-1} \quad (9\text{-}106)$$

where ρ_{w} is the formation water resistivity and ρ_{sh} the resistivity of the dispersed shale.
Then, the conductivity of the dispersed shale model results in:

$$\sigma_{\text{disp},t} = \rho_{\text{disp},\,t}^{-1} = (\rho_{\text{slurry}} \cdot F_{\text{im}})^{-1} = \frac{\Phi_{\text{im}}^{m}}{a} \cdot S_{\text{im}} \cdot \left(\frac{S_{\text{im}} - q}{\rho_{\text{w}}} + \frac{q}{\rho_{\text{sh}}} \right) \quad (9\text{-}107)$$

The water saturation as a fraction of the true effective porosity is

$$S_{\text{w}} = \frac{S_{\text{im}} - q}{1 - q} \quad (9\text{-}108)$$

Winsauer and McCardell (1953), Guyod and Whichman (1953, 1972), and de Witte (1955) gave a physical explanation of the clay conductivity based on the double layer and counterion mobility.

In 1967, Waxman and Smits published their theory and model for shaly sands. In this model, "the electrical current associated with the clays travels along the same tortuous paths as the current attributed to ions in the pore water". This results in a relationship with an unique formation factor F

$$\sigma_0 = \frac{1}{F} \cdot (\sigma_{\text{w}} + B \cdot Q_v) \quad (9\text{-}109)$$

or in terms of specific resistivities

$$\rho_0 = F \cdot \rho_{\text{w}} \cdot (1 + B \cdot Q_v \cdot \rho_{\text{w}})^{-1} \quad (9\text{-}110)$$

The empirical parameter B (counterion mobility) after Waxman and Thomas (1974) describes the slight interface conductivity dependence on salinity

$$B = 3.83 \cdot [1 - 0.83 \cdot \exp(-0.5/\rho_{\text{w}})] \quad (9\text{-}111)$$

The shalyness parameter Q_{v} is related to the cation exchange capacity-*CEC* by

$$Q_v = \frac{1-\Phi}{\Phi} \cdot \frac{d_s}{100} \cdot CEC \qquad (9\text{-}112)$$

where d_s is the grain solid density (in g/cm^3) and Φ the porosity; the cation exchange capacity *CEC* is given in milliequivalents per 100 g of rock. *CEC* is measured with laboratory methods.

The *cation exchange capacity* is a typical property of clays and relates to the clay mineral structure. The structures are sheetlike as a result of the geometric nature of these fundamental structural units:

- one type consists of octahedral units of oxygen or hydroxyl around a central atom (usually aluminium),
- the other type is a tetrahedral unit consisting of a central silicon atom surrounded by oxygen.

Figure 9.32 shows five types of clay minerals as a stacking of these fundamental units.

The particular *physical properties* of clays are related to two phenomena:

➢ the presence of water trapped between the "plates" with its influence on conductivity and porosity measurements, but not as a part of the effective porosity,

➢ the presence of a substantial negative surface charge with ability to absorb ions (for example radioactive ions) and to create a double layer if the pore fluid is an electrolyte.

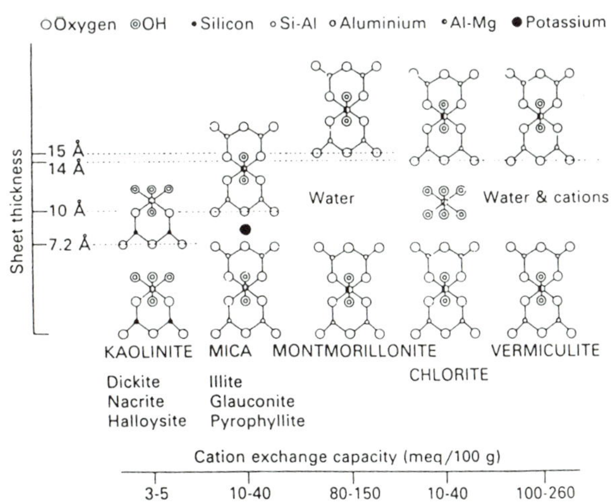

Figure 9.32. Schematic representation of the structure and composition of clay minerals (Ellis, 1987)

This ability of clay minerals to form the *electric double layer* is the basis for the cation exchange phenomenon. During the formation of clay through weathering, positive charges (cations) are adsorbed on the clay surface. These cations (for example Ca, Mg, H, K, Na) are loosely held to the surface and can subsequently be exchanged for other cations or essentially go into solution should the clay be mixed with water. For this reason, they are called exchangeable ions and the *cation exchange capacity* is a measure of the number of cations that are required to neutralise the clay particle as a whole i.e. the weight of ions in milliequivalents adsorbed per 100 grams of clay. Along the bottom of Figure 9.32 typical value ranges for the CEC are given.

Waxman and Smits (1968) extended equation 9-109 to the case of partial water saturation. It is assumed that the exchange ion population remains unchanged as water is replaced by oil. Thus, these ions become more concentrated in the remaining water-filled pore space, which requires that Q_v be replaced by Q_v/S_w (Hearst and Nelson, 1985):

$$\sigma_t = \rho_t^{-1} = \frac{S_w^n}{F}\left(\sigma_w + \frac{BQ_v}{S_w}\right) \qquad (9\text{-}113)$$

Tables 9.10 and 9.11 show some data for surface area and exchange capacity of various materials.

Table 9.10. Surface area and exchange capacity of various materials; References: Olhoeft 1981 - O; Keller and Frischknecht, 1966 - K

Material	Surface area m^2/g	Cation exchange meq/100g	Anion exchange meq/100g	Ref.
silica	300...973	10 ...400		O
zeolithes		100...620		O
organic matter	10...260	130...350		O
vermiculite		100...150	4	O, K
montmorillonite	82...767	29 ...126	23...41	O
		80 ...150		K
halloysite (2 H_2O)	43	5 ...10		O, K
halloysite (4 H_2O)	43	40...50		O, K
chlorite	42	4 ...47		O
		10...40		K
illite	97...113	10...40		O, K
glauconite		11...20	7...20	O
kaolinite	15...23	3 ...15		O, K

Table 9.11. Cation exchange capacity (CEC) and colloid content of five soils of different textures (after Webber, see McNeill, 1980)

Soil type	Organic matter in %	Clay in %	CEC in meq/100 g
sand	1.7	7	6.3
sandy loam	3.2	13.2	13.7
loam	4.9	16.8	20.2
silt loam	5.4	18.4	24.0
clay loam	5.5	31.2	27.2

Patchett, 1975 has found a correlation between CEC and the specific surface area (Figure 9.33a):

$$CEC = \alpha_e \cdot S_m \tag{9-114}$$

where α_e is the equivalent surface charge density (in meq/m^2). Börner (1983) and Schön (1984) found a comparable correlation for unconsolidated sediments (Tertiary and Pleistocene, Germany) as shown in Figure 9.33b. Table 9.12 gives some examples for α_e , including data of Figure 9.33.
The correlation between CEC and the specific surface area brings a link to grain size as a controlling factor of the second conductive component in rocks.

Table 9.12 Surface charge density for some rock types; References: P - Patchett (1975); PW - Pape and Worthington (1983); BS - Börner (1983), Schön (1984)

Rock type	α_e in meq/m^2	Ref.
Unconsolidated sediments: Tertiary, clay, silt	6.9×10^{-3}	BS
Pleistocene, sand, silt	12×10^{-3}	
Clastic sediments (consolidated)	5.32×10^{-3}	PW
Clastic sediments (consolidated), clay minerals	1.67×10^{-3}	P

The additional conductive component was also detected in *porous rocks other than in shaly sands and clay*. For example, Repsold (1976) found a dependence of the formation factor on water conductivity in sand, particularly in fine sand. Studies by Rink and Schopper (1974), Pape and Worthington (1983), Börner and Schön (1987), and Schön (1986) of various consolidated and unconsolidated sediments leads to an explanation of the second conductive component of rocks based on the specific internal surface, pore wall morphology, and surface charge density.

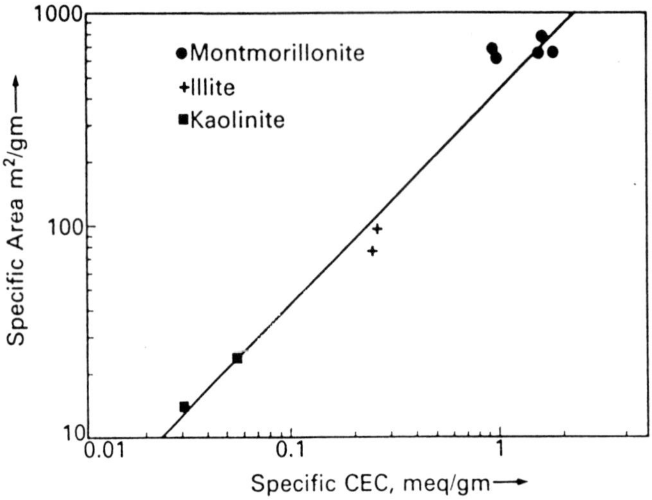

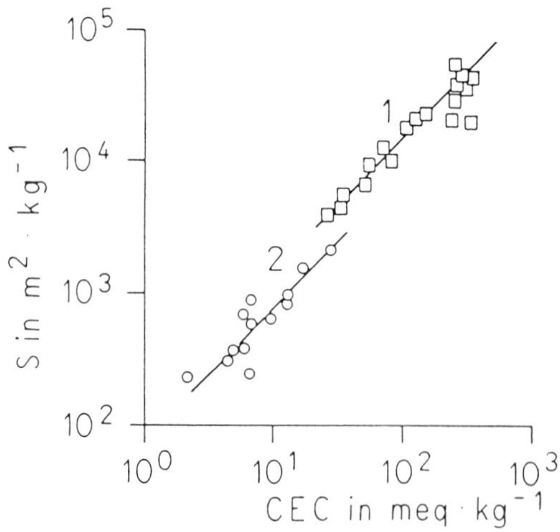

Figure 9.33. Correlation between CEC and specific internal surface
a) Clay (Patchett, 1975; Ellis, 1987)
b) Unconsolidated sediments (Börner, 1983, Schön, 1984); 1 - Tertiary 2 - Pleistocene

For the water saturated case, Rink and Schopper, 1974 published the equation

$$\sigma_0 = \frac{\sigma_w}{F} + \sigma_q = \frac{\sigma_w}{F} + \frac{\alpha\beta}{\lambda^2 \Lambda^2 F} \cdot S_{por} \tag{9-115}$$

where α is the surface charge density (in C/m^2)

β is the charge mobility (counterion mobility) in the double layer (in $V^{-1}s\text{-}1\ m^2$)

λ^2 is the tortuosity of the grain surface, exceeding pore tortuosity

Λ^2 is the pore tortuosity

S_{por} is the specific internal surface related to the pore volume (in m or in µm).

Pape and Worthington (1983) found to a first approximation that α and β are constant: $\alpha = 0.51\ C\ m^{-2}$ $\beta = 2.69 \times 10^{-8}\ V^{-1}s\text{-}1\ m^2$.

Börner (1991, 1992) has investigated the electrical behaviour of porous rocks with correlations to salinity and to pore geometrical properties like specific internal surface in the frequency range between 10^{-3} and 10^4 Hz. Figure 9.34 shows a significant correlation between the real part of the interface conductivity and the specific surface area (the correlation between the imaginary part of the interface conductivity and the specific internal surface area will be discussed later).

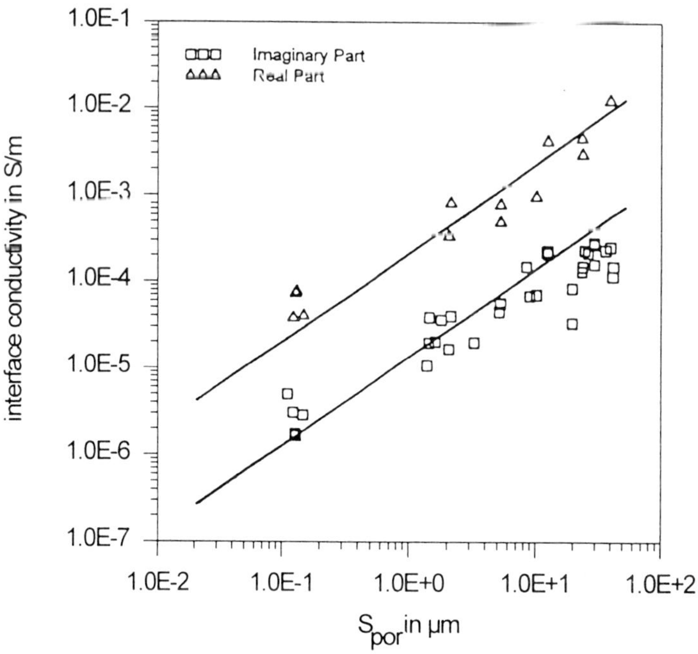

Figure 9.34. Correlation between the real and imaginary component of the interface conductivity and the specific surface area, measured at sandstone samples; frequency range $10^{-3}...10^4$ Hz; Börner (1991, 1992), Börner and Schön (1991).

This correlation between the real interface conductivity component and pore or grain geometrical properties was also found for *unconsolidated sediments*. Figure 9.35 shows results of laboratory investigations of water saturated sand and gravel samples (Pleistocene, Tertiary). These measurements were made with various concentrated brines (i.e. various σ_w). From the sediment conductivity σ_0 versus brine conductivity σ_w graph (Figure 9.35a), the two properties

- formation factor F
- interface or grain-surface conductivity σ_q

were derived using the first expression of equation 9-115:

$$\sigma_0 = \frac{\sigma_w}{F} + \sigma_q \qquad (9\text{-}116)$$

In a formation factor vs. porosity graph (Figure 9.35b) and a surface conductivity vs. grain size plot (Figure 9.35c), good correlation result and confirm that, dominantly

- the volume or electrolytic (Archie) term of conductivity is controlled by a pore *volume effect*
- the interface conductivity term is controlled by a pore (or grain) *surface effect*.

The correlation between interface conductivity σ_q and mean grain diameter D reflects the influence of grain surface area. The specific grain surface area increases with decreasing grain size.

From Figure 9.35, a first approximation for the formation factor vs. porosity relationship results in

$$F = \frac{1}{\Phi^{1.4}} \qquad (9\text{-}117)$$

and for the interface conductivity vs. mean grain diameter relationship

$$\sigma_q = 9 \cdot 10^{-4} \cdot (D)^{-0.4} \qquad (9\text{-}118)$$

where σ_q is in S/m and D in mm. Thus, a rock conductivity relationship incorporating both porosity and grain size effects results:

$$\sigma_0 = \frac{1}{\Phi^{1.4}} \cdot \sigma_w + 9 \cdot 10^{-4} \cdot (D)^{-0.4} \qquad (9\text{-}119)$$

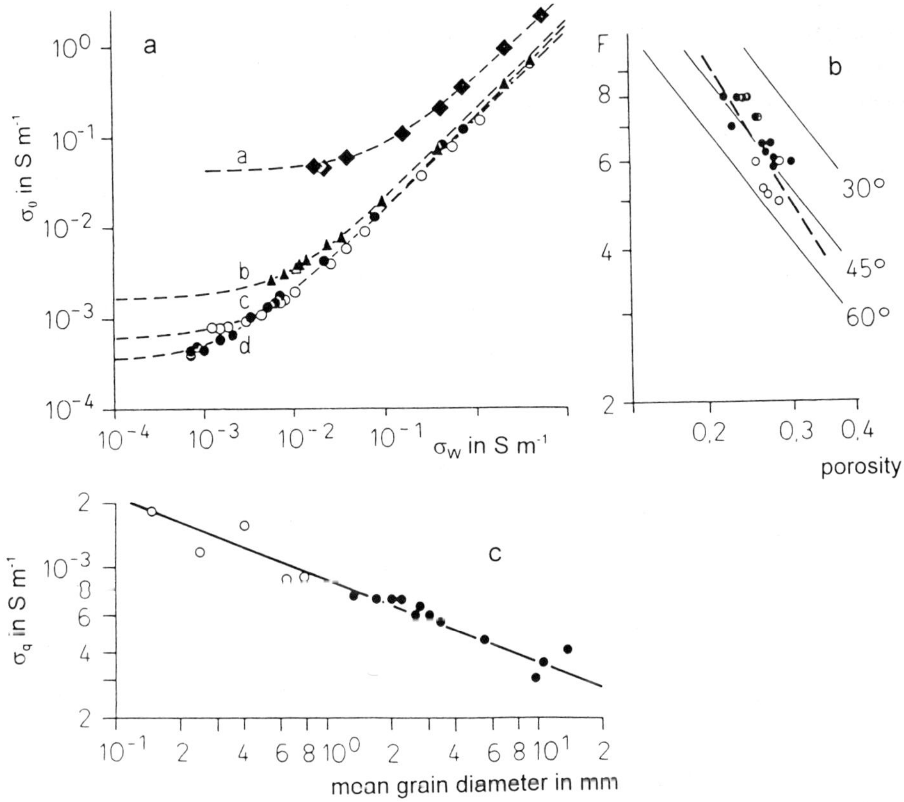

Figure 9.35. Electrolytic and interface conduction components of unconsolidated sediments (Börner, 1983, Börner and Schön, 1984, Schön, 1984)

a) Examples of the dependence of rock conductivity σ_0 upon the brine conductivity σ_w (a - clay, b, c, d - sand, gravel)

b) Formation factor F versus porosity Φ (the three lines are calculated after the structured model with the angle α as parameter)

c) Interface conductivity σ_q versus mean grain diameter D

The study of interface phenomena opens the possibility for determination and characterisation of *pore geometry properties* using geophysical methods. In particular, specific internal surface and grain size (and porosity) are properties related to hydrodynamic permeability (see chapter 2), which is of fundamental importance for all fluid motion processes.

The *selection of the interface conductivity* term from the whole rock conductivity based on equation 9-116 needs measurements made with different brine conductivities σ_w. However, complex conductivity measurements offer a further method for separation of electrical volume and interface effects, such as

- Vinegar and Waxman (1984) and Börner and Schön (1991) for the low frequency range (f < 10 kHz) and following for the Induced Polarisation (IP),
- Kulenkampff and Schopper (1988) for frequencies above 1 kHz.

The frequency dependency of rock conductivity (9-10) can be approximated by a power law for the frequency range $10^{-3}...10^{4}$ Hz (Börner and Schön, 1991; Börner, 1992)

$$\sigma*(\omega) = \sigma*'(\omega) + i \cdot \sigma*''(\omega) = \sigma_n \cdot (i \cdot \omega_n)^{1-p} \qquad (9\text{-}120)$$

where $\sigma*(\omega)$ is the frequency dependent conductivity,
σ'_n is the conductivity amplitude at $\omega = 1\ s^{-1}$; it has a real and an imaginary component; σ'_n is the real component of the conductivity at $\omega = 1\ s^{-1}$; it is the volume (electrolytic) conductivity combined with the real component of the interface conductivity at y $\omega = 1\ s^{-1}$; σ''_n is the imaginary component of the conductivity at $\omega = 1\ s^{-1}$; it is the imaginary component of the interface conductivity at $\omega = 1\ s^{-1}$
$\omega_n = \dfrac{\omega}{\omega = 1 s^{-1}}$; *1-p* is the frequency exponent (Börner, 1991; $p = 0.95 ... 1$).

This power law (9-120) is identical to the frequency independent phase angle and typical for water adsorbing systems (Dissado and Hill, 1984).
The investigations of sandstones in the frequency range between 10^{-3} and 10^{4} Hz confirm the strong correlation between the *imaginary part of complex conductivity and specific surface area* (Figure 9.34). A phenomenological explanation is that, in the considered frequency range, the only source of an internal capacity of the porous rock is the interface region. The double layer mechanism acts as a capacitor.
Thus, a rough rule shows that:

► real conductivity is influenced by the volume (electrolytic) and the interface properties
► imaginary conductivity part is influenced by the interface effects alone.

This suggests that the determination of porosity and specific surface area of reservoir rocks directly from complex electrical measurements can be made.

From 9-120 results

$$\sigma_n' = \sigma_n \cdot \cos[\pi/2 \cdot (1-p)] \qquad \text{real part} \qquad (9\text{-}121)$$

$$\sigma_n'' = \sigma_n \cdot \sin[\pi/2 \cdot (1-p)] \qquad \text{imaginary part} \qquad (9\text{-}122)$$

Thus, σ_n', σ_n'' are related to the frequency exponent 1-p and the phase angle φ according to:

$$\tan\varphi = \frac{\sigma_n''}{\sigma_n'} = \tan[\pi/2 \cdot (1-p)] \qquad (9\text{-}123)$$

The real part can be substituted by equation 9-109 or 9-116. The imaginary part in equation (9-122) is caused solely by the capacitive behaviour of the interface and it is proportional to Q_v (Vinegar and Waxman, 1984) and to S_{por} (Börner and Schön, 1991).

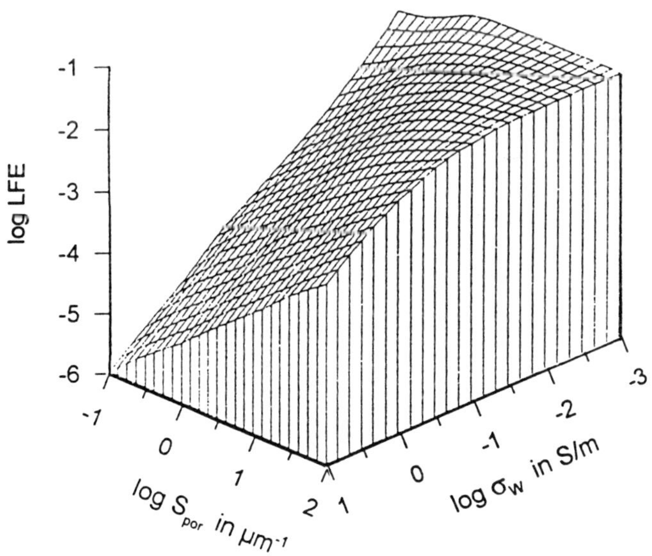

Figure 9.36. Variation of frequency dependence of logarithmic frequency effect LFE = 1 - p with water conductivity σ_w and surface-to-porosity-ratio S_{por}; calculated using equation 9-124 with l = 0.01, Börner and Schön (1995)

Thus, follows a complex conductivity model to explain the phase angle or the frequency exponent:

$$\tan\varphi = \tan[\pi/2(1-p)] = \frac{F^{-1}\cdot f(\sigma_w)\cdot S_{por}\cdot l}{F^{-1}[\sigma_w + f(\sigma_w)\cdot S_{por}]} = \frac{l}{1+\frac{\sigma_w}{f(\sigma_w)\cdot S_{por}}} \qquad (9\text{-}124)$$

where l is the ratio of the capacitive and the ohmic interface contribution and $f(\sigma_w)$ is a general function considering salinity dependence of the interface conductivity (see equation. 9-111). The parameter l is slightly different in various formations ($l \approx 0.01...0.15$) and is nearly independent of salinity. The overlay of the effects of water conductivity and surface-to-porosity-ratio on the frequency dependence of $1 - p$ = LFE is shown in Figure 9.36; LFE is the "logarithmic frequency effect" (Börner, 1992). Thus, complex conductivity measurements or spectral IP measurement (see section 9.3.5) allow the derivation of the parameters in these equations, following Figure 9.37.

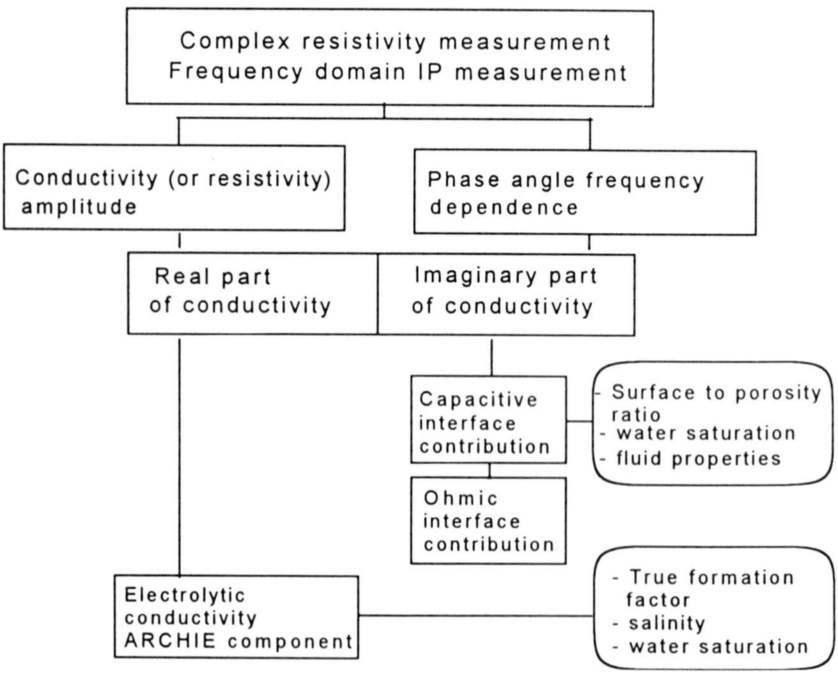

Figure 9.37 Scheme to derive pore and fluid properties from complex electrical data or frequency domain IP data, respectively, using the constant phase angle model (Börner, 1992, personal communication).

Two remarks may be added:

1) The low frequency complex electrical conductivity is influenced by changes of the interface properties as a result of *contamination*. Olhoeft (1984, 1985) based on complex resistivity spectrum measurements between 10^{-3} and 10^{6} Hz found significant indications of different electrochemical reactions, particularly clay-organic reactions.
Börner et al., (1993) studied the influence of various contaminants (hexane, dichlormethane, benzene, methanol, nitrate, and various chlorides) on complex electrical conductivity. The interface is very sensitive to changes of the pore fluid. The experimental results show an influence of the contaminant with respect to the level of conductivity components and to the shape of the frequency dependence. Figure 9.38 shows an example.
The sensitivity of electrical properties particularly in the low frequency range is the physical basis for the detection of contamination with induced polarisation measurements (see also section 9.3.4.).

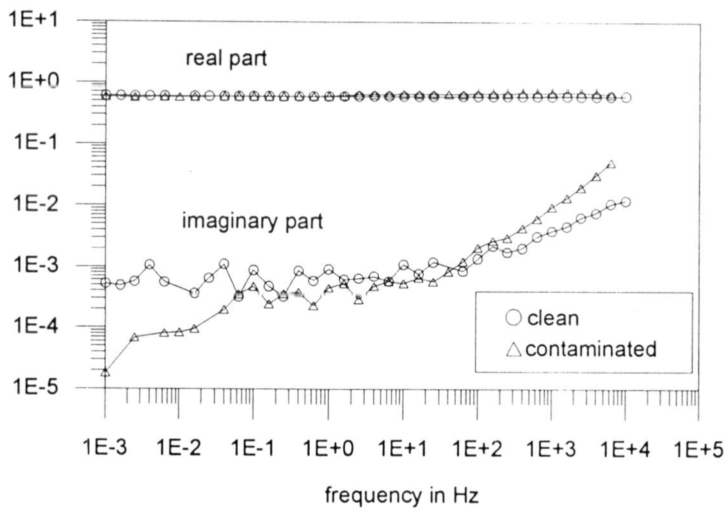

Figure 9.38. Frequency dependence of the real and imaginary part of the complex conductivity for a clay sample, prepared with a NaCl-solution (σ_w = 1 S/m) and organic contaminants.

2) The low frequency power law dependence of conductivity (equation 9-120) changes for *frequencies* above 10 kHz to another power law with an exponent $1 - \alpha$ which is higher than $1 - p$. An experimental study of the broad band (10^{-3} to 10^{6} Hz) complex conductivity of porous rocks by Kulenkampff et

al. (1993) resulted in two distinct ranges with respect to the frequency dependence of complex conductivity σ^* (Figure 9.39).
For both ranges, the frequency dependence can be described by a power law with different exponents:

$$\sigma^* = \sigma_0\left[\left(i\frac{\omega}{\omega_0}\right)^{1-p} + \left(i\frac{\omega}{\omega_0}\right)^{1-\alpha}\right] \qquad (9\text{-}125)$$

The low frequency part is characterised by a small frequency exponent $(1 - p) = 0...0.05$ and the high frequency part by an exponent of $(1 - \alpha) = 0.5...1$ (Börner, 1992, Kulenkampff et al., 1993). ω_0 is an equivalent relaxation frequency, which controls the transition between the two frequency ranges. It seems to be a rock-specific property and is of the order 10^5 to 10^9 Hz . A theoretical explanation can be given based on fractal properties of the interface region:

► The low frequency dependence of the conductivity is controlled by "large structures" of the pore space (tortuosity, constrictivity) and thus, related to permeability.

► The high frequency dependence of the conductivity is controlled by "small structures" of the pore space (pore wall morphology, clay particles).

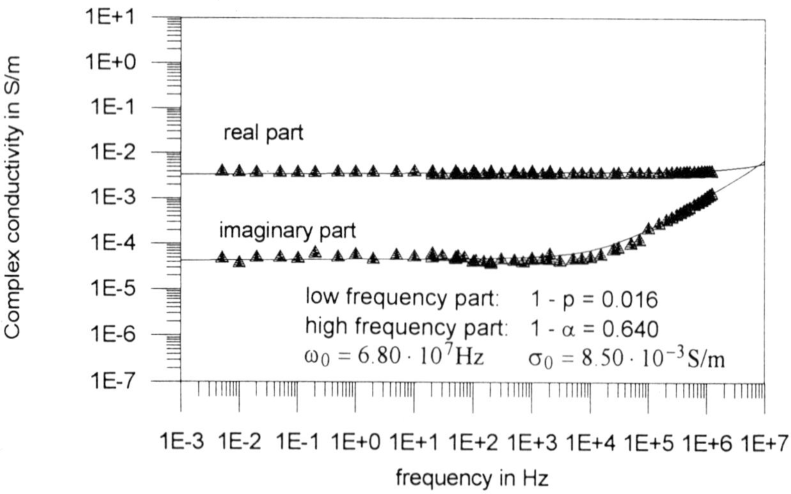

Figure 9.39. Real and imaginary part of the complex conductivity for the frequency range between 10^{-3} and 10^7 Hz; Cretaceous sandstone (Elbe River, Germany); Börner, personal communication (1994)

9.3.4. Induced Polarisation

9.3.4.1. Origin and mechanisms

Induced polarisation is a current-stimulated electrical phenomenon observed as a *delayed voltage response* in earth materials (Ward, 1990). This delay effect can be described and measured in the *time-domain* and in the *frequency-domain*. The connection between the two domains is given by the Fourier-transformation. The fundamentals of induced polarisation (or overvoltage phenomenon) are described by Wait (1959, 1971), Marshall and Madden (1959), Ward and Frazer (1967), and Sumner (1976).

Induced polarisation effects have two physical mechanisms:
a) *electrode polarisation* or metallic polarisation (for example pyrite)
b) *membrane polarisation* or boundary layer polarisation (for example clay).
In both cases the polarisation is attributed to the presence of interfaces

- ➢ between zones of electronic and ionic conduction in case of electrode polarisation) or
- ➢ between zones of unequal ionic transport properties in case of membrane polarisation (Hearst and Nelson, 1985).

Induced polarisation and complex conductivity behaviour are caused by the same capacitive phenomena of the multiphase system "wet rock" with its complicated internal structure of the phase boundary layers.
In the following a brief phenomenological description is given, which mainly follows a description given by Ward (1990).

Electrode polarisation

Electrode polarisation occurs in an electric field if an electronic conducting mineral (for example pyrite) is in contact with ionic conducting pore fluid. The mode of current conduction changes from ionic to electronic (or metallic) and again to ionic. The current flows across the interface and its energy barrier constitutes an electrical impedance. Figure 9.40a shows a simplified model of this process: Cations and anions in the electric field move in opposite directions. But the interface between the electronic conducting mineral and the ionic conducting fluid is a barrier for current flow because

- electrons cannot go into solution, and
- ions cannot creep into the solid mineral lattice.

The charges are exchanged by electrochemical (oxidation, reduction) processes.

Thus, in the electrolyte these contact areas produce clouds of ions with different polarities. Within the mineral, charges with opposite polarity are produced. After switching off the primary field, this accumulation of charges fades away by diffusion.

The current conduction at an interface cannot be expressed by a simple combination of resistors and capacitors; the frequency dependence is more complicated. Customarily the interface properties are described by a particular interface impedance, the Warburg-impedance. Its magnitude varies inversely with the square root of the frequency.

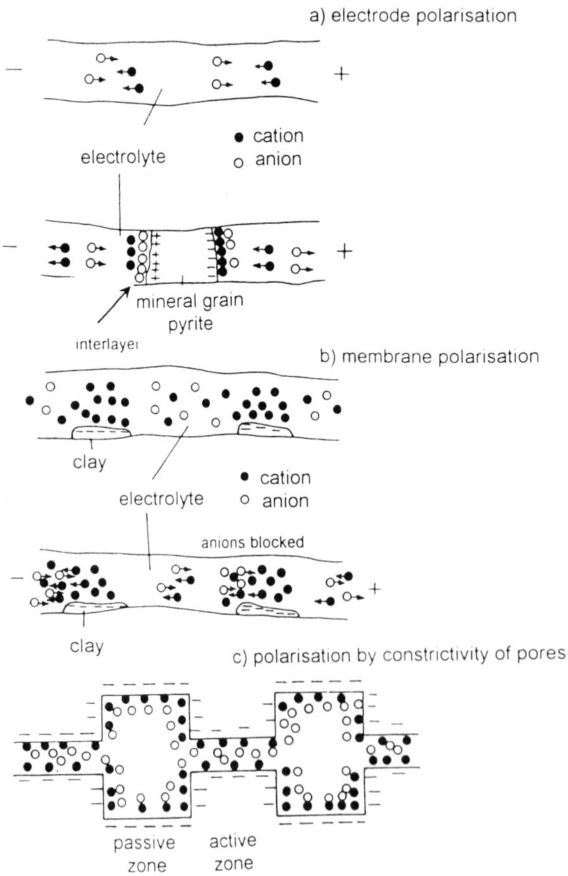

Figure 9.40. Induced polarisation - models and mechanisms:
a) electrode polarisation (metallic polarisation) b) membrane polarisation c) polarisation by constrictivity of pores.

Membrane polarisation

Membrane polarisation is of importance in rocks containing a few percent clays distributed throughout the rock matrix. It arises chiefly in porous rocks in which clay particles partially block ionic solution paths (Figure 9.40b).

The diffuse cloud of cations (double layer) in the vicinity of a clay surface is characteristic of clay-electrolyte systems (Ward, 1990). Under the influence of an electric field, positive charge carriers (cations) easily pass through the cationic cloud but negative charge carriers (anions) accumulate. The clay particle and its vicinity acts as an ion-selective membrane: it is a barrier to anions.

Upon elimination of the electrical field, all charges return to equilibrium positions by a diffusion process. The described process results in an inhomogeneity of charge distribution (surplus of charges at one end of the membrane zone, and a deficiency at the other end). The resulting ion concentration gradient oppose the current flow, thereby, reducing the overall mobility of ions. The ion mobility reduction has a stronger influence for slow potential variations (for example 0.1 Hz) related to the time of diffusion of ions between adjacent membrane zones. For fast potential variations (for example 1 kHz), the mobility reducing effect is much smaller due to the "inertia" of the membrane system with respect to the frequency. Therefore, the conductivity of a membrane system increases with frequency (Ward, 1990).

Two remarks may be added:

1) The surface of silicate minerals often carry negative charges (see for example Pape et al., 1987); it is therefore a phenomenon not only of clay minerals. The effect of the interface on physical properties of porous rocks (electrical and hydraulic) is controlled by the ratio of the interface (or double layer) thickness and the pore radius. For large pore radii, the influence of the double layer is small, and the ion motion is nearly undisturbed. But for small pore radii (or sections with small pore radius), the double layer can touch each other and acting as a barrier to anions. Under the influence of an electric field, a varying concentration of anions and cations result. This behaviour can be described by a model of *passive and active pore sections* (Figure 9.46c), where the notation "active" and "passive" refers to the polarisation origin within the porous rock.

2) Ward (1990) notes that, most of the anomalous charge adsorbed adjacent to an interface is contained within a distance d_{int} from the surface. Following the Debye-Hückel theory (see Falkenhagen, 1991) this distance is

$$d_{\text{int}} = \sqrt{\frac{\varepsilon_{\text{fl}} \cdot \text{k} \cdot T}{2 \cdot n \cdot \nu^2 \cdot \text{e}^2}}. \qquad (9\text{-}126)$$

where n normal ion concentration of electrolyte
ν valence of the normal ions
ε_{fl} dielectric permittivity of the fluid
e elementary charge
k Boltzmannn constant
T temperature

For environmental problems it is important that the *double layer thickness* increases with the relative dielectric permittivity, but also decreases with the normal ion concentration of electrolyte and the valence of the normal ions. In the presence of a contaminant, the fluid can markedly change d_{int}. The greater the thickness of d_{int} the greater the blocking effect (totally or partially) and the probability of detection by induced polarisation or complex conductivity measurements. Practical applications have been published by Olhoeft (1985), Cahyna et al., 1990, Soininen and Vanhala (1992).

In summary, in wet or water saturated rocks induced polarisation results from interface or boundary layer effects occurring in rocks containing clay as well as in rocks without clay. If electronic conducting minerals are also present, their stronger (electrode) polarisation will be superimposed on the boundary layer polarisation. The *separation* of the various origins of induced polarisation effects seems to be the fundamental problem of the data interpretation related to practical applications. For discrimination, two techniques have been utilised (Song and Vozoff, 1985):

- the Complex resistivity method or CR (Zonge, 1973)
- the Spectral IP method or SIP (Pelton et al., 1978).

These methods are based on different time constants related to the various electrochemical processes and the accompanying non-linear behaviour of electrochemical parameters at high current density.

9.3.4.2. Simple parameters for description of IP effects

Figure 9.41 schematically shows the time-domain (or transient) and frequency domain responses of the IP effect.

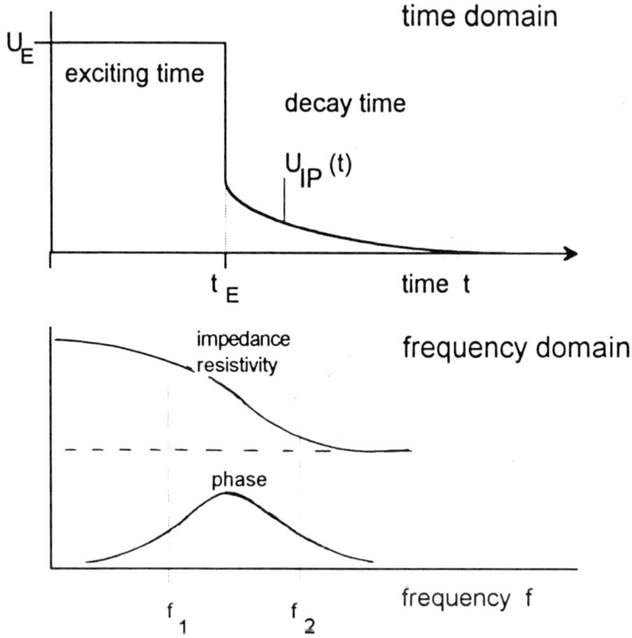

Figure 9.41. IP response in the time domain and frequency domain

From the figure the typical parameters for describing and measuring IP-effects can be derived:

a) in the time domain

In the simplest case there is a primary voltage U_E during an excitation time t_E. After this time, the current is switched off and the decay voltage $U_{IP}(t)$ is measured as the induced polarisation effect.

A measure of the IP response, called *chargeability* is the ratio of the secondary voltage at a time t after switching off to the primary voltage

$$P(t) = \frac{U_{IP}(t)}{U_E} \tag{9-127}$$

Chargeability as defined above is dimensionless; usually the secondary voltage is in mV range and the primary voltage in V. Thus, the chargeability is given in mV/V (sometimes percent of this value is used).

Such a definition of chargeability has an infinite number of possibilities to define the time t. Modern IP methods are directed on a study of the decay curve shape. One simple step is to integrate under a portion of the decay curve between the decay times t_1 and t_2:

$$M_{t_1}^{t_2} = \frac{1}{U_E} \int_{t_1}^{t_2} U_{IP}(t)dt \tag{9-128}$$

In this case, the units are time (mostly given in ms from (mVs/V). A division by the integration-time results in a relationship (see Reitmayr, 1985) for a dimensionless parameter:

$$M_{t_1}^{t_2} = \frac{1}{U_E \cdot (t_2 - t_1)} \int_{t_1}^{t_2} U_{IP}(t)dt \tag{9-129}$$

The *shape* of the decay curve U_{IP} or the time dependence of the chargeability *M(t)* is described as an exponential function of the form

$$M(t) = a \cdot \exp\left(-\frac{t}{\tau}\right) \tag{9-130}$$

where τ is the time constant of the relaxation mechanism.
Frequently, experimental data are fitted by a summation of exponential terms (for example Bertin and Loeb, 1976)

$$M(t) = \sum_i a_i \cdot \exp\left(-\frac{t}{\tau_i}\right) \tag{9-131}$$

The coefficients a_i and τ_i can be determined by least square analysis.

b) in the frequency-domain

The induced polarisation generally results in a decrease of resistivity or increase of conductivity with increasing frequency.
One measure of the IP effect is the change in resistivity measured at two frequencies f_1 and f_2 ($f_1 < f_2$) of a harmonic alternating current. This parameter is called frequency effect *FE*:

$$FE_{f_1}^{f_2} = \frac{\rho(f_1) - \rho(f_2)}{\rho(f_1)} \tag{9-132}$$

The frequency effect is dimensionless, and frequently is given as "percent frequency effect" *PFE*

$$PFE = FE \cdot 100\% \tag{9-133}$$

For the low frequency f_1, Bertin and Loeb (1976) give a range between 0.05 and 0.5 Hz, and for the high frequency f_2, a range between 1 and 10 Hz. Snyder et al. (1977) recommend a decade frequency difference ($f_2 = 10 \cdot f_1$). The *"metal factor"* is the ratio of the *PFE* to the specific rock resistivity.

Modern techniques allow measurements over a broad frequency spectrum. The *spectral* IP is the frequency-domain equivalent of the decay-curve analyses in the time-domain. The frequency effect is related to the phase angle and the frequency exponent (1-*p)* of the low frequency power law dependence following equation 9 - 121.

9.3.4.3. Some experimental results - examples

In 1959, Collett published a classic paper with fundamental laboratory data. measurements in both domains.

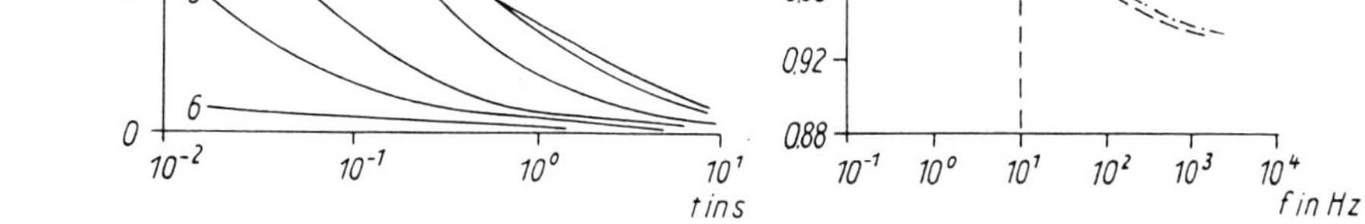

Figure 9.42. IP effects of some metallic and non-metallic minerals, after Collett, 1959

1 - graphite	2 - pyrite	3 - chalkopyrite
4 - galena	5 - magnetite	6 - malachite

a) Decay curves in the time-domain (duration of the exciting primary current 21 sec)

b) Frequency characteristic (frequency-domain); the conductivity σ_f at the frequency f is normalised by its value σ_{10} at the frequency 10 Hz

Figure 9.42a shows the decay curves and Figure 9.42b the frequency characteristic for some metallic and nonmetallic minerals. The matrix material was an andesite with grain diameter between 0.84 and 2.0 mm. The pore fluid is a 5 percent 0.01N NaCl solution. The metallic and non-metallic mineral content is 3 percent (by solid volume) with the same grain diameter as the matrix material. Note that the same sequence of minerals were used for
Figure 9.43 shows the influence of *grain size* and metallic component concentration upon the chargeability.

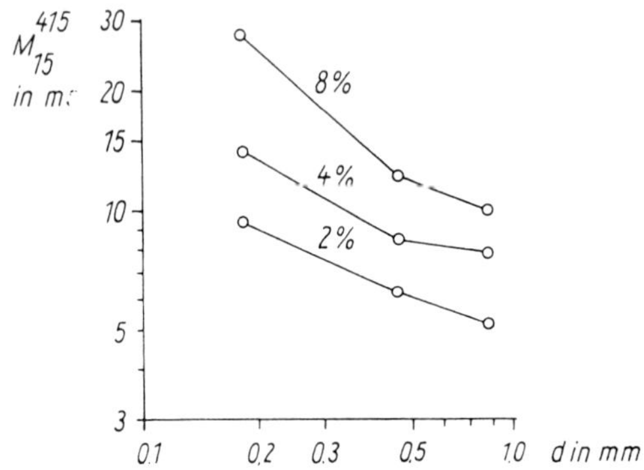

Figure 9.43. Chargeability as a function of grain size *d* and concentration of the sulfidic ore particles (in percent as curve parameter); t_1 = 15 ms t_2 = 415 ms; duration of the exciting primary current 5 s; electrolyte: 0.3 mN NaCl-solution, after Bertin and Loeb (1976)

In rocks without electronic conducting minerals, the IP effects are mainly influenced by the content and distribution of clay, the pore geometric properties, and the properties of the electrolyte. The chargeability of sedimentary rocks was studied in early papers by Vaquier et al. (1957), Marshall and Madden (1959).
These and subsequent investigations are directed mostly at correlations between IP parameters and *reservoir properties* (clay content, hydrodynamic permeability) and, now to contamination problems. The complexity of the various influencing properties results in general difficulty in determining the individual causes of IP effects. The following examples portray this situation and some of its tendencies: Figure 9.44 shows the chargeability of four classes of unconsolidated sediments.
From sand to sandy loam, the chargeability increases. This may be caused by the increase of clay within the sandy sediment skeleton and, thus, produces an increase of membrane barriers. A further increase of clay content obviously results in a decrease. In this case, there are a decreasing number of "membrane

polarisation creating combinations" of a free pore channel section and a clay particle region; it dominates a clay behaviour. Maximum membrane effects are observed at clay contents between 3 and 8 % (Ogilvy and Kuzmina, 1972). The optimal clay content depends on the clay mineral type; it is relatively low for montmorillonite and higher for illite and kaolinite (Telford et al., 1976).

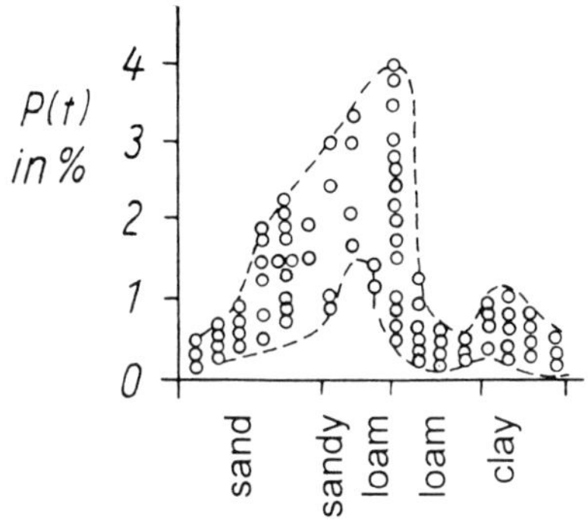

Figure 9.44. Chargeability for unconsolidated sediments, Ogilvy and Kuzmina (1972)

Figure 9.45 shows the same tendencies in the IP-behaviour of sediments. A some more clear classification gives a IP vs. resistivity plot. In this figure the apparent values from field measurements were used.

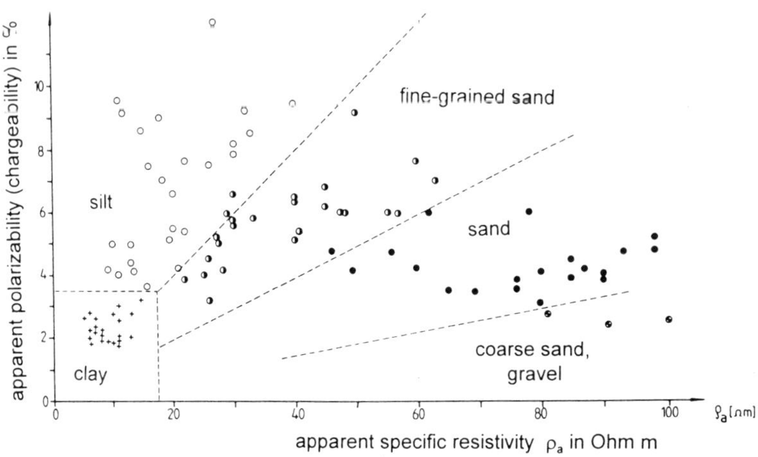

Figure 9.45 Apparent resistivity versus apparent chargeability for unconsolidated sediments (ELGI, Budapest)

The decrease of chargeability with increasing permeability (Figure 9.46) could be explained by a clay effect below the optimal clay content (decreasing clay content results in an increase of permeability and a decrease of chargeability) and by a pore radius effect (increasing pore radius results in an increase of permeability and a decrease of chargeability).

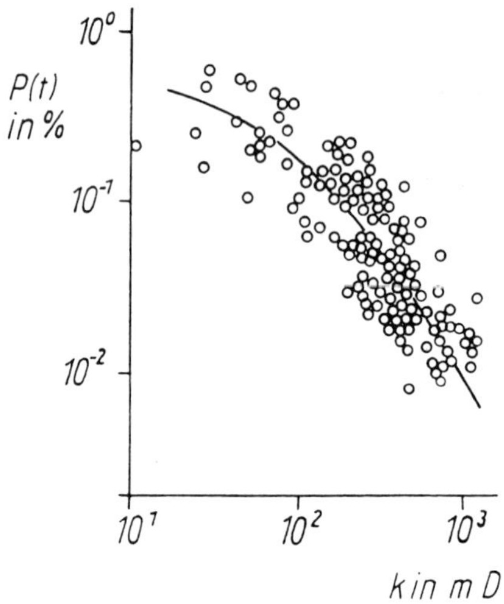

Figure 9.46. Chargeability of sedimentary rocks versus permeability, after Dachnov (1975)

Iliceto et al. (1982) have investigated similar material as Kuzmina and Ogilvy

- sand with grain size > 74 μm
- silt and loam with an average grain size between 74 μm and 2..3 μm
- clay with particle size < 2...3 μm.

The decay curves were fitted using equation 9-131, where the parameters a_i and τ_i are determined by least-square technique. The analyses of the results show

- for the dependence of the chargeability on water content and lithotype, a broad scatter and an overlap of values on chargeability versus water content plot,
- a clearer differentiation between lithotypes utilising the ratio of two parameters from the decay curve approximation a_2/a_3 (Figure 9.47), where a_2 is the coefficient from the time range 0.5...0.8 and a_3 is the coefficient from the time range 2..3 s (values for an excitation time 4 s).

Japsen, 1980 (see Schön, 1983) found comparable results using three terms in the equation 9-131.

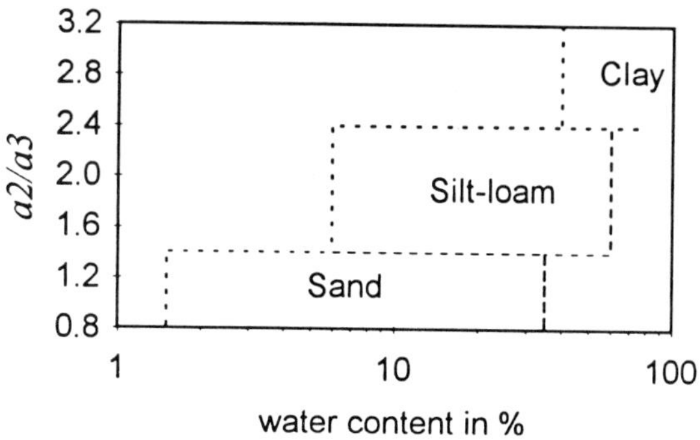

Figure 9.47. Ratio of the two coefficients a_2 and a_3 (derived from a decay curve analysis) versus water content for three lithotypes, after Iliceto et al. (1982)

A *power function* is also used to fit experimental data (see for example Vogelsang et al. 1992):

$$M(t) = c \cdot t^{-\varsigma} \tag{9-134}$$

where c and ς are empirically. Vogelsang et al. (1992) note that this function may be chosen, because most of the relaxation processes in solid bodies can be described by power laws (Jonscher, 1981).

Figure 9.48 shows an example of a well log decay curve (stationary measurement) and the result of fitting with the two relationships 9-131 and 9-134; both functions are considered to be equivalent.

curve A: $$M(t) = 120.8 \cdot \exp\left(-\frac{t}{0.098}\right) + 85.8 \cdot \exp\left(-\frac{t}{1.244}\right) \tag{9-135}$$

curve B: $$M(t) = 40.1 \cdot t^{-0.462} \tag{9-136}$$

Based on a study of *fractal properties* of porous rocks (Papé et al. 1987), the behaviour of the decay curve was analysed by Vogelsang et al. (1992). The decay curve, calculated by fractals, supplies the following information from exponential analyses (equation 9-131):

- the smallest relaxation term (beginning of the decay curve) is controlled by the double layer thickness (in the case of boundary layer polarisation) and by the mineral grain size (in the case of electrode polarisation);

- the greatest relaxation time (end of the decay curve) is controlled by the thickness of the whole layer of diffusion;
- the gradient of the transient decay curve is a function of the fractal dimension of the polarised surface; this indicates the influence of pore wall texture and constrictivity of the pores.

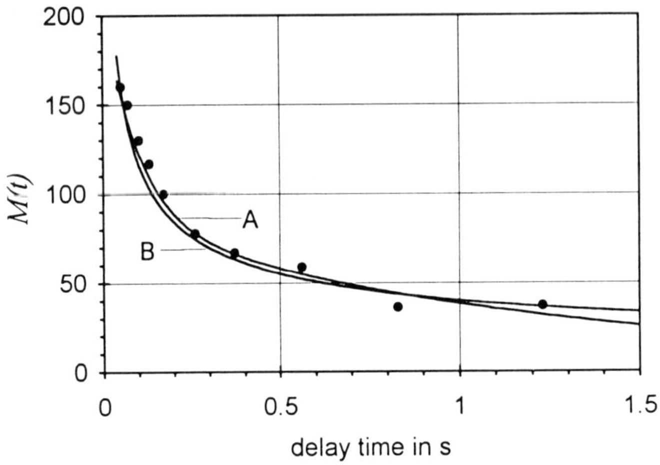

Figure 9.48. Decay curve recorded in the borehole KTB-Oberpfalz VB in the depth 580 m, after Vogelsang et al. (1992); exciting time 8 s.

In summary, induced polarisation is influenced by the following parameters:

- type and volume fraction of electronic conducting minerals,
- size and shape of grains,
- size and shape of pores,
- type, distribution, and volume fraction of clay minerals,
- type and salinity of the electrolyte in the pores,
- grade of water saturation,
- temperature,
- porosity and permeability.

These parameters influence the *amount* of chargeability and, also, the *shape* of the transient decay curve. The two effects are classified by Sumner (1976) as effects of first and second order.

9.3.5. Permittivity of rocks

9.3.5.1. Experimental results

Permittivity of rocks and soils is of interest particularly for interpretation of georadar and borehole electromagnetic propagation measurements (a review about borehole dielectric logging published Wright and Nelson, 1993). Table 9.13 shows some characteristic values and dependencies of physical properties for georadar:

- ➢ permittivity or (relative) dielectric number
- ➢ propagation velocity of the electromagnetic wave
- ➢ attenuation of the electromagnetic wave.

Table 9.13 Characteristic values of physical properties of media important for georadar (van Overmeeren, 1994)

Medium	Dielectric number ε_r	propagation velocity in cm/ns	attenuation in dB/m
air	1	30	0
dry sand	4	15	0.01
water saturated sand	25	6	0.03...0.3
clay	5...40	4.7...13	1...300
peat	60...80	3.4...3.9	0.3
water (fresh)	80	3.4	0.1
water (saline)	80	3.4	1000

The following discussion is related to the behaviour of the real part of the relative dielectric permittivity. The dielectric permittivity of *water bearing (porous or fractured) rocks* is controlled dominantly by their water content. The extreme difference between the permittivity range for the rock forming minerals ($\varepsilon_r \approx 3...7$) and the permittivity of water ($\varepsilon_r = 81$) results in a strong dependence of the rock permittivity on water content: as water content increases the permittivity also increases.

This behaviour is similar in tendency to the dependence of the rock conductivity. But unlike conductivity with its strong dependence on the salinity of the pore water, permittivity is relatively unaffected by the salinity (compare section 9.2.2).

With decreasing *frequency* the imaginary part of the permittivity increases depending on the conductivity σ (see equation 9-16, 9-17). Therefore, the *effective permittivity* of water bearing rocks shows an increase with decreasing frequency.

The following figures illustrate this behaviour. For the general tendencies in case of reservoir rocks mean values (compare also Tables 9.1 and 9.2) may give a first orientation (after Knight and Nur, 1987):

Quartz $\varepsilon_r' = 4.5...4.7$	Calcite $\varepsilon_r' = 7...8$	Shale $\varepsilon_r' = 13...15$
Gas $\varepsilon_r' = 1$	Oil $\varepsilon_r' = 2.2$	Water $\varepsilon_r' = 80$

Figure 9.49a shows the dependence of the real part of the effective permittivity of fresh water saturated sandstone samples of various porosity as a function of frequency. With increasing *frequency*, permittivities decrease and tend towards asymptotic values for high frequencies at about 0.5 GHz. This value for high frequencies increases with increasing *porosity* as demonstrated by the five samples. Figure 9.49b shows this permittivity value for all samples with a clear correlation between ε_r' and porosity Φ.

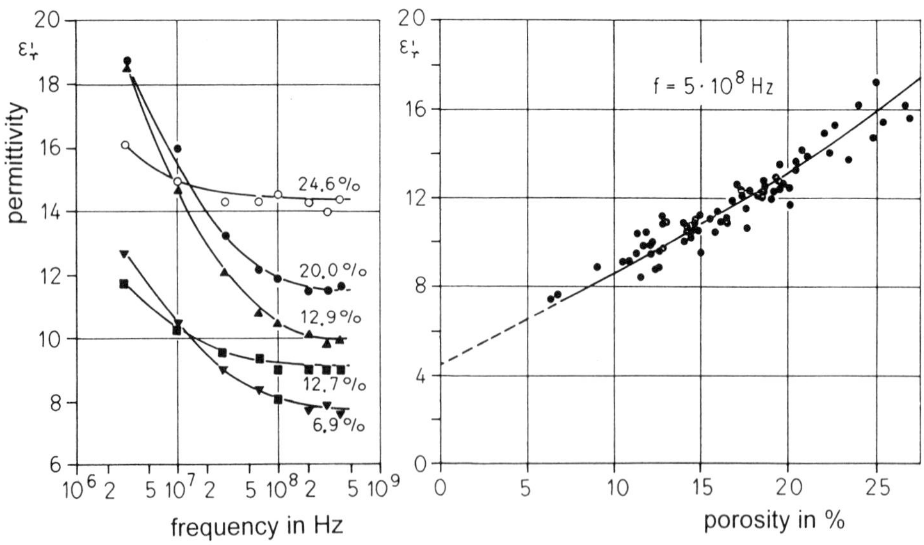

Figure 9.49. Permittivity as a function of frequency and porosity, measured on sandstones with fresh water saturation (Pooley et al., 1978)
a) Real part of effective dielectric permittivity as function of frequency for samples of different porosity
b) ε_r' as function of porosity at the frequency 0.5 GHz

Permittivity increases with increasing *water content* or water saturation S_w. Knight and Nur (1987) published results of systematic laboratory studies of a tight gas sandstone behaviour. Figure 9.50 shows an experimental example with a typical "*hysteresis*" for the imbition-drying cycle. For the interpretation the authors may be cited: "In the region of $S_w < 0.03$, there is little change in ε_r' with S_w; we interpret this region as corresponding to the presence of a monolayer of water on the surface of the pore space. In the region of $0.03 < S_w < 0.12$, there is a rapid increase in ε_r' with increasing S_w; we interpret this region as the wetting of the pore surfaces by two or three monolayers of water, creating both water-grain and water-gas capacitors. In the region of $S_w > 0.12$ dielectric hysteresis occurs; we interpret this region as corresponding to the filling and emptying of the central volume of the pore space with water, the hysteresis being a result of the changing geometry of the liquid and gas phases."

Endres and Knight (1992) investigated this effect of *microscopic fluid distribution* on the dielectric properties of partially saturated rocks also. The results (experimental and model calculations) show that not only do the volume fraction and properties of the individual components influence permittivity, but also the internal geometry, particularly the pore geometry and pore fluid configuration within the pore. The consequence is a difference of permittivity for imbition and drainage conditions.

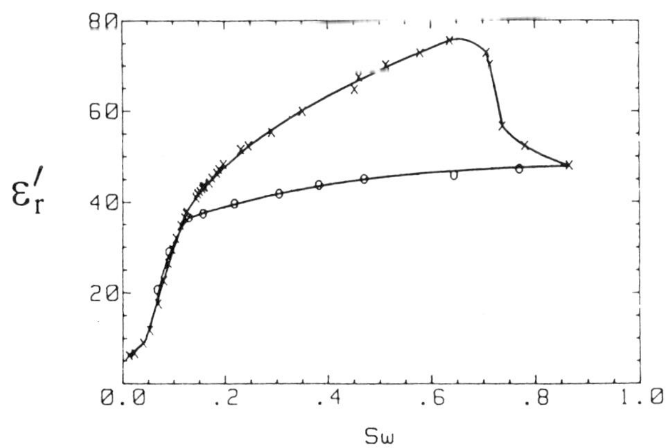

Figure 9.50 ε_r' versus S_w measured during increasing S_w through adsorption and imbition (x data points), and during decreasing S_w through drying (○ data points); frequency 30 kHz; tight gas sandstone (Spirit River Formation, Alberta Basin), porosity $\Phi = 0.07$, surface area $S_{por} = 1.04 \times 10^6\ \text{cm}^{-1}$; gas permeability 7.18 μDarcy (after Knight and Nur, 1987)

An example of *unconsolidated sand* is depicted in Figure 9.51.

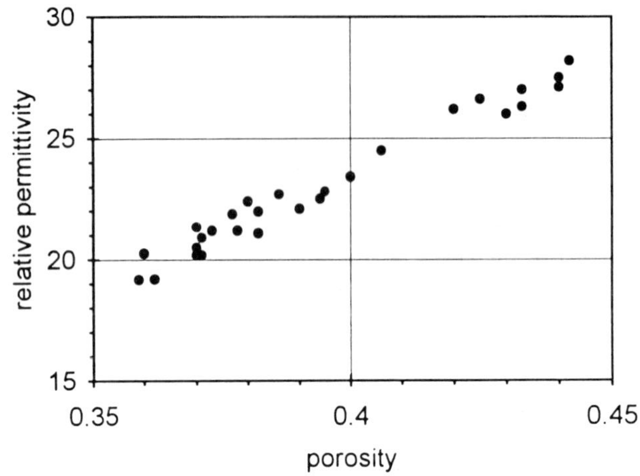

Figure 9.51. Dielectric permittivity versus porosity for water saturated quartz sand (grain diameter 0.25...0.5 mm) after Cerniak (1964)

The presence of *clay* in the rock skeleton results in a tendency of increasing permittivity. This effect is due to bound water at the clay particle surface. In dry or hydrocarbon bearing rocks (Figure 9.52) this effect is more

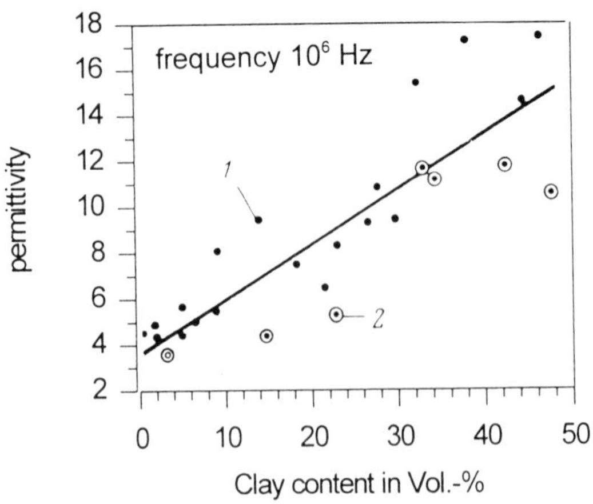

Figure 9.52. Dependence of dielectric permittivity ε_r of gas saturated sandstones (Mesozoic sandstone from West Siberia) on the shale content V_{sh}, after Brylkin et. al., (see Kobranova, 1989)
1 - polymictic sandstone and aleurotite 2 - rocks with increased hydromica cement content

pronounced than in water saturated rocks; the very strong influence of water upon permittivity masks the clay effect.

The *combined effect of water content and clay content* upon the relative dielectric permittivity and the loss tangent (and specific electrical resistivity) is demonstrated in Table 9.14 and the graphic presentation in Figure 9.53.

Table 9.14. Relative dielectric permittivity and loss tangent of unconsolidated sediment mixtures of various composition (Olhoeft, 1986); Frequency 100 MHz; specific electrical resistivity (in Ohm m) is DC resistivity;
Sand is pure silica sand, clay is the Clay Mineral Society CMS-SWy-1 Na-montmorillonite (Wyoming), water is a distilled, de-ionized water

Sand	Clay weight - %	Water	Relative permittivity	Loss tangent	Specific resistivity
83.6	0	16.4	12.1	0.088	269
80.7	0	19.3	14.0	0.14	125
79.5	4.9	15.6	16.4	0.91	15.7
76.7	1.9	21.4	18 1	0.40	33.5
77.5	4.0	18.5	20.2	0.84	13.4
58.2	10.1	31.7	32.3	1.09	6.6
29.3	13.9	56.8	49.7	1.00	4.7
17.9	16.5	66.4	52.8	1.00	4.4
0	22.9	77.1	62.6	1.14	3.4

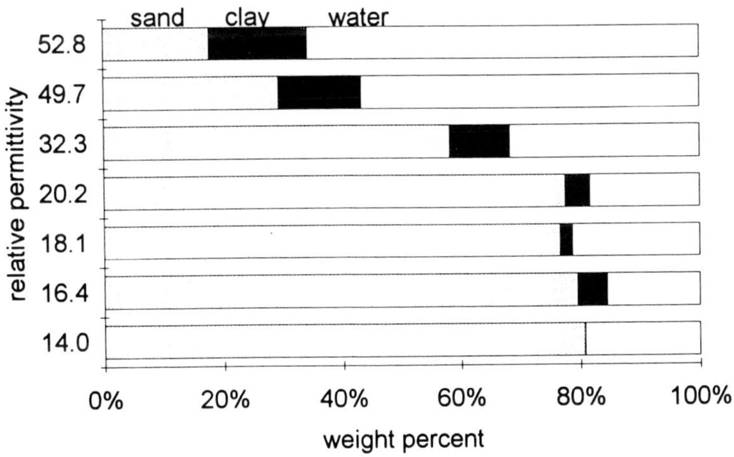

Figure 9.53. Relative dielectric permittivity for the sediment compositions of Table 9.14.

A study of the dielectric properties of *soils* was published by Wang and Schmugge (1980). They found a general increase of permittivity with increasing water content. At low water content, the observed increase was relatively small up to a transition point; beyond this transition point permittivity increases rapidly with moisture content. The transition point is at higher water contents for clayey soils (volumetric water content about 0.26...0.33) than for sandy soils (volumetric water content about 0.17...0.28).

Garrouch and Sharma (1994) investigated the combined influence of clay content, porosity, and frequency on the dielectric properties of shaly sands in the frequency range between 10 Hz and 10 MHz. The samples are Berea sandstone and Ottawa sand-bentonite packs. They found an increase of dielectric number with

- increasing porosity of the brine-saturated rock,
- increasing cation exchange capacity CEC, describing the clay effect and ionic double-layer polarisation phenomena at the solid-fluid interface.

A regression analysis suggests an empirical correlation with a proportionality

$$\varepsilon_r \sim \exp(0.15 \cdot CEC + 0.23 \cdot \Phi + 11.1) \tag{9-137}$$

where the porosity is as a fraction and CEC in meq/100 g.
The frequency dependence was described using the Lima-Sharma model (Lima and Sharma, 1992).

The influence of *contaminants* in the pore space upon permittivity and resistivity of sand and clay is illustrated with the data in Table 9.15.

Table 9.15 Influence of contaminants on permittivity (100 MHz) and resistivity (DC) of sand and clay; Olhoeft, 1987

composition (weight %)					properties	
sand	clay	water	methanol	benzene	ρ Ohm m	ε_r
79	-	21	-	-	132	19.1
82	-	-	18	-	447	9.8
81	-	-	-	19	> 5000	3.2
100	-	-	-	-	2500	2.4
-	28	72	-	-	3.5	53.4
-	40	-	60	-	26.6	21.5
-	36	-	-	64	1340	2.3
-	100	-	-	-	> 5000	1.7

Impressive are the results of the "Borden experiment" (Greenhouse et al., 1993); Table 9.16 presents some physical properties. The contaminant of the water bearing sand with a porosity of 40 percent was perchlorethylene (PCE); it is a so-called DNAPL (dense nonaqueous phase liquids).

Table 9.16 Soil properties as a function of perchlorethylene (PCE) saturation (Borden experiment); after Greenhouse et al.. (1993)

PCE saturation % pore volume	Density g/cm³	Electrical conductivity mS/m	Relative dielectric permittivity	Electromagnetic wave reflection coefficient
0	2.03	14.5	29.7	0
15	2.06	10.5	24.3	0.05
30	2.10	7.1	19.4	0.11
45	2.14	4.4	14.9	0.17
60	2.18	2.3	11.0	0.24
75	2.22	0.9	7.6	0.33
100	2.28	0	3.4	0.49
Components				
PCE	1.62	0	2.3	
water	1.00	50.0	83.6	
sand	1.71	0	4.5	

Figure 9.54 shows an example for the *temperature dependence* of the relative permittivity.

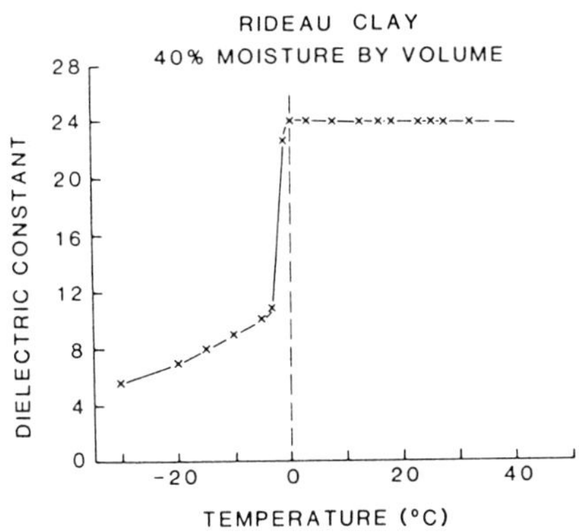

Figure 5.54 Dependence of relative permittivity on temperature for clay; Annan and Davis, 1978; see also Scott et al. (1990)

Ice has a low permittivity (about 3...5). This results in a sharp decrease soil (or wet rock) permittivity as temperature decreases through 0°C. The drop must not be exactly at 0°C, because a large quantity of unfrozen water can occur at temperatures as low as -5 °C (Scott et al., 1990).

9.3.5.2. Models and mixing formulas

In order to predict and interpret dielectric permittivity values of rocks (or other composite media), various models and theories are derived. They are related either

- by the "pure" dielectric behaviour (non conductive material)
- or by the behaviour of materials with dielectric and conductive properties in complex models.

The following discussion is related to the first type and gives a short overview of some theories and equations. Models, derived theories and mixing formulas are reviewed by Wang and Schmugge (1980) and Olhoeft (1985).

The earliest mathematical derivation of a mixing formula is credited to Mossotti in 1846; it is usually called the "Clausius-Mossotti approximation" (Olhoeft, 1985):

$$\frac{\bar{\varepsilon}-\varepsilon_1}{\bar{\varepsilon}+2\varepsilon_1} = V_2 \cdot \frac{\varepsilon_2-\varepsilon_1}{\varepsilon_2+2\varepsilon_1} \qquad (9\text{-}138)$$

where $\bar{\varepsilon}$ is the (effective) dielectric permittivity of the mixture,

ε_1 and ε_2 are the dielectric permittivities of the two components,

V_2 is the volume fraction of component 2.

Also from the early history of material sciences is the Rayleigh formula (Rayleigh, 1892; see also Wang and Schmugge, 1980):

$$\frac{\bar{\varepsilon}-1}{\bar{\varepsilon}+2} = V_1 \cdot \frac{\varepsilon_1-1}{\varepsilon_1+2} + V_2 \cdot \frac{\varepsilon_2-1}{\varepsilon_2+2} + \ldots \qquad (9\text{-}139)$$

The Maxwell-formula for the permittivity of a material consisting of suspended spherical (component *2*) particles in a host substance *1* is (compare equation 8-56) is:

$$\bar{\varepsilon} = \varepsilon_2 \cdot \frac{2\varepsilon_2+\varepsilon_1-2V_1(\varepsilon_2-\varepsilon_1)}{2\varepsilon_2+\varepsilon_1+V_1(\varepsilon_2-\varepsilon_1)} \qquad (9\text{-}140)$$

In particular, suspended grains (ε_m) in water (ε_w) result in

$$\bar{\varepsilon} = \varepsilon_m \cdot \frac{(2\varepsilon_m + \varepsilon_w) - 2\Phi \cdot (\varepsilon_m - \varepsilon_w)}{(2\varepsilon_m + \varepsilon_w) + \Phi \cdot (\varepsilon_m - \varepsilon_w)} \tag{9-141}$$

and for distributed water drops (pores) within a solid matrix material

$$\bar{\varepsilon} = \varepsilon_w \cdot \frac{3\varepsilon_m - 2\Phi \cdot (\varepsilon_m - \varepsilon_w)}{3\varepsilon_w + \Phi \cdot (\varepsilon_m - \varepsilon_w)} \tag{9-142}$$

The Hanai-Bruggeman relationship (compare with equation 9-87) for permittivity is:

$$\left(\frac{\bar{\varepsilon} - \varepsilon_2}{\varepsilon_1 - \varepsilon_2}\right) \cdot \left(\frac{\varepsilon_1}{\bar{\varepsilon}}\right)^{\psi} = V_1 \tag{9-143}$$

where ψ is the depolarisation factor. It depends on the particle shape; for spheres ψ is 1/3.

For porous water saturated rock, the resulting equation is

$$\left(\frac{\bar{\varepsilon} - \varepsilon_m}{\varepsilon_w - \varepsilon_m}\right)\left(\frac{\varepsilon_w}{\bar{\varepsilon}}\right)^{\psi} = \Phi \tag{9-144}$$

Sen (1980) derived an equation similar to the Bruggeman formula using a self-consistent method.

Compilations of the various *mixing formulas* are given by Olhoeft (1985), Parchomenko (1965), and Wang and Schmugge (1980). These theories and formulas are based on simple mean equations, for example

- *arithmetic mean*

$$\bar{\varepsilon} = \sum_{i=1}^{n} V_i \cdot \varepsilon_i \tag{9-145}$$

or for a two component porous rock

$$\bar{\varepsilon} = (1 - \Phi) \cdot \varepsilon_m + \Phi \cdot \varepsilon_w \tag{9-146}$$

A more generalised form of the arithmetic mean equation was applied by Meador and Cox, 1975 (see also Hearst and Nelson, 1985) for a porous rock filled with water and hydrocarbon:

$$\bar{\varepsilon}^{\gamma} = (1 - \Phi) \cdot \varepsilon_m^{\gamma} + \Phi \cdot S_w \cdot \varepsilon_w^{\gamma} + \Phi \cdot (1 - S_w) \cdot \varepsilon_{hydrocarbon}^{\gamma} \tag{9-147}$$

where S_w is the water saturation and the exponent γ varies between 0 and 2, depending on rock type.

- *geometric mean*

$$\bar{\varepsilon} = \prod_{i=1}^{n} \varepsilon_i^{V_i} \tag{9-148}$$

This is identical the so called *Lichtenecker-formula* (see for example Olhoeft, 1985; Kobranova, 1986):

$$\lg \bar{\varepsilon} = \sum_{i=1}^{n} V_i \cdot \lg \varepsilon_i \tag{9-149}$$

or for a two component porous rock

$$\bar{\varepsilon} = (\varepsilon_m)^{1-\Phi} \cdot (\varepsilon_w)^{\Phi} \tag{9-150}$$

The propagation time of an electromagnetic wave - for methods such as ground penetrating radar (GPR) and electromagnetic borehole tools - is

$$\Delta t_0 = \Delta t_{\text{vacuum}} \cdot \sqrt{\varepsilon_r} \tag{9-151}$$

This relationship is valid for a lossless formation. Using the "philosophy of the time-average relationship" (see chapter 6.4.) a mean formula can be derived for the propagation time in a porous formation

$$\Delta t_{\text{formation}} = (1 - \Phi) \cdot \Delta t_m + \Phi \cdot \Delta t_w \tag{9-152}$$

Thus, results in the permittivity relationship of (also called Complex Refractive Index Method or CRIM-formula, see Baker et al., 1985; Kenyon and Baker, 1986)

$$\bar{\varepsilon} = [(1 - \Phi) \cdot \sqrt{\varepsilon_m} + \Phi \cdot \sqrt{\varepsilon_w}\,]^2 \tag{9-153}$$

This equation corresponds to equation 9-147 for $\gamma = 1/2$ and $S_w = 1$.

Figure 9.55 shows a relative dielectric permittivity versus porosity plot for a water saturated porous rock calculated using some selected relationships. In Figure 9.56, these curves are compared with some experimental data; for this particular case, a good fit is given by the CRIM formula.

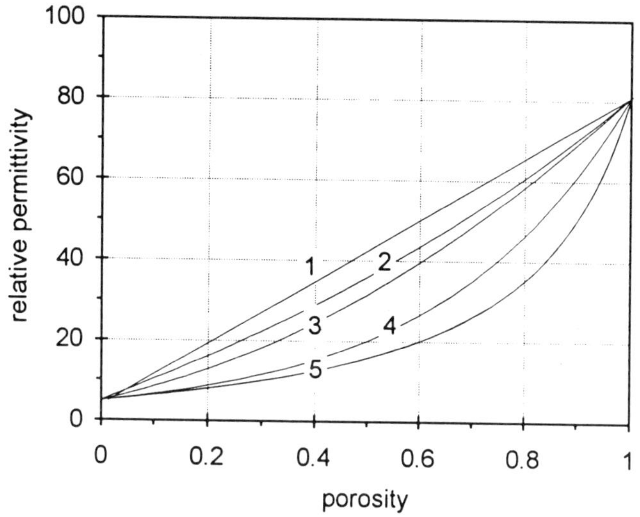

Figure 9.55 Relative dielectric permittivity as a function of porosity for a water saturated rock using some mixing equations; $\varepsilon_m = 5$ $\varepsilon_w = 81$
curve 1: arithmetic mean (9-146) curve 2: Maxwell equation (9-142)
curve 3. CRIM equation (9-153) curve 4: Lichtenecker equation (9-150)
curve 5: Bruggeman equation (9-144), Maxwell equation (9-141)

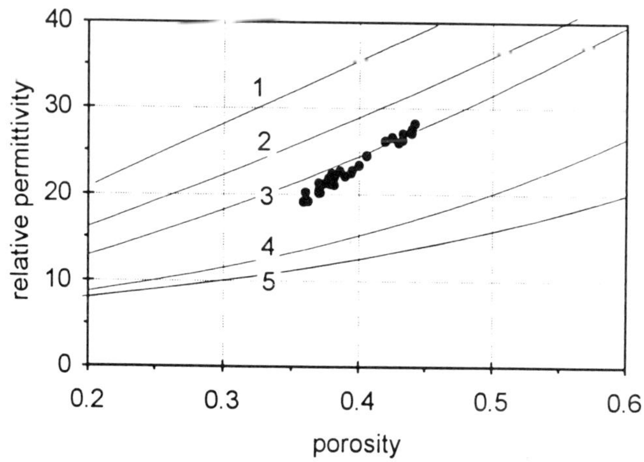

Figure 9.56 Relative dielectric permittivity as a function of porosity for a water saturated rock; curves: calculated using some mixing equations (see Figure 9.55); points: experimental data for sand (see Figure 9.51)

Coutanceau-Monteil and Jacquin (1993) investigated the permittivity of sedimentary rock samples in the frequency range from 20 to 1000 MHz. A comparison of the experimental results with model equations shows that the CRIM model fits the data for water and brine-saturated quarry limestones, whereas, the Bruggeman-Hanai-Sen equation fits the data for water and brine saturated Fontainebleau sandstones fairly well.

9.3.6. Anisotropy of electrical properties

The anisotropy of the electrical conductivity or of the resistivity in natural rocks is caused by

- ► preferred orientation of conducting minerals or accessories (for example graphite, ores),
- ► the preferred orientation of the wet or water saturated pores or cracks,
- ► fine layering of rock components with different conductivity (clay - sand "sandwich layering").

Whereas, the first two types are intrinsic caused anisotropy, the third type is a macroanisotropy.

Figure 9.57 shows as example a azimuthal plot of the specific resistivity measured at a carbonate rock with low porosity; anisotropy is caused by fracturing of the water saturated sample.

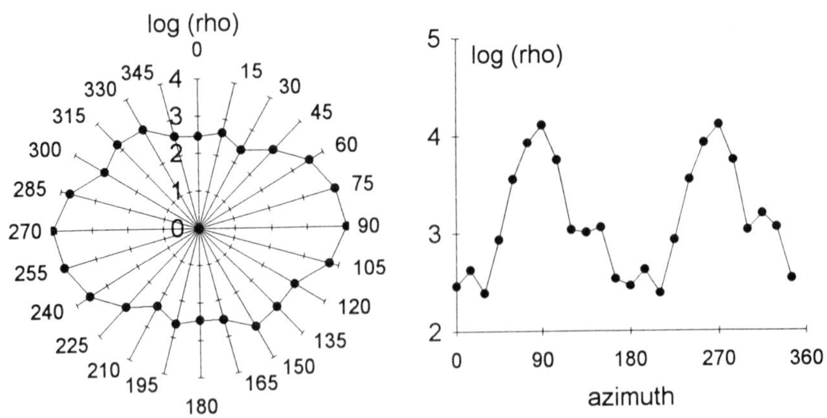

Figure 9.57 Azimuthal plot of the logarithm of specific resistivity; low porous carbonate rock, Imsterberg, Tyrol Austria; microfracturing has the orientation of the smallest resistivities

Keller and Frischknecht (1966) defined the *coefficient of anisotropy* for layered rocks

$$A_\rho = \left(\frac{\rho_\perp}{\rho_{//}}\right)^{1/2} \tag{9-154}$$

where the resistivity perpendicular to the layer plane is

$$\rho_\perp = \frac{\Sigma(\rho_i \cdot h_i)}{\Sigma h_i} \tag{9-155}$$

and the resistivity parallel to the layer plane is

$$\rho_{//} = \frac{\Sigma h_i}{\Sigma\left(\frac{h_i}{\rho_i}\right)} \tag{9-156}$$

with the resistivity ρ_i of the i-th layer and its thickness h_i.

Table 9.17 contains anisotropy coefficient ranges for some rock types.

Table 9.17. Range of the anisotropy coefficient for some rock types, after Keller and Frischknecht (1966)

Rock type	A_ρ
Volcanic tuff (Eocene and younger), Nevada	1.10 ... 1.20
Sections of Alluvium, south-western US	1.02 ... 1.10
Interbedded limestone, limey shales (north-eastern Colorado)	2.0 ... 3.0
Interbedded anhydrite and shale (north-eastern Colorado)	4.0 ... 7.5
Massive shale beds	1.01 ... 1.05
Interbedded shale and sandstone	1.05 ... 1.15
Baked shale or low-rank slate	1.10 ... 1.60
Slates	1.40 ... 2.25
Bituminous coal and mudstone	1.7 ... 2.6
Anthracite coal and associated rocks	2.0 ... 2.6
Graphitic slate	2.0 ... 2.8

Foliated rocks (gneisse) show an anisotropy closely correlated to the foliation structure. The lowest resistivities are measured parallel to the strike of foliation, the highest values perpendicular to that direction. Rauen and Lastovickova (1995) investigated horizontal core cuts from rock samples of the KTB (Continental Deep Drilling Project) and defined a "horizontal anisotropy" (Table 9.18):

$$FA = \left(\frac{\rho_{max}}{\rho_{min}}\right)^{1/2} \tag{9-157}$$

Gneiss with a pronounced foliation shows the highest anisotropy values.

Table 9.18 Summary of laboratory measurements of the horizontal anisotropy, KTB samples, Rauen and Lastovickova (1995)

Rock	horizontal anisotropy FA	number of samples
gneiss	2.840	9
amphibolite	1.335	28
metagabbro, lamprophyre	1.241	4

Most studies of the electrical resistivity of *sedimentary rocks* are based on Archie's equation. Barker and Worthington (1973) compared values of the formation factors $F_{\perp}$ (current direction perpendicular to bedding plane) and $F_{//}$ (current direction parallel to bedding plane) of a Triassic sandstone from north-west England:

$$F_{\perp} = \frac{1.54}{\Phi^{1.27}} \qquad F_{//} = \frac{1.43}{\Phi^{1.26}} \tag{9-158}$$

This results in the ratio

$$A_F = \frac{F_{\perp}}{F_{//}} = 1.077 \cdot \Phi^{-0.11} \tag{9-159}$$

Worthington (1981) noted that electrical anisotropy is greater in shaly formations than in clean formations. This may be caused by the spatial orientation of the "second conducting component clay or surface conductivity".

Dielectric permittivity of rocks also exhibits anisotropic behaviour (Table 9.19).

Table 9.19. Anisotropy of the dielectric permittivity, after Parchomenko (1965)

Rock type	$\varepsilon_{\perp}$		$\varepsilon_{//}$		$A_{\varepsilon} = \varepsilon_{\perp}/\varepsilon_{//}$	
	dry	wet	dry	wet	dry	wet
Marble	8.5	8.7	8.7	8.8	0.98	0.99
Limestone	8.3	10.4	8.5	12.2	0.98	0.85
Limestone	7.8	10.6	7.8	12.0	1.00	0.88
Keratophyre	5.7	5.9	5.5	5.9	1.04	1.00
Eclogite	12.7	17.0	9.7	14.1	1.31	1.20
Granite	5.9	8.4	5.9	8.7	1.00	0.97
Granite	5.7	7.1	7.0	7.2	0.81	0.99

CHAPTER 10

Some Relationships between Various Physical Properties of Rocks

"The future of rock physics will be to help ... in integration of geophysical, core, and engineering data through a better understanding of the relationships between different rock properties an how those properties are controlled by the physical characteristics (lithology, pore geometry, fluid type) of the rock." *(David P. Yale, 1985)*

10.1. Overview

The previous chapters show that the investigated petrophysical properties (density, magnetic susceptibility, natural radioactivity, velocity and absorption coefficient of elastic waves, specific electric resistivity, dielectric permittivity, and thermal conductivity) depend on different parameters. They are summarized in section 1.3.

The physical basis for the existence of *relationships* between the various physical properties - for example, between the velocity of elastic waves and the thermal conductivity - is given by

a) relationships between the properties of the individual *rock constituents* (for example between the elastic and thermal properties of minerals),
b) the correlation of the considered parameters with *"parameterinvariant"* quantities like porosity, rock composition etc..

The knowledge of such relationships is the foundation for an integrated interpretation of geophysical data.

In regard to their application, we can distinguish between two types of relationships:

1) Relationships between properties determined by geophysical methods in the narrow sense (for example, between the velocity of elastic waves and the thermal conductivity of rocks) including influences like rock composition or lithology, and porosity. In many cases, these relationships and algorithms also utilise other measured parameters, such as by nuclear methods (various cross sections, neutron porosity). They are included in the following discussions. This type of relationship is the general basis for interpretation; such as in well log analysis and for conversions.
2) Relationships between properties determined by geophysical methods and other particular physical properties (for example, geotechnical properties, hydraulic permeability).

These are described by some selected examples in the following sections.

10.2. Relationships between the properties determined by geophysical methods

10.2.1. Relationships based on solid state physics theory

In section 6.3 (equation 6-23) Birch's empirical relationship between seismic wave velocities and density was discussed. A relationship between elastic properties and density was also derived by Anderson (1967); the simplified form of his "seismic equation of state" is

$$d = a \cdot m_A \cdot \Psi^n \tag{10-1}$$

where d is the density, m_A is the mean atomic mass, the exponent n is of the order of 1/4 to 1/3 and is related to the Grüneisen constant, and Ψ is the seismic parameter:

$$\Psi = v_p^2 - \frac{4}{3} \cdot v_s^2 = \frac{k_{ad}}{d} \tag{10-2}$$

Figure 10.1 shows data for thirty-one selected minerals and rocks in a plot $\ln\left(\frac{d}{m_A}\right)$ versus $\ln \Psi$. The solid line is a least squares fit to the data following equation 10-1

$$\frac{d}{m_A} = 0.048 \cdot \Psi^{0.323} \pm 0.12 \tag{10-3}$$

where density is in g/cm^3 and velocity in km/s.

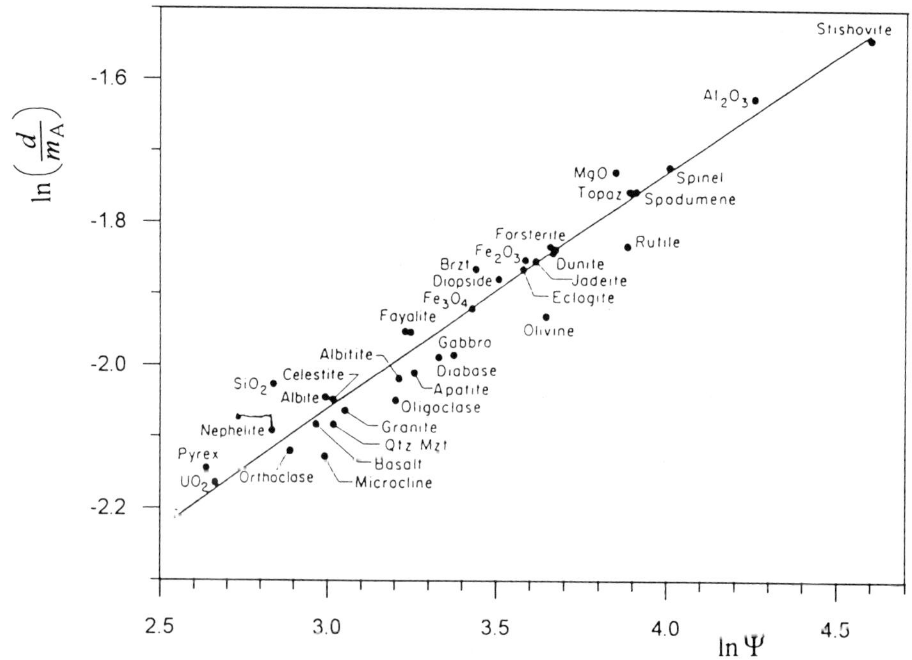

Figure 10.1 Correlation between the ratio d/m_A versus seismic parameter Ψ for selected rocks and oxides in a logarithmic plot; solid line is a least square fit (equation 10-3), after Anderson (1967)

The Debye-theory of lattice dynamics gives a relationship between thermal conductivity λ, specific heat c, density d, and mean sound velocity v_m:

$$\lambda \sim c \cdot d \cdot v_m \tag{10-4}$$

where

$$\frac{3}{v_m^3} = \frac{1}{v_p^3} + \frac{2}{v_s^3} \tag{10-5}$$

Such relationships established from the solid state physics were combined or modified with *empirical relations*. Horai and Simmons (1969, 1970) developed relationships between the compressional and shear wave velocity and thermal conductivity based on experimental data from silicates (Figure 10.2). Straight lines fitted to the experimental data give the following linear equations

$$v_p = (405.0 \pm 47.7) \cdot \lambda + (5930 \pm 170) \tag{10-6}$$

$$v_s = (214.8 \pm 47.7) \cdot \lambda + (3310 \pm 160) \tag{10-7}$$

where v_p, v_s is in m s^{-1} and λ in W m^{-1} K^{-1}.

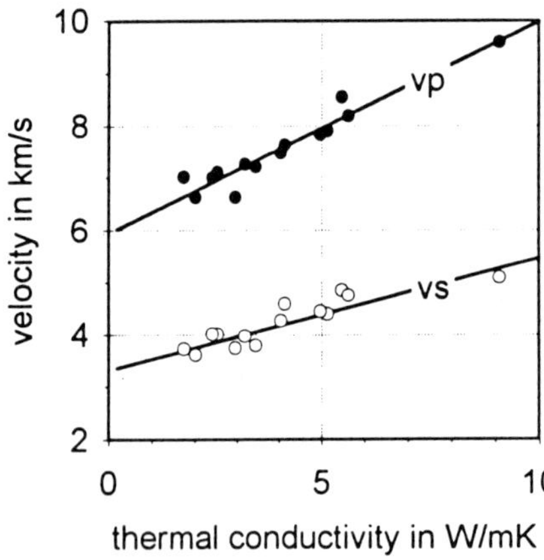

Figure 10.2 Correlation of thermal conductivity and seismic wave velocities for silicates; data after Horai and Simmons, 1970 (converted into SI units)

Referring to the version 6-30 of Birch's law and equation 8-7 a more general equation including the influence of the mean atomic mass is given:

$$v_p = a_1 \cdot \lambda + b_1(m_A) \tag{10-8}$$

$$v_s = a_2 \cdot \lambda + b_2(m_A) \tag{10-9}$$

Rybach and Buntebarth (1982, 1984) derived various relationships between properties (density, compressional wave velocity, and heat generation) based on an *"ionic model for minerals"*. They start with the discussion that the chemical composition (and the mean atomic mass) alone is not sufficient to describe mineral assemblages. Thus, structural aspects like the packing density of ions must be considered. To quantify these aspects, the fraction of cations present in the mineral structures seems to be appropriate. The cation fraction can be defined by the *cation packing index K*, which is the ratio of the number of cations per molar volume, normalised by Avogadro's number. This cation packing index was determined from modal analyses on thin sections of rock samples (from rhyolitic/granitic to ultrabasic composition). Table 10.1 shows some selected values.

Table 10.1. Cation packing index K, density d, compressional wave velocity (at 50 MPa) v_p, and heat generation A of some rock types (mean values, standard deviation, range); selected values after Rybach and Buntebarth (1982, 1984)

Rock type	Number of samples	K in 10^{-2} mole/cm^3	d in g/cm^3	v_p in km/s	A in µW/m^3
Granites	17	4.91 ± 0.21 4.59... .40	2.63 ± 0.03 2.57...2.69	5.84 ± 0.16 5.52...6.04	2.81 ± 1.00 0.806...4.65
Granodiorites	11	5.05 ± 0.16 4.82...5.40	2.70 ± 0.06 2.61...2.86	5.93 ± 0.18 5.66...6.30	2.50 ± 1.25 0.628...4.41
Diorites	7	5.55 ± 0.10 5.34...5.63	2.86 ± 0.08 2.74...2.98	6.26 ± 0.27 5.84...6.57	0.877 ± 0.30 0.472...1.32
Amphibolites	6	5.87 ± 0.07 5.80...5.98	2.97 ± 0.05 2.89...3.03	6.30 ± 0.30 5.92...6.66	0.367 ± 0.194 0.154...0.701
Peridotites	6	6.72 ± 0.12 6.57...6.84	3.17 ± 0.10 3.05...3.29	7.71 ± 0.20 7.47...7.99	0.017 ± 0.01 0.0019...0.0265
Serpentinites	6	7.79 ± 0.32 7.38...8.19	2.69 ± 0.04 2.64...2.75	6.49 ± 0.64 5.72...7.44	0.012 ± 0.016 0.0007...0.0450

Density d (in g/cm^3), compressional wave velocity v_p (in km/s, measured at 50 MPa), and heat generation A in µW/m^3) correlate with the cation packing index K (in mole/cm^3):

$$d = 1.27 + 28.3 \cdot K \quad \text{with a correlation coefficient of } r = 0.982 \tag{10-10}$$

$$v_p = 1.58 + 84.1 \cdot K \quad \text{with a correlation coefficient of } r = 0.933 \tag{10-11}$$

$$\ln A = 12.20 - 230.5 \cdot K \quad \text{with a correlation coefficient of } r = -0.935 \tag{10-12}$$

The resulting relationships between density, compressional wave velocity (measured at 50 MPa), and heat generation are

$$d = 0.74 + 0.34 \cdot v_p \qquad (10\text{-}13)$$

$$\ln A = 16.5 - 2.74 \cdot v_p \qquad (10\text{-}14)$$

$$\ln A = 22.5 - 8.15 \cdot d \qquad (10\text{-}15)$$

These correlations are valid mostly for minerals and rocks as a solid material under the conditions of the deeper crust and the earth interior. For natural rocks of the upper part of the earth's crust, the effects of cracks, fractures, pores, grain contacts etc. dominantly influence the physical properties; the rock cannot be considered as an undisturbed solid material without defects. Therefore, other models must be used which include these defects.

10.2.2 Volumetric models - a system of linear equations for rock composition and porosity determination

The volumetric model was introduced in connection with the definition and discussion of density (chapter 3). The density of a porous rock is (see equation 3-4):

$$d = (1 - \Phi) \cdot d_m + \Phi \cdot d_{fluid} \qquad (10\text{-}16)$$

or in general for a rock consisting of n components

$$d = \sum_{i=1}^{n} V_i \cdot d_i \qquad \sum_{i=1}^{n} V_i = 1 \qquad (10\text{-}17)$$

where V_i is the volume fraction and d_i the density of the i-th component. This equation expresses that the effective value of the rock density is unambiguously determined by the individual densities of the components and their volume fractions.

In a more *general formulation*, the effective value of a physical property X is given by the sum over the products of the property X_i and volume fraction V_i of the individual components:

$$X = \sum_{i=1}^{n} V_i \cdot X_i \qquad \sum_{i=1}^{n} V_i = 1 \tag{10-18}$$

This relationship and the volumetric model is *valid for all scalar properties* (for example density, nuclear cross sections). Applied to various properties, it forms a *system of linear equations*, which are a fundamental part of many log interpretation algorithms and strategies. It is the basics of compositional analysis.

In addition to the density - determined by a gamma-gamma-density measurement - neutron porosity Φ_N (as a second scalar property) is also used. Both form this system of two linear equations:

$$d_{GG} = \sum_{i=1}^{n} V_i \cdot d_{GG,i} \tag{10-19}$$

$$\Phi_N = \sum_{i=1}^{n} V_i \cdot H_{N,i} \tag{10-20}$$

d_{GG} is the density measured by a gamma-gamma tool. If the tool is calibrated using a material with a ratio of atomic number (Z) to mass number (A) of $Z/A = 0.5$, then the relationship between the gamma-gamma density d_{GG} and the "mass-density" d as defined by equation 3-1 is

$$d_{GG} = d \cdot \frac{(Z/A)_{eff}}{0.5} = 2 \cdot d \cdot (Z/A)_{eff} \tag{10-21}$$

where $(Z/A)_{eff}$ is the mean effective ratio for the rock. Table 10.2 shows mean values for some minerals.

Φ_N is the neutron-porosity. A neutron-neutron (or neutron-gamma) tool reads the slowing down and neutron capture properties of rock. Hydrogen has an extraordinarily high slowing down cross section for neutrons. Components with high hydrogen content (water, oil) have the dominant effect on this measurement. Neutron tools, therefore, reflect the liquid-filled porosity in clean formations. The influence of gaseous components (with a relatively low hydrogen concentration compared with liquids) and most minerals is small. If the minerals contain hydrogen or water within their crystal lattice or have "bound water", then this acts in the same manner on the neutron flux as the free water within the pore space. This causes neutron effects of gypsum and clay.

The neutron effect of the rock components is described by a "hydrogen index" H. The hydrogen index for fresh water is defined $H_{\text{fresh water}} = 1$ by the tool calibration. All other materials are related to this reference value. Thus, the hydrogen index of any material is the fresh water equivalent with respect to the neutron response. In practice, the neutron tool is usually calibrated in porous limestone saturated with fresh water. Using this calibration, the hydrogen index of calcite is defined as zero.

Table 10.2 shows some mean values; the relatively high values for clays result from interlayered and bound water and the high value for the non-porous gypsum is produced by hydrogen in the crystal lattice.

Table 10.2. Mean values for the density d (in g/cm^3), gamma-gamma-density d_{GG} (in g/cm^3), effective atomic number Z_{eff} and ratio $(Z/A)_{eff}$, hydrogen index H, and photoelectric cross section U (in barns/cm3) or P_e (in barns/electron); data after Schlumberger (1989), Hearst and Nelson (1985).

Substance	d	d_{GG}	Z_{eff}	$(Z/A)_{eff}$	H	U	P_e
Quartz	2.654	2.650	11.8	0.499	-0.02	4.79	1.81
Calcite	2.710	2.708	15.7	0.500	± 0	13.77	5.08
Dolomite	2.870	2.864	13.7	0.499	0.02	9.00	3.14
Halite	2.165	2.074	14.0	0.479	-0.02	9.45	4.65
Anhydrite	2.960	2.957	15.6	0.499	-0.01	14.95	5.05
Gypsum	2.320	2.372	16.4	0.511	0.5	9.37	3.99
Barite	4.500	4.011	47	0.446	-0.01	1070	267
Kaolinite	2.44	2.44	14.22	0.500	0.34	4.44	1.83
Illite	2.65	2.63	11.85	0.496	0.2	8.73	3.45
Montmorillonite	2.12	2.12	12.19	0.500	0.4	4.04	2.04
Average shale	2.65	2.645	14.0	0.499	0.3...0.4	9.05	3.42
Anthracite coal	1.70	1.75	6.0	0.515	0.37	0.28	0.16
Bituminous coal	1.40	1.47	6.2	0.525	0.5	0.26	0.18
Water, fresh	1.000	1.110	7.5	0.555	1.000	0.40	0.36
Oil	0.850	0.948	5.5	0.558	~ 1	0.12	0.13

For *water* dissolved NaCl (or other ions) takes up space and thereby reduces the hydrogen density and hydrogen index (see Table 10.3):

$$H_{solution} = 1 - 0.4 \cdot C_{NaCl} \qquad (10\text{-}22)$$

where C_{NaCl} is the concentration in ppm; the relationship is given for a temperature of 75° F = 24 ° C (Schlumberger, 1989).

For an estimation of the hydrogen index of *liquid hydrocarbons*, the following determination can be applied (Doveton, 1986, Schlumberger, 1989): For a relative measure of the hydrogen content of water, the expression

$$H^{*}_{\mathrm{H_2O}} = d_{\mathrm{H_2O}} \cdot \frac{2}{2+16} = d_{\mathrm{H_2O}} \cdot \frac{1}{9} \qquad (10\text{-}23)$$

can be used. The atomic weight of hydrogen and carbon are approximately 1 and 12, respectively). For hydrocarbon with the general formula $n(\mathrm{CH_2})$, the corresponding expression would be

$$H^{*}_{n(\mathrm{CH_2})} = d_{n(\mathrm{CH_2})} \cdot \frac{2}{2+12} = d_{n(\mathrm{CH_2})} \cdot \frac{1}{7} \qquad (10\text{-}24)$$

For the hydrogen index of the hydrocarbon results in

$$H_{n(\mathrm{CH_2})} = H^{*}_{n(\mathrm{CH_2})} / H^{*}_{\mathrm{H_2O}} = d_{n(\mathrm{CH_2})} \cdot \frac{9}{7} = 1.286 \cdot d_{n(\mathrm{CH_2})} \qquad (10\text{-}25)$$

For an oil density of 0.85, the hydrogen index is therefore 1.09. Equation 10-25 is recommended for heavier hydrocarbons. For light hydrocarbons (> 0.25) the equation $H \approx 2.2 \cdot d$ is proposed (Schlumberger, 1989).

The index for *gaseous hydrocarbons* depends on gas composition, pressure and temperature; for illustration Table 10.3 shows some examples.

Table 10.3. Hydrogen index of some pore fluids (after Western Atlas, 1992)

Substance	Hydrogen Index	
	T = 60°F = 15.6°C p = 14.7 psi = 0.101 MPa	T =200°F = 93.3 °C p = 7000 psi = 48.3 MPa
Pure water	1	1
Water (2 x 10^5 ppm NaCl)	0.92	0.90
N-hexane C_6H_{14}	0.96	0.92
N-octane C_8H_{18}	1.00	0.96
N-decane $C_{10}H_{22}$	1.02	0.98
Average natural gas	0.0017	0.54
Methane CH_4	0.0015	0.49
Ethane C_2H_6	0.0023	0.74

The *neutron porosity* Φ_N for a porous, partial saturated rock is

$$\Phi_N = \Phi \cdot H_{fluid} + (1 - \Phi) \cdot H_{matrix} \qquad (10\text{-}26)$$

where Φ is the "true" porosity as defined by equation 2-1. H_{matrix} is the hydrogen index of the matrix, and H_{fluid} the hydrogen index of the pore fluid. For a gas and water bearing rock, H_{fluid} is

$$H_{fluid} = H_w \cdot S_w + H_{gas} \cdot (1 - S_w) \approx S_w \qquad (10\text{-}27)$$

These volumetric models and the resulting linear equations are used as fundamentals for various log interpretation methods for *lithology and porosity estimation*.

Frequently, the sonic log and its parameter - the transit time Δt - is included with the linear time-average relationship (see section 6.4.3.1):

$$\Delta t = (1 - \Phi) \cdot \Delta t_{ma} + \Phi \cdot \Delta t_{fluid} \qquad (10\text{-}28)$$

or in its more general form for n components

$$\Delta t = \sum_{i=1}^{n} V_i \cdot \Delta t_i \qquad (10\text{-}29)$$

Despite the common application of the time average relationship, it must be remarked that elastic (seismic) properties as tensorial properties do not exactly follow simple volumetric models without consideration of spatial distribution of the components, their bonding etc. (see section 6.4.). The inclusion of the seismic traveltime - on the other hand - is a very powerful component of integrated interpretation methods. The techniques are described in detail in textbooks of well logging (such as Serra, 1984; Jorden and Campbell, 1984; Ellis, 1987; Hearst and Nelson, 1985) and manuals (for example, Schlumberger, 1989; Western Atlas, 1992). The following is a brief discussion of some principles only; for practical log interpretation the reader should refer to these special papers.

The following main types of techniques - based on volumetric models - are used:

a) Lithology- or matrix identification and porosity determination using two measured properties (logs)

As an example, Figure 10.3 shows the sonic-neutron crossplot calculated with the properties given in Table 10.4.

Table 10.4. Mean values for crossplot and model calculations

Substance	d_{GG} in g/cm³	H	Δt in μs/m
Quartz	2.65	-0.02	180
Calcite	2.71	± 0	156
Dolomite	2.87	0.02	143
Water	1.00	1.00	620

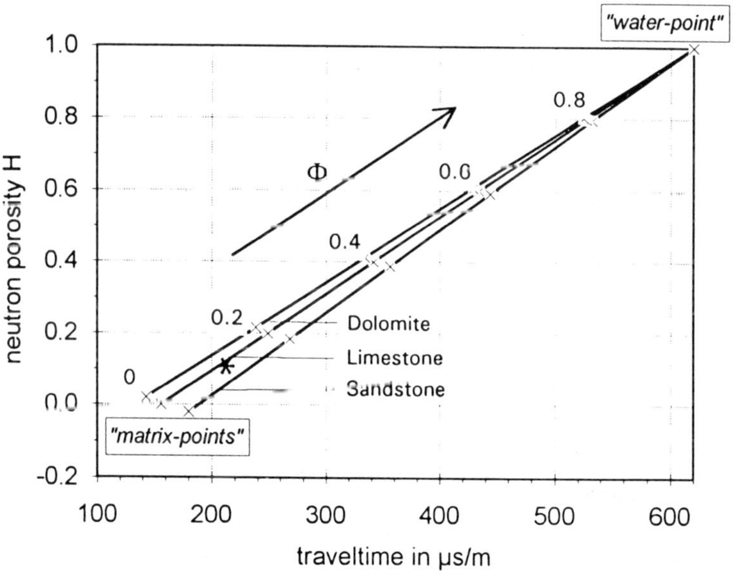

Figure 10.3 Sonic-neutron crossplot for sandstone and carbonate rocks

The three monomineralic porous (water saturated) rocks are represented by three straight lines with a linear porosity scale. The pure substances (quartz, calcite, dolomite, and water) determine the end points of these lines.
For practical applications usually

- the left part (porosity below 0.3) is plotted and
- the curves are empirically modified.

Any set of measured sonic traveltime and neutron porosity corresponds to one point in the crossplot and determines the lithotype and porosity.

The asterix in Figure 10.3 describes one example (the result is a limestone with a porosity $\Phi = 0.10$).

b) Lithology- or matrix identification and porosity determination using a condensed presentation of three measured properties (logs)

When three porosity logs (such as sonic, neutron, and density) are available, a three dimensional crossplot simultaneously incorporates all three. The difficulties of the handling of such a plot are avoided by the M-N plot (Schlumberger), derived by Burke et al. (1969) as "an ingenious solution to this three-dimensional problem that reduces it to two dimensions by the elimination of one of the unknowns" (Doveton, 1986).

Burke et al. (1969) have defined two parameters *M* and *N*. *M* combines the sonic and density properties, *N* combines the neutron and density properties. For the forward calculation based on *known matrix and fluid properties* they are:

$$M = \frac{\Delta t_{\mathrm{fl}} - \Delta t_{\mathrm{m}}}{d_{\mathrm{GG,m}} - d_{\mathrm{GG,fl}}} \cdot 100 \qquad (10\text{-}30)$$

$$N = \frac{H_{\mathrm{fl}} - H_{\mathrm{m}}}{d_{\mathrm{GG,m}} - d_{\mathrm{GG,fl}}} \qquad (10\text{-}31)$$

Using matrix and fluid values (for example Table 10.3), the porosity-independent parameters *M* and *N* can be calculated for the various matrix types and presented in a *M-N*-plot (Figure 10.4).

A second version of the two parameters is used for the interpretation:

$$M = \frac{\Delta t_{\mathrm{fl}} - \Delta t}{d_{\mathrm{GG}} - d_{\mathrm{GG,fl}}} \cdot 100 \qquad (10\text{-}32)$$

$$N = \frac{H_{\mathrm{fl}} - \Phi_{\mathrm{N}}}{d_{\mathrm{GG}} - d_{\mathrm{GG,fl}}} \qquad (10\text{-}33)$$

where $\Delta t, d_{\mathrm{GG}}$, and Φ_{N} are the *measured values*. These determine the position of the measured profile point or section in the calculated *M-N* plot and give a lithology estimate. It can be shown that the two pairs of equations (10-30, 10-31) and (10-32, 10-33) are identical by insertion of the linear response equations (10-16, 10-26, 10-28) into (10-30) and (10-31).

In the following step, porosity can be calculated based on the determined matrix type and the corresponding matrix properties.

M, N data pairs between calculated matrix points in the *M-N* plot represents a *composite matrix*. For more detailed studies, log interpretation textbooks (for example Doveton, 1986; Serra, 1984) are recommended.

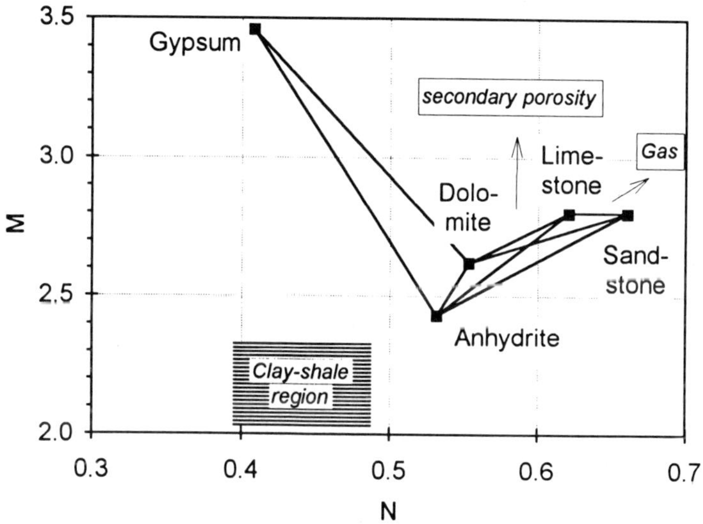

Figure 10.4 Simplified M-N plot for some common minerals or sedimentary rock types; M is in μs kg^{-1} m^2 and N is in 10^{-3} kg $^{-3}$

c) Numerical solution of the equation system for lithology and porosity estimation

The main steps follow an excellent description by Doveton (1986) of the philosophy of this method, physical and mathematical fundamentals of the matrix algebra methods.

For a water saturated porous rock with a composite matrix, consisting of calcite (C), dolomite (D), and quartz or chert (Q), respectively, the response equations with the properties of Table 10.4 are:

$$d_{GG} = 2.71 \cdot V_C + 2.87 \cdot V_D + 2.65 \cdot V_Q + 1.00 \cdot \Phi \qquad (10\text{-}34)$$

$$\Phi_N = 0.00 \cdot V_C + 0.02 \cdot V_D - 0.02 \cdot V_Q + 1.00 \cdot \Phi \qquad (10\text{-}35)$$

$$\Delta t = 156 \cdot V_C + 143 \cdot V_D + 180 \cdot V_Q + 620 \cdot \Phi \qquad (10\text{-}36)$$

and the volume balance equation

$$1 = V_C + V_D + V_Q + \Phi \qquad (10\text{-}37)$$

where V_C, V_D, and V_Q are the volume fractions of the mineral components. The system of the four linear equation can be written as a matrix equation:

$$\begin{bmatrix} d_{GG} \\ \Phi_N \\ \Delta t \\ 1 \end{bmatrix} = \begin{bmatrix} 2.71 & 2.87 & 2.65 & 1.00 \\ 0.00 & 0.02 & -0.02 & 1.00 \\ 156 & 143 & 180 & 620 \\ 1.00 & 1.00 & 1.00 & 1.00 \end{bmatrix} \begin{bmatrix} V_C \\ V_D \\ V_Q \\ \Phi \end{bmatrix} \qquad (10\text{-}38)$$

Expressed in symbols

$$\mathbf{M}=\mathbf{RV} \qquad (10\text{-}39)$$

where **M** is the matrix of the measured properties, **R** the response matrix, and **V** the matrix of the volume fractions. The matrix of volume fractions (usually the "goal of interpretation") is given by

$$\mathbf{V} = \mathbf{R}^{-1}\mathbf{M} \qquad (10\text{-}40)$$

where $\mathbf{R}^{-1}$ is the inverse coefficient matrix. In this case, the matrix equation in expanded form is

$$\begin{bmatrix} V_C \\ V_D \\ V_Q \\ \Phi \end{bmatrix} = \begin{bmatrix} -12.68 & 8.19 & -0.0665 & 45.74 \\ 7.63 & 3.02 & 0.0216 & -24.03 \\ 5.11 & -11.91 & 0.0445 & -20.78 \\ -0.05 & 0.70 & 0.0004 & 0.0651 \end{bmatrix} \begin{bmatrix} d_{GG} \\ \Phi_N \\ \Delta t \\ 1 \end{bmatrix} \qquad (10\text{-}41)$$

The system can be determined, if the

number of logs = number of components - 1.

Using this technique, the composition of any part of the profile can be immediately found by premultiplying a column vector of the log readings and one by the inverse of the coefficient matrix (Doveton, 1986).

From this foundation two development directions are important:

1) Inclusion of additional properties and corresponding linear equations. Most of these properties are determined by nuclear techniques. A first step is the measurement of the photoelectric cross section. The photoelectric cross section is controlled by the gamma ray interaction with a material at the low energy level, where the photoelectric effect takes place. Tools based on this have a high Z (atomic number) response, which can be used for a lithology estimation. The photoelectric cross section is described either by a parameter P_e (in barn/electrons) or by a parameter U (in barn/cm^3). The linear equation for U is

$$U = \sum_{i=1}^{n} V_i \cdot U_i \qquad (10\text{-}42)$$

Table 10.2 contains some mean values for U (note the high value for barite). Other methods are based on various spectral techniques with an elemental selectivity. The system of equations results in a "geochemical log" (Schlumberger), such as was applied for the determination of mineralogical composition to the KTB-profile (Draxler, 1990).

2) Consideration of the case of an underdetermined system: *(number of logs) < (number of components - 1)* and the case of an overdetermined system: *(number of logs) > (number of components - 1)*. In particular, overdetermined systems are solved with the criterion of a minimum error of the final solution as the most probable answer (for details see Doveton, 1986).

As first step all these techniques require a formation model. It comprises the constituents of the formation, the tool responses, and physical constraints. Quirein et al., 1986 provide a coherent framework for developing multiple models for formation evaluation.

10.2.3. A simple model for fractured rocks

In section 6.4.3.4, a simple fractured rock model for calculation of elastic wave velocities was introduced and was also applied to thermal properties in section 8.4.2.3. The resulting equations are

for velocity $$v = (1 - D)^{1/2} \cdot v_{\mathrm{m}} \quad (10\text{-}43)$$

and for thermal conductivity $$\lambda = (1 - D) \cdot \lambda_{\mathrm{m}} \quad (10\text{-}44)$$

where the parameter D expresses the effect of defects like microcracks, fractures etc. on the two properties; the subscript m denotes the properties of the unfractured solid matrix substance.

The application of the same model on two properties offers the possibility to derive relationships between them; the *connecting model part* is the "defect parameter D". This results in

$$\frac{\lambda}{v^2} = \frac{\lambda_{\mathrm{m}}}{v_{\mathrm{m}}^2} \quad (10\text{-}45)$$

or

$$\lambda = \left(\frac{\lambda_{\mathrm{m}}}{v_{\mathrm{m}}^2} \right) \cdot v^2 = A_{\lambda v} \cdot v^2 \quad (10\text{-}46)$$

This means that as a result of defects in the fractured rock, a proportionality between the thermal conductivity and the square of the elastic wave velocity exists:

$$\lambda \sim v^2 \quad (10\text{-}47)$$

The parameter $A_{\lambda v} = \frac{\lambda_{\mathrm{m}}}{v_{\mathrm{m}}^2}$ is controlled by the matrix properties.

Figure 10.5 shows an example of a bilogarithmical plot thermal conductivity versus velocity. Equation 10-46 results in parallel straight lines with the slope 2 and the curve parameter $A_{\lambda v}$. Some typical values of $A_{\lambda v}$ for minerals are shown on the upper axis. The experimental data for gneiss, schist, melaphyre, and basalt are arranged in two clusters, which correspond to

$A_{\lambda v} = 0.09...0.13$ for gneiss and phyllite with a mean of $A_{\lambda v} \approx 0.1$

$A_{\lambda v} = 0.04...0.06$ for basalt and melaphyre with a mean of $A_{\lambda v} \approx 0.045$.

In their $A_{\lambda v}$-values the gneiss and phyllite trend towards the acid mineral components (quartz), whereas, basalt and melaphyre trend to basic mineral components. As a first approximation, two equations result for the correlation between thermal conductivity and elastic wave velocity:

$\lambda \approx 0.1 \cdot v^2$ for gneiss and phyllite
$\lambda \approx 0.045 \cdot v^2$ for basalte and melaphyre.

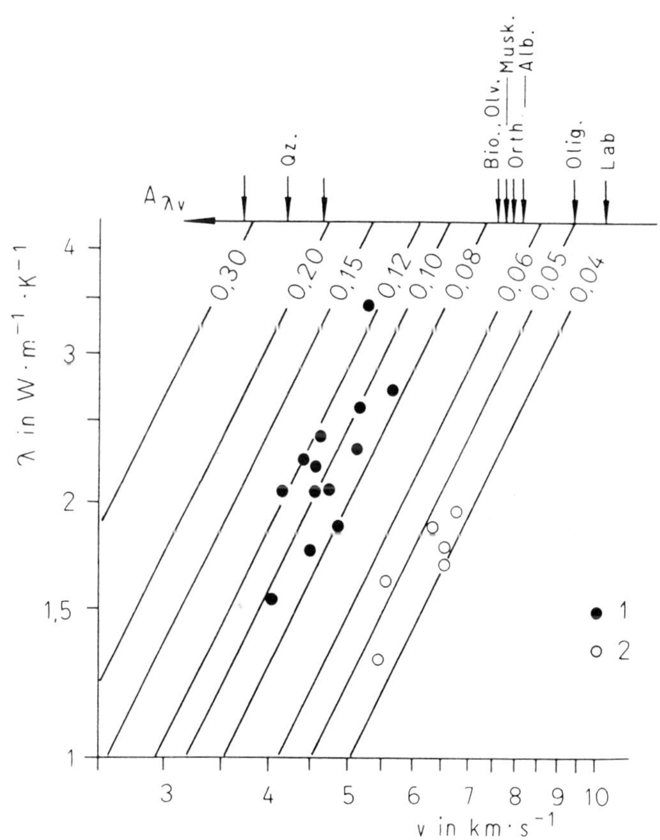

Figure 10.5 Thermal conductivity versus elastic wave velocity plot
Curves: Calculated after equation 10-46 with $A_{\lambda v}$ as curve parameter
Points: Experimental data after Hurtig (1966); 1 - gneiss, phyllite 2 - basalt, melaphyre
The arrows along the upper scale denotes values for some selected minerals: Qz - quartz, Bio - biotite, Musk - muscovite, Olv - olivine, Orth - orthoclase, Alb - albite, Olig - oligoclase, Lab - labradorite

10.2.4. Models with a variable internal structure

In section 6.4.8, a model with a variable internal structure - expressed by the "structure angle α" was introduced. This model was applied to various properties:

- elastic properties (section 6.4.8.)
- thermal properties (section 8.4.5.)
- electrical properties (section 9.3.3.3.)

The resulting equations show the same architecture for the calculated macroproperty (related to the model or rock as a whole)

macroproperty = f (microproperties, volume fractions, structure, bonding)

In this general relationship, the microproperties (for example thermal properties of the minerals and pore constituents) are related to the macroproperty (for example thermal conductivity of the rock); this is also valid for the bonding properties.
But the volume fraction and structure are always the same property for elastic, electrical, thermal, and further physical rock properties. Therefore, volume fraction and structure are acting as "joint connections" between the various physical rock properties.

It is advantageous to use dimensionless (normalized) parameters for the derivation of the principal correlations. The properties are divided by the *"microproperty of dominant influence"*, these are

- for elastic wave velocities, the velocity of the solid matrix material (v_m)
- for thermal conductivity, the conductivity of the solid matrix material (λ_m)
- for electrical rock conductivity, the conductivity of the pore water ($\sigma_W = \rho^{-1}{}_W$); normalization results in the formation factor.

Figure 10.6 shows an example plot of normalized velocity (compare section 6.4.8)

$$v_{p,\text{norm}} = \left(\left[\frac{G}{1-\Phi}\right] \cdot s_{33}^{33}(\alpha, f)\right)^{1/2} \tag{10-48}$$

versus normalized thermal conductivity

$$\lambda_{3,\text{norm}} = \left(\frac{\lambda_3}{\lambda_s}\right) \tag{10-49}$$

The normalized velocity, in this case, is related to the longitudinal wave (vertical propagation) in a dry porous rock. It includes the effects of porosity, internal structure, and bonding properties; the actual velocity value for a given rock results from multiplication with a factor which includes the matrix properties. The normalized thermal conductivity is the rock (or model) conductivity divided by the conductivity of the solid matrix material.

The plot shows a *network* of two curve families: one is given by the *porosity* and the other by the *structure angle* as parameter. This means that any correlation between the velocity and thermal conductivity is controlled by

- porosity and internal structure (network parameters),
- contact or bonding properties ($f = 4$ and $\lambda_c/\lambda_s = 0.5$),
- grain and pore geometric properties (network is calculated for $\alpha_g/\alpha_p = 0.5$),
- matrix properties (factor of normalization).

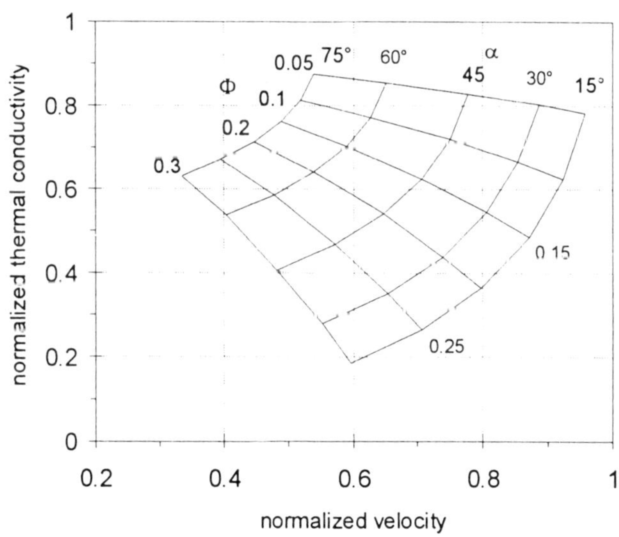

Figure 10.6 Normalized thermal conductivity versus normalized longitudinal wave velocity with porosity Φ and structure angle α as parameters. Grain and pore shape properties $\alpha_g/\alpha_p = 0.5$; elastic bonding property $f = 4$; ratio of the thermal conductivity of the contact region and solid matrix substance $\lambda_c/\lambda_s = 0.5$

In the following examples, this "network" is used for an *analysis of experimental results*.

Figure 10.7 shows the plot converted into a *velocity versus thermal conductivity* plot by empirical factors:

$$v_p = (4700\ \text{m/s}) \cdot v_{p,\,norm} \qquad \lambda = (2\ \text{W/mK}) \cdot \lambda_{3,\,norm}$$

The relatively low value of $\lambda_s = 2$ W/mK is caused by the clay content.

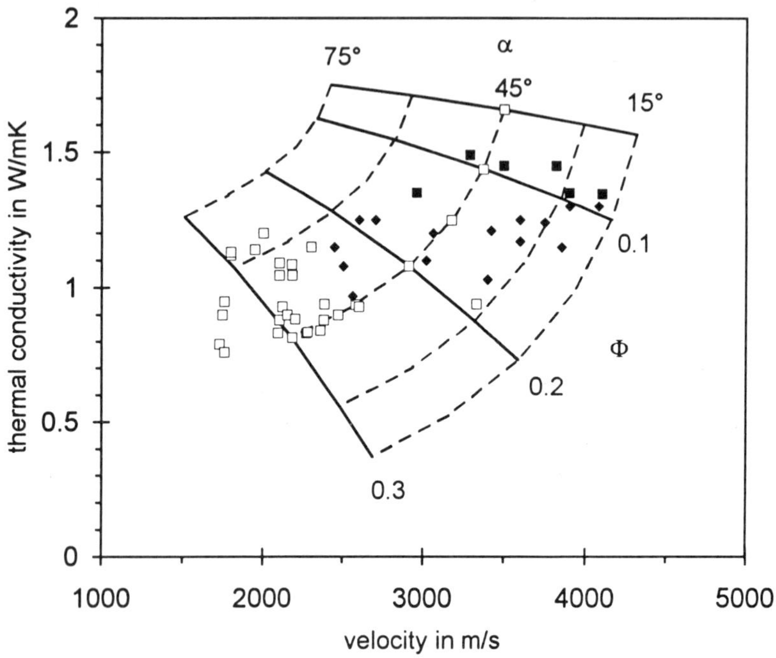

Figure 10.7. Thermal conductivity versus longitudinal wave velocity - analyse of experimental data by calculated plots.
Points: Experimental data after Moisseenko et al. (1970); Shaly sandstone and shale (Siberia), dry: ■ $\Phi < 0.1$ ◆ $0.1 < \Phi < 0.2$ □ $\Phi > 0.2$
Curves: Calculated for the dry rock model with the following parameters:
$\alpha_g/\alpha_p = 0.5$ $f = 4$ $\lambda_c/\lambda_s = 0.5$ $v_p = (4700\ \text{m/s}) * v_{p,\,norm}$ $\lambda = (2\ \text{W/mK}) * \lambda_{3,\,norm}$

The experimental data are arranged within the network and within the "porosity ranges". The scatter of data can be interpreted as the combined influence of porosity and structural changes. Each measured point is correlated to both porosity and structural values.

For water saturated porous rock, the thermal conductivity can be calculated by the equations in section 8.4.5 with $\lambda_p = \lambda_{water}$ and using the Biot-Geertsma-Smith-equation for the calculation of velocity for the water saturated rock (computed from the velocity of the dry rock). Figure 10.8 shows an example of this case: thermal conductivity and longitudinal wave velocity are determined on cores of unconsolidated sediments from a borehole in Lake Zürich, Switzerland. Calculated curves give a good approximation and show that the "structure" of the sediment is relatively constant ($\alpha = 60°$), but the porosity changes significantly (as a result of the overburden pressure).

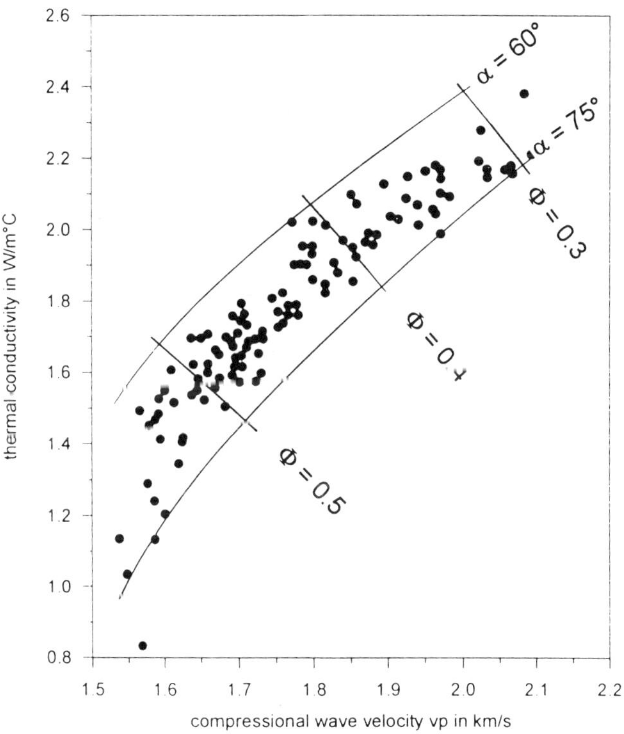

Figure 10.8 Thermal conductivity versus longitudinal wave velocity - analyse of experimental data by calculated plots.
Points: Experimental data after Finkh (1983); Sediments (water saturated) from a borehole in Lake Zürich, Switzerland
Curves: Calculated for the rock model with the following parameters:
$\alpha_g/\alpha_p = 1$ $f = 10$ $\lambda_c/\lambda_s = 0.5$ $v_p = (2000\ \text{m/s})\ v_{p,\text{norm}}$ $\lambda_s = 3.7$ W/mK
Biot-Geertsma-Smith-calculation with:
$d_s = 2.65\ 10^3$ kg m^3 $d_p = 1.00\ 10^3$ kg m^3 $k_s = 3.7\ 10^{10}$ Pa $k_p = 2.2\ 10^9$ Pa.

In a comparable manner, relationships can also be calculated for thermal conductivity versus formation factor and formation factor versus velocity.

Figure 10.9 shows as example the *thermal conductivity versus formation factor* correlation. A "network" of curves with constant porosity and curves with constant structure angle also results for this case. The grid was applied to experimental data determined on Cambrian sandstones with a relatively high quartz content (69...87 %). The scatter of the experimental data can be interpreted by a variation of porosity and structure. The high value for $\lambda_s = 5.4$ W/mK is caused by the high quartz content.

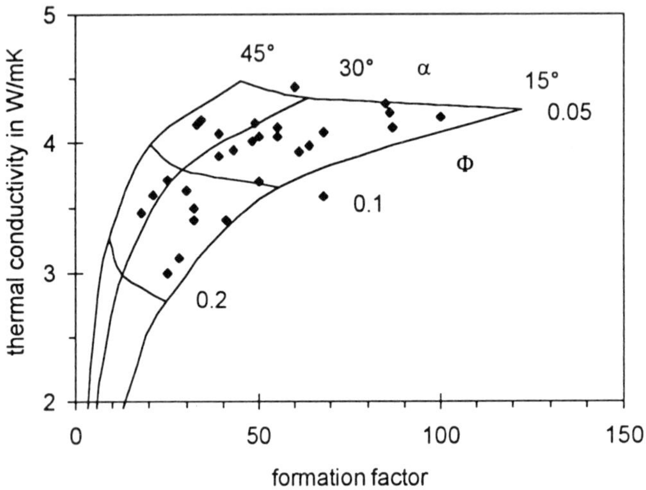

Figure 10.9 Thermal conductivity versus formation factor - analysis of experimental data by calculated plots.
Points: Experimental data after Plewa, 1976; Cambrian sandstones (water saturated), Poland
Curves: Calculated for the rock model with the following parameters:

$\alpha_g/\alpha_p = 1$ $\lambda_c/\lambda_s = 0.5$ $\lambda_s = 5.4$ W/mK

All calculations based on the structured model result in normalized crossplots. They are a network of two curve families; one is given by porosity and the other by the structure angle. This means that the correlations are controlled by

- porosity and internal structure (network parameters),
- contact or bonding properties,
- grain and pore geometric properties,

and the matrix properties as factor of normalization.

10.3. Relationships between properties determined by geophysical methods and geotechnical properties

"Engineering recognizes geophysics as an tool which can often give an important information about a site as effectively and more cheaply than a very large number of boreholes"
(D. H. Griffith and R.F. King, 1965)

10.3.1 Introduction

One direction of the expanding application of geophysical methods is concentrated towards geotechnical problems. As in other applications, there are two main questions:
- structure or "architecture" of the underground (bed boundaries, faults etc.)
- properties of the underground.

With respect to the *properties*, we can distinguish between
- ➢ a general description or classification of the rock material related to its geotechnical behaviour (for example rock type, degree of fracturing, density etc.) and
- ➢ a direct estimation of geomechanical properties (modulus of deformation, strength properties).

In practice, a *combination of geophysical and geotechnical methods* is advantageous. A model of the underground in terms of physical properties can be derived with geophysical methods; this allows the definition of "homogeneous underground sections". Then parts can be selected for the detailed geotechnical characterisation by geotechnical tests (in situ or on samples in the laboratory).
The extension of geophysical measurements or their repetition gives valuable information about a *change of properties*. Observations of the same subject at different times can signal dangerous changes of strength properties or - in the case of environmental investigations - changes of the pore content.

The direct *correlation* between geophysically determinable properties (e.g. velocity, resistivity) and geomechanical properties (e.g. modulus of deformation) can be related to the same problems and is based on the same principles as the correlation between various geophysical properties:

- the properties are of *different physical features*; for example, there is no direct connection between the elastic wave velocity and the strength of a rock material,
- both groups of properties depend on some "*joint influencing parameters*" like porosity or fracturing. Based on this, a correlation for individual rock types is given and is physically reasonable.

In the following sections, three problems are discussed:
(1) the characterisation of fractured rocks,
(2) the relationship between "statically and dynamically determined" elastic properties,
(3) the correlation of strength properties with geophysical properties.

The discussion includes a very brief description of the geomechanical problem. A deeper insight into this problem is given in textbooks and special literature about geomechanics (for example Jaeger and Cook, 1976, Kezdi, 1969, 1976, Fjær et al., 1992).

10.3.2. Fracture properties derived from seismic measurements

Fractures and cracks result in a decrease of seismic velocities (see chapter 6.); the theoretical basis for this phenomenon is discussed in section 6.4. For practical applications, the strong influence of fracturing on the elastic wave velocities is used to derive "measures of fracturing":

- a fracture coefficient ζ can be determined by the time average relationship

$$\frac{1}{v}=\frac{1-\zeta}{v_{solid}}+\frac{\zeta}{v_{fracture}} \tag{10-50}$$

$$\zeta=\frac{v_{solid}-v}{v_{solid}-v_{fracture}} \tag{10-51}$$

where v is the measured rock velocity, v_{solid} is the velocity of the non-fractured solid matrix material, and $v_{fracture}$ is the velocity of the fracture filling fluid.

- a measure of the "defects" as introduced in section 6.4.3.4. with the defect parameter D.

These parameters are primarily a measure of the fracturing *effect on the geophysical parameter* (velocity, transit travel time) only. A correlation with parameters used in *geotechnical practice* is difficult; such parameters are the frequency of joint occurrence or the number of cracks per length (e.g., per meter) and the Rock Quality Designation (RQD) index. The RQD index represents the percentage of rock core in each drill run that exceeds 4 inch (= 10.16 cm) in length without the presence of a discontinuity, discounting any mechanical fractures or breaks as is given in Table 10.5a (Carmichael, 1969).

A detailed study of the correlation between the parameters RQD index, number of cracks per meter, and longitudinal wave velocity for Scandinavian magmatites and metamorphics was published by Sjøgren et al. (1979). Table 10.5b shows the mean values for the three parameters.

Table 10.5a Rock Quality Designation (RQD) index, after Carmichael (1989)

Rock quality	RQD index in %
very poor	0...25
poor	25...50
fair	50...75
good	75...90
excellent	90...100

Table 10.5b. Longitudinal wave velocity and rock mechanic parameters (number of cracks per meter and Rock Quality Designation (RQD) index), Scandinavian magmatites and metamorphics, after Sjøgren et al. (1979)

Velocity v_p in m/s	Number of cracks per meter C	Rock Quality Designation (RQD) index in %
5500	3.5	94
5000	4.0	88
4500	6.5	78
4000	9.5	63
3500	13.5	45
3000	19.0	25

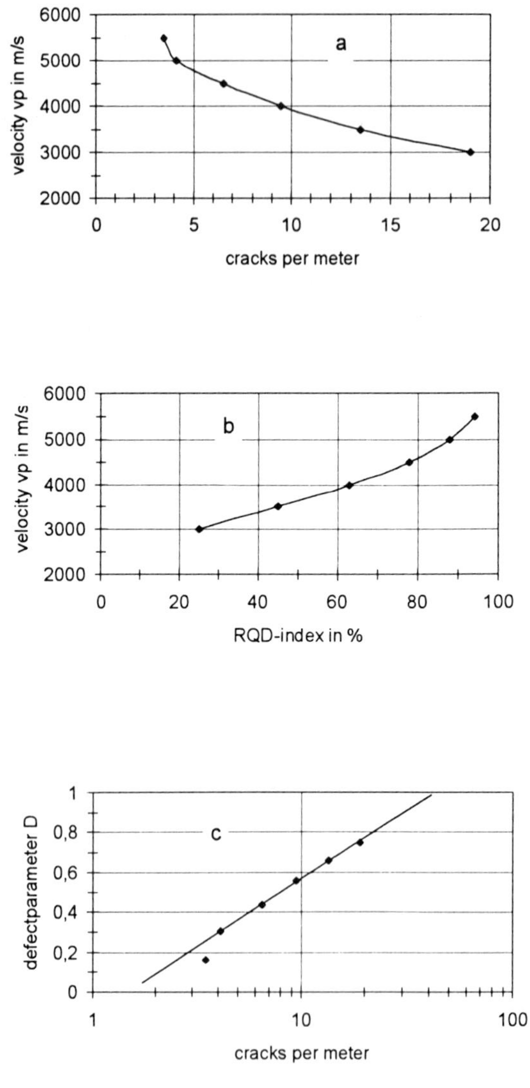

Figure 10.10. Correlation between geotechnical parameters and longitudinal wave velocity (data after Table 10.5.)
a) Correlation between the number of cracks per meter C and longitudinal wave velocity v_p
b) Correlation between the RQD index and longitudinal wave velocity v_p
c) Derived correlation between defect parameter D and number of cracks per meter C.

Figures 10.10a and 10.10b show the significant dependence of velocity v_p on both geotechnical properties. In Figure 10.9c, the parameter "cracks per meter" is compared with the "defect parameter" D, as introduced in section 6.4.3.4.. The measured velocities (Table 10.4.) were converted into D-values, where a velocity of the unfractured solid material of $v_m = 6000$ m/s was assumed.
It results in a correlation between the defect parameter D and the logarithm of the parameter C (number of cracks per meter):

$$D = 0.70 \cdot \log C - 0.123 \qquad (10\text{-}52)$$

Jamscikov et al. (1985) and Savic et al. (1969) applied velocity measurements for fracture characterisation of rocks. A relationship between elastic wave velocities and the number of cracks per meter C was also found by Idziak (1981) for sedimentary rocks (limestone, dolomite) of the Uppersilesian Coal Basin, Poland. The C values are in the range between 3 and 11 cracks per meter. The experimental data were fitted by regression equations of the type

$$v_{\text{rock}} = v_{\text{m}} \cdot (1 + a \cdot C^{m})^{-1} \qquad (10\text{-}53)$$

Results for the compressional or longitudinal wave velocity are

$$v_{\text{p}} = 7670 \cdot (1 + 0.252 \cdot C^{3/2})^{-1} \qquad (10\text{-}54)$$

and for the shear or transverse wave velocity are

$$v_{\text{s}} = 4260 \cdot (1 + 0.245 \cdot C^{3/2})^{-1} \qquad (10\text{-}55)$$

King et al. (1986) also reported about seismic measurements for the characterisation of fractured (columnar-jointed) rocks and discussed the effect of anisotropy on velocities.

10.3.3. Static and dynamic moduli

In chapter 6 (Elastic Properties), the elastic moduli were introduced. They are related to the velocities of elastic waves by equations derived from the classic *theory of elasticity*. For the isotropic homogeneous medium, these are (compare equation 6-5 to 6-10):

$$v_p = \left[\frac{E}{d} \cdot \frac{1-\sigma}{(1-2\sigma)(1+\sigma)} \right]^{1/2} \quad (10\text{-}56)$$

$$v_s = \left[\frac{\mu}{d} \right]^{1/2} = \left[\frac{E}{d} \cdot \frac{1}{2(1+\sigma)} \right]^{1/2} \quad (10\text{-}57)$$

From these equations, the elastic parameters E (Youngs modulus), μ (shear modulus), and σ (Poissons ratio) directly result for known or separately determined rock densities d (for example, by a gamma-gamma-density measurement).

$$\sigma = \frac{\Theta - 2}{2(\Theta - 1)} \quad (10\text{-}58)$$

where $\Theta = (v_p/v_s)^2$

$$\mu = v_s^2 \cdot d \quad (10\text{-}59)$$

$$E = v_p^2 \cdot d \cdot \left[\frac{(1-2\sigma)(1+\sigma)}{1-\sigma} \right] \quad (10\text{-}60)$$

The other elastic parameters can also be derived using Table 6.2. These elastic parameters are determined by seismic or ultrasonic measurements at frequencies in the range of 10 Hz up to MHz. The determination is made with a *dynamic measurement.*

This type of a dynamic measurement is different from most geotechnical test methods; these are based on a *static* or quasistatic loading and *measurement* of the deformation as a function of pressure following Hooke's definition. The *non-linearity* between stress and strain results in a formulation of Hooke's law in terms of differentials such as for the Young modulus

$$E(p) = \frac{dp}{d\varepsilon} \quad (10\text{-}61)$$

where the modulus itself is a function of pressure. For *engineering use* the deformation properties are described by a number of moduli such as (McCann and Entwisle, 1992):

(a) *Tangent Modulus*, measured at a stress level which is some fixed percentage of the ultimative strength (50 % of the uniaxial strength),

(b) *Average Modulus*, determined from the average slope of the straighter portion of the stress-strain curve,

(c) *Secant Modulus*, taken as the gradient of the line joining the origin to some fixed percentage of the ultimative uniaxial compressive strength on the stress-strain curve (50%).

Natural rocks show the phenomenon of deformation hysteresis: loading and unloading curves differ as a result of non-elastic deformation. The modulus determined by a static technique frequently is called "*static modulus*" (the term "static modulus of elasticity" is not correct, because this modulus also includes the non elastic rock behaviour).

The statically determined moduli are necessary for many calculations in geotechnique, soil mechanics and foundation problems - this has created the special testing techniques used in these disciplines. The static measurement includes non-elastic (e.g. viscous) deformations, as well as, elastic deformations. In general, the statically determined modulus is lower than the dynamic in real rocks. Only for an ideally elastic material are both moduli equal (Figure 10.11).

The *relationships between static and dynamic moduli* have been studied by Onodera (1963), Simmons and Brace (1965), Langer, 1965, Linovski (1969), Belikov et al. (1970), King, 1970, Stötzner, 1975, Linovski and Stötzner, 1976, Cheng and Johnston (1981), McCann and Entwisle (1992), King (1983), Schön (1983), Militzer et al. (1968, 1986), Jizba and Nur (1990) et al..

They confirm the general result that

- the static modulus is less than the dynamic modulus
- the difference increases with fracturing and porosity; extremely high differences occur in unconsolidated rocks,
- the difference decreases with increasing confining stress.

These tendencies are expressed by Figure 10.11: For low values of the modulus (as an indicator for unconsolidated or fractured rocks), we find big differences. For high values of the modulus (as an indicator for an undisturbed compact rock), the difference is small.

The following gives some comparative *examples* of the two moduli for igneous, consolidated sedimentary, and unconsolidated rocks. Figure 10.12 confirms the predicted trends of Figure 10.11 with experimental data.

The strong correlation between the ratio of dynamically and statically determined moduli (in this case the bulk modulus) and *pressure* on the one hand and the *crack porosity* on the other hand was studied by Cheng and Johnston (1981) for various rock types (Navajo and Berea sandstone, Westerly granite, Ammonia Tanks tuff). The ratio increases with increasing crack porosity. The defects affect the deformation of the entire rock specimen during the static measurement more than the propagation characteristics of the ultrasonic waves during the dynamic measurement. The closure of cracks with increasing pressure, therefore, results in a decrease of the ratio. A comparable result for Boise sandstone was published by King (1970).

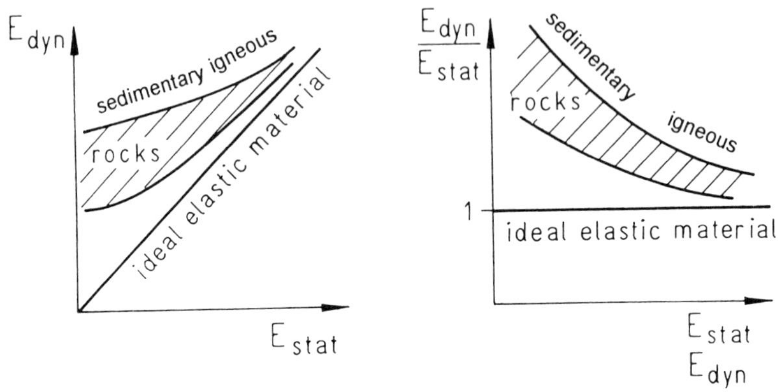

Figure 10.11 Static and dynamic modulus - schematic

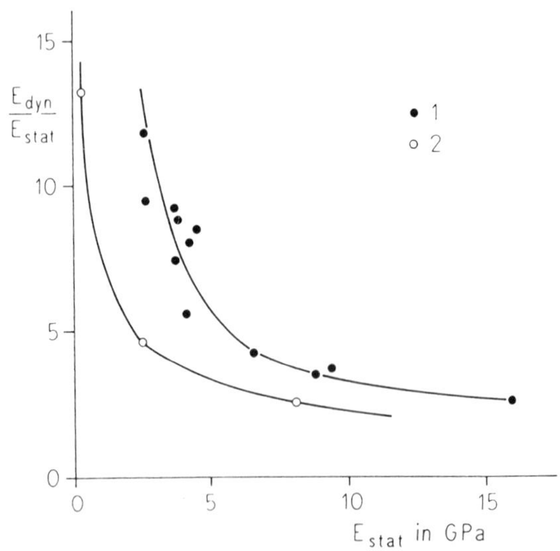

Figure 10.12 Ratio of the dynamically and statically determined Young's modulus as a function of the static Young's modulus (data after Thiel et al., 1970): 1 - limestone 2 - schist

Figure 10.13 gives a summary picture of measurements on a microcline-granite:

- both moduli decrease with increasing crack porosity, but the static modulus shows a stronger decrease than the dynamic modulus (a);
- therefore, the ratio E_{dyn}/E_{stat} increases with increasing crack porosity (b);
- relatively compact (slightly fractured) rocks have high moduli and low ratios E_{dyn}/E_{stat} (near 1), but fractured rocks have low moduli and high ratios E_{dyn}/E_{stat} (d).

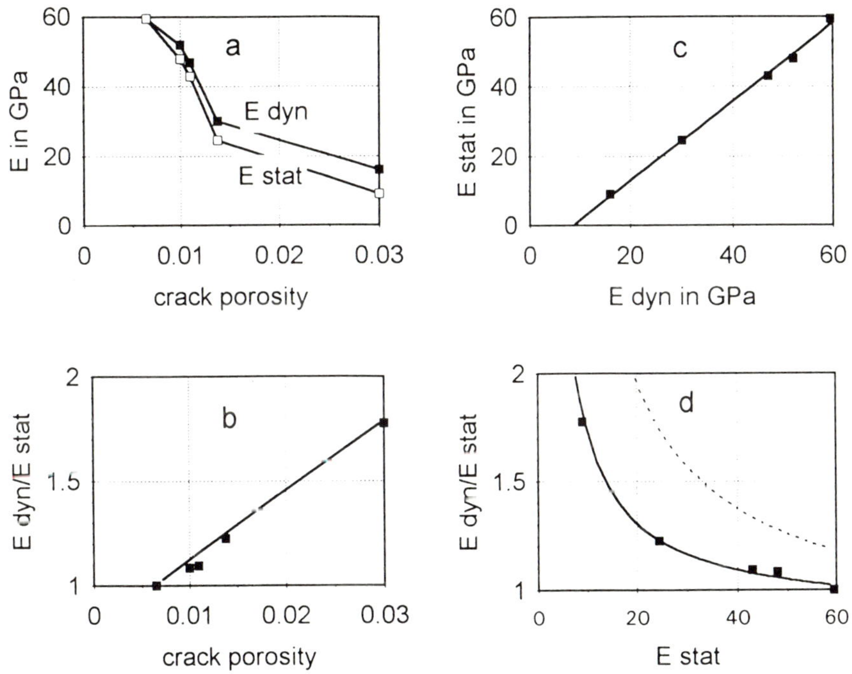

Figure 10.13 Statically and dynamically determined Young's modulus for microcline-granite (data after Belikov et al., 1970)

a) E_{stat} and E_{dyn} as function of the crack porosity Φ_c

b) ratio E_{dyn}/E_{stat} as function of the crack porosity Φ_c

c) E_{stat} versus E_{dyn} ; the curve represents the linear correlation (10-62)

d) ratio E_{dyn}/E_{stat} as function of the static modulus E_{stat},

solid curve calculated after equation 10-62 dashed curve calculated after equation 10-63

E_{stat} versus E_{dyn} is plotted in Figure 10.13c. The curve represents a linear correlation as a first approximation to fit the data of

◆ the mean values for the ratio E_{dyn}/E_{stat} for unconsolidated rocks is as great as 5 for cohesionless and 20 for cohesive rocks; McCann and Entwisle (1992) reported values of 100...200 for soft mudrocks and alluvial materials.

Due to these conditions, a strong correlation between the two moduli cannot be expected. However, one physical *basis for a correlation* is the fact that both moduli depend on the *porosity* and show a remarkably similar *pressure dependence*.

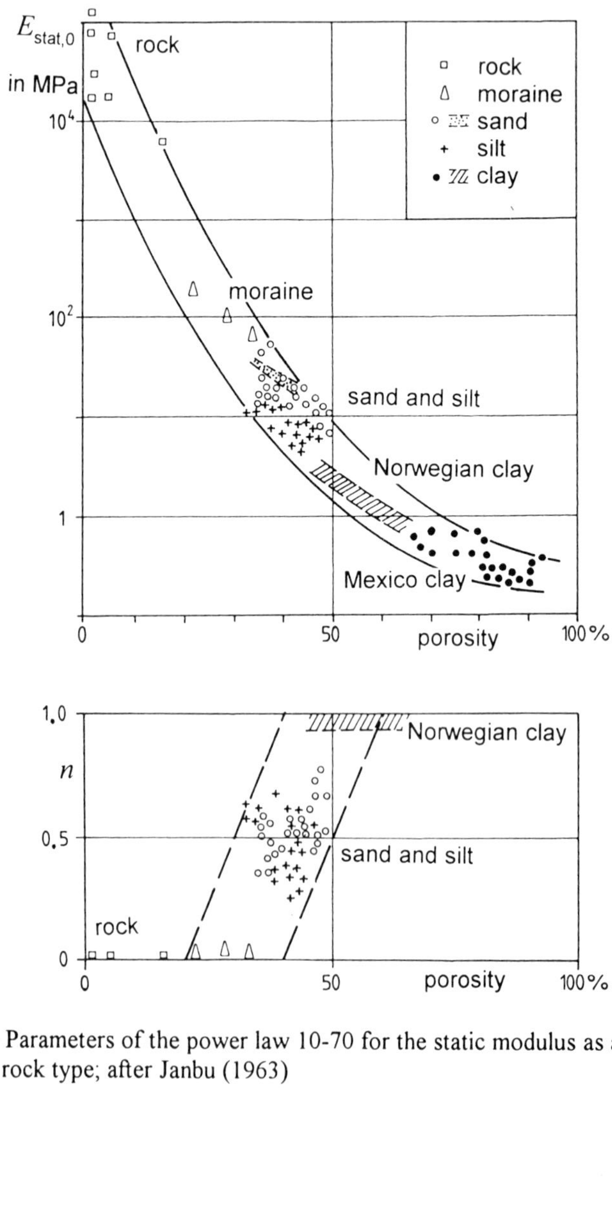

Figure 10.15 Parameters of the power law 10-70 for the static modulus as a function of rock porosity and rock type; after Janbu (1963)

In section 6.3.4.4, a power law for the pressure dependence of velocity was found (equation 6-80):

$$v = v_0 \cdot \left(\frac{p}{p_0}\right)^m \qquad (10\text{-}68)$$

From this, the equation for the *dynamic modulus* can be derived:

$$E_{\text{dyn}} = E_{\text{dyn},0} \cdot \left(\frac{p}{p_0}\right)^{2m} \qquad (10\text{-}69)$$

Janbu (1963) also has derived a power law for the *static modulus*:

$$E_{\text{stat}} = E_{\text{stat},0} \cdot \left(\frac{p}{p_0}\right)^{n} \qquad (10\text{-}70)$$

The exponent n depends on the rock type (see Figure 10.15) and shows - similarly to the exponent for the dynamic modulus (equation 6-81) - the tendency to increase from sand to clay. Figure 10.15 also confirms the influence of porosity, mainly on the value $E_{\text{stat},0}$.

Gorjainov and Ljachowickij (1979) have determined the dynamic and static Young's modulus from shallow seismic measurements and soil mechanic tests for a depth up to 10 m and applied a linear regression to the data - similar to equation 10-62 - of the form

$$E_{\text{stat}} = a \cdot E_{\text{dyn}} + b \qquad (10\text{-}71)$$

Some mean values for the coefficients a and b are given in Table 10.7.

Table 10.7 Mean values for the coefficients of the regression equation 10-71; after Gorjainov and Ljachowickij (1979); E_{stat} and E_{dyn} in MPa

Rock type	a	b in MPa
sand, wet	0.085	3
clay	0.033	6.5
soil, wet, sandy	0.061	2.85

In view of the magnitude of the data scatter, these and comparable relationships yield only a *very raw approximation*; for a general and practical application they are not transferable, but must be derived in each case for the individual rock type. It can be expected that correlations between dynamic properties derived from shear wave velocities give a better correlation, because shear wave velocities, particularly for unconsolidated rocks, are controlled by the "skeleton properties of the rock" (see section 6.3.4.). These "skeleton properties" dominantly control the (static) mechanical properties. This will also be confirmed in the following section about the correlation between geophysical properties and the strength of rocks.

10.3.4. Correlation between geophysical properties and strength properties of rocks

10.3.4.1. Strength properties

Strength properties are necessary to calculate the stability of soil and rock construction, natural slopes, excavation and for drilling processes. The investigation of failure criterion, the failure processes, and the determination of characterising parameters are the subjects of research and applications in rock and soil mechanics. For a detailed study the reader may be referred to the specific literature (for example Jaeger and Cook, 1976).

The simplest and most frequently used criterion was introduced by Coulomb in 1773. He suggested that the shear stress tending to cause failure across a plane (Figure 10.16) is resisted by the cohesion c of the material and by a parameter μ times the normal stress across the plane. This criterion is given by the equation

$$\tau = \mu \cdot \sigma_n + c \tag{10-72}$$

where τ is the shear stress and σ_n the normal stress across the plane, c is the cohesion and μ the coefficient of internal friction. The coefficient of internal friction is related to the angle of internal friction φ by

$$\mu = \tan \varphi \tag{10-73}$$

Using the principal stress components (σ_3 axial principal stress, $\sigma_1 = \sigma_2$ radial principal stress), the criterion is

$$\sigma_1 = \sigma_3 \cdot \tan^2(45 - \varphi/2) - 2c \cdot \tan(45 - \varphi/2) \tag{10-74}$$

In the τ versus σ_n plot, this form of the criterion (Mohr's criterion) results in circles (Mohr's circles) with the envelope as graphic expression of the failure criterion (the region of failure is above the line).

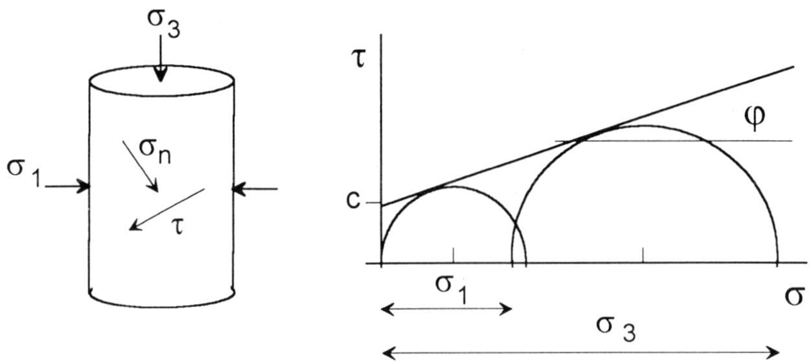

Figure 10.16 The Coulomb-Mohr failure criterion
a) A cylindrical sample under the influence of the principal stresses $\sigma_1 = \sigma_2 \neq \sigma_3$ with the shear (τ) and normal stress (σ_n) across the failure plane
b) Coulomb diagram and Mohr's circle and envelope

Table 10.8 gives mean values of μ and c for some materials.

Table 10.8 Values of μ and c for some materials; after Jaeger and Cook (1976)

Rock type	μ	c in MPa
granite	0.64	0.31
gabbro	0.66	0.38
trachyte	0.68	0.41
sandstone	0.51	0.28
marble	0.75	1.10

The strength properties of rocks depend mainly on

- the bonding type and quality of the solid particles (solid bonds in the case of igneous rocks, cementation for consolidated sediments, cohesion for clay, friction for cohesionless unconsolidated sediments like sand and gravel),
- the internal structure of the rock skeleton.

For many practical problems, the "*uniaxial compressive strength*" σ_c is used. This strength parameter is defined as the axial strength (σ_3) of an unconfined ($\sigma_1 = \sigma_2 = 0$) cylindrical sample. Thus, equation 10-74 becomes

$$\sigma_3 = \sigma_c = \frac{2c}{\tan(45 - \varphi/2)} \qquad (10\text{-}75)$$

Table 10.9 gives mean values for the uniaxial compressive strength of some rocks.

Table 10.9 Mean values for the uniaxial compressive strength of some rocks; References: K - Kezdi, 1969, J - Jaeger and Cook, 1976, H - Howarth et al., 1986, M - Müller, 1978, R - Rshewski and Novik, 1978; Rshewski and Jamscikov, 1973 (rocks from the formerly Sovjet Union)

Rock type	σ_c in MPa	Ref.
granite	100...250	R
granite (Westerly)	229	J
syenite	100...340	R
quartzite	290...300	R
quartzite, Cheshire	460	J
diabase, Frederick	487	J
marble	60...190	R
marble, Tennessee	152	J
Carrara	89	J
Carrara ($\Phi = 0.69 \pm 0.23$)	93.6 ±15.3	H
greywacke, Thuringia, Germany($\Phi = 0.4...1.5\%$)	34...123	M
sandstone	35...150	R
sandstone, Gosford($\Phi = 12.7 \pm 2.4$)	44.1 ± 6.7	H
Saxonia, Germany, Cretaceous($\Phi = 19...29\%$)	23.5...75.0	M
limestone	90...120	R
limestone, Solenhofen	224	J
Harz, Germany ($\Phi = 0.4...3.2\%$)	37...181	M
rock salt	35	R
sand	0.004...0.012	K
silt, clay - soft	...0.05	K
- medium	0.05...0.1	K
- hard	0.1...0.4	K

An engineering classification for intact rocks is based on the uniaxial compressive strength (Table 10.11).

Table 10.10 Engineering classification of intact rocks based on compressive strength; after Deere and Miller, 1966 (in: Carmichael, 1989)

Class	Type	σ_c in MPa
A	very high strength	> 220
B	high strength	110...220
C	medium strength	55...110
D	low strength	28...55
E	very low strength	< 28

The strength properties of *igneous, metamorphic and consolidated sedimentary rocks* are influenced strongly by fracturing and porosity. Rshewski and Novik (1978) recommend a regression equation of the form

$$\sigma_c = a \cdot (1 - b \cdot \Phi)^2 \qquad (10\text{-}76)$$

For limestone they found $a \approx 277$ MPa and b between 2 and 5.

Unconsolidated sediments have the weakest strength values. For *cohesionless* sediments (sand, gravel) the strength is controlled by friction at the grain to grain contacts. The coefficient of internal friction is between 0.5 and 1 (Kezdi, 1964) and the cohesion acts only for the case of partial saturation as a result of capillary forces. The coefficient depends mainly on grain shape and porosity (see Feda, 1982). Figure 10.17 demonstrates the influence of porosity on the coefficient of internal friction of sand.

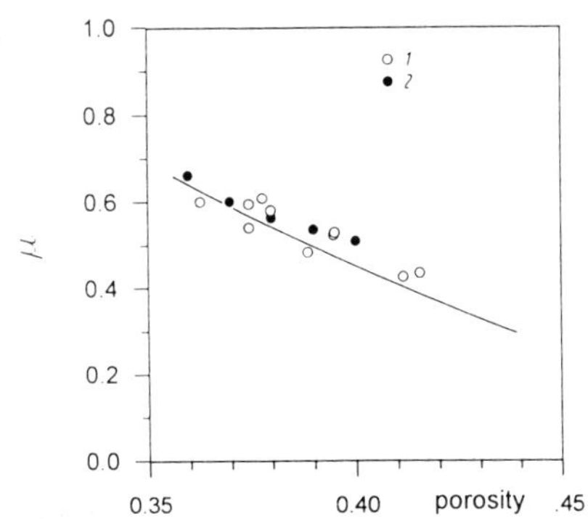

Figure 10.17 Coefficient of internal friction μ as a function of porosity Φ for unconsolidated non-cohesive sediments

1- glass spheres 0.5...1 mm diameter, (Schön, 1969, 1983)

2 - Ottawa sand (Taylor, see Herbst and Winterkorn, 1962)

As a result of this experimental test series, Lundgren (1960) derived an empirical formula for the angle of internal friction of gravel and sand including such influences:

$$\varphi = 36 + \varphi_1 + \varphi_2 + \varphi_3 + \varphi_4 \tag{10-77}$$

where φ_1 is +1° for angular grains and up to -6° for rounded grains
φ_2 is ±0 for sand, +1° for fine gravel, and +2° for medium and coarse gravel
φ_3 is -3° for uniform grain size distribution and +3° for non-uniform grain size distribution
φ_4 is -6° for loose, ±0° for medium, and +6° for dense packing.

For *cohesive* unconsolidated rocks (clay), the strength is controlled mainly by cohesion. The values are between about 10^{-3} MPa for soft marine clay and 1 MPa for consolidated clay such as the London clay (Hamilton, 1970). The magnitude depends mainly on the degree of consolidation, water content, and consistency.

10.3.4.2. Correlations

Properties like porosity, bonding type, water content and others have an influence on elastic wave velocities and on strength properties. Correlations between wave velocities and strength properties are based on this fact. As in case of the problem of static and dynamic moduli, the complexity of influences and dependencies, as well as, the fundamental difference between the two physical properties are the cause of a broad data scatter and a high uncertainty in derived relationships.

Figure 10.18 shows one example for the correlation between the uniaxial compressive strength and longitudinal wave velocity. For the sandstone samples a linear regression equation was derived (Freyburg, 1972):

$$\sigma_c = 0.035 \cdot v_p - 31.5 \tag{10-78}$$

where v_p is in m/s and σ_c in MPa.

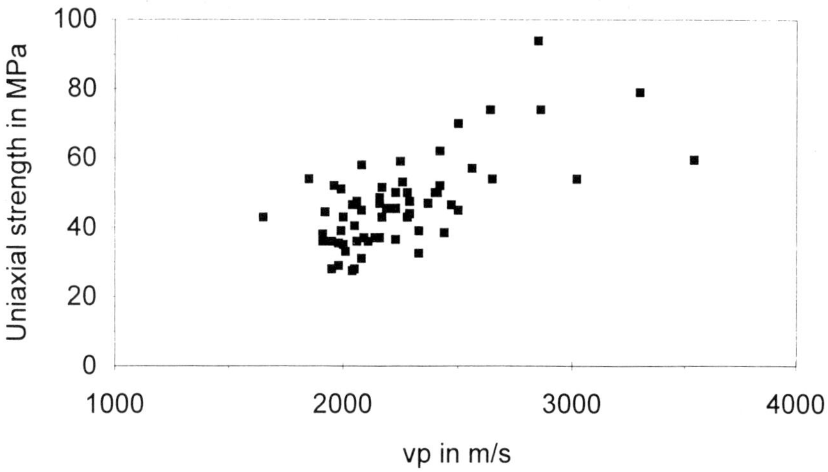

Figure 10.18 Uniaxial compressive strength versus longitudinal wave velocity for sandstone (Thuringia, Germany) after data from Freyburg, 1972

In the following, some examples of empirical relationships (after conversion of the different units) between seismic velocity (in km/s) and uniaxial compressive strength (in MPa) are given:

- Gorjainov and Ljachovickij (1979) derived a polynomial equation for sandy and shaly rocks:

$$\sigma_c = -0.98 \cdot v_p + 0.68 \cdot v_p^2 + 0.98 \qquad (10\text{-}79)$$

- for limestone, Militzer and Stoll (1973) found the following relationship

$$\sigma_c = 2.45 \cdot v_p^{1.82} \qquad (10\text{-}80)$$

- Golubev and Rabinovich (1976) used a logarithmic relationship

$$\lg \sigma_c = 0.358 \cdot v_p + 0.283 \qquad \text{for limestone} \qquad (10\text{-}81)$$
$$\lg \sigma_c = 0.444 \cdot v_p + 0.003 \qquad \text{for schist} \qquad (10\text{-}82)$$

McNally (1987) studied the correlation between the uniaxial compression strength and the travel time measured with a sonic logging tool. For fine to medium-grained sandstone from the German Creek formation (Queensland, Australia), he derived the relationship on 142 samples with a correlation coefficient of 0.91; in this relationship, σ_c is in MPa and Δt in μs/ft. Conversion to velocities (in m/s) results in

$$\sigma_c = 1277 \cdot \exp\left(-\frac{11200}{v_p}\right) \tag{10-84}$$

Figure 10.19 shows the experimental data.

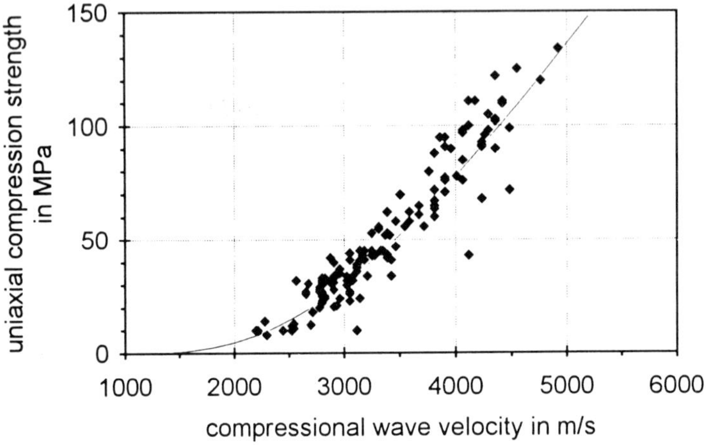

Figure 10.19 Uniaxial compression strength versus compressional wave velocity, sandstone German Creek formation, Queensland, after McNally (1987)
points: experimental data curve: regression equation 10-84

Fjær (1995) investigated the correlation between shear wave velocity or dynamic shear modulus and the shear strength on natural and synthetic sandstone (Figure 10.20).

In actual *practice*, some further parameters in addition to the uniaxial compression strength are used for a characterisation of the mechanical properties. The "*Schmidt-hammer test*" is such a "simple to use" method. Frantz (1990) has found a correlation between the parameter determined by this test and the compressional wave velocity for sandstones.

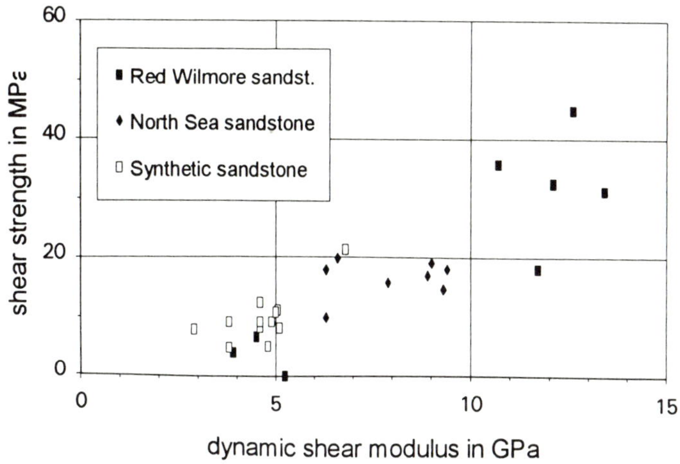

Figure 10.20 Correlation between dynamic shear modulus and the shear strength, sandstone; Fjær (1995)

The correlation between seismic velocities and strength properties (including fracture effects) is the basis for a "*Seismic rippability chart*" as published by Bison Instruments Inc.. This chart gives a classification of three material groups (rippable, marginal, non-rippable by tractors) by velocity ranges. For the main rock groups, these are given in Table 10.12.

Table 10.12 Seismic rippability classification (after data from Bison Instruments Inc.)

Rock group	v_p in km/s		
	rippable	marginal	non-rippable
Igneous rocks	< 2.1	2.1...2.6	> 2.6
Sedimentary rocks	< 2.4	2.4...3.2	> 3.2
Metamorphic rocks	< 2.3	2.3...2.9	> 2.9
Minerals and ores	< 2.4	2.4...3.2	> 3.2

Also, correlations between elastic wave velocities and parameters of the rock *drillability* have been investigated. For example, Somerton et al. (1969) reported in a study that sonic velocity is a good indicator of rock drillability for a given rock type (in this case limestone and sandstone) and the type of drilling tool. Howarth et al. (1989) has measured various physical properties (density,

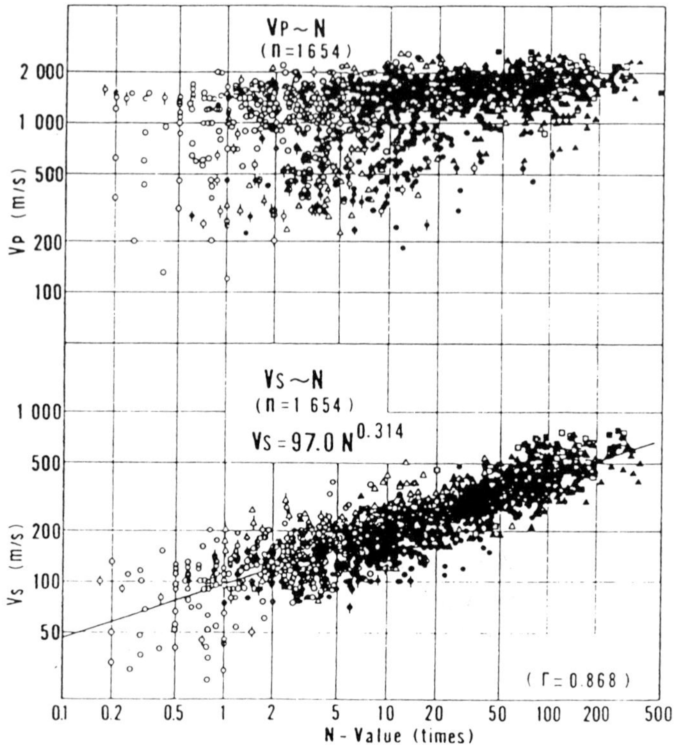

Figure 10.22 Compressional and shear wave velocity versus *N*-value for unconsolidated rocks; after Suyama et al. (1984).

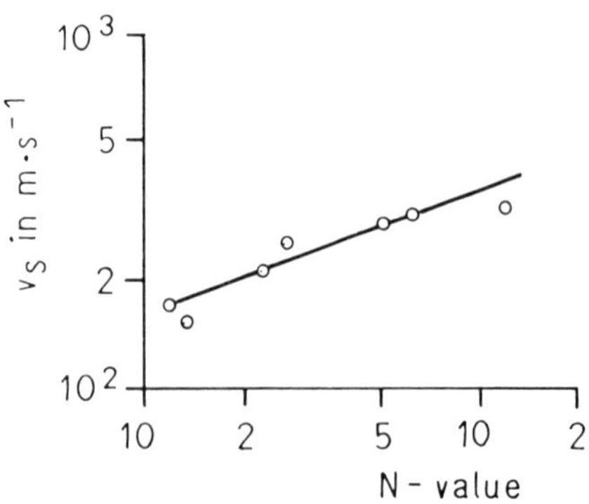

Figure 10.23. Shear wave velocity versus N-value for Oxford clay; after Davis (1989).

10.3.4.3. A model consideration

The following contains the test of the simple model for a *fractured rock* in order to derive a general form for a correlation between velocities and uniaxial compressive strength. The objective of this is not to find a numerically exact solution, but a basic formulation of the relationship.

Assuming that the strength of a material is controlled by the material section without defects, the simple relation for the model results is

$$\sigma_c = \sigma_{c,m} \cdot (1 - D) \tag{10-88}$$

where $\sigma_{c,m}$ is the strength of the intact matrix material. Comparison with equation 6-136 results in a relationship between the *strength and the square of the velocity*

$$\sigma_c = v_p^2 \cdot \left(\frac{\sigma_{c,m}}{v_m^2}\right) = A_{\sigma v} \cdot v_p^2 \tag{10-89}$$

The parameter $A_{\sigma v} = (\sigma_{c,m}/v_m^2)$ is controlled by the matrix material properties.

Figure 10.24 shows a comparison between this equation with a proportionality $\sigma_c \sim v^2$ and experimental data.

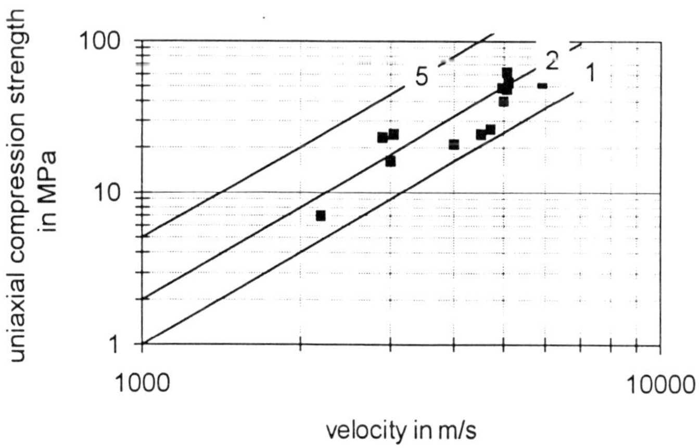

Figure 10.24 Comparison between the model-calculated correlation σ_c vs. v_p and experimental data for limestone; data after Stoll (1971); curves: calculated after equation 10-89 with curve parameter $A_{\sigma v}$; σ_c in MPa, v_p in m/s.

There is a reasonable degree of confidence between the calculated dependence and the measured data and also for the empirical relationships 10-81 and 10-83 (Figure 10.25).

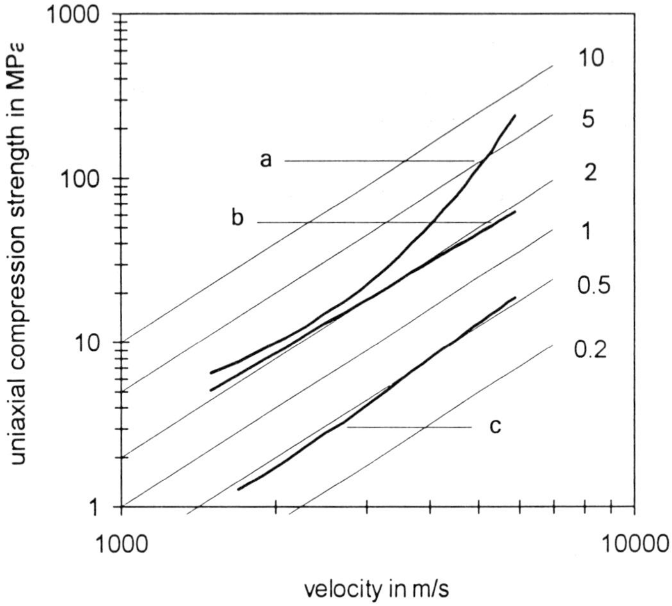

Figure 10.25 Comparison between the model-calculated and empirically determined correlations σ_c vs. v_p
empirical equations: a - equation 10-84, b - equation 10-83, c - equation 10-81
curves: calculated after equation 10-89 with curve parameter $A_{\sigma v}$; σ_c in MPa, v_p in m/s.

10.3.4.4. Remarks about changes of physical parameters connected to the failure process

The failure is connected with changes of the internal rock constitution. The pre- and post-failure changes are subject of studies about the failure process and also the problem of an earthquake prediction. Despite numerous problems and open questions many experimental observations show significant *pre-failure changes* such as

- v_p/v_s-ratio (Semenov, 1969, Nur, 1972, 1975, Aggarwal et al., 1973, Whithcomb et al., 1973, Stiller et al., 1979, Feng and Yu, 1990)
- specific electrical resistivity (Brace and Orange, 1968, Barsukov, 1970, Mazella and Morrison, 1974, King and Luo, 1990)
- Radon emission (King and Luo, 1990).

The analysis made in section 6.4.8.2 show that a model using internal structure is one way to interpret changes of the internal constitution by the categories "*structure and bonding*". A hint of a variation of structure and bonding in connection with uniaxial pressure on a microfractured sample was given in Figure 6.74. This was based on a v_p/v_s versus v_s plot.

Volarovich and Budnikov (1979) have investigated changes of compressional and shear wave velocities during uniaxial pressure experiments on plagiogranite and gneiss blocks. The observed velocities are plotted in Figure 10.26 in a v_p/v_s versus v_s grid.

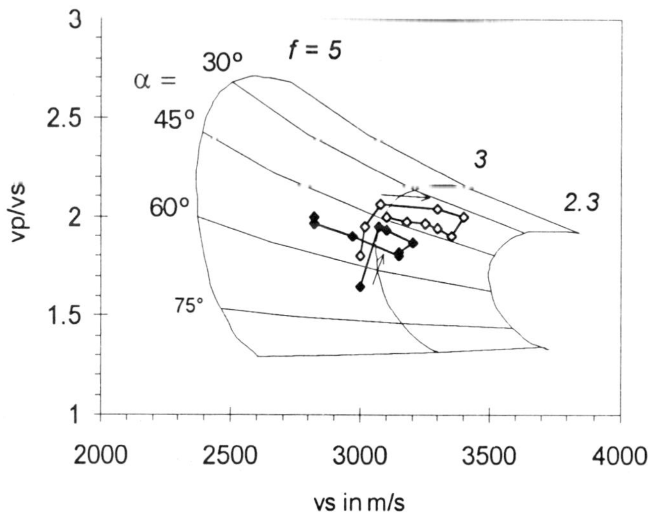

Figure 10.26 Analyse of changes of the v_p/v_s-ratio during a failure process
grid: calculated for the model with internal structure for different structure angles (α) and bonding parameters (*f*); The empirical factor *A* (see section 8.4.8.2.) is *A* = 7800 m/s
points: experimental data (after Volarovich and Budnikov, 1979); the arrow indicates the direction of increasing pressure (from 0 to about 0.8 GPa at the failure)
open symbols - plagiogranite, solid symbols - gneiss

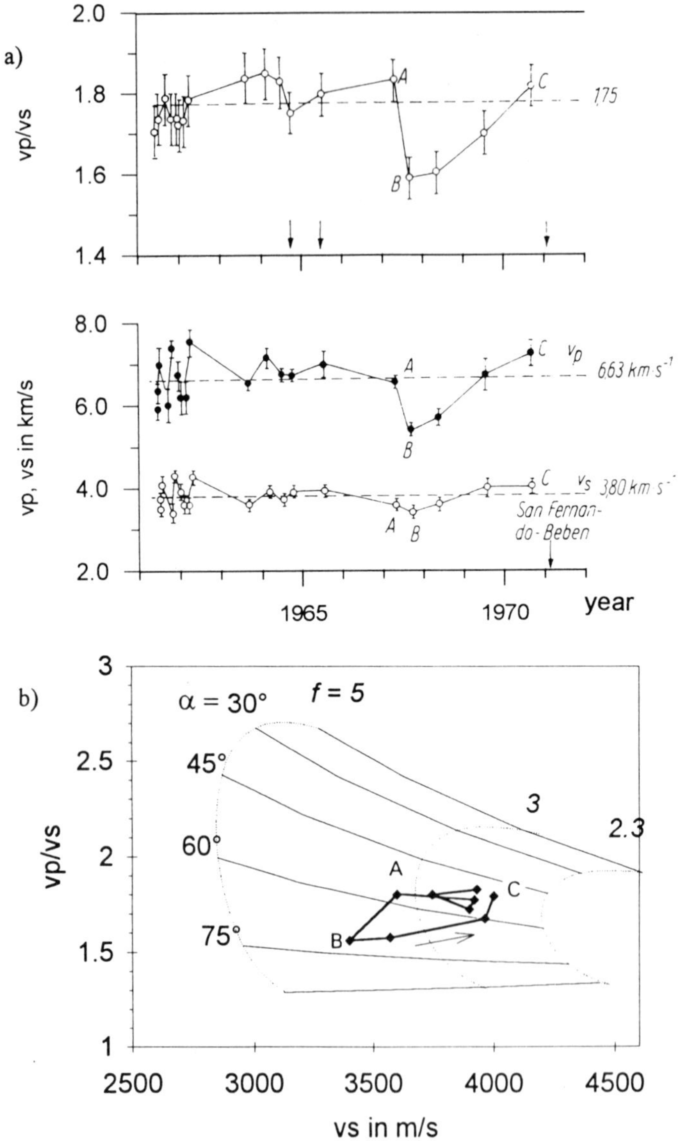

Figure 10.27 Changes of the longitudinal and transverse wave velocities in an tectonic active area of California, data from Aggarwal et al. (1973)
a) Observed v_p and v_s data and ratio; the arrows indicate events with a magnitude M > 4
b) Calculated v_p/v_s versus v_s plot and observed data from Figure 10.27a (only the two loops are plotted), A = 8000 m/s

In the pre-failure phase, the data follow a loop of dominant changes of structure and bonding. The change of structure may be a change in the main *direction of active fracture systems* under these stress conditions. The change of bonding with increasing values of the parameter f can be interpreted as a *loss of mechanical connection* between the grains and minerals along fractures and other defects. This phenomenon is probably related to dilatancy. This example shows that the concept of such a model including the categories structure and bonding may be suitable for an analysis of failure phenomena.

The *seismological observations* of the longitudinal and transverse wave velocities (Figure 10.27a) over a period of eleven years are shown in the v_p/v_s versus v_s plot (Figures 10.27b and 10.27c). Also shown are typical *loops for earthquakes*. The smaller loop is related to the event of the lower magnitude in 1964 and the large loop is related to the San Fernando Earthquake in 1971.

It may also be noted that changes of the electrical resistivity can be interpreted to be due to a change of the main active fracture orientation with this type of model (Schön, 1983).

References

Adams, J. S., P. Gasparini (1970) Gamma-ray spectrometry of rocks, Elsevier, Amsterdam

Adams, J. S., R. Weaver (1958) Thorium to uranium ratio as indicator of sedimentary processes: examples of concept of geochemical facies, Amer. Assoc. Petrol. Geol., Bull. 42, 2.

Aggarwal, Y. P., Sykes, L. R.. Armbruster, I., L. M. Sbar (1973) Premonitory changes in seismic velocities and prediction of earthquakes. Nature 241, 101-104

Ahorner, L., M. Budny (1985) Seismische Bestimmung der bodendynamischen Kennwerte von oberflächennahen Schichten in Erdbebengebieten der Niederrheinischen Bucht, In: Heitfeld, K.-H. (ed.) Ingenieurgeologische Probleme im Grenzbereich zwischen Locker- und Festgesteinen, Spriger Verlag Berlin Heidelberg, 560-582

Alexandrov, K. S., L. A. Eisenberg (1966) Sposob vycislenia fiziceskich konstant polikristalliceskich materialov, Dokl. A. N. SSSR, 1967 No. 5

Alexandrov, K. S., B. P. Belikov, T. V. Rysova (1966) Vycislenie uprugich parametrov gornich porod po mineralnomu sostavu. Izv. A. N. SSSR, Ser. Geol. Moskva, 2, 3-19

Amaefule, J., M. Altunbay, D. Tiab, D. Kersey, D. Keelan (1993) Enhanced reservoir description: Using core and log data to identify hydraulic flow and predict permeability in uncored intervals/wells; SPE 68th Annual Conference, Houston TX, Paper SPE 26436, p. 1-16

Anderson, O.L., J. E. Nafe (1965) The Bulk Modulus-Volume Relationship for Oxide Compounds and Related Geophysical Problems, Journ. Geophys. Res., 70, 3951-3963

Anderson, D. L. (1967) A seismic equation of state. Geophys. Journ., 13, 9-30.

Angenheister, G., H. Soffel (1972) Gesteinsmagnetismus und Paläomagnetismus. Studienhefte zur Physik des Erdkörpers, Borntraeger Berlin/Stuttgart

Annan, A. P., J. L. Davis (1978) High-frequency electrical methods for the detection of freeze-thaw interfaces, Proc. 3rd Internat. Conference on Permafrost Geophysics, Nat. Res. Council, Canada, 495-500

Anstey, N. A. (1990) Velocity in thin section, First Break , 9, 10, 439-457

Archer, J. S., X. D. Jing (1991) The influence of reservoir condition measurements of shaly sand electrical properties on equity studies. Trans. 14th European Formation Evaluation Symposium (SPWLA), London, paper L, 1-24

Archie, G. E. (1950) Introduction to petrophysics of reservoir rocks. Bull. AAPG, 34, 943-961

Archie, G. E. (1942) The electrical resistivity log as an aid in determining some reservoir characteristics. Trans. Americ. Inst. Mineral. Met., 146, 54-62

Arestad, J. F., B. W. Mattocks, T. L. Davis, R. D. Benson (1995) 3-D, 3-C Seismic Characterization of the Nisku Carbonate Reservoir, Joffre Field, South-Central Alberta, Canad. SEG Recorder, April 1995, 5-10

Arestad, J. F., B. Mattocks (1993) Progress Report on the Characterization of Nisku Carbonate Reservoirs, Reservoir Characterization Project - Sponsor Meeting Colorado School of Mines, paper A

Arps, J.J. (1953) The effect of temperature on the density and electrical resistivity of sodium chloride solutions, Journal Petroleum Technology, Technical Note 195, 17-20

Arshavskaya, N. I. (1973)Teplovie potoki iz Kory i verchney mantii zemli, in: Vlodavec, V. I., Ljubimova, E. A. (eds.) Verchnaja Mantia No. 12, Izdat.Nauka, Moskva, p. 26-31.
Artemieva, I. M., E. M. Cesnokov (1991) Thermal characteristics of anisotropic media with inclusions, Geophys. Journ. Int., 107, 557-562
Arts, R. J., P. N. J. Rasolofason (1992) Approximation of velocity and attenuation in general anisotropic rocks, SEG MNeeting 1992, Expand. Abstr., paper RP1.7, 640-643
Arts, R. J., P. N. J. Rasolofason, B. E. Zinszner (1992) Experimental determination of the complete anisotropic viscoelastic tensor in rocks, SEG Meeting 1992, Expand. Abstr., paper RP1.6, 636-640
Athy, L. F. (1930) Density, porosity and compaction of sedimentary rocks. Bull. Am. Ass. Petrol. Geol., 14, 1.
Atkins, E. R., G. H. Smith (1961) The significance of particle shape in formation resistivity factor-porosity relationships. J Petrol. Techn., 13, 3, 285-291
Attewell, P. B., Y. V. Ramana (1966) Wave attenuation and internal friction as functions of frequency in rocks. Geophysics, v. 31, 1049-1056.
Auberger, M., J. S. Rinehardt (1961) Ultrasonic velocity and attenuation of elastic waves in rocks, J. Geoph. Res. 66, 191-199.
Avcan, G. M., A. A. Matveenko, Z. B. Stefankiewicz (1968) Vlijanie vsestoronnego davlenija na fiziceskie svojstva pescanikov. Prikladn. Geofiz. Moskva, 55 (1969), 149-164
Avcan, G. M., A. A. Matveenko, S. B. Stefankevic (1968) Vlianie davlenija i temperatury na fiziceskie svoistva osadocnich gorni porod nasyscennych sidkostju; in: Fiz.-mech. svoistva gornich porod (ed. Rosanov, J. A.) Izdat. Nauka Moskva, 173-177,
Babuska, V. (1977) Elastic anisotropy of igneous and metamorphic rocks. Stud. geoph. et geod., Praha 12, 3, 291-303
Babuska, V. (1977) Elastic properties and velocity anisotropy in ultrabasic rocks. In: Volarovic, M. P., H. Stiller, T. S. Lebedev (ed.) Issledovania fiziceskich svoistv mineralnogo vecestva zemli pri vysokich termodinamiceskich parametrach, Izdat. Naukova Dumka, Kiev, p. 67-72
Bachman, R.T. (1983) Elastic Anisotropy in Marine Sedimentary Rocks, Journ. Geophys. Res., 88, B1, 539-545
Bachmann, R. T. (1969) Acoustic anisotropy in marine sediments and sedimentary rocks. J. Geophys. Res. 84 B, 13, 7661-7663.
Backus, G. E. (1962) Long-wave anisotropy produced by horizontal layering. Journ. Geophys. Res. 67, 4427-4440.
Bacri, J-C, D. Salin (1986) Sound velocity of a sandstone saturated with oil and brine at different concentrations, Geophys. Res. Letters, 13, 4, 326-328
Baermann, A., S. Hübner (1984) Ingenieurgeologische und geophysikalische Untersuchungen an Geschiebemergeln im Norddeutschen Raum. Geologisches Jahrbuch; Hannover, C 37, 17-57.
Bajuk, E. I., F. M. Levitova (1990) Uprugaja anisotropija ultrabasitov pri vysokich davlenii i temperature, Geofiziceski Zurnal (Kiev), 4, 12, 38-44
Bajuk, E. I. (1982) Uprugaja anisotropija gornich porod pri vysokich davleniach, Izdat. Nauka Moscow

Bajuk, E. I., P. W. Tedev (1978) Vlijanie temperatury na skorost poperecnich voln v gornich porodach pri vycokich davleniach, Fiz. Svoistva gornich porod i mineralov pri vysokich temperaturach. Izdat. Nauka, Moskva, 101-111.

Baker, P. L., W. E. Kenyon, J. M. Kestner (1985) EPT Interpretation using a textural model, SPWLA 26-th Annual Logging Symposium 1985, paper DD, 19 p.

Bakhmutov, V. G., G. F. Zagniy (1990) Secular variation of the geomagnetic field: data from the varved clays of Soviet Karelia, Physics of the Earth and Planetary Interiors, 63, 121-134

Baldridge, W. S., G. Simmons (1971) Thermal expansion of Lunar rocks., Proc. 2nd Lunar Science Conf., vol. 3, 2317-2321, Mass. Inst. Technol. Press 1971.

Baldwin, B., C. O. Butler (1985) Compaction Curves, AAPG Bulletin, 69, 622-626

Balling, N., J. Kristiansen, N. Breiner, K.D. Poulsen, R. Rasmussen (1981) Geothermal measurements and subsurface temperature modelling in Denmark, Geolog. Skrifter, 16

Balsley, J. R., A. F. Buddington (1958) Iron-titanium oxide minerals, rocks, and aeromagnetic anomalies of the Adirondack area, Econ. Geol (New York), 53, 777-805

Barker, R. D, P. F. Worthington (1973) The hydrological and electrical anisotropy of the Bunter Sandstone of northwest Lancashire, Quart. Jour. Eng. Geol., 6, 169-175

Barker, R. D., P. F. Worthington (1973) Some hydrogeophysical of the Bunter Sandstone of Northwest England, Geoexploration, Amsterdam, 11, 151-170.

Barsukov, O. M. (1970) O svjasy elektriceskogo soprotivlenia gornich porod s tektoniceskomi prozessami, Izvest. A. N. SSSR, Fiz. Zemli, 1, 84-89

Batzle, M., Z. Wang (1992) Seismic properties of pore fluids, Geophysics, 57, 11, 1396-1408

Beblo, M., J. R. Schopper, V. Haak (1982) Electrical properties; In: Landolt-Börnstein Numerical Data and Functional Relationships in Science and Technology (K.-H.Hellwege, ed.), New Series; Group V. Geophysics and Space Research, Vol.1 Physical Properties of Rocks, Subvol.b, 239-307. Springer-Verlag Berlin Heidelberg, New York 1982

Beck, A. E. (1976) An improved method of computing the thermal conductivity of fluid-filled sedimentary rocks, Geophysics, 41, 1, 133-144.

Beck, A. E. (1977) The use of thermal resistivity logs in stratigraphic correlation. Log Analyst, XVIII, 1, 17-22

Belikov, B. P., K. S. Alexandrov, T. W. Rysova (1970) Uprugie svoistva porodo-obrasujscich mineralov i gornich porod, Izdat. Nauka, Moskva

Berg, R. R. (1970) Method for determining permeability from reservoir rock properties, Trans. Gulf Coast Ass. of Geol. Soc., vol. 20, 303-317

Bergmann, L. (1954) Der Ultraschall. S. Hirzel, Zürich

Berktold, A. (1982) Elektrische Leitfähigkeit von Lösungen reiner Salze und von natürlichen Wässern; in: Landolt-Börnstein Zahlenwerte und Funktionen aus Naturwissenschaft und Technik, Neue Serie V 1b, Springer Verlag Berlin Heidelberg New York

Berryman, J. G., S .C. Blair (1986) Use of Digital Image Analysis to Estimate Fluid Permeability of Porous Materials: Application of two-point correlation functions. Journ. Appl. Phys., 60, 1930-1938.

Bertin, J., J. Loeb, (1976) Experimental and theoretical aspects of induced polarization, Geopublication Ass. (Vol. 1 and 2), Gebr. Borntraeger, Berlin-Stuttgart

Brinch Hansen, J. H., H. Lundgren (1960) Hauptprobleme der Bodenmechanik, Springer Verlag Berlin, Göttingen, Heidelberg
Brockhaus, abc - Physik. (1972), VEB F. A. Brockhaus Leipzig
Bruggeman, D. A. G. (1935) Berechnung verschiedener physikalischer Konstanten von heterogenen Substanzen. I: Dielektrizitätskonstanten und Leitfähigkeiten der Mischkörper aus isotropen Substanzen. Ann. d. Phys. 24, 636-679
Bruggeman, D. A. G. (1937) Berechnung verschiedener physikalischer Konstanten von heterogenen Substanzen. III: Die elastischen Konstanten der quasi-isotropen Mischkörper aus isotropen Substanzen. Ann. d. Phys. 29, 5, 160-178
Buchsteiner, H., N. R. Warpinski, M. J. Economides (1993) Stress induced permeability in fissured reservoirs, SPE paper 26513, 10 p.
Budiansky, B., R. J. O'Connell (1976) Elastic moduli of a cracked solid, Int. Journ. Solids Struct., 12, 81-97
Budny, M. (1991) Seismic reservoir interpretation of deep Permian carbonates and sandstones (NW German gas province) First Break, 9, 2, 55-61
Bulatova, S. M., E. A.Volkova, E. Dubrov (1970) Akusticeski karotas. Izdat. Nedra Leningrad
Bulin, N. K. (1968) Velicina parametra v zemnoj kore pri naliciiosadocnogo sloja. Prikladn. Geofiz. Moskva 52, 73- 82.
Bullard, E.C., A.A. Day (1961) The flow of heat through the floor of the Atlantic Ocean, Geophys. Journ. Royal Astron. Soc., 4, 282-292
Bunch, A. W. H., P. W. Dromgoole (1995) Lithology and fluid prediction from seismic and well data, Petroleum Geoscience , Vol. 1, 49-57
Burke, J. A., A. W. Schmidt, R. L. Campbell (1969) The Litho-porosity cross plot. Log Analyst, Nov.-Dez., 25-43.
Burkhardt, H., J. Paffenholz, R. Schütt (1986) Measurements of Q on natural rock samples in different frequency ranges, DGMK Report (Absorption seismischer Wellen), Hamburg 1986, 203- 236
Burkhardt, H., R. Mörig, R. Schütt (1990) Experimental and theoretical investigations of seismic wave absorption mechanisms in sedimentary rocks. DGMK Report 386 (Appl. of the Absorption of Seismic Waves in Hydrocarbon Exploration), Hamburg 1990
Burkhardt, H., R. Mörig, R. Schütt (1991) Laboratory investigations on rock sample to establish further fundamentals for the lithological interpretation of seismic measurements. DGMK Report (Appl. of the Absorption of Seismic Waves in Hydrocarbon Exploration), Hamburg 1991
Bussian, A. E. (1983) Electrical conductance in a porous medium, Geophysics, 48, 9, 1258-1268
Cadoret, T., D. Marion, B.E. Zinszner (1992) 1 kHz Elastic wave velocities in partially saturated limestones: evidence of fluid distribution effect, Expand. Abstracts SEG Ann. Meeting 1992, paper RP 2.4, 658-661
Cahyna, F., O. Mazac, D. Venhodova (1990) Determination of the extent of cyanide contamination by surface geoelectrical methods, in: St. Ward (ed) Geotechnical and environmental geophysics - Investigations in Geophysics No. 5 SEG, 97-99
Campbell, J. L. P. (1951) Radioactivity well logging anomalies, Petroleum Engineer, 23, 6, B7-B12

Cardoret. T, D. Marion, B. Zinszner (1993) Sonic wave velocity and X-ray tomography images for partially saturated rocks. Evidence of microscopic fluid distribution effect on acoustic properties, in: Worthington, P.F., C. Chardaire-Riviere (eds.) Advances in Core evaluation: Reservoir Management, Proc. 3rd European Core Analysts Symposium Paris, 275-290

Carman, P.C. (1956) Flow of Gases Through Porous Media, Academic Press Inc., Butterworth Sc. Publ., London

Carmichael, R. S. (1989) Practical Handbook of Physical Properties of Rocks and Minerals, CRC Press, Boca Raton Ann Arbor, Boston

Carslaw, H. S. and Jaeger, J. C. (1959) Conduction of heat in solids. Clarendon Press, Oxford.

Castagna, J. P., M. L. Batzle, R. L. Eastwood (1985) Relationships between compressional-wave and shear-wave velocities in clastic silicate rocks, Geophysics, 50, 571-581

Cattaneo, C. (1938) Sul Contatto di due Corpi Elasticti. Accad. Lincei, 27, 342-348, 434-436- 474-478

Cermak, V. (1976) Heat flow investigation in Czechoslovakia; in: Adam, A. (ed.) Geoelectric and geothermal studies (KAPG Monograph) Akad. Kiado, Budapest, 414-424.

Cermak, V. (1967) Coefficient of thermal conductivity of some sediments, its dependence on density and on water-content of rocks. Chemie der Erde, Jena, 26, 4, 271-278.

Cermak, V., L. Rybach (1982) Thermal properties, In: Landolt-Börnstein Numerical Data and Functional Relationships in Science and Technology (K.-H.Hellwege, ed.), New Series; Group V. Geophysics and Space Research, Vol.1 Physical Properties of Rocks, Subvolume a, 305-371. Springer-Verlag Berlin. Heidelberg. New York 1982

Cerniak, G. J. (1964) Dielektriceskie metody issledovania vlasnych gruntov, Izdat. Nedra, Moskva.

Cesnokov, E. M. (1977) Seismiceskaja anisotropija verchnei mantii zemli, Izdat. Nauka Moscow

Chapman, R. E. (1981) Geology and water: An introduction to fluid mechanics for Geologists, M. Nijhoff/W. Junk, The Hague

Chen, C. T., L .S. Chen, F. J. Millero (1978) Speed of sound in NaCl, $MgCl_2$, Na_2SO_4 and $MgSO_4$ aqeous solutions as functions of concentration, temperature, and pressure, Journ. Acoust. Soc. Am., 63, 1795-1800

Cheng, C. H., Toksöz, M. N. (1979) Inversion of seismic velocities for the pore aspect ratio spectrum of a rock, Journ. Geophysical Research 84, 7534-7543

Cheng, C. H., D. H. Johnston (1981) Dynamic and static moduli, Geophys. Res. Letters, 8, 1, 39-42

Chilingar, G. V. (1969) Deltaic and Shallow Marine Deposits, L.M.J.U. van Straaten (ed.) Elsevier Amsterdam

Choquette, P. W., L. C. Pray (1970) Geologic Nomenclature and Classification of Porosity in Sedimentary Carbonates, Bull. Am. Ass. Petr. Geol., 54 (2), 207-250

Christensen, N. J. (1966) Elasticity of ultrabasic rocks. Journ. Geoph. Res. Wash., 71, 5921-5931.

Christensen, N. J., U. H. Salisbury (1975) Structure and constitution of the lower oceanic crust. Rev. Geophys. Space Phys., 13, 57-86.

Christensen, N. I., R. S. Crosson (1968) Seismic Anisotropy in the Upper Mantle, Tectonophysics 6 (2)

Christensen, N. .I., Z. Wang (1985) The influence of pore pressure and confining pressure on dynamic elastic properties of Berea sandstone, Geophysics, 50, 207-213

Chung, D. H. (1963) Elastic moduli of single crystal and polycristalline MgO, Phil.Mag., 8, 833-841

Clark jr., S. P., Ed.(1966): Handbook of Physical Constants; The Geological Soc. Am., Inc.; New York

Clavier, C., G. Coates, J. L. Dumanoir (1977) The theoretical and experimental basis for the dual water model for the interpretation of shaly formations, SPE of AIME Fall Meeting 1977, paper SPE 2961

Coates, G. R., J. L. Dumanoir (1974) A new approach to improved log derived permeability, The Log Analyst (Jan.-Febr. 1974), 15, 1, 17-29

Cole, K. S., R. H. Cole (1941) Dispersion and Absorption in Dielectrics, Journ. Chem. Phys., 9, 341-351

Collett, L. S. (1959) Laboratory investigation of overvoltage; in: Wait, J.R. (ed.) Overvoltage Research and Geophysical Applications, Pergamon Press

Cornell, D., D. L. Katz (1953) Flow of Gases through consolidated porous media, Ind. Eng. Chem., 45, 2145

Correns, C. W. (1968) Einführung in die Mineralogie, 2-nd edition, Springer Verlag, Berlin. Heidelberg New York

Coutanceau-Monteil, N., Ch. Jacquin (1993) Improvements of the coaxial line technique for measuring complex dielectric permittivities of centimetric samples in the 20 to 1,000 MHz range: Application to sedimentary rocks, The Log Analyst, Sept-Oct. 1993, 21-33

Crampin, S. (1984) Effective anisotropic elastic constants for wave propagation through cracked solids Geophys. Journ. Roy. Astr. Soc., 76, 135-145

Crampin, S., J.H. Lovell (1991) A decade of shear-wave splitting in the Earth's crust: what does it mean? what use can we make of it? and what should we do next? Geophysical Journ. Internat., 107, 3, 387-408

Crampin, S., Booth, D. C. (1989) Shearwave splitting showing hydraulic dilation of pre-existing joints in granite; Scientific Drilling, 1, 21-26

Crosson, R. S., Jia-Wen Lin (1971).Voigt and Reuss Prediction of Anisotropic Elasticity of Dunite. Journ. Geophys. Res.,76, 2, 570-578

Dachnov, W. N., D. I. Djakonov (1952) Termiceskie issledovanija skvashin, Goztoptechizdat. Moskow

Dachnov, W. V. N., M. G. Latysova, V. A. Ryapalov (1967) Well Logging by means of induced polarization, The Log. Analyst., 3-18.

Dachnov, W. N. (1962) Interpretacia rezultatov geofiziceskichissledovanii razrezov skvazin, Izdat. Gostoptechizdat, Moskva.

Dachnov, W. N. (1975) Geofiziceskie metody opredelenija kollektorskich nefte-gasonasyscenija gornich porod, Izdat. Nedra, Moskva

Damm, V. (1988) Susceptibilitätsanisotropien in Sedimentiten und Magmatiten aus dem Gebiet der DDR und in Metamorphiten der Schirmacher-Oase (Antarktika), Veröff. Zentralinst. Physik der Erde, Potsdam, 118 p.

Darcy, H .(1856) Les fontaines publiques de la Ville de Dijon, V. Dalmont, Paris

Datta, S. (1968) Absorption of dilatational waves in rocks, Geoexploration 6, 3, 127-139.

David, C., M.Darot (1989) Permeability and conductivity of sandstones, in: Maury and Fourmaintraux (eds.) Proc. Symp. "Rock at great depth", Rotterdam, 203-209

Davis, A. M. (1989) Determination of dynamic elastic parameters from crosshole testing. Scientific Drilling, 1, 54-62

Davis, T. L. (1995) The New Wave, Canad. SEG Recorder, April 1995, 3

De Witte, L. (1955) A study of electric log interpretation methods in shaly formations, Trans. AIME 204, 103-110.

de Vries (1952) Het Wärmtegeleidingsvermögen van Grond; Doctoral Thesis, Rijksuniversiteit Leiden

de Waal, J. A., R. M. M. Smits, J. D. de Graaf, B. A. Schipper (1989) Measurement and evaluation of resistivity index curves, Trans. 30th Ann. Logging Symp. SPWLA, paper II

de la Cruz, V., T. J. T. Spanos (1985) Seismic wave propagation in a porous medium, Geophysics, 50, 1556-1565

Debschütz, W., Krückel U., J. R. Schopper, J. R. (1989) Der Einfluß des Überlagerungs- und Porendrucks auf die Permeabilität, den Klinkenberg-Faktor und andere Strömungsparameter; in: Maury and Fourmaintraux (eds.) Proc. Symp. "Rocks at Great Depth", Balkema Rotterdam, 179 - 185

Deresievics, H. (1958) Mechanics of Granular Matter. Advances in Applied Mechanics, Acad. Press. Inc. New York Vol. V, 233-306

Desai, K. P., D. P. Helander, E. J. Moore (1969) Sequential measurement of compressional and shear velocities of rock samples under Triaxial pressure. Soc. Petr. Engrg. J. New York, 94, 378-394.

Desbrandes, R. (1968) Theorie et interpretation des diagraphies, Edit. Technip. Paris

Diederix, K. M. (1982) Anomalous relationships between resistivity index and water saturations in the Rotliegend Sandstone (The Netherlands), Trans. 23rd Ann. Log. Symp. SPWLA, paper X

Digby, P. J. (1981) The effective elastic moduli of porous granular rocks, Journ. Appl. Mech. ASME, 48, 803-808

Dissado, L. A., R. M. Hill (1979) Non-exponential decay in dielectrics and dynamics of correlated systems, Nature, 279, 685-689

Dissado, L. A., R. M. Hill (1984) Anomalous Low-frequency dispersion, Journ. Chem. Soc. Faraday Trans., 80, 291-319

Dissado, L.A., R.M. Hill (1987) Small-signal AC frequency response functions, Solid State Ionics, 22, 331-336

Djakonov, D. I., B. A. Jakovlev (1969) Opredelenie i ispolsovanie teplovych svoistv gornich porod i plastovych sidkostei neftanych mestorosdenii, Izdat. Nedra, Moskva

Dobrynin, W. M. (1970) Deformacii i izmenenia fiziceskich svoist kollektorov nefti i gaza, Izdat. Nedra, Moskva

Dobrynin, W. M. (1963) Izmenenie fiziceskich svoistv pescanikov pod deistveiem vsestoronnego davlenija, Trudy MINChGP, vyp. 41, Gostop Moscow

Dodson, C. R., M. B. Standing (1945) Pressure-volume-temperature and solubility relations for natural-gas-water mixtures; in: Drilling and Production Practices, 1944, Am. Petr. Inst.

Doering, M. A., D. P. Smith (1974) Locating extraneous water sources with the gamma ray log, SPE Fall Meeting, paper SPE 5116
Domenico, S. N. (1976) Effect of brine gas mixture on velocity in unconsolidated sand reservoir, Geophysics 41, 882-894.
Domenico, S. N. (1977) Elastic properties of unconsolidated porous sand reservoirs, Geophysics 42, 1339-1368.
Domenico, S. N. (1984) Rock lithology and porosity determination from shear and compressional wave velocity, Geophysics, 49, 1188-1195
Donaldson, E. C., T. K. Siddiqui (1987) Relationship between the Archie saturation exponent and wettability, SPE paer 16790, 439-443
Dorofeyeva, R. P. (1990) Teplogeneracia gornich porod i radiogennoe teplo Pribaikala i Sabaikala, Fizika Semli, 1, 82-90
Dorsey, N. E. (1940) Properties of ordinary water substance, Reinhold Publ. New York
Dortman, N. B, M. Magid (1969) Zavisimost skorosti uprugich voln i plotnosti gornich porod ot kristallochimiceskich osobennostej minerlalov, Geologia i geofizika No. 4, 93-105.
Dortman. N. B. (1976) Fiziceskie svoistva gornich porod i polesnich iskopamych. Izdat. Nedra, Moskva
Doveton, J. H. (1986) Log analysis of subsurface geology, John Wiley & Sons New York
Doyen, P. M. (1988) Permeability, conductivity, and pore geometry of sandstone, Journ. Geophys. Res., 95, B5, 6993-7005
Draxler, J. (1990) Untertage-Spektroskopie und Scanner-Verfahren, Freiberger Forschungshefte C 447, Leipzig 1990, 43-68
Draxler, J, K.R. Hänel (1988) Grundlagenforschung und Bohrlochgeophysik, Bohrlochmessungen in der KTB-Oberpfalz VB, Intervall 1529. 4-3009 m, KTB-Report 88-7, Hannover
Dresser Atlas (1985) A Guide To Formation Damage Prevention In Fine-Grained Sandstones. Dresser Atlas Ind. Inc., Rept 10/85 9531
Dresser Atlas (1982) Well logging and interpretation techniques (The course for home studies), Dresser Ind. Inc., Houston
Dresser Atlas (1985) A Guide to Formation Damage Prevention in Fine Grained Sandstones, Dresser Atlas Rep. 10/85 - 9531
Dresser, Ind., Spectralog, Houston/USA.
Drury, M. J., A. M. Jessop (1983) The estimation of rock thermal conductivity from mineral content - an assessment of techniques, Zentralbl. f. Geol. und Paläontol., 1-2, 35-48
Duffy, J., R. D. Mindlin (1959) Stress-Strain Relations and Vibrations of Granular Medium Journ. of Applied. Mech. ASME, New York, 24, 585-593.
Duffy, S. Th., D. L. Anderson (1989) Seismic Velocities in Mantle Materials and the Mineralogy of the Upper Mantle, Journ. Geophys. Res., 94, B2, 1895-1912.
Eastwood, R. L., J. P. Castagna (1983) Basis for interpretation of vp/vs ratios in complex lithologies, Trans. 24th Ann. Logging Symp. SPWLA, paper G 1-17
Eberhart-Phillips, D., D-H. Han and M. D .Zoback (1989) Empirical relationships among seismic velocity, effective pressure, porosity, and clay content in sandstone, Geophysics, 54, 1, 82-89.

Ebert, H. (ed.) (1976) Physikalisches Taschenbuch, Vieweg und Sohn, Braunschweig, 2. Aufl., 5. Aufl. 1976.

Economides, M., P. Ungemach, eds. (1987) Applied Geothermics. John Wiley & Sons, Chichester, New York, Brisbane, Toronto, Singapore.

Ehrlich, R., S. K. Kennedy, S. J. Crabtree, R. L.Cannon (1984) Petrographic Image Analysis, Journ. of Sediment Petrology, vol. 54, 1365-1378.

Eichelberger, J. F. (1989) Research drilling in young volcanoes, Scientific Drilling, 1, 90-102

Eissa, E. A., A. Kazi (1988) Relation between static and dynamic Young's moduli of rocks, Int. Journ. Rock Mech. Sci. & Geomech. Abstr., Vol. 25, 479 - 482

Elliott, S. E., B. F.Wiley, (1975) Compressional velocities of partially saturated, unconsolidated sands, Geophysics, 40, 949-954

Ellis, D. V. (1987) Well Logging For Earth Scientists, Elsevier Sc. Publ., New York, Amsterdam, London.

Elphick, R. Y. (1993) From the editor, The Log Analyst, Nov-Dec., 2

Endres, A L., R. Knight (1992) A theoretical treatment to the effect of microscopic fluid distribution on the dielectric properties of partially saturated rocks, Geophysical Prospecting, 40, 307-324

Endres, A., R. Knight (1989) The effect of microscopic fluid distribution on elastic wave velocities, The Log Analyst, Nov-Dec., 437-445

Engelhardt, W. V. (1960) Der Porenraum der Sedimente, Springer-Verlag Berlin, Göttingen, Heidelberg

Eötvös Lorand Geophysical Inst. -ELGI: Water

Ermankov, W. A., A. I. Levykin, T. K. Slobin (1989) Sostav i stroenie semnoi kory i verchnej mantii Kurilskich Ostrovov, Fizika Semli, 10, 18-35

Eshelby, J .D. (1957) The determination of the elastic field of an ellipsoidal inclusion and related problems, Proc. Roy. Soc., Ser. A, 221, 376-396

Falkenhagen, H. (1971) Theorie der Elektrolyte, Hirzel Verlag Leipzig

Falkenhagen, H. (1932) Elektrolyte, Hirzel-Verlag Leipzig

Fatt, I. (1953) The Effect of Overburden Pressure on Relative Permeability, Petr. Trans., AIME, 198, 325 -326

Fatt, I., D. H. Davis (1952) Reduction in Permeability with Overburden Pressure. Trans., AIME, 195, 329.

Faust. L. (1975) Velocity function including lithologic variation, Geophysics, 183, 271-287.

Feda, J. (1982) Mechanics of particulate materials, Academia Press Prague

Feng, D., Yu, X. (1990) v_p/v_s variations before reservoir-induced seismic events, Gerlands. Beitr. zur Geophysik, Leipzig, 99, 3, 259-264

Ferrand, L. A., M. A. Celia, W. E. Soll (1990) Percolation-based Models for Pore-to Lab Scale Calculations in Multifluid Porous Media. In: J. H.Cushman (ed.) Dynamics of Fluids in Hierarchical Porous Media , 663-683. Academic Press, London.

Fertl, W. H. (1979) Gamma ray spectral data assists in complex formation evaluation, Transact. 6-th Europ. SPWLA (London), Paper Q

Fertl, W. H. (1983) Gamma ray spectral logging: A new evaluation frontier (part I....VII), World Oil, series March 1983...November 1983

Fertl, W. (1984) Well Logging and Its Applications in Cased Holes, Journ. Petr. Technol., 250-266

Finckh, P (1983) On the calibration of lacustrine heat flow density measurements in a borehole in Lake Zürich, Switzerland, Zbl. Geol. Paläont. (Stuttgart), Teil I, H. 1/2, 102-117

Fjaer, E., R. M. Holt, P. Horsrud, A. M. Raaen, R. Rines (1992), Petroleum Related Rock Mechanics, Elsevier Science Publ., Amsterdam

Fjær, E. (1995) personal communication

Francl, J. W., D. Kingery (1954) Thermal conductivity: IX. Experimental investigation of effect of porosity on thermal conductivity, Journ. Amerc. Ceram. Soc., 37, 2, 99-107.

Frantz, E. (1990) Untersuchungen zum Verwitterungsverhalten von Sandsteinen des Mittleren und Oberen Buntsandsteins; Ber.-,Reports, Geol.-Paläont. Inst. Univ. Kiel, Nr. 37,

Freund, D. (1987) K voprosu o svjasi pronizaemosti i elektriceskoi provodimosti maloporistych osadocnich porod pri vysokom davlenii, Geofiziceski Journ., 9, 5, 58-61, Kiew

Freund, D., G. Nover (1995) Hydrostatic pressure tests for the permeability - formation factor relation on crystalline rocks from the KTB Drilling Project; Surveys in Geophysics, 16, No. 1, 47-62

Freyburg, E. (1972) Der untere und mittlere Buntsandstein SW-Thüringens in seinen gesteinstechnischen Eigenschaften, Ber. dte. Ges. geol. Wiss. A; Berlin, 17 6, 911-919.

Fricke, H. (1931) The electric conductivity and capacity of disperse systems, Physics, 1, 106-115

Frisillo, A. L., L. Thomsen (1991) Effect of fluid micro-distribution on velocity and attenuation in porous rock, 53rd Meeting EAEG, 1991 Florence, abstracts paper C019, 246-247

Fröhlich, L. (1975) Einsatzmöglichkeiten und Ergebnisse einer geophysikalischen Komplexmethodik zur Strukturuntersuchung und Kennwertbestimmung, Diss. Univ. Leipzig (unpublished)

Fröhlich, L., G. Schwinge (1978) Geophysikalische Messungen zur Lösung ingenieurgeologischer Bauwerksüberwachung für Wasserspeicher im Lockergesteinsbereich, Zeitschr. f. geol. Wiss. Berlin, 6, 8, 1017-1024.

Frölich, F. (1959) Magnetismus der Erdkruste. In: Fanselau (ed.) Geomagnetismus und Aeronomie. VEB Deutscher Verlag der Wissenschaften, Berlin Bd III.

Fuller, B. D., Ward, S. H. (1970) Linear system description of the electrical parameters of rocks, IEEEE Trans. Geosc.Electronics GE-8, 7-8

Galdin, N. E. (1966) Vlianie vysokogo davlenia na skorost prodolnych voln v gneisach i kristalliceskich slanzach, in: Elektriceskie i mechaniceskie svoistva gronich porod pri vysokich davleniach, AN SSSR. Trudy Inst. Fiz. Semli No. 37 (204), Izdat. Nauka Moskva, 98-117.

Garanin, V. A., O. J. Rogosa, F. M. Sibagarullina (1965) O pogloscajuscich svoistvach vodonasyscennich i neftegasonasyscennych kollektorov, Prikladnaja Geofizika, Moskva, 44, 99-100.

Garanin, V. A. (1968) O zavisimosti pogloscajuscich i skorostnych svojstv dvuchfaznych sred ot vjazkosti prorozapolnitelja, Prikladnaja Geofizika, Moskva, 52, 83-92.

Gardner, G. H. F., M. R. J. Wyllie, U. D. Droschak (1964) Effects of pressure and fluid saturation of elastic waves in sand, Journ. Petrol. Technol. 16, 2, 189-198.

Gardner, G. H. F., L. W. Gardner, A. R. Gregory (1974) Formation velocity and density. The diagnostic basis for stratigraphic traps, Geophysics, 39, 770-780

Garrouch A A., M. M. Sharma (1994) The influence of clay content, salinity, stress, and wettability on the dielectric properties of brine-saturated rocks: 10 Hz to 10 MHz, Geophysics, 59, 6, 909-917

Gassmann, F. (1951) Elastic waves through a packing of spheres, Geophysics, Tulsa, 16, 673-685.

Gassmann, F. (1951) Über die Elastizität poröser Medien, Vierteljahresschr. d. Naturforsch. Ges. Zürich, 96, 1-22.

Gassmann, F. (1964) Introduction to seismic traves time methods in anisotropic media, Pure and Appl. Geophys. 58 II, 63-113.

Gavrilenko, P. Y. Gueguen (1989) Pressure dependence of permeability: a model for cracked rocks. Geophys. Journ. Internat., 98, 159-172.

Gearhart Ind. (1978) Formation evaluation data handbook, Edition Forth Worth/USA.

Gebrande, H., H. Kern, F. Rummel (1982) Elasticity and Inelasticity. In: Landolt-Börnstein Numerical Data and Functional Relationships in Science and Technology (K.-H.Hellwege, ed.), New Series; Group V. Geophysics and Space Research, Vol.1 Physical Properties of Rocks, Subvolume b, 1-233. Springer-Verlag Berlin. Heidelberg. New York

Geertsma, I., D. C. Smith (1961) Some aspects of elastic wave propagation in fluid saturadet porous solids. Geophysics, Tulsa, 26, 169-181.

Geertsma, A, J. (1961) Velocity log interpretation: The effect of rock bulk compressibility; Soc. Petr. Eng. Journ., 1, 235

Geertsma, A. J. (1974) Estimating the coefficient of inertial resistance in fluid flow through porous media, Journ. Petroleum Techn., 211-216

Georgi, D.T., S.K. Menger (1994) Reservoir quality, porosity and permeability relationships, Trans. 14. Mintrop-Seminar (Beilage) DGMK und Ruhr-Universität Bochum 163/1 - 163/35

Gerland, S., M. Richter, H. Villinger, G. Kuhn (1993) Non-destructive porosity determinations of Antarctic marine sediments derived from resistivity measurements with an inductive method, Marine Geophysical Researches, 15, 201-218

Giesel, W. (1963) Elastische Anisotropie in tektonisch verformten Sedimentgesteinen, Geophysical Prospecting, 11, 4, 423-458

Glover, P.W.J., F.J. Vine (1995) Beyond KTB - Electrical Conductivity of the Deep Continental Crust, Surveys in Geophysics (Kluwer Acad. Publ.), 16, 5-36

Golubev, *A*. A, V. I. Oparin, G. A. Rabinovic (1971) Izucenie fiziko mechaniceskich svoistv gornych porod metodom akusticeskogo karotaza. Razvedocnaja Geofizika, Moskva, 46, 116-120.

Golubev, A., G. J. Rabinovic (1976) Resultaty primenenia apparatury akusticeskogo karotasa dlja predelenia procnostych svoistv gornych porod na mestorosdeniach tverdych iskopaemych. Prikladnaja Geofizika Moskva, 73, 109-116.

Gorbusina, L. W., D. F. Simin, W. W. Naglaja(1974) Radiometria i jadernaja geofizika, Izdat. Nedra, Moskva

Gordon, R. B., L. A. Davis (1968) Velocity and attenuation of seismic waves in imperfectly elastic rock, Journ. Geophys. Research, 73, 3917-3935.

Gorjainov, N. N., F. M. Ljachovickij (1979) Seismiceskie metody v insenernoi geologii, Izdat. Nedra, Moskva
Goss, R., J. Combs, A.Timur (1975) Predicition of thermal conductivity in rocks from other physical parameters and from standard geophysical well logs, SPWLA 16-th. Ann. Log Symp. New Orleans
Graham, G., S. Crampin, L. M. Fernandez (1991) Observations of shear-wave polarizations from rockbursts in a South African gold field: an analysis of acceleration and velocity recordings, Geophysical Journ. Internat., 107, 3, 661-672
Graton Fraser (1935) Systematic packing of spheres with particular relation to porosity and permeability, Journ. Geol. 43, 8, 785-909
Gregory, A. R. (1976) Fluid saturation effects on dynamic elastic properties of sedimentary rocks, Geophysics, 41, 895-921
Gregory, A. R. (1977) Aspects of rock physics from laboratory and log data that are important to seismic interpretation, in: Seismic Stratigraphy - application to hydrocarbon exploration, AAPG Memoir 26
Gregory, A. R. (1990) Ultrasonic velocity and attenuation measurements in water-based drilling muds, Paper presented at ASME 1990 ETCE Conference, New Orleans, La ; Jan. 1990
Gregory. R. (1963) Shear wave velocity measurements of sedimentary rock samples under compression, Proc. 5-th. Symp. Rock Mech., Perg.-Press New York 439-471.
Griffith, A. (1924) The theory of rupture. Proc. 1-st Int. Congr. Appl. Mech., Delft, 55-63.
Griffith, D. H., R. F. King (1965) Applied Geophysics for Engineers and Geologists, Pergamon Press
Griffiths, C. M., N. R. Brereton, R. Beausillon, D. Castillo (1992) Thermal conductivity prediction from petrophysical data: a case study, in: Geological Applications of Wireline Logs II (ed.: Hurst, A., C. M. Griffiths, P. F. Worthington), Geological Society Special Publication No. 65, pp. 299-315
Grim, R. E. (1968) Clay Mineralogy, 2nd ed., McGraw-Hill Book, New York
Gueguen, Y., Ch. David, M. Darot (1986) Models and time constants for permeability evolution, Geophys. Res. Letters, 13, 5, 460-463.
Gueguen,Y., J. Dienes (1989) Transport Properties of Rocks from Statistics and Percolation. Mathematical Geology, 21, 1, 1-13
Guyod, H., P. A. Wichman (1953) Ionic double-layer conductivity in reservoir rock, Trans. Am. Inst. Mining and Metall. Engrs.(Petrol. Trans. AIME) 198, 129-134.
Guyod, H., P. A. Wichman (1972) Theoretical notes on the resistivity of shaly sands, Log Analyst 13, 3, 11-16.
Haack, U. (1982) Radioactivity of rocks; in: Landolt-Börnstein Numerical Data and Functional Relationships in Science and Technology (K.-H.Hellwege, ed.), New Series; Group V. Geophysics and Space Research, Vol.1 Physical Properties of Rocks, Subvolume b, 433-481. Springer-Verlag Berlin Heidelberg New York
Haalck, H. (1953) Lehrbuch der angewandten Geophysik, 2. Aufl., Gebr. Borntraeger Berlin
Haalck, H. (1959) Physik des Erdinnern, Akad. Verl. Ges. Geest u. Portig Leipzig
Hagiwara, T. (1986) Archie's "m" for permeability, The Log Analyst, Jan-Febr., 39-42
Hamdi, F., D. T. Smith (1982) The influence of permeability on compressional wave velocity in marine sediments, Geophys. Prospecting, 30, 622-640

Hamilton, R. U. (1965) Temperature variation at constant pressures of the electrical conductivity of periclase and olivin, Journ. Geoph. Res. 70, 22, 5679-5692.

Hamilton, E. L. (1979) Sound velocity gradients in marine sediments, Journ. Acoust. Soc. Am. 6, 909-922.

Hamilton, E. L. (1972) Compressional-wave attenuation in marine sediments. Geophysics, 37, 4, 620-646.

Hamilton, N., W. H. Owens, A. I. Rees (1968) Laboratory experiments on the production of grain orientation in shearing sand, Journ. Geol., 76, 465-472.

Hamilton, E. L. (1970) Sound velocity and related properties of marine sediments. North Pazific; Journ. Geoph. Res., 75, 4423-4446.

Hamilton, E. L. (1971) Elastic properties of marine sediments. Journ. Geoph. Res., 76, 579-604.

Hamilton, E. L. (1976) Shear wave velocity vs. depth in marine sediments: a review. Geophysics, 41, 5, 985-966

Hamilton, E. L. (1976) Variations of Density and Porosity with depth in deep-sea sediments. Journ. Sed. Petrol. 46, 280-300.

Hamilton, E. L. (1978) Sound velocity-density relations in sea floor sediments and rocks. Journ. Acoust. Soc. Am. 63, 366-377.

Hamilton, E. L. (1979) vp/vs and Poissons ratios in marine sediments and rocks, Journ. Acoust Soc. Am. 66, 1093-1101.

Hamilton, E.L. (1976) Attenuation of shear waves in marine sediments, Journ. Acoust Soc. Am. 60, 2, 334-338.

Hamilton, E.L. (1976) Sound attenuation as function of depth in the sea floor, Journ. Acoust. Soc. Am. 59, 3, 528-535

Hamilton, E. L., R. T. Bachman, W. H. Berger, T. C. Johnson, L. A. Mayere (1982) Acoustic and related properties of calcareous deep-sea sediments, Journ. Sedimentary Petrol., 52, 0733-0753

Han, D., A. Nur, D. Morgan (1986) Effects of porosity and clay content on wave velocities in sandstones, Geophysics, 51, 11, 2093-2107

Hanai, T. (1960) Theory of the dielectric dispersion due to the interfacial polarization and its application to emulsions, Kolloidzeitschrift, vol. 171, 23-31

Hara, G. (1935) Theorie der akustischen Schwingungsausbreitung in gekörnten Substanzen und experimentelle Untersuchungen an Kohlepulver, Elektr. Nachrichtentechnik, 12, 7, 191-200.

Hashin, Z. (1964) Theory of mechanical behavior of heterogeneous media, Appl. Mech. Reviews 17, 1, 1-9.

Hashin, Z., S. Shtrikman (1962) On some variational principles inanisotropic an nonhomogenous elasticity, Journ. Mech. and Phys. Solids, 10, 335-342.

Hashin, Z., S. Shtrikman (1962) A variational approach tt the theory of elastic behavior of polycrystals, Journ. Mech. Phys. Solids, 10, 343-352

Hashin, Z. (1962) The elastic moduli of heterogeneous materials, Journ. Appl. Mech., 29, 143-150

Hazen, A. (1893) Some Physical Properties of Sands and Gravels with Special Reference to their Use in Filtration, 24nd Ann. Rept., Mass. State Board of Health 1892, Pub. Doc. 34, 539-556.

Hughes, D . S., J. H. Cross (1951) Elastic wave velocities in rocks at high pressures and temperatures. Geophysics 16, 4, 577-593.
Hunt, G. R., G. R. Johnson , G. R. Olhoeft, D. E. Watson, K. Watson (1979) Initial Report of the Petrophysics Laboratory, Geol.Survey Circ.789, 74 p., U.S.Govern. Print.Off: 1979-677- 026/67
Hunter, A. N., R. Legge, E. Matsukawa (1961) Measurements of acoustic attenuation and velocity in sand. Acoustica 11, 1, 26-31.
Hurtig, E. (1965) Untersuchungen der Wärmeleitfähigkeitsanisotropie von Sandsteinen, Grauwacken und Quarziten. Pure and Appl. Geoph. 60, 1, 85-100.
Hurtig, E. (1963) Untersuchungen über die Fortpflanzungsgeschwindigkeit von P-Wellen in Sedimentsgesteinsproben. Geophysik und Geologie, Leipzig 5, 38-73.
Hurtig, E. (1968) Zum Problem der Anisotropie petrophysikalischer Parameter in geologischen Köı ɔern. Geophysik und Geologie, Leipzig, Folge 12, 3-36.
Hurtig, E. (1968) Ergebnisse von Wärmeleitfähigkeitsmessungen an Gesteinsproben aus der DDR im Rahmen komplexer petrophysika-lischer Untersuchungen. Freiberger Forschungshefte, C 238, Leipzig 101-118.
Hurtig, E., H. Brugger (1970) Wärmeleitfähigkeitsmessungen unter einaxialem Druck, Tectonophysics, 10, No.1-3, 67-77.
Hurtig, E., P. Schlosser (1976) Geothermal studies in the GDR and relations to the geological structure. In: Adam, A. (ed.) Geoelectric and geothermal studies (KAPG Monograph) Akad.Kiado, Budapest, 384-394.
Hurtig, E., P. Schlosser (1976) Vertical changes of the heat flow in boreholes in the North-German sedimentary basin. In: Adam, A. (ed.) Geoelectric and geothermal studies (KAPG Monograph) Akad. Kiado, Budapest, 395-401.
Hutt, J. R., J. W. Berg (1968) Thermal and eletrical conductivities of sandstone rocks and ocean sediments. Geophysics, 33, 489-500
Idziak, A. (1981) Seismic Study of Fractured Sedimentary Rocks, Trans. 26-th Geophysical Symposium, Leipzig 1981, 116-123
Iliceto, V., G. Santarato, S. Veronese (1982) An approach to the identification of fine sediments by induced polarization laboratory measurements, Geophysical Prospecting, 30, 331-347
Imai, T.Fumoto, K. Yokota (1975) The relation of mechanical properties of soils to Pand S-wave Velocities in Japan, Proceedings Fourth Japan Earthquake Engineering Symposium, 1975, 13 p.
International Atomic Energy Agency (IAEA), (1976): Radiometric Reporting Methods and Calibration in Uranium Exploration, Techn. Rept. Ser. 174, Vienna, Austria,
Ising, G. (1942) On the magnetic properties of varved clay. Arkiv Astron. Fysik, 29 A (5), 1-37
Itenberg, S. S. (1978) Interpetacija resultatov karotasa skvasin. Isdat. Nedra, Moskva
Iverson, W. P., R. Satchwell (1989) Permeability: An Elusive Goal of Production Geophysics, Extend. Abstr. SEG-Ann. Meeting Dallas paper RP 1.5, 576-578
Jackson, P. D., D. Taylor, P. M. Smith (1978) Standard Resistivity- porosity-particle shape relationships for marine sands. Geophysics, 43, 6, 1250-1268
Jaeger, J.C., N.G.W. Cook (1976) Fundamentals of rock mechanics, Chapman and Hall, John Wiley & Sons, New York

Jahns, E., S. Siegesmund, T. Chloupac (1994) In situ seismic velocities versus laboratory measurements: an example from the KTB, Scientific Drilling, 4, 215-226
Jahren, C. E. (1963) Magnetic susceptibility of bedded iron-formation, Geophysics, 28, 5, 756-766
Jakosky, J. J. (1950) Exploration Geophysics, 2nd ed.
Jamscikov, W. S., W.L. Schkuratnik, A.B. Bobrov (1985) O kolicestvennoi ozenke mikrotrecinobatocti gornich porod ultrasvukovim velocimmetriceskim metodom, Akad. Nauk SSSR, Sibirskoe otdelenie, No. 4, 110- 114
Janbu, N. (1963) Soil compressibility as determined by Oedometer and Triaxal tests, Proc.Europ. Conf. Soil Mech. Found. Engrg. Wiesbaden, Vol. II, 19-25.
Japsen, P. (1980) IP-measurements on sand samples, Master-thesis, Univ. of Copenhagen, Geophys. Inst.(unpublished)
Jelic, K. (1984) Odnos gustoce i poroznosti s dubinom litostratigrafskih formacija savske i dravske potoline. Nafta (Zagreb), 35, 12, 637-643
Jessop, A. M. (1990) Thermal Geophysics, Elsevier Amsterdam -Oxford - New York
Jizba, D., A. Nur (1990) Static and dynamic properties of sandstones, Expanded Abstracts EAEG/SEG Research Workshop "Estimation and Practical Use of Seismic Velocities", Cambridge England, 326-332
Jödicke, H. (1991) Zonen hoher elektrischer Krustenleitfähigkeit im Rhenoherzynikum und seinem nördlichen Vorland, Diss. Univ. Münster, Hochschulschriften Bd. 24
Johnston, D. H., M. K. N. Toksöz, A. Timur (1979) Attenuation of seismic waves in dry and saturated rocks (II. mechanisms.), Geophysics, 44, 4, 691-711.
Johnston, D.H. (1981) Attenuation: A state-of-the-art summary, in: Toksöz, M. N., D. H. Johnston (eds.) Seismic Wave Attenuation, SEG reprint series No. 2, 123-139
Johnston, D. H., M. N. Toksöz (1981) Chapter 1 - Definitions and terminology. In: SEG Geophysics reprint series No. 2 Seismic Wave Attenuation, pp.1-5, SEG Tulsa/Oklahoma
Jonscher, A. K. (1977) A many-body model of dielectricpolarization in solids, Phys. Stat. Sol., 84, 159-186
Jonscher, A. K. (1981) Review: A new understanding of the dielectric relaxation of solidsj Journ. Material Science, 16, 2037-2060
Jorden, J. R., F. L. Campbell (1984) Well logging, Henry L.Doherty Mem. Fund of AIME, Soc.of Petr. Engrs. of AIME; New York-Dallas 1984
Just, G. (1980) Bestimmung von Thorium, Uran und Kalium in Magmatiten als Beitrag zur petrophysikalisch - petrochemischen Interpretation regionaler geophysikalischer Arbeiten. Zeitschr. geol. Wiss., Berlin, 8, 5, 525-544.
Kaelin, B., L. Rybach, E. H. K. Kempter (1992) Rates of Deposition, Uplift and Erosion in the Swiss Molasse Basin, Estimated from Sonic- and Density-Logs; Bull. Swiss. Ass. of Petroleum Geol. and Eng., 58, 133, 9-22
Kanamori, H., N. Fujii, H. Mizutani (1969) Thermal diffusity measurements of rockforming minerals from 300 to 1100 K., Journ. Geoph. Res. 732, 595-605.
Kapicka, A. (1990) Variations of the mean susceptibility of rocks under hydrostatic and non-hydrostatic pressure. In: McClelland, E., L. Pesonen, R. Enkin, V. Shcherbakov, V. Kropacek (ed.), Special Issue "Paleomagnetism: Database and experimental techniques", Physics of the earth and planetary interiors, 63, 1-2., 78-84.

Kappelmeyer, O., R. Haenel (1974) Geothermics. Geoexploration Monographs, Series 1 - No. 4 (O. Rosenbach, C. Morelli, eds.), Gebr. Borntraeger, Berlin, Stuttgart.
Käppler, R., J. Schön (1979) Untersuchungen zum Absorptionskoeffizienten elastischer Wellen in kohäsionslosen Sedimenten, Neue Bergbautechnik, 9, 12, 667-669.
Karato, S., H. A. Spetzler (1990) Defect microdynamics in minerals and solid-state mechanisms of seismic wave attenuation and velocity dispersion, Reviews of Geophysics, 28, 4, 399-421
Karus, E. B., I. P. Pasecnik (1954) Izucenie uprugich i poglascajuscich svoistv gornich porod v ich estestvennom zaleganii metodami seismoakustiki, Izvestija. AN SSSR, Ser. Geofiz., 6.
Katsube, T. J., J. P. Hume (1987) Permeability determination in crystalline rocks by standard geophysical logs, Geophysics, 52, 342-352
Katsube, T. J., L. S. Collett (1976) Electromagnetic propagation characteristic of rocks. In: Strens (edit.) The physics and chemistry of minerals and rocks, J. Wiley Sons, London N. York Sydney, 279-296.
Katz, A. J., A. H. Thompson (1985) Fractal Sandstone Pores: Implications for Conductivity and Pore Formation, Phys. Rev. Letters, 54, 1325-1328.
Kaye, G. W. C., T. H. Laby (1968) Tables of physical and chemical constants, Longmanns, London
Keller, G. V. (1989) Electrical properties, in: Practical handbook of Physical Properties of Rocks and Minerals (Ed.: Carmichael, R.S.) CRC Press, 1989, Boca Raton
Keller, G. V., F. C. Frischknecht (1966) Electrical methods in geophysical prospecting, Pergamon Press 1966
Kenyon, W. E., P. L. Baker (1986) EPT Interpretation in Carbonates Drilled With Salt Muds, SPE Formation Evaluation, October 1986; SPE 13 192, paper presented at the 1984 SPE Ann. Techn. Conf. Houston
Kermabon, A., C. Gehin, P. Blavier (1969) A deep-sea electrical resistivity probe for measuring porosity and density of unconsolidated sediments, Geophysics, 34, 4, 554-571.
Kern, H. (1990) Laboratory seismic measurements: an aid in the interpretation of seismic field data, Terra Nova, 617-628
Kezdi, A.(1969) Handbuch der Bodenmechanik, VEB Verlag für Bauwesen, Berlin, Akademiai Kiado Budapest
Kezdi, A. (1976) Handbuch der Bodenmechanik, Verlag Bauwesen, Berlin
Kezdi, A. (1964) Bodenmechanik, Verlag Bauwesen, Berlin
Khusnullin, M. K. (1973) Application of gamma method for determination of waterflooded strata, Geol. Neft i Gaza, 12, Dec. 1973, 63-67
King, M. S. (1984) Elastic-wave velocities in quartz monzonite at different levels of water saturation Int. Journ. Rock Mech.& Geomech. Abstr. , 1, 35-38
King, M. S., B. N. P. Poulsson (1981) Acoustic velocities in heated block of granite subjected to uniaxial stress, Geophys. Res. Letters, 8, 7, 699-702
King, Chi-Yu, G. Luo (1990) Variations of electric resistance and H^2 and Rn emissions of concrete blocks under increasing uniaxial compression, PAGEOPH, 134, 1, 45-51
King, M. S., Stauffer, M. R., Yang, H. J. P., Hajnal, Z. (1988) Elastic wave and related properties of clastic rocks from the Athabasca Basin, Western Canada, Can.Journ. Expl. Geophys., 24, 2, 110-116

Mann, U., J.D. Neisel, W.-G. Burchard, V. Heinen, D. H. Welle (1994) Fluid-rock interfaces as revealed by cryo-scanning electron microcopy, First Break, 12, 3, 131-136

Marion, D., A. Nur (1990) Pore-filling material and its effect on velocity in rocks, Expanded Abstracts EAEG/SEG Research Workshop "Estimation and Practical Use of Seismic Velocities", Cambridge England, 333-342

Marion, D., D. Jizba (1992) Acoustic properties and their dependence on porosity, mineralogy, and saturation: Application to field-scale measurements, in: Worthington, P. F., C. Chardaire-Riviere (eds.) Advances in core evaluation - Reservoir management; Proc. 3-rd Core Analysis Symp. Paris

Marks, S. G., C. McCann, J. S. Oliver, J. Sothcott (1992) Compressional wave quality factors of reservoir sandstone at 50 Hz and 1 MHz, 54th Meeting EAEG, Paris 1992, Abstracts D026, 338-339

Marle, Chr. (1978) Geophysik und subsequenter Vulkanismus (dargestellt am Beispiel des NW-sächsischen Vulkanitkomplexes), Thesis Universität Leipzig (unpublished)

Marshall, D. J., T. R. Madden (1959) Induced polarization, a study of its causes, Geophysics, 24, 4, 790-816

Martin, R. J., R. W. Haupt, M. N. Toksöz (1992) Modulus dispersion and attenuation in sandstone and granite, Trans. 54th EAEG Meeting, Paris, paper D028

Mauritsch, H., M. Becke,V. Kropacek, T. Zelinka, P. Hejda (1987) Comparison of the hysteresis characteristics of synthetic samples with different magnetite and hematite contents, Phys. Earth and Planet. Interiors, 46, 93-99

Mavko, D., A. Nur (1975) Melt squirt in the astenosphere, Journ. Geoph. Res., 80, 1444-1448.

Mavko, G., A. Nur (1979) Wave attenuation in partially saturated rocks. Geophysics, 44, 2, 161-178.

Mavko, G. M., Nur, A. (1978) The effect of nonelliptical cracks on the compressibility of rocks, Journ. Geophys. Res., 83, No B9, 4459-4468

Mavko, G., E. Kjartansson (1979) The effect friction on velocity and attenuation, Geophysics, 44, 3, 334-335.

Mavko, G., D. Jizba (1990) Effects of grain-scale pore fluid flow on velocity dispersion in rocks, Expanded Abstracts EAEG/SEG Research Workshop "Estimation and Practical Use of Seismic Velocities", Cambridge England, 298-307

Mavko, G., D. Jizba (1991) Estimating grain-scale fluid effects on velocity dispersion in rocks, Geophysics, 56, 1940-1949

Maxwell, I. C., P. Verrall (1954) Low porosity may limit oil in deep sands, World Oil, 138, 5, 106-113, 6, 102-104.

Maxwell, J. C. (1964) Influence of Depth, Temperature and Geologie Age on Porosity of Quartzose Sandstones, Bull. A.A.P.G., 48, 697-709.

Mazella, A., H. F. Morisson (1974) Electrical resistivity variation associated with earthquake on the San Andreas fault, Science, 185, 4154, 855-857.

McCann, D. M., D. C. Entwisle (1992) Determination of Young's modulus of the rock mass from geophysical well logs; in: Hurst, A., Griffiths, C. M., Worthington, P. F. (eds.) Geological applications of wireline logs II, Geological Society Special Publication No. 65, 317-325

McCann, C., D. M. McCann (1969) The attenuation of compressional waves in marine sediments, Geophysics, Tulsa, 34, 6, 882-892.

McCann,C., S. Assefa, J. Sothcott, D. M. McCann, P.D. Jackson (1989) In-situ borehole measurements of compressional and shear wave attenuation in Oxford Clay, Scientific Drilling, 1, 11-20

McClelland, E., L. Pesonen, R. Enkin, V. Shcherbakov, V. Kropacek (ed., 1990) Special Issue "Paleomagnetism: Database and experimental techniques", in: Physics of the earth and planetary interiors, vol. 63, No.1-2.

McDonal, F. J., F. A. Angona, R. L.Mills, R. L.Sengbush, R. G. van Norstrand, J. E. White (1958) Attenuation of shear and compressional waves in Pierre shale, Geophysics, 23, 421-439.

McNally, G. H. (1987) Estimation of coal measures rock strength using sonic and neutron logs, Geoexploration, 24, 381-395

Meador, R.A., P.T. Cox (1975) Dielectric Constant Logging: A Salinity Independent Estimation of Formation Water Volume, SPE paper 5504, 50th Ann. Fall Meeting 1975

Mehta, C. H. and B.M. Verma (1991) Porosity interpretation through seismics, Geoexploration, 28, 91-106.

Meinke, W., E. Hurtig, J. Weiner (1967) Temperaturverteilung, Wärmeleitfähigkeit und Wärmefluß im Thüringer Becken, Geophysik und Geologie (Leipzig), Folge 11, 40-71

Melnikov, N. W., W. W. Rshewski, M. M. Prodotjakonov (1975) Spravocnik (kadastr.) fiziceskich svoistv gornich porod, Izdat. Nedra, Moskva

Merkulova, W. M. (1968) Pogloscenie ultrasvukovych voln v gornych porodach v oblasti castot 10-160 kHz, Izvestija A.N.SSSR Fizika Semli 22, 6, 20-25.

Merkulova, W. M. (1966) Castotanaja zavisimost zatuchania ultrasvuka v gornich porodach dlja megagerzevoi oblasti, Izvestija A.N.SSSR Fizika Semli 7, 47-60.

Merkulova, W. M. (1965) Izmerenie skorostei ultrasvukovych voln v nekotorych gornich porodach, Izvestija A.N.SSSR, Fizika Semli, 5, 75-81.

Merkulova, W. M., E. G. Pigulevskii, V. M. Zaplev (1972) Izmerenie pogloscenia zvuka v gornych porodach pri otnoosnom ssati. Izvestija A.N.SSSR Fizika Zemli, Moskva 3, 59-61.

Merkulova, V. U., E. A. Wasilcov (1967) Izmerenie pogloscenia v gornych porodach sposobom izgibnych kolebanii, Izvestija A.N.SSSR Fizika Semli, Moskva 4, 75-77.

Militzer, H., R. Scheibe (1981) Grundlagen der angewandten Geomagnetik, Freiberger Forschungshefte C 352, (Verlag Grundstoffindustrie Leipzig)

Militzer, H., R. Stoll (1973) Einige Beiträge der Geophysik zur Primärdatenerfassung im Bergbau, Neue Bergbautechnik, Leipzig, 3, 1, 21-25.

Militzer, H., J. Schön, U. Stötzner, R. Stoll (1978) Angewandte Geophysik im Ingenieur- und Bergbau, VEB Deutscher Verlag f. Grundstoffind., Leipzig

Militzer, H., J. Schön, U. Stötzner (1986) Angewandte Geophysik im Ingenieur- und Bergbau, Ferd. Enke Verlag Stuttgart and VEB Deutscher Verlag für Grundstoffindustrie, Leipzig

Mindlin, R. D. (1949) Compliance of elastic bodies in contact, Journ. Appl. Mech. Trans., ASME, 71, 259-268

Mjelde, R. (1991) Analysis of shear waves with respect to splitting near the San Andreas Fault, Geophysical Journ. Internat., 107, 3, 649-660

Ohta , Y., N. Goto (1978) Empirical shear wave velocity equations in terms of characteristic soil indexes, Earthq. Engrg. Struct. Dynamics, 6, 167-187

Olevskj W. A. (1990) Svjas mezdu fiziko-mechaniceskimi i petrochimiceskimi charakteristikami glavnich tipov magmaticeskich gornich porod, Fizika Zemli (Moscow), 11, 95-99

Olhoeft, G. R. (1978) Electrical Properties of Permafrost, Proc. 3rd Intern. Conf. on Permafrost, Nat. Res. Counc. Canada, 127-131

Olhoeft, G. R. (1992) Geophysical Detection of Hydrocarbon and Anorganic Chemical Contamination, Proc. Symposium on the Application of Geophysics to Engineering and Environmental Problems, April 26-29, 1992, 587-595

Olhoeft, G. R. (1976) Electrical properties of rocks, in: Strens (edit.) The physics and chemistry of minerals and rocks. J. Wiley a. Sons. London-N. York-Sydney, 261-278.

Olhoeft, G. R. (1981) Electrical properties of rocks, in: Y. S. Touloukain, W. R. Judd, R. F. Roy (eds.) Physical Properties of Rocks and Minerals, McGraw Hill, New York

Olhoeft, G. R. (1985) Low frequency electrical properties, Geophysics, 50, 2492-2503

Olhoeft, G. R. (1985) Electrical properties of rocks and minerals, Short Course Notes, US Geological Survey, Denver

Olhoeft, G. R. (1984) Clay-organic reactions measured with complex resistivity, SEG Ann. Intern. Meeting, Atlanta , Expandet abstracts, 356-357

Olhoeft, G. R., R. L. Reynolds, J. D. Friedman, G. R. Johnson, G. R. Hunt (1981) Physical properties of the June 1980 dacite dome, in: Lipman, P.W., D.R. Mullineaux (eds.) The 1980 eruptions of Mt. St. Helens, Washington, U.S. Geol. Survey Prof. Paper 1250, 549-556

Olhoeft, G. R., G. R. Johnson (1989) Densities of Rocks and Minerals, in: Carmichael (ed.) Practical Handbook of Physical Properties of Rocks and Minerals, section II, CRC Press, Boca Raton Florida

Olhoeft, G. R. (1987) Electrical properties from 10-3 to 10+9 Hz - physics and chemistry. Proc. 2nd Int. Symp. Physics and Chemistry of porous Media; Schlumberger-Doll, Ridgefield, p. 281-298

Olhoeft, G. R. (1979) Electrical properties, Initial Report on the Petrophysics Laboratory, U.S. Geol. Survey, Circ. 789.

Onodera, T. F. (1962) Dynamic investigation of foundation rocks in situ, Proc. 5-th. Symp. on Rock Mech. (ed. Fairhurst, C.), Minnesota Perg. Press, 1963, 517-533.

Owen, P. F. (1993) Porosity, permeability and rock strength parameters in siliciclastic reservoirs, First Break, 11, 9, 375-389

Owens, W. H.(1974) Mathematical model studies on factors affecting the magnetic anisotropy of deformed rocks, Tectonophysics, 24, 115-131

Palmer, T., N. Pallatt (1991) The effect of overburden pressure on formation factor for shaly sands, Trans. 14th European Formation Evaluation Symposium, London, paper M, 1-19

Pandit, B. I., M. S. King (1979) A study of the effects of pore-water salinity on some physical properties of sedimentary rocks at permafrost temperatures, Canad. Journ. Earth Sci., 16, 1566-1580

Pape, H., P. F. Worthington (1983) A surface-structure model for the electrical conductivity of reservoir rocks, Trans. 8th Europ. Logging symp. SPWLA, papaer Z

Pape, H., L. Riepe, J. R. Schopper (1987) Theory of self-similar network structures in sedimentary and igneous rocks and their investigation with microscopical and physical methods, Journ. Microcopy, 148, 2, 121-147

Pape, H., L. Riepe, J. R. Schopper (1982) A Pigeon-Hole Model for relating Permeability to Specific Surface, Log Analyst, 23,No.1, 5-13 (errata see No.2, 50)

Pape, H., L. Riepe, J. R. Schopper (1984) Calculating permeability from electrical logging data, Trans. 9-th Int. Form. Eval. Symp., Paris, paper 22

Parasnis, D. S. (1971) Physical guide for rocks and minerals, ABEM Geophysical Memorandum 4/71

Parasnis, D. S. (1973) Mining Geophysics, Elsevier Publ. Comp. Amsterd. New York, London

Parchomenko, E. J. (1965) Elektriceskie svoistva gornich porod, Izdat. Nauka, Moskva

Parchomenko, E. J. (1977) Electrical properties of minerals and rocks, in: Volarovic, M. P., Stiller, H. (ed.) Issledovanija fiziceskich svoistv mineralnogo vecestva zemli pri vysokich termodinamiceskich parametrach, Izdat. Naukova Dumka, Kiev, 145-146.

Parrot, J. E., A.- D. Stuckes (1977) Thermal conductivity of solids, Pion, London 1975.

Patchett, J. G. (1975) An investigation onf shale conductivity, SPWLA Ann. Symposium, New Orleans, paper U

Patchett, J. G.ed. (1982) Shaly Sand Reprint Volume, Society of Professional Well Log Analysts (SPWLA)

Patnode, W. H., M. R. J. Wyllie (1950) The presence of conductive solids in reservoir rocks as factor in electric log interpretation, Petroleum Transactions, AIME 189, 47-52

Pelton, W. H., H. Wards, P. G. Hallhof, W.-R. Sill, P. H. Nelson (1978) Mineral Discrimination and Removal of Inductive Coupling with Multifrequency I.P., Geophysics, 43,3, 588-609

Perez-Rosales, C. (1976) Generalization of the Maxwell Equation for formation resistivity factors, Journ. Petrol. Technol., 819-824.

Peselnick, L., J. Zietz (1959) Internal frition of fine-grained limestones at ultrasonic frequencies, Geophysics, 24, 285-296.

Petersen, N. (1985) Gesteinsmagnetismus, in: Bender, F. (ed.) Angewandte Geowissenschaften, Vol. II, Ferd. Enke Verlag, Stuttgart

Petersen, N., Bleil, U. (1982) Magnetic properties of rocks.in: Landolt-Börnstein Numerical Data and Functional Relationships in Science and Technology (K.-H. Hellwege, ed.), New Series; Group V. Geophysics and Space Research, Vol. 1 Physical Properties of Rocks, Subvolume b, 366-432 Springer-Verlag Berlin. Heidelberg. New York 1982

Pettijohn, F. J. (1975) Sedimentary Rocks, 3rd edition, Harper & Row Publ. Inc., New York

Pickett, G. R. (1963) Acoustic character log and their applications in formation evaluation; Journ. Petr. Technol., 659 - 667

Pickett, G. R. (1960) The use acoustic logs in the evaluation of sandstone reservoirs. Geophysics, 25, 1, 250-269.

Pickup, G. E., P. S. Ringrose, P. W. M. Corbett, J. L. Jensen, K. S. Sorbie (1995) Permeability tensors for sedimentary structures, Mathematical Geology, 26, 2, 227-250

Pickup, G. E., P. S. Ringrose, P. W. M. Corbett, J. L. Jensen, K. S. Sorbie (1995) Geology, geometry and effective flow, Petroleum Geoscience, 1, 37-42

Pierce, C. I., R. B. Loewe (1958) Some tests related to resistivity - water saturation measurements of Appalachian sandstone specimens, U.S.Bureau Mines Rept. Inv. 5389, 21 p.

Pirson, S. J. (1966) Handbook of well log analysis; Prent-Hall N.York 1963

Platou, S. W. (1968) On the petrophysical properties of granitic rocks, Geologiska Föreningens i Stockholm Förhandlingar, 90, 427-433

Plewa, S. (1976) Correlation between thermal conductivity and other physical parameters of rocks, in: Adam, A. (ed.) Geoelectric and geothermal studies KAPG Geophysical monograph, Akad. Kiado, Budapest, 48-52.

Plona, T. J. (1980) Observation of a second bulk compressional wave in a porous medium atultrasonic frequencies, Appl. Phys. Letters, 36, 259-261

Plona, T. J., L. Tsang (1979) Characterisation of the average microscopic dimension in granular media using ultrasonic pulses: theory and experiments, Geophysics, 44, 3, 334-335.

Podio, A. L., A. R. Gregory (1990) Ultrasonic velocity and attenuation measurements in water-based drilling muds, paper presented at ASME 1990 ETCE Conference, New Orleans, 6 p.

Polak, L. S., M. B. Rapoport (1961) Osvjasi skorosti uprugich prodolnych voln s nekotorymi fiziceskimi svoistvami osadocnych porod, Prikladnaja Geofizika, Moskva, 29, 12-19.

Poley, I. Ph., I. I. Nooterboom, P. J. De Waal (1978) Use of V.H.F. dielectric measurements for formation analysis, Log Analyst XIX, 3, 8-30.

Porter, C. R., J. E. Carothers (1971) Formation factor - porosity relation derived from well log data, The Log Analyst, Jan-Febr., 16-26

Postma, G. (1955) Elastic waves in a stratified medium, Geophysics, 20, 780-806.

Poulsen, K. D., S. Saxov, N. Balling, J.I. Kristiansen (1982) Thermal conductivity measurements on Silurian limestones from the Island of Gotland, Sweden, Geologiska Föreningens i Stockholm Förhandlingar, 103, Pt. 3, 349-356

Poupon, A., W. R. Hoyle, A. W. Schmidt (1971) Log analysis in formation with complex lithologies. Journ. Petrol. Technol. 23, 8, 995-1005.

Prasad, M., R. Meissner (1992) Attenuation mechanisms in sands: Laboratory versus theoretical (Biot) data, Geophysics, 57, 5, 710-719

Press, F., W. F. Brace (1966) Earthquake prediction. Science 152 , 3729, 1575-1584.

Pros, Z. (1977) Studies of anisotropy in elastic properties of rocks at uniform pressures on spherical samples, in: Volarovic, M. P., Stiller, H., Lebedev, T. S. (eds.) Issledovania fiziceskich svoistv mineralnogo vecestva zemli pri vysokichtermodinamiceskich parametrach. Izdat. Naukova Dumka, Kiew, 56-66.

Pros, Z., J. Vanek, K. Klima (1962) The velocity of elastic waves in diabase and greywacke under pressures up to 4 kilobars. Studia Geoph. et Geodetica, Praha, vol. 6, 347-367.

Prutkina, M. I., W. L. Saskin (1975) Spravocnik po radiometriceskomu analizu, Atomizdat, Moskva

Pusch, G., J. R. Weber (1993) Influence of deviatoric stress conditions on micro fracture permeability of crystalline rock, KTB-Report 93-2, Hannover 1993 (ed. by Emmermann et al.), 289-293

Puzankov, J. M., V. A. Bobrov, A. D. Duchkov (1977) Radioaktivnye elementy i teplovoy potok zemnoi kory Poluostrova Kamchatki, Izdat. Nauka Sibir. otd., Novosibirsk

Puzicha, K. (1977) Der Magnetismus der Gesteine als Funktion ihres Magnetitgehaltes, Beiträge z. angew. Geophysik, Leipzig, 9, 2, 158-186.
Qaisar, M. (1979) Attenuation properties of viscoelastic material, PAGEOPH, 131, 4, 704-713
Quiel, F. (1975) Thermal-IR in Geology, Photogrammetric Engrg. and Remote Sensing, 341-346.
Quirein, J., S. Kimminau, J. La Vigne, J. Singer, F. Wendel (1986) A coherent framework for developing and applying multiple formation evaluation models, SWPLA 26-th Annual Logging Symposium, paper DD
Quirein, J. A., J. L. Baldwin, R. L. Terry, M. Hendrieks (1981) Estimation of clay types and volumes from well log data - An extension of the GLOBAL method, Trans. 22nd Ann. Log. Symp. SPWLA, paper Q
Rabbel, W. (1992) Seismic anisotropy at the KTB deep drilling site, KTB Report, Hannover, 92-5, 275-289
Raiga-Clemenceau, J. (1977) The cementation exponent in the formation factor - porosity relation: The effect of permeability, SPWLA 18-th Ann.Logging Symp., paper R, 1-13
Raiga-Clemenceau, J., J. P. Martin, S. Nicoletis (1988) The concept of acoustic formation factor for more accurate porosity determination from sonic transit time data, The Log Analyst, Jan-Febr. 54-59
Rao, A. A. (1967) Some studies on attenuation of elastic waves in rocks, Bull. NGRJ, New Dehlhi, 5, 4, 183-194.
Ratcliffe, E.H. (1960) The Thermal Conductivities of Ocean Sediments, Journ. Geophys. Research, 65, No. 5, 1535-1541
Rauen, A.S. Keyssner, Ch. Bücker (1990) Magnetische Anomalien in der KTB-Vorbohrung. KTB-Report 90-4, 461, Hannover
Rauen, A. (1991) Untersuchungen des komplexen elektrischen Widerstandes, insbesondere dessen Anisotropie und Frequenzabhängigkeit, von Proben des Kontinentalen Tiefbohrprogrammes der Bundesrepublik Deutschland (KTB), Thesis (Dissertation) University of Munich (Ludwig-Maximilians-Universität München)
Rauen, A., E. Lippmann, E. Huenges, Ch. Bücker, J. Wienand, K. E. Wolter, K. M. Streit (1990) Tiefbohrung KTB Oberpfalz VB, Ergebnisse der geowissenschaftlichen Bohrungsbearbeitung im KTB-Feldlabor,Teufenbereich von 3509 bis 4000 m.KTB-Report 90-2, D1-D60, Hannover
Rauen, A., M. Lastovickova (1995) Investigation of Electrical Anisotropy in the Deep Borehole KTB, Surveys in Geophysics (Kluwer Acad. Publ.), 16, 37-46
Rayleigh (1892) On the influences of obstacles arranged in rectangular order on properties of a medium, Phil. Mag., 34, 481-502
Raymer, D. S., E. R. Hunt and J. S. Gardner (1980) An improved sonic transit time-to-porosity transform, Proc. SPWLA 21st Ann. Meeting, paper P.
Reeves, D. R. (1981) Coal Interpretation Manual, BPB Instruments, East Leake, England
Reich, H. (1933) Angewandte Geophysik für Bergleute und Geologen, Leipzig
Reitmayr, G. (1985) Induzierte Polarisation, in: Bender, F.(ed.)Angewandte Geowissenschaften, Vol . II, 339-351, Ferd. Enke, Stuttgart
Repsold, H. (1976) Über das Verhalten des Formationsfaktors in Lockersedimenten bei schwach mineralisierten Porenwässern, Geologisches Jahrbuch E 9, Hannover, 19-39.

Reuss, A. (1929) Berechnung der Fliegrenze von Mischkristallen, Angew. Mathem. u. Mech., 9, 49-58.
Rider, M. H. (1986) The Geological Interpretation of Well Logs, Blackie -Glasgow, London; John Wiley and Sons - New York
Riepe, L., J. R. Schopper, R. Wehr (1986), Permeability estimations from nuclear measurements (NMR-, spectral GR-data), Trans. 10-th Europ. Form. Eval. Symp., Aberdeen, Scotland, paper X
Riepe, L. (1984) Theoretische und experimentelle Untersuchungen über den Einfluß der spezifischen inneren Oberfläche auf petrophysikalische und bohrlochgeophysikalische Parameter von Sedimentgesteinen, Clausthaler Geowissenschaftliche Dissertationen H. 13, Clausthal-Zellerfeld
Rink, M., J. R. Schopper (1974) Interface conductivity and its implication to electric logging, Trans. 15th Ann. Logging Symp. Houston, 1-15
Rink, M., J. R. Schopper (1978) On the Application of Image Analysis to Formation Evaluation, Log Analyst, Jan-Feb, 12-22.
Riznichenko, J. W. (1949) O seismiceskoi kwasianisotropy, Izvestija Akad. Nauk SSSR, Ser. Geogr. i Geofiz., 13, 6, 518-544
Rösler, R. (1959) Die galvanisch induzierte Polarisation an Dispersionen rotationsellipsoidischer Teilchen, Freiberger Forschungshefte, C 67, Berlin
Rösler, H. J., H. Lange (1972) Geochemical Tables, Elsevier, Amsterdam-London-New York
Rösler, R. (1958) Betrachtungen zu den Spannungs-Dehnungs- Beziehungen nach Nakamura, Gerlands Beitr. z. Geophys, 67, 31-48.
Roy, R. F., A. E. Beck, Y. S. Touloukian (1981) Thermophysical Properties of Rocks, in:Y.S. Touloukian, C.Y. Ho (ed.) Physical properties of rocks and minerals, McGraw-Hill/CINDAS Data Series on Material Properties, Vol. II-2
Rudzki, M. P. (1911) Parametrische Darstellung der elastischen Welle in anisotropen Medien, Anz. d. Akad. d. Wiss. Krakau, 503-536
Ruffet, C., Y. Gueguen, M. Darot (1991) Complex conductivity measurements and fractal nature of porosity, Geophysics, 56, 6, 758-768
Ruffet, C., M. Darot, Y. Gueguen (1995) Surface Conductivity in Rocks: a Review, Surveys in Geophysics (Kluwer Acad. Publ.), 16, 83-105
Rybach, L., G. Buntebarth (1984) The variation of heat generation, density and seismic velocity with rock type in the continental lithosphere, Tectonophysics, 103, 335-344
Rybach, L., Cermak (1982) Radioactive heat generation in rocks, Chapter 4.4 in: Haack, U. (1982) Radioactivity of rocks; In: Landolt-Börnstein Numerical Data and Functional Relationships in Science and Technology (K.-H.Hellwege,ed.), New Series; Group V. Geophysics and Space Research, Vol.1 Physical Properties of Rocks, Subvol. b, 433-481. Springer-Verlag Berlin Heidelberg New York
Rybach, L., G. Buntebarth (1982) Relationships between the petrophysical properties density, seismic velocity, heat generation, and mineralogical constitution, Earth and Planetary Science Letters, 57, 367-376
Rybach, L. (1976) Radioactive heat production; a physical property determined by the chemistry of rocks, in: Strens (edit.): The physics and chemistry of minerals and rocks. J. Wiley u Sons, London-N. York-Sydney, 309-318.

Rysova, T. W. (1964) Uprugie svoistva porodoobrasyjuscich mineralov plagioklasy, Izvestija. A. N. SSSR, Ser. Geofiz, 7.
Rzewski, V., G. Novik (1971) The physics of rocks, Mir Publishers Moscow (translated from the Russian by A.K. Chatterjee)
Rzhewski, W. W., G. J. Nowik (1978) Osnovy fiziki gornich porod, Izdat. Nauka, Moskva
Rzhewski, W. W., W. S. Jamscikov (1973) Akusticeskie metody issledovania i kontrolja gornich porod v massive, Izdat. Nauka, Moskva
Saint Amant, M., D. W. Strangway (1970) Dielectric properties of dry geologic materials, Geophysics, 35, 4, 624-645.
Saporosec, W. M. (1978) Sakvsinnaja jadernaja geofizika, Izdat. Nedra, Moskva
Sarma, V. V. J, V. B. Rao (1963) Variation of electrical resistivity of river sands, calcite and quartz powders with water content, Geophysics, 27, 470-479, 28, 310-3413.
Savic, A. J, W. J. Koptev, W. N. Nikitin, Z. D. Jascenko (1969) Seismoakusticeskie metody izucenia massivov skalnych porod, Izdat. Nedra, Moskva
Scheidegger, A. E. (1960) The Physics of Flow through Porous Media, Univ. of Toronto Press
Schlumberger (1989) Log Interpretation Principles/Applications, Schlumberger Ed. Serv.
Schlumberger (1989) Log Interpretation Charts. Schlumberger Ed. Serv.
Schlumberger Oilfield Review (Schlumberger Publ.), Archie III: Electrical Conduction in Shaly Sands, Vol. 1, Number 3, 43-53
Schlumberger (1972) Log interpretation principles; Log interpretation charts. Schlumberger Ltd.
Schlumberger (1982) Natural Gamma Ray Spectrometry, Essentials of N.G.S. Interpretation, Schlumberger Ed. Serv.
Schneider, F., M. Bouteca, G. Vasseur (1994) Validity of the porosity/effective-stress concept in sedimentary basin modelling, First Break, 12, 6, 321-326
Schoenberg, M. (1994) Transversely isotropic media equivalent to thin isotropic layers, Geophysical Prospecting, 42, 885-915
Schoenberg, M., Levin, F.K. (1974) Apparent attenuation due to intrabed multiples, Geophysics, 43, 4, 730-737
Schoenberg, M., Levin, F.K. (1974) Apparent attenuation due to intrabed multiples I, Geophysics, 39, 3, 278-291
Scholz, C. H., L. R. Sykes, Y. P. Aggarwal (1973) Earthquake Prediction. A physical basis, Science 1814102, 803-810.
Schön, J., Börner, F.(1985) Untersuchungen zur elektrischen Leitfähigkeit von Lockergesteinen - der Einfluß matrixbedingter Leitfähigkeitsanteile, Neue Bergbautechnik 15 (1985) 6, S 222-224 Leipzig
Schön, J. (1983) Petrophysik, Akademie Verlag Berlin and Ferd. Enke Verlag Stuttgart
Schön, J. (1969) Die Ausbreitungsgeschwindigkeit elastischer Wellen in Lockerböden und ihre Beziehung zu bodenmechanischen Kennwerten. Freiberger Forschungshefte, C 250, Verl. Grundstoffind. Leipzig
Schön, J. (1986) Ein Beitrag zu den elektrischen Eigenschaften von Lockergesteinen, Freiberger Forschungshefte C 416, 70-79, Leipzig

Schön, J. (1984) Electrical conductivities in unconsolidated sediments - their correlation to other physical and petrographic properties, Transact. 9-th. European Formation Evaluation Symposium, Oct. 1984 Paris, paper. 23

Schön, J. (1968) Ein Beitrag zur Ausbreitungsgeschwindigkeit elastischer Wellen in Lockerböden. Teil II, Bergakademie, Leipzig, 20, 381-384.

Schön, J. (1968) Ein Beitrag zur Ausbreitungsgeschwindigkeit elastischer Wellen in Lockerböden, Teil I. Bergakademie, Leipzig, 20, 6, 332-336.

Schön, J., Börner, F.: Experimental investigation of the electrical conductivity of unconsolidated rocks - the relationship between interphase conductivity and lithologic character, Proc. 32-nd. Geophys. Symp. Nov. 2 - 5. 1987, Dresden

Schön, J. (1987) Note on the paper "Estimation of mud filtrate resistivity in fresh water drilling muds" by T. A. Loewe and H. F. Dunlap, The Log Analyst, Jan-Feb., 56

Schön, J. (1973) Vergleichende Betrachtungen zur Druckabhängigkeit der Ausbreitungsgeschwindigkeit elastischer Wellen an lunaren und terrestrischen Gesteinsproben, Gerlands Beitr. z. Geoph., 82, 6, 472-476.

Schön, J. (1979) Die Ausbreitungsgeschwindigkeit elastischer Wellen in granularen Medien - speziell im Regolith, Gerlands Beiträge z. Geoph., 88, 5, 387-393.

Schön, J. (1964) Modellseismische Untersuchungen im Hinblick auf die Schallgeschwindigkeit in Lockergesteinen. Freiberger Forschungshefte, C 173, Verlag Grundstoffindustrie Leipzig

Schön, J. (1974) Ein Gesteinsmodell zur Berechnung petrophysikalischer Parameter, Freiberger Forschungshefte, C 299, Verlag Grundstoffindustrie, Leipzig

Schopper, J. R. (1982) Porosity and permeability, in: LANDOLT-BÖRNSTEIN Numerical Data and Functional Relationships in Science and Technology, New Series; Group V. Geophysics and Space Research, Vol.1 Physical Properties of Rocks, Subvol. a Springer-Verlag Berlin. Heidelberg. New York

Schopper, J. R. (1967) A theoretical study on the reduction of statistical pore system parameters to measurable quantities, Geophys. Prospecting, 15, 2, 262-287

Schopper, J. R. (1966) A theoretical investigation on the formation factor-permeability-porosity relationships using a network model, Geophys. Prospecting, 14, 3, 301-341.

Schopper, J. R. (1983) Inelastic absorption in porous rocks related to common petrophysical quantities, Trans. SPWLA 24-th Ann. Symp. paper L, 14 p.

Schopper, J. R., L. Riepe (1986) Experimetal and theoretical petrophysical investigations on relations between the inelastic absorption in porous rocks and other petrophysical and reservoir parameters, DGMK Report (Absorption seismischer Wellen), Hamburg 1986, 239-275

Schröder, D. (1991) Der Einfluß lithologischer Eigenschaften auf das seismische Wellenfeld in Bohrlöchern, Diss. TU Braunschweig 1991

Schuch, M. (1980) Physik des Torfes und der Moorböden, in: K. Göttlich, Moor- und Torfkunde, Stuttgart 1980

Schultz-Krutisch, Th., F. Heller (1985) Measurement of magnetic susceptibility anisotropy in Buntsandstein deposits from southern Germany, Journ. Geophys. Res, 56, 51-58

Sclater, J. C., P. A. F. Christie (1980) Continental streching: an explanation of the post mid Cretaceous subsidence of the central North Sea basin, Journ. Geophys. Res., 85, 3711-3739

Scott, W. J., A .E. Kay (1988) Earth Resistivities of Canadian Soils, Canad. Elect. Assoc., Montreal, Vol. 1 (Main Report), Vol. 2 (Site data)
Scott, W. J., P. V. Sellmann, J. A. Hunter (1990) Geophysics in the Study of Permafrost, in: St. Ward (ed) Geotechnical and environmental geophysics - Investigations in Geophysics No. 5 SEG, 355-384
Seeburger, D. A., A. Nur (1984) A pore space model for rock permeability and bulk modulus, Journ. Geophys. Res., 89, 527-536
Seipold, U. (1990) Pressure and temperature dependence of thermal transport properties for granites, High Temperatures - High Pressures, 22, 541-548
Seipold, U. (1993) Determination of the thermal transport properties of amphibolithes at high pressure and high temperature, KTB Report 93-2, ed. by Project Management of the Continental Deep Drilling Programme FRG, 305-308
Seipold; U. (1995) The Variation of Thermal Transport Properties in the Earth's Crust, Journ. Geodynamics, 20, 2,145-154
Semenov, A. N. (1969) Izmerenie otnosenia vremen probega poperecnych i prodolnych voln pered silnymi semletraseniami, Izvestija A.N. SSSR, Fizika Zemli 4, 72-74.
Sen, P.N., P.A. Goode, A. Sibbit (1988) Electrical conduction in clay-bearing sandstones at low and high salinities, Journ. Appl. Phys, 63, 4832-4840
Sen, P. N., C. Straley, W. E. Kenyon, M. S. Whittingham (1990) Surface-to-volume ratio, charge density, nuclear magnetic relaxation, and permeability in clay-bearing sandstones, Geophysics, 55, No.1, 61-69.
Sen, P. N. (1981) Relation of certain geometrical features to the dielectric anomaly of rocks, Geophysics, 46, 1714-1720
Sen, P. (1980) The dielectric and conductivity response of sedimentary rocks, SPE 9379, Soc. of Petroleum Engineers of AIME, 11 p.
Serra, O. (1984) Fundamentals of well-log interpretation, Elsevier Amsterdam Oxford New York Tokyo
Shapiro, S. A., P. Hubral, H. Zien (1994) Frequency-dependent anisotropy of scalar waves in a multilayered medium. Journ. Seismic Exploration, 3, 37-52
Sharma, S. K., U. V. S. Nohwar, H. Raj, J. Koithara, G. V. Chilingar (1980) Interrelationships among various petrophysical parameters of carbonate reservoir rocks in the Bombay High Oil Field, India, Energy Sources, 5, 1, 53-69
Sharma, P. V. (1975) Geophysical methods in geology, Elsevier Publ. Comp., Amsterdam
Sheng, P. (1989) Effective medium modeling of electrical and elastic properties of sedimentary rocks, SEG-Ann. Symp. extendet abstr. RP 1.8, 585-588
Shumway, G. (1960) Sound speed and absorption studies of marine sediments by a resonance method, Part I, Geophysics, 25, 2, 451-567; Part II, Geophysics 25, 3, 659-682.
Shumway, G. (1956) A resonant chamber method for sound velocity and attenuation measurements in sediments, Geophysics, 21, 305-319
Siegesmund, S., A. Vollbrecht, At. Chlupac, G. Nover, H. Dürrast, J. Müller, K. Weber (1993) Fabric-controlled anisotropy of petrophysical properties observed in KTB core samples, Scientific Drilling, 4, 31-54
Silaeva, O. O., O. G. Samina (1969) Pogloscenie ultrasvuka v granitach, Izvestija Akad. Nauk SSSR, Ser. geofiz., 9, 1354-1359.

Simmons, G. (1964) Velocity of compressional waves in various minerals at pressure to 10 kbars, Journ. Geoph. Res., 69, 6, 1117-1121.
Simmons, G., W. F. Brace (1965) Comparison of static and dynamic measurements of compressibility of rocks, Journ. Geoph. Res., 70, 22, 5649-5656.
Simmons, G. (1967) Hashin bounds for aggregates of cubic crystal, Journ. of the Graduate Res. Center; South. Method. Univ. Press, Dallas, XXXVI 1.
Simmons, G., T. Todd, W. S. Baldridge (1975) Toward a quantitative relationships between elastic properties and cracks in low porosity rocks, Am. Journ. of Science, 275, 318-345.
Simmons, G., A. W. England (1969) Universal equations of state for oxidees and silicates, Phys. Earth Planet. Interiors, 2, 69-76
Simmons, G., D. Richter (1976) Microcracks in rocks, in: Strens (ed.), The physics and chemistry of minerals and rocks. J. Wiley u. Sons, London-N. York-Sydney 105-137.
Sjogren, B., A. Ofsthus, I. Sandberg (1979. Seismic classification of rock mass qualities, Geophys. Prospecting, 27, 2, 409-442.
Skopec, R.A. (1992) Recent advances in rock characterization, Log Analyst, May-June, 270-285
Snyder, D. D., R. H. Merkel, J. T. Williams (1977) Complex formation resistivity - the forgotten half of the resistivity log, SPWLA 18-th. Ann. Logg. Symp. paper Z.
Soininnen, H. T., H. Vanhala (1992) Spectral induced polarization method in mapping soils polluted by organic chemicals, 54th EAEG Meeting, Paris June 1992, Abstracts of papers, D040, 366-367
Somerton, W.H. (1958) Some thermal characteristics of rocks, Trans. AIME, 213, 375 378
Somerton, W. H., G. D. Boozer (1969) Thermal characteristics of porous rocks at elevated temperatures, Journ. Petr. Techn., 12, 6, S. 77-81.
Somerton, W. H., J. A. Keese, S. C. Chu (1974) Thermal behavior of unconsolidated oil sand, Soc. of Petroleum Engineering Journ., 14, 5, 513-521
Somerton, W. H, F. Esfandiari, A. Singhal (1969) Further studies of the relation of physical properties of rock to rock drillability, Soc. of Petr. Engr. of AIME, Paper No SPE 2390, Dallas
Somerton, W.H. (1958) Some thermal characteristics of porous rocks, Journ. Petr. Techn., 10, 5, 61-64
Song, L., K. Vozoff (1985) The complex resistivity spectra of models consisting of two polarizable media of different intrinsic properties, Geophys. Prospecting, 33, 1029-1062
Spangenberg, E. (1995) Ein fractales Modellkonzept zur Berechnung physikalischer Gesteinseigenschaften und dessen Anwendung auf die elastischen Eigenschaften poröser Gesteine, Dissertation (Ph.D. Thesis) TU Braunschweig
Spravocnik Geofizika(1966), Izdat. Nedra, Moskva .
Srebredolskij, D. M., G. P. Matcinova (1977) Svjaz estestvennoi radioaktivnosti s glinostju gornich porod, Neftegaz. Geol. i Geofiz., Moskva, 9, 32-34.
Stacey, F. C., S.. K. Banerjee (1974) The Physical Principles of Rock Magnetism, Elsevier Amsterdam
Stegena, L. (1964) The structure of the earths crust in Hungary, Acta Geologica, Budapest 8, 413-431.

Stieber, S. J. (1980) Pulsed neutron capture log evaluation in the Louisiana Gulf Coast, SPE of AIME Fall Meeting 1980, paper SPE 6859

Stiller, H., F. C. Wagner, H. Vollstädt (1979. Influence of pressure and joints on the velocity of elastic waves in jointy rocks in the case of earthquakes, in: Stiller, H.; Volarovic, M. P.(ed.), Theoretical and experimental investigations of physical properties of rocks and minerals under extreme p, T-conditions. Akademie-Verlag, Berlin, 177-197.

Stiller, H., M. P. Volarovic (ed.) (1979) Theoretical and experimental investigations of Physical properties of rocks and minerals under extreme p, T-conditions, Akademie-Verlag, Berlin

Stoll, R., G. M. Bryan (1970) Wave attenuation in sediments, Journ. Acoust. Soc. Am., 47, 1440-1447.

Stoll, R. (1971) Vibrotorseismische Untersuchungen in realen Medien zur Nahortung und Kennwertbestimmung für ingenieurgeophysikalische Belange, Freiberger Forschungshefte, C 274, Verl. Grundstoffindustrie, Leipzig

Stötzner, U. (1975) Ingenieurgeophysikalische Untersuchungsmethodik und Komplexinterpretation zur Lösung felsmechanischer Aufgaben, Freiberger Forschungshefte C307, Verlag Grundstoffindustrie, Leipzig

Strens, R. G. I. (edit.) (1976. The Physics and Chemistry of Minerals and Rocks, J. Wiley u. Sons, London-New-Yoek-Sydney

Stümpel, H., S. Kähler, R. Meissner, B. Milkereit (1984) The use of seismic shear waves and compressional waves for lithological problems of shallow sediments, Geophysical Prospecting, 32, 662-675

Sugawara, A., Y. Yoshizawa (1961) An investigation on the thermal conductivity of porous materials and its application to porous rock, Austral. Journ. Physics, 14, 4

Sumner, J. S. (1976) Principles of induced polarization for geophysical exploration, Elsevier Scientific Publ.

Sundberg, K. (1932) Effect of impregnating waters on electrical resistivity of soils and rocks, Trans. AIME. Geoph. Prosp., 367.

Suyama, K., T Imai, K Ohtomo, T Takahasi (1984) Delineation of structures in Alluvium and Dilluvium using SH-wave reflection and VSP methods, Trans. 45th SEG Meeting, 165-179

Swanson, B.F. (1980) Rationalizing the influence of crude wettingon reservoir fluid flow with electrical resistivity behaviour, Journ. Petrol. Tech., 32, 1459-1464

Tagiev, K. K., K. A. Mustafaev (1975) Mechanism pogloscenia energii seismiceskich voln v nefti i vode, Prikladnaja Geofizika, Moskva, 80, 114-119.

Tao, G., M. S. King, M. Nabi-Bidhendi (1995) Ultrasonic wave propagation in dry and brine-saturated sandstones as a function of effective stress: laboratory measurements and modelling, Geophysical Prospecting, 43, 299-327

Tao, G., M. S. King (1993) Porosity and pore structure from acoustic well logging data, Geophysical Prospecting, 41, 435-451

Tarling, T.H., D.H. Hrouda (1993) The magnetic Anisotropy of Rocks, Chapman & Hall, London . Glasgow . New York . Melbourne . Madras

Tarling, T.H. (1983) Palaeomagnetism, Chapman & Hall, London New York

Tatham, R. H. (1982) v_p/v_s and lithology, Geophysics, 47, 336-344

Tatham, R. H. (1985) Shear waves and lithology in: Seismic Shear Waves, Part B: Applications, 87-133; (ed. Dohr,G.P.), Geophysical Press, London Amsterdam.

Tatham, R.H., M.D. McCormack (1991) Multicomponent seismology in petroleum exploration, Invest. in Geophysics No. 6, SEG

Taylor, J. M. (1950) Pore space reduction in sandstone, Am. Assoc. Petrol. Geol. Bull. 34, 701

Telford, W. M., L. P. Geldart, R. E. Sheriff, D. S. Keys (1976) Applied geophysics, Cambridge Univ. Press.

Terzaghi, K. V. (1955) Influence of Geological Factors on the Engineering Properties of Sediments, Economic Geology, 15-th Ann. Vol., 557-618.

Thiel. K., H. Linovski, R. Lukaszek (1970) Verformbarkeitsprüfungen eines heterogenen Gebirgsmassivs, Proc. 2nd Congr. Int. Soc. Rock Mech., Belgrad, 2-34.

Thomas, E. C. (1992) 50th Anniversary of the Archie Equation: Archie Left More Than Just an Equation, Log Analyst, May-June, 199-205

Thomas, E. C., S. J. Stieber (1975) The distribution of shale in sandstones and its effect upon porosity, SPWLA 16-th. Annual Logging Symp. New Orleans-Transactions paper T.

Thompson, A. H., A. J. Katz, C. E. Krohn (1987) The microgeometry and transport properties of sedimentary rock, Advances in Physics, 36, 5, 625-694

Thomsen, L. (1986) Weak elastic anisotropy, Geophysics, 51, 10, 1954-1966

Thomson, L. (1985) Biot-consistent elastic moduli of püorous rocks: Low frequency limit, Geophysics, 50, 12, 2797-2807

Timur, A. (1968) An investigation of permeability, porosity, and residual water saturation relationships for sandstone reservoirs, Log Analyst, July-August, 8-17

Timur, A. (1969) Pulsed nuclear magnetic resonance studies of porosity, movable fluid and permeability of sandstones, Journ. Petr. Techn., 21, 775-786

Tixier, M. P. (1949) Evaluation of permeability from electric-log resistivity gradients, Oil and Gas Journ.,June 16,

Todd, T., G. Simmons, W. S. Baldridge (1973) Acoustic double refraction in low-porosity rocks, Bull. Seismol. Soc. Am., 63, 6, 2007-2020

Todd, T., G. Simmons (1972) Effect of Pore Pressure on the Velocity of Copressional Waves in Low-Porosity Rocks, Journ. Geoph. Res., 70, 20, 3731-3743.

Toksöz, M. N., D.H. Johnston (1981) Attenuation Mechanisms, Preface Chapter 3 SEG Geophysics reprint series No. 2: Seismic Wave Attenuation, pp.136-139. SEG Tulsa/Oklahoma,

Toksöz, M. N., D. H. Johnston, A. Timur (1979) Attenuation of seismic wave in dry and saturated rocks: I. Laboratory measurements, Geophysics, 44, 4, 681-690

Toksöz, M.N., D.H. Johnston (1981) Preface. In: SEG Geophysics reprint series No. 2 Seismic Wave Attenuation, pp.1-5. SEG Tulsa/Oklahoma

Toksöz, M. N., C. H. Cheng, A. Timur (1976) Velocities of seismic waves in porous rocks, Geophysics, 41, 4, 621-645.

Toksöz, M. N.and D. H. Johnston (1981) Attenuation Mechanisms, Preface Chapter 3 SEG Geophysics reprint series No. 2 Seismic Wave Attenuation, pp.136-139. SEG Tulsa/Oklahoma,

Toksöz, M. N., D. H. Johnston, A. Timur (1979) Attenuation of seismic wave in dry and saturated rocks: II. Mechanisms, Geophysics, 44, 4, 691-711

Tosaja, C. and A. Nur (1982) Effects of diagenesis and clays on compressional velocities in rocks, Geophys. Res. Letters, 9, No.1, 5-8.

Toulhoat, T., S. Bigot, J. Chupeau (1989) In situ fixation of U, Ca, and K pumping through multilayered aquifers in sandstone-clay sequential successions, Panel Proc. IAEA, Vienna, Mar. 1989, 77-91

Tucker, M.E. (1991) Sedimentary Petrology (Sec. Edition), Blackweel Sc. Publ., Oxford

Tunn, W. (1966) Ermittlung einiger wichtiger gesteinsphysikalischer Daten von Erdöl-und Erdgas- Speichergesteinen, Erdöl-Erdgas-Zeitschrift, Wien 82, 10, 404-413.

Uhrig, L. F., F. A. van Melle (1955) Velocity anisotropy in stratified media, Geophysics, 20, 774-779

Underwood, E. E. (1970) Quantitative Stereology, Reading, Mass., Addison-Wesley

Urban, C., W. Debschütz, J. Kulenkampff, U. Yaramanchi (1994) Der Einfluß von Kontamination auf Gesteinseigenschaften, paper presented on Colloquium "Gesteinsmodelle-Petrophysik und Umwelt" Bucha/Universität Leipzig-TU Braunschweig

Uyeda, S., M. D. Fuller, J. C. Belshe, R. W. Girdler (1963) Anisotropy of magnetic susceptibility of rocks and minerals, Journ. Geophys. Res., 68, 279-292.

v. Herzen, R. P., E. A. Maxwell (1959) The measurement of thermal conductivity of deep sea sediments by a needle-probe method, Journ. Geoph. Res., 64, 1557-1563.

v. Hippel, A. R. (1954) Dielectrics and Waves, M I T Press, Cambridge Mass.

van Overmeeren, R. A. (1994) Georadar for hydrogeology, First Break, 12, 8 401-408

Vaquier, V., C. R. Holmes, P. R. Kitzinger, M. Lavergne (1957) Prospecting for groundwater by induced electrical polarization, Geophysics, 22, 660-687

Vassiliev, Y. I, G. I. Gurevich (1962) On the ratio between attenuation decrements and propagation velocities of longitudinal and transversal waves, Izvestija Akad. Sci. USSR, Geophys. Ser. No. 12, 1061-1074 (Engl. Transl.)

Vendelsteijn, B. J., W. W. Larionow (1964) Ispolzovanie dannych promyslovoi geofiziki pri pogscete zapasov nefti i gaza, Izdat. Nedra, Moskva

Vinegar, H. J., M. H.Waxman (1984) Induced Polarization of Shaly Sands, Geophysics, 49, 1267-1287

Vogelsang, D., M. Grinat, H. Pape (1992) Logging of induced polarizationin the KTB Oberpfalz VB interpreted by a fractal model, Scientific Drilling, 3, 105-114

Voigt, W. (1910) Lehrbuch der Kristallphysik, Teubner-Verlag, Leipzig

Voigt, H.-D. (1979) Berechnung der Erdgasvorräte nach der Druckabfallmethode bei deformierbarem Porenraum, Zeitschr. Angew. Geol., 25, 1, 15-20

Volarovic, M. P., A. K. Kurskeev, A. I. Levykin (1970) Issledovanie plotnosti gornich porod Zentralnogo Kasachstan pri vysokich davleniach, Izvvestija A.N. SSSR, Fizika Zemli 1, 46-51.

Volarovic, M. P. (1968) Issledovanie uprugich svoist v gornich porod pri vysokich davleniach v svjasi c izuceniem zemnoi kory i verchnej mantii. Fiz.- mechan. svoistva gornich porod, Izdat. Nauka, Moskva 30-36.

Volarovic, M. P., E. I. Parchomenko, A. T. Bondarenko (1966) Issledovanie elektrosoprotivlenija osnovnych ultrarasnovnych i scelocnych gornich porod i mineralov. Trudy Inst. Fiz. Semli No. 37 (204) Izdat. Nauka, Moskva, 168-179.

Volarovic, M. P., E. I. Bajuk, G. S. Schaganjan (1978) Izucenie skorosti prodolnych voln u ich pogloscenia v gornich porodach i mineralach v ustanovke vysokogo davlenia, in: Fiz. Svoistva gornich porod i mineralov pri vysokich davleniach i temperaturach Izdat. Nauka, Moskva, 9-23.

Volarovic, M. P., E. I. Bajuk (1977) Elastic properties of rocks, in: Volarovic, M. P., Stiller, H., Lebedev, T. S. (eds.) Issledovanie Fiziceskich svoit sv mineralnogo vescestva zemli pri vysokich termodinamiceskich parametrach, Izd. Nakova dumka, Kiev 43-49.

Volarovic, M. P., E. I. Bajuk, T. M. Salechli, W. A. Pavlogradski (1968) Vlianie vsestoronnego davlenija na skorosti prodolnych voln nasysceniach kerosinom i vodio obrazach osadocnich porod, in: Fiz.-mechan. Svoistva gornich porod, Izdat. Nauka, Moskva, 123-127.

Volarovic, M. P., V. A. Budnikov (1979) Geschwindigkeiten elastischer Wellen in Gesteinsproben im Prozess ihrer unelastischen Deformation unter Druckbedingungen, Gerlands Beiträge. z. Geoph.. 88, 3, 193-200.

Volarovic, M. P. (1979) Investigation of physical properties of rocks at high pressure and temperature as related to geodynamic processes, in: Stiller, H., Volarovic, M. P. (ed.), Theoretical and experimental investigations of physical properties of rocks and minerals under extreme p, T-conditions. Akademie-Verlag Berlin,

Volarovic, M. P., H. Stiller, T. S. Lebedev (1977) Issledovanie fiziceskich svoistv mineralnogo vescestva zemli pri vysokich termodinamiceskich parametrach, Izdat. Naukova dumka, Kiev

Volarovic, M. P., E. I. Bajuk, T. M. Salechli, F. G. Guseinov (1966) Skorosti prodolnych voln v osadocnich gornych prodach Aserbaidshana pri vcestoronnych davleniach go 4000 kg/cm^2. Trudy Akad. Nauk SSSR, Inst. Fiz. Zemli No. 37 (204), Izdat. Nauka, Moskva, 130-139.

Volarovich, M. P., Parchomenko, E. I. (1976) Electrical properties of rocks at high temperatures and pressures, in: Geoelectric and geothermal studies - KAPG Geophysical Monograph (Ed. by A. Adam), Akademia Kiado Budapest, 319-369

Waff, H. S. (1974) Theoretical considerations of electrical conductivity in a partially molten mantle, and implications of geothermometry, Journ. Geophys. Research, 79, 4003-4010

Wait, J. R. (1959) Overvoltage Research and Geophysical Applications, Pergamon Press, London, New-York, Paris

Wait, J. R. (edit.) (1971) Electromagnetic probing in geophysics, Golem Press, Boulder/Color.

Walls, J. B, A. Nur (1979) Pore pressure and confining pressure dependence of permeability in sandstone, presented at: 7th Formation Evaluation Symposium Canadian Well Logging Society

Walsh, J. B. (1966) Seismic wave attenuation in rock due to friction, Journ. Geoph. Res. 71, 2591-2599.

Walsh, J. B. (1965) The effect of cracks on the Poisson ratio, Journ. Geoph. Research, 70, 20, 5249-5257.

Walsh, J. B, M. A. Grosenbaugh (1979) A new model for analyzing the effect of fractures on compressibility, Journ. Geophys. Research, 84, 3532-3536

Walsh, J. B. (1965) The effect of cracks on the uniaxial elastic of rocks. Journ. Geoph. Research, 70, 2, 399-441.

Walsh, J. B. (1965) The effect of cracks on the compressibility of rocks, Journ. Geoph. Res., 70, 2, 381-389.

Walsh, J. B. (1968) Attenuation in partially melted material, Journ. Geophys. Research, 73, 2209-2216
Walsh, J. B., W. F. Brace (1964) A fracture criterion of brittle anisotropic rock, Journ. Geophys. Research, 69, 16, 3449-3456.
Walther, H. C. (1968) Saturation from logs - laboratory measurements of logging parameters, Journ. Petr. Technol., 20, 251-258
Wang, Z. (1990) Laboratory measurements and applications of seismic velocities, Expanded Abstracts EAEG/SEG Research Workshop "Estimation and Practical Use of Seismic Velocities", Cambridge England, 298-307
Wang, J. R., Th. J. Schmugge (1980) An empirical model for the complex dielectric permittivity of soils as a function of water content, IEEE Trans. Geosci. and Remote Sensing N. York GE-18, 4, 288-295.
Wang, Z., A. Nur (1988) Effect of Temperature on Wave Velocities in Sands and Sandstones With Heavy Hydrocarbons, SPE Reservoir Engineering, Vol. 3, 1, 158-164
Wang, Z., A. Nur (1988) Velocity dispersion and the "local flow" mechanism in rocks, 58-th Int. SEG meeting, Expand. abstracts 928-930.
Wang, H., T. Todd, D. Weidner, G. Simmons (1971) Elastic properties of Apollo 12 rocks. Proc. 2-nd Lunar Science Conf., vol. 3, Mass. Inst. Technol. Press, 2327-2336.
Wang,Y., M. M. Sharma (1988) A network model for the resistivity behaviour of partially saturated rocks, Trans. 29th Ann. Logging Symp. SPWLA, San Antonio, paper S
Wänstedt, S. (1992) Geophysical borehole logging applied to the location and characterisation of base metal ores, Licentiate Thessis, Tekniska Högskolan i Lulea, Sweden
Ward, S. H., D.C. Fraser (1967) Conduction of electricity in rocks, in: Mining geophysics, vol. II, Tulsa, Society of Exploration Geophysicists (SEG)
Ward, St. H. (1990) Resistivity and Induced Polarization Methods, in: St. Ward (ed) Geotechnical and environmental geophysics - Investigations in Geophysics No. 5 SEG, 169 - 189
Warrick, R. E. (1974) Seismic investigation of a San Francisco Bay Mud Site, Bull. Seismol. Soc. Am., 64, 375-385
Waxman, M. H., L. J. M. Smits (1968) Electrical conductivity in oil bearing sandstones, Soc. Petroleum Engrg. J., 8, 2, 107-122.
Waxman, M. H., E C. Thomas (1974) Electrical conductivity in shaly sands, Journ. Petroleum Technol., 213-225
Wedepohl, K. H. (1969) Handbook of Geochemistry, Springer Verlag, Berlin. Heidelberg, New York
Wehr, R. G. (1989) Experimentelle und theoretische Untersuchungen zur Bestimmung der Permeabilität aus Bohrlochmessungen - Zusammenhänge zwischen natürlicher Gamma-Radioaktivität, innerer Oberfläche und Permeabilität, PhD-Thesis, TU Clausthal, Germany
Weinberg, A. K. (1967) Spravnitelnaja charakteristika raslicnich sposobov opredelenija konzentracii magnetita v gornich porodach i rudach, Vopros. razved. geofiz., vyp. 6. Izdat. Nedra Moskva, 156-167.

Wenk, H.-R., E. Wenk (1969) Physical Constants of Alpine Rocks (density, Porosity, Specific Heat, Thermal Diffusivity and Conductivity), Beiträge zur Geologie der Schweiz Nr. 45, Leemann AG. Zürich, 343-357

Western Atlas (1992) Introduction to Wireline Log Analysis, Western Atlas International, Inc., Houston TX

Wheathcraft, S. W., G. A. Sharp, S. W. Tyler (1990) Fluid Flow and Solute Transport in Fractal Heteogeneous Porous Media, in: Dynamics of Fluids in Hierarchical Porous Media (J. H. Cushman, ed.) 306-326. Academic Press, London.

Whitcomb. J. H., J. D. Garmany, D. L. Anderson (1973) Earthquake, prediction: Variation of seismic velocities before San Fransisco Earthquake, Science, 180, 632-623.

White, J. E., F. A. Angona (1955) Elastic wave velocities in laminated media, Journ. Acoust. Soc. Am. 27, 310-317

White, J. E., R. L. Sengbush (1953) Velocity measurements in near surface formations, Geophysics, 18, 54-69.

White, J. E. (1975) Computed seismic speeds and attenuation in rocks with partial gas saturation, Geophysics, 40, 224-232

White, J. E. (1965) Seismic waves: Radiation, transmission, and attenuation, McGraw-Hill Book Co., Inc. New York.

White, J. E., L. Martineau-Nicoletis, C. Monash (1983) Measured anisotropy in Pierre shale, Geophys. Prospecting, 31, 5, 709-725

White, J. E. (1983) Underground Sound, Elsevier Publ. Amsterdam, Oxford, New York

White, J. E., Wang, KeXie (1989) Waves in a gel-saturated porous solid, Journ. Acoust. Soc. Am., 86, (3), 1149-1154

White, J. E. (1966) Static friction as a source of seismic attenuation, Geophysics, 31, 2, 333-339.

Wienand, J., A. Rauen, E. Huenges, Ch. Bücker, K. E. Wolter (1989) Tiefbohrung KTB-Oberpfalz VB, Ergebnisse der geowissenschaftlichen Bohrungsbearbeitung im KTB-Feldlabor: 3000-3500 m, KTB-Report 89-5, D1-D50, Hannover

Wieseneder, H. (1976) Verhalten der Gesteine unter extremen Druck- und Temperaturbedingungen. Erdöl-Erdgas Ztschr., 92, 411-413.

Wilkens, R., G. Simmons, L. Caruso (1984) The ratio v_p/v_s as a discriminant of composition for siliceous limestones, Geophysics, 49, 1850-1860

Wilson, D. A., W. M. Hensel (1982) The Cotton Valley Sandstones of East-Texas: A log - core study, SPWLA 23-rd Ann. Logging Symp., paper R

Winkler, K., A. Nur (1979) Pore fluids and seismic attenuation in rocks, Geophys. Research Letters, 6,1-4

Winkler, K. W., A. Nur (1982) Seismic attenuation: Effects of pore fluids and frictional sliding, Geophysics, 47, 1-15

Winsauer, W. O., H. M. Shearin, P. H. Mason, M. Williams (1952) Resistivity of brine-saturated sands in relation to pore geometry, Am. Ass. Petrol. Geol. Bull., 36, 253-277

Winsauer, W. O., W, M. Mc Cardell (1953) Ionic double layer conductivity in reservoir rocks, Trans. AIME 198, 129-134.

Winterkorn, H.F. (1962) Behaviour of moist soils in a thermal energy field, Clays and Clay Minerals, Pergamon Press, 9, 85-103

Winterstein, D. F. (1990) Velocity anisotropy terminology for geophysicists, Geophysics, 55, 8, 1070-1088

Wohlenberg, J. (1982) Density, in: LANDOLT-BÖRNSTEIN Numerical Data and Functional Relationships in Science and Technology, New Series; Group V. Geophysics and Space Research, Vol.1 Physical Properties of Rocks, Subvol. a Springer-Verlag Berlin. Heidelberg. New York

Wong, J. (1979) An electrochemical model of induced - polarization phenomen in disseminated sulfide ores, Geophysics 44, 1245-1265.

Wood (1930) A textbook of sound, G. Bell and Son, London

Woodside, W., J. H. Messmer (1961) Thermal conductivity of porous media (I. Uncons. porous sands, II Consolidated rocks.), Journ. Appl. Phys., 32, 9, 1688-1699.

Worthington, P. F., T. J. Toussaint-Jackson, N. Pallatt (1988) Effect of sample preparation upon saturation exponent in the Magnus Field, U.K. North Sea, The Log Analyst, Jan-Febr., 48-53

Worthington, P .F., N. Pallatt (1989) Effect of variable saturation exponent upon the evaluation of hydrocarbon saturation, SPE paper 20538, 101-110

Worthington, P. F. (1991) The Direction of Petrophysics: A Five-Year Perspective, The Log Analyst, March-April 1991, 57-62

Worthington, P. F. (1981) The influence of formation anisotropy upon resistivity - porosity relationships, SPWLA Twenty-second Ann. Logging Symp. paper AA, 25 p.

Worthington, P.F. (1973) Estimation of the permeability of a Bunter sandstone aquifer from laboratory investigations and borehole resistivity measurements, Water and Water Engineering, July 1973, 251-257

Worthington, P. F. (1986) The relationship of aquifer petrophysics to hydrocarbon evaluation, Quarterly Journ.. Engineering Geology, London, 19, 97-107

Worthington, P. F., N. Pallatt, T. J. Toussaint-Jackson (1989) Influence of microporosity on the evaluation of hydrocarbon saturation, SPE Formation Evaluation 4, 203-209

Worthington,A. E., J. E. Hedges, N. Pallatt (1990) SCA Guidelines for Sample Preparation and Porosity Measurement of Electrical Resistivity Samples - Part I, The Log Analyst, Jan-Feb. 1990, 20-28

Worthington,P. F. (ed.) Advances in Core Evaluation, Gordon and Breach Publ., New York

Wright, D. L., P. H. Nelson (1993) Borehole dielectric logging: a review and a laboratory experiment, Proc. 5th Int.Symp.on Geophysics for Minerals, Geotechnical, and Environmental Applications (MGLS), 24-28 October, Tulsa, 23 p.

Wyble, D. O. (1958) Effect of applied pressure on conductivity, porosity and permeability of sandstones, Trans. AIME 21, 430.

Wyllie, M. R. J., A. R. Gregory (1953) Formation Factors of Unconsolidated Porous Media - Influence of Particle Shape and Effect of Cementation. Journ. Trans. AIME 198, 103-110.

Wyllie, M. R. J., G. H. Gardner, A. R. Gregory (1962) Studies of elastic wave attenuation in porous media, Geophysics, 27, 5, 569-589.

Wyllie, M. R. J., W. D. Rose (1950) Some theoretical considerations related to the quantitative evaluation of the physical characteristics of reservoir rock from electrical log data, Journ. Petr. Techn. (April 1950), 189

Wyllie, M. R. J., M. B. Spangler (1952) Application of electrical resistivity measurements to problem of fluid flow in porous media, Bull. Am. Ass. Petr. Geol., 36, 359-403

Wyllie, M. R. J., A. R. Gregory, L. W. Gardner (1956) Elastic wave velocities in heterogeneous and porous media, Geophysics, 21, 41-70
Wyllie, M. R. J, A. R. Gregory, G. H. F. Gardner (1958) An experimental investigation of factors affecting elastic wave velocities in porous media, Geophysics, 23, 459-493
Wyss. M. (1975) Earthquake prediction and rock mechanism, Birkhäuser Verlag, Basel/Stuttgart
Xu, S., R. E. White (1995) A new velocity módel for clay-sand mixture, Geophysical Prospecting, 43, 91-118
Xu, H., R. Desbrandes (1991) Formation Evaluation Using In-Situ Measurements of Formation Thermal Properties, The Log Analyst, March-April 1991, 144-157
Xu, S., M.S. King, (1989) Shear-wave bifrigerence and directionalm permeability in fractured rock, Scientific Drilling, 1, 27-33
Yale, D. P (1985) Recent advances in rock physics, Geophysics, 50, 12, 2480-2491
Yale, D. P, E. S. Sprunt (1989) Prediction of Fracture Direction Using Shear Acoustic Anisotropy, The Log Analyst, March-April, 65-70
Yoon, Y. K. Lee, H. K. (1993) The effect of the loading condition and rock joint roughness on the permeability of rocks, Journ. Korean. Inst. Min. and Energy Res. Engrs. Seoul, 30, 238-247
Zang, A., K. Wolter, H. Berkhemer (1989) Strain recovery, microcracs and elastic anisotropy of drill cores from KTB deep well, Scientific Drilling, 1, 3, 115-126
Zareva, N. W. (1956) Rasprostranenie uprugich voln v peske, Izvestija Akad. Nauk SSSR, Ser. geofiz., Moskva, 9, 1044-1053.
Zimmerman, R. W., M. S.King (1986) The effect of the extent of freezing on seismic velocities in unconsolidated permafrost, Geophysics, 51, 6, 1285 - 1290
Zoback, M. D. (1975) High pressure deformation and fluid flow in sandstone, granite, and granular material, Ph.D. Thesis Stanford University
Zonge, K. L. (1973) In situ mineral discrimination using a complex resistivity method, Geophysics, 38, 197
Zoth, G. (1989) Die Bestimmung der Wärmeproduktionsrate aus Bohrlochmessungen. KTB-Report 89-1, 45-51, Hannover
Zwikker, C. (1954) Physical properties of solid materials, Pergamon Press. London

INDEX